Reproductive Biology of Angiosperms

The science of understanding plant reproduction is more than four hundred years old. Today, with integration of molecular biological tools, plant reproductive biology has catapulted into an exciting field of research. It has become an integral part of evolutionary biology, conservation biology, climate change studies, population biology, genetics, horticulture and many more fields. Considering the widening scope of plant reproductive biology, this book focusses on teaching the core concepts of plant reproduction supplemented with latest findings in the field. Uniquely, this book addresses both theoretical and practical perspectives by providing easy protocols of experiments related to the content of each chapter, thus, making the book useful for an entire spectrum of students, teachers and researchers.

The content of the book is designed for the undergraduate syllabi of embryology, reproductive biology of flowering plants, reproductive ecology of flowering plants, and plant breeding, taught in universities across the country. The content is well-supplemented with photographs and illustrations to enhance the understanding of the structures and processes involved in plant reproduction. Interesting information, which may incite the curiosity of learners, appears at appropriate places in the narrative in a box format. Detailed comparisons of similar/related concepts which commonly are difficult to comprehend for first-time learners are especially brought out in the text. The key concepts are revisited at the end of each chapter in the form of glossary. Practice questions at the end of each chapter have been added as part of pedagogical approach of conceptual learning. To help readers develop a complete understanding of the subject, step-by-step description of experiments related to the content are also provided.

Yash Mangla is Assistant Professor in the Department of Botany at Acharya Narendra Dev College, University of Delhi. His specialization is developmental, molecular, and reproductive biology of angiosperms.

Priyanka Khanduri is Assistant Professor in the Department of Botany at Vidyasagar Metropolitan College, University of Calcutta. Her research interests include developmental and reproductive biology of angiosperms, and phylogenetics.

Charu Khosla Gupta is Professor of Botany at Acharya Narendra Dev College, University of Delhi, with teaching and research experience spanning 24 years. She specializes in reproductive biology of angiosperms. She is the recipient of Teaching Excellence Award for innovation (2015) from the University of Delhi and Meritorious Teacher Award (2020) from the Directorate of Higher Education, Government of NCT of Delhi.

Reproductive Biology of Angiosperms

Concepts and Laboratory Methods

Yash Mangla

Priyanka Khanduri

Charu Khosla Gupta

CAMBRIDGE
UNIVERSITY PRESS

CAMBRIDGE
UNIVERSITY PRESS

University Printing House, Cambridge CB2 8BS, United Kingdom

One Liberty Plaza, 20th Floor, New York, NY 10006, USA

477 Williamstown Road, Port Melbourne, VIC 3207, Australia

314–321, 3rd Floor, Plot 3, Splendor Forum, Jasola District Centre, New Delhi–110025, India

103 Penang Road, #05–06/07, Visioncrest Commercial, Singapore 238467

Cambridge University Press is part of the University of Cambridge.

It furthers the University's mission by disseminating knowledge in the pursuit of education, learning and research at the highest international levels of excellence.

www.cambridge.org
Information on this title: www.cambridge.org/9781009160407

First published 2022

Printed in India by Nutech Print Services, New Delhi 110020

Library of Congress Cataloging-in-Publication Data

Names: Mangla, Yash, author. | Khanduri, Priyanka, author. | Khosla, Charu, author.
Title: Reproductive biology of angiosperms : concepts and laboratory methods / Yash Mangla, Priyanka Khanduri, Charu Khosla Gupta
Description: New York, NY : Cambridge University Press, 2022. | Includes bibliographical references and index.
Identifiers: LCCN 2021055613 (print) | LCCN 2021055614 (ebook) | ISBN 9781009160407 (paperback) | ISBN 9781009160414 (ebook)
Subjects: LCSH: Angiosperms—Reproduction.
Classification: LCC QK495.A1 M35 2022 (print) | LCC QK495.A1 (ebook) | DDC 583–dc23/eng/20211221
LC record available at https://lccn.loc.gov/2021055613
LC ebook record available at https://lccn.loc.gov/2021055614

ISBN 978-1-009-16040-7 Paperback

To Our Gurus ...

Our Constant Source of Inspiration

Contents

Foreword

Plants in general and flowering plants (angiosperms) in particular are the essential components for sustenance of life of all non-photosynthetic organisms on our planet. Plants reproduce by asexual as well as sexual means. Asexual reproduction is not congenial for long-term sustenance and evolutionary processes of the species because of genetic uniformity of the progeny. Sexual reproduction which permits genetic recombination is the dominant mode. Although Angiosperms were the last to evolve as land plants, they soon became the most successful and dominant group amongst land plants. Their success is largely due to the mode of their reproduction through the evolution of the flower and the consequent advantages it brought in. For human beings, flowering plants provide most of their essential needs – food, fibres, shelter, medicines, clean air and water. Reproduction is the basis for sustenance of any species. Thus, understanding reproductive biology of flowering plants is important not only from the fundamental point of view but also for their manipulation for human welfare. Reproductive biology of angiosperms is more complex when compared to other groups of plants because of the involvement of the flower. The progress in understanding the structural and functional aspects of reproduction has been very slow.

Initial studies on reproductive biology of angiosperms were largely confined to examining embryological details using fixed and sectioned materials. Enormous data accumulated over the years on the developmental details of the pollen grains, ovules and female gametophyte, double fertilization, embryo and endosperm, seed and fruit development. These advances were taught to the undergraduate and postgraduate students under the title embryology of angiosperms as a part of their curriculum. Following the development of electron microscopy and histochemistry, embryological details were further elaborated by using these techniques. Development of aseptic culture techniques broadened scope for experimental studies on embryological processes leading to a slow but steady understanding of the functional details of embryological structures. These developments were incorporated in some of the books of embryology under a chapter on experimental embryology. However, there was hardly any integrated account of embryological processes in relation to the structure with their function. Pre-fertilization aspects of reproductive biology covering the details of pollen, pistil, and pollen–pistil interactions, which are unique to angiosperms and play a critical role in their successful evolution, were the last to enter the field of embryology of angiosperm. Surprisingly, pollination on which plenty of literature has long been available and which is a critical requirement for angiosperm reproduction was not a part of embryology.

Now enormous data has accumulated on all aspects of reproductive biology of angiosperms through interdisciplinary studies and it is high time to teach this subject as "Reproductive Biology" (rather than embryology) integrating all these advances starting with flower development until fruit and seed maturation and their dispersal.

Continued research and teaching of reproductive biology of angiosperms has become highly relevant in the light of human-induced environmental changes (intensification of agriculture, habitat loss and degradation, overexploitation of bio-resources, introduction of alien species and climate change) in recent decades and their impact on the sustenance of biodiversity and crop productivity. Human activities have not only accelerated extinction of species but also have pushed a large number of species to endangered status leading to the sixth mass extinction crisis. In the absence of effective remedial measures, a large proportion of biodiversity is likely to become extinct by the end of this century. Therefore, conservation of biodiversity is likely to become one of the most important agendas of the world in the coming decades. Reproduction is the basis for the sustenance of any species. In flowering plants, recruitment of new individuals which is the final step in a series of sequential reproductive events is the basis of species sustenance. Environmental changes have induced severe constraints on many reproductive events, particularly pollination and seed dispersal, leading to recruitment constraint. For any effective conservation of angiosperm species, it is important to understand reproductive biology of endangered species to identify reproductive and/or recruitment constraints and to apply effective measures to overcome such constraints. Plant conservation attempts so far, particularly in developing countries, have not been very successful; one of the reasons being our ignorance of the reproductive biology of the species to be conserved. Knowledge about reproductive biology is important not only for effective conservation but also to monitor the success of conservation measures.

Present book, *Reproductive Biology of Angiosperms: Concepts and Laboratory Methods*, by Yash Mangla, Priyanka Khanduri and Charu Khosla Gupta is going to be an important contribution to the field. It begins with a brief introduction that gives an account of different aspects of reproductive biology and highlights the importance of the subject to other areas of plant biology such as conservation biology, crop production, and evolutionary biology. Historical account, that comes next, elaborates briefly the development of "classical embryology" through various stages into an integrated discipline of "reproductive biology" and highlights the contributions of several well-recognized investigators. Various chapters starting from the flower until seed dispersal give a reasonably comprehensive account of different aspects of reproductive biology. A chapter on genetic transformation is also included at the end. The book is illustrated with photographs and diagrammatic sketches of high quality.

An attempt is made in all the chapters to give an integrated account of reproductive structures (as revealed through light and electron microscopy) and their functions. The deep understanding involved was realized through the application of the techniques from different disciplines such as histochemistry, physiology, biochemistry, molecular biology, and evolutionary biology. Such an integrated account is hardly given in any of the existing books in this field. Boxes and tables have been used profusely to provide clarity on many aspects. Glossary of a large number of technical terms and key questions are given at the end of each chapter. The details of the practical exercises given for each chapter are going to be very useful for the teachers as well as students. To my knowledge there is no such book available so far in this field. The present book is thus a very welcome addition to the literature on reproductive biology of angiosperms. I congratulate the authors for compiling such a book. I am confident that the volume is going to be very useful not only for teachers and students of Reproductive Biology but also those interested in any aspect of reproduction of angiosperms.

K. R. Shivanna
(Former Professor and Head, Department Botany, University of Delhi)
INSA Honorary Scientist
Ashoka Trust for Research in Ecology and the Environment,
Bengaluru, India

Preface

The inception of interest in understanding mechanisms of plant reproduction is as old as inception of interest in biology. The seminal work and critical observations by Charles Darwin can be regarded as a foundation for establishing a wide interest in pollinators and reproductive biology of angiosperms as a formal subject. In the last few decades, systematic field investigations, advancement of microscopy tools, and molecular techniques have taken the reproductive biology of angiosperms to a new zenith. The scope of the subject is no longer limited to just studying embryo-endosperm development and taxonomic studies but is extended to study the effect of climate change, evolution, conservation of threatened taxa, raising commercial plantations and orchards, pollinator management, seed development, population biology, phyto-geography, and much more. The reproductive biological studies are also closely linked with the understanding of, physiology, genetics and epigenetics of plants.

For a thorough understanding of the subject, a textbook summarizing the basic concepts of plant reproduction integrated with current research, is the need of the hour for both students and instructors. The aim of the present book is to provide a comprehensive account of basic concepts and recent developments in the field of reproductive biology of flowering plants with essential practical exercises. The book extensively covers all the topics from structure of a flower to seed dispersal and presents the concepts with accompanying color photographs and illustrations wherever necessary, to enhance the level of a student's perception. The new, advanced and interesting information is also provided in a box format in each chapter to reinforce learning. An elaborate glossary and questions are provided with each chapter for quick revision and concept enhancement. Boxes summarizing differences between two terms/concepts which students otherwise usually find difficult to comprehend have also been furnished in the book. This book is a blend of theoretical concepts and details of hands-on exercises in the field and laboratory. Methods for field observations, sample observation tables, and suggestions for plant materials to be used for classroom studies/demonstrations pertaining to each concept have also been provided. In addition, the observation sections under practicals are supplemented with the photographs. This should surely help the instructors to demonstrate and students to grasp the concept effectively

The current book has been specially structured keeping in mind the syllabi of the leading universities in the country. The content is all inclusive of different curricula frameworks implemented by the University Grants Commission across the universities. It can be used as a text book as well as a reference book by both graduate and undergraduate students along with researchers in the field. The glossary of technical terms at the end of each chapter should help students for quick revision for their competitive and entrance examinations. We hope the present edition of the book will certainly turn into a knowledge resource for the young minds and learners in the field.

Acknowledgments

We want to thank all those related directly or indirectly to this book. We dedicate this book to our mentors Late Professor H. Y. Mohan Ram, Professor K. R. Shivanna and Professor Rajesh Tandon who have been a constant source of inspiration and motivation for all three of us. We hold a deep sense of gratitude and reverence towards them. We are also thankful to Professor S. C. H. Barrett and Professor Odair José Garcia de Almeida for readily providing us with some wonderful photographs. We are equally indebted to all the researchers in the field of reproductive biology who supplemented our effort by providing illustrations and photographs from their original research. Our friends from the botany fraternity, Dr. P. Chitralekha, Dr. Sudip Kumar Roy, Dr. Chandan Barman, Dr. Vineet Kumar Singh, Mr. Arjun Adit, and Mr. Ashish Jangam very graciously shared their research work for which we are deeply obliged. We are also thankful to the International Association of Sexual Plant Reproduction Research (IASPRR) and the University of Wisconsin, Stevens Biology Lab for consenting to our use of images from their work. The permissions to use images from the Copyright Clearance Centre, Indian National Science Academy, Indian Academy of Sciences, Phytomorphology, Journal of Indian Botanical Society, Brazilian Journal of Botany and various journals have been invaluable and duly acknowledged. Our thanks are also due to all the creators whose work (text, figures, and illustrations) licensed under Creative Commons and free and open access journals, has been used in this book. The impetus for writing and publishing this book provided by Professor Ravi Toteja, Officiating Principal Acharya Narendra Dev College, University of Delhi and Dr. Ram Swarup Gangopadhyay, Principal, Vidyasagar Metropolitan College, University of Calcutta is deeply appreciated. Professor Toteja was also kind enough to run the plagiarism checks for our chapters. Hence, we take this opportunity to thank him and the University of Delhi for providing access to plagiarism software "Urkund", which helped immensely in making the content original. We are also thankful to the anonymous reviewers for their constructive comments and suggestions which encouraged us to not skip even the minutest details of the subject.

We are indebted to Dr. Vaishali Thapliyal, Senior Commissioning Editor at Cambridge University Press, for helping us bring out this comprehensive volume. She has been instrumental right from inception of proposal till its printing. Support provided by Cambridge University Press and staff members, especially Mr. Aniruddha De and Mr. Vikash Tiwari, is duly recognized.

Last but not the least we really thank our families and friends who supported us in this herculean venture.

Image Sources

Chapter 2

Figure 2.1 A. *Ranunculus*: 'Renoncules sauvages Bouton-d'or (Ranunculus acris)' by Giancarlo – Foto 4U is licensed under CC BY 2.0.

H. *Fuchsia* sp.: Arjun Adit, Research Scholar, Department of Botany, University of Delhi.

U. *Yucca* sp.: Hemant Bisht.

V. *Cynotis* sp.: Ashish Jangam, Research Scholar, Department of Botany, Kolhapur University.

Figure 2.4 A and *Figure 2.5* B. *Phalaenopsis*, *Anthurium* sp. respectively: Arjun Adit, Research Scholar, Department of Botany, University of Delhi.

Figure 2.6 B and F. *Solanum* sp., *Ricinus communis*: Arjun Adit, Research Scholar, Department of Botany, University of Delhi.

C. *Dianthus* sp.: Hemant Bisht.

Chapter 3

Rudolf Jakob Camerarius, German botanist, 1665–1721. Artist unknown: http://www.biologie.uni-hamburg.de/b-online/e08/08.htm, Wikipedia public domain.

Josef Gottlieb Koelreuter: Wikipedia public domain.

Charles Darwin: By unknown author, originally published in *The Hornet* magazine; this image is available on University College London Digital Collections (18886), Public Domain, https://commons.wikimedia.org/w/index.php?curid=23436.

Amici: By Michele Gordigiani, http://catalogo.museogalileo.it/galleria/RitrattoGiovanni BattistaAmiciDepGAMFirenze.html, Public Domain, https://commons.wikimedia.org/w/index.php?curid=16010822.

Strasburger: By unknown; published in Munich. by J.F. Lehmann, http://ihm.nlm.nih.gov/images/B22559, Public Domain, https://commons.wikimedia.org/w/index.php?curid=18716986.

William Hofmeister: By Universitätsbibliothek Heidelberg, CC BY-SA 4.0, https://commons.wikimedia.org/w/index.php?curid=71111381.

Sergei Navashin: http://molbiol.ru/forums/index.php?showtopic=105009.

P. Maheshwari, BM Johri, SC Maheshwari, HY Mohan Ram, NS Rangaswamy: Indian National Science Academy, New Delhi.

Professor KR Shivanna: Professor KR Shivanna.

Chapter 4

Figure 4.1 B and *Figure 4.18*: University of Wisconsin-Stevens Biology Lab and IASPRR.
Figure 4.5 B: Dr Arun Kumar Mourya, Assistant Professor, Multanimal Modi College, Modinagar, Uttar Pradesh.
Parts of *Figure 4.15*, *4.16* and *4.23*: Halbritter et al. 2018.

Chapter 5

Figure 5.22 A–E: University of Wisconsin, Stevens Biology Lab.
F: https://commons.wikimedia.org/wiki/File:Lilium_embryo_4_nuclei.jpg.
Figure 5.19 A–C: Professor Odair José Garcia de Almeida, UNESP – Universidade Estadual Paulista, IB/Campus do Litoral Paulista.

Chapter 6

Figure 6.3 B. *Gloriosa superba*: Dr Rajesh Chaudhary, Associate Professor, Department of Biomedical Sciences, Acharya Narendra Dev College, University of Delhi.
Figure 6.8 D. Moth pollination: Dr Chandan Barman, Assistant Professor, Department of Botany, University of Gaur Banga, Malda, West Bengal.
Figure 6.10. Bat pollination: Dr Sudip Kumar Roy, Assistant Professor, Department of Botany, Charuchandra College Kolkata, West Bengal.
Figure 6.11. Snail pollination: Professor Rajesh Tandon, Professor, Department of Botany, University of Delhi.

Chapter 7

Figure 6.3 C. Dr Chandan Barman, Assistant Professor, Department of Botany, University of Gaur Banga, Malda, West Bengal.
F. Dr Vineet Kumar Singh, Assistant Professor, Department of Botany, Acharya Narendra Dev College, University of Delhi.

Chapter 8

Figure 8.1 B. Tristyly in *Lythrum salicaria*: Professor SCH Barrett, Professor Emeritus, University of Toronto, Canada.
Figure 7.8 C. Self-pollinated stigma showing deposition of callose on stigma and pollen tubes: Professor Rajesh Tandon, Professor, Department of Botany, University of Delhi.

Chapter 9

Figure 9.19: Dr P. Chitralekha, Associate Professor, Department of Botany, Dyal Singh College, University of Delhi.

Chapter 10

Figure 10.17: University of Wisconsin, Stevens Biology Lab.

Chapter 12

Figure 12. 1 A: 'Lodoicea maldivica. Coco de mer. Half used by locals for boat balers' by Mary Gillham Archive Project is licensed under CC BY 2.0, https://search.creativecommons.org/photos/d2a16ded-66bf-4fc5-b7b3-3376af3eff6f.
B: Arjun Adit, Research Scholar, Department of Botany, University of Delhi.
Figure 12.9 B: 'Magnolia Arils' by Editor B is licensed under CC BY 2.0, https://search.creativecommons.org/photos/e92bcbf9-77a9-4087-801e-fed2c8a3aae6.
D. Oil palm: Operculum: Professor S. Natesan, Tamil Nadu Agricultural University.
F: 'Nutmeg' by Giselleai is licensed under CC BY 2.0, https://search.creativecommons.org/photos/493b35d8-b8b3-49dc-951a-69a4c05ea66a.
I: 'Acacia tetragonophylla open seed pods' by John Tann is licensed under CC BY 2.0, https://search.creativecommons.org/photos/0907e44e-dfef-4bca-964e-60d1e8f151c6.
'Acacia tetragonophylla seeds' by John Tann is licensed under CC BY 2.0, https://search.creativecommons.org/search?q=Acacia%20tetragonophylla%20seeds&license=by&license_type=commercial.
Figure 12.11 B: 'File:Alsomitra macrocarpa seed (syn. Zanonia macrocarpa).jpg' by Scott Zona from Miami, Florida, USA, is licensed under CC BY 2.0, https://search.creativecommons.org/photos/3b0545dd-7586-4a25-8112-e81b1a894374.
C: Dr. Sudip Kumar Roy, Assistant Professor, Department of Botany, Charuchandra College Kolkata, West Bengal.
F: 'Poppycock' by Jenny Downing is licensed under CC BY 2.0, https://search.creativecommons.org/photos/ac555fb7-107c-46bc-b00f-6727ddeab3b6.
Figure 12.12 A: COCONut 'Dauin beach' by bortescristian is licensed under CC BY 2.0, https://search.creativecommons.org/photos/ceebca24-633a-430e-a57d-38bed0eca5a9.
B: *Encyclopædia Britannica*, https://www.britannica.com/plant/coconut-palm/images-videos#/media/1/123794/238237.
C: *Encyclopædia Britannica*, https://www.britannica.com/science/seed-plant-reproductive-part/Dispersal-by-water#/media/1/532368/217444.
D: 'Lotus seed / 蓮の実(はすのみ)' by TANAKA Juuyoh (田中十洋) is licensed under CC BY 2.0,
https://search.creativecommons.org/photos/8ac504c4-8475-4614-8a9a-b62bc26525b6.
Figure 12.14 A: Winter berries by qmnonic is licensed under CC BY 2.0, https://search.creativecommons.org/photos/04fb465e-d1db-4562-9267-31593a0719f3.
B: *Encyclopædia Britannica*, https://www.britannica.com/science/seed-plant-reproductive-part/Dispersal-by-animals#/media/1/532368/155752.
C: Novemberry by Wildlife Terry is marked with CC0 1.0, https://search.creativecommons.org/photos/fbee2f01-41cb-471b-8b76-04dbd91f30bb.
D: King of the Stump by Peter G Trimming is licensed under CC BY 2.0, https://search.creativecommons.org/photos/4d61cec8-9286-4d9c-9bae-b7e28f378e41.

1

An Introduction to the Reproductive Biology of Flowering Plants

Reproduction is a very important stage in the life-history of a species, being essential for its survival and sustenance. Different organisms adopt different strategies as they attempt to maximize their reproductive success and produce a favourable number of new individuals. Reproduction in plants can be achieved by either vegetative or sexual means or a combination of both. The seeds and propagules produced by asexual and sexual modes of reproduction have differing implications on the perpetuation of the species. Asexual means (such as vegetative reproduction) in plants is a quicker reproductive strategy that leads to production of new individuals genetically identical to parents. However, there is a limitation of genetic variability in vegetative reproduction and this may affect the long-term survival of a species. On the other hand, reproduction by sexual means brings genetic heterogeneity in progeny resulting in their wider adaptability and better survival. Sexual reproduction in angiosperms is a complex process involving several sequential events which take place in different organs of a flower. Thus, flower is a unit of sexual reproduction in angiosperms.

Plant reproductive biology is the study of the mechanisms of both sexual and asexual reproduction in plants. It involves the study of interactions of plants with biotic factors (such as pollinators, seed dispersal agents) and abiotic components (such as soil, space, climate) in the environment. With the integration of the many aspects of ecology, reproductive biology of flowering plants is now also known as *Reproductive Ecology of Flowering Plants*.

1.1 Different aspects of Reproductive Biology of Flowering Plants

Study of reproductive biology of plants broadly includes observations on phenology, structural and functional floral biology, sexual system, pollination biology, mating system, pollen–pistil interactions, fertilization, embryo-endosperm development, seed formation, seed dispersal and seed recruitment. These events may also be considered as the series of steps neccessary for the formation of a perfect new sporophyte. These aspects being interconnected, each of these is discussed sequentially in the subsequent sections. ·

- **Phenology:** Phenology is the timing of recurring biological phases in response to seasonal variations. In the life-cycle of flowering plants various events such as appearance of leaves, onset of flowering, fruit initiation and seed dispersal occur in consonance with seasonal changes and are termed as *phenoevents*. The timing of these recurring and periodic life-cycle events plays a significant role in interaction with other species in the ecosystem. Therefore, the phenological behavior of any species is important for insights into its long-term survival under its ecological conditions. Variations in the phenoevents are very good indicators of climate change; and can be used for predicting such changes in coming years. For these predictions, recordings of phenoevents is conducted over several years in a row at various levels, namely, individual flowers, individual plants as well as the population as a whole. At the level of flower, the timing of flower opening (floral anthesis), anther dehiscence, stigma receptivity, and pollination are recorded. This leads to a detailed understanding of pollination syndromes and breeding strategies of a particular species. All these events are species specific and may happen at any time in a twenty-four hour cycle.

- **Floral Biology and Sexual System:** Floral biology studies provide an understanding of the structural and functional features of a flower. The structural (morphological as well as anatomical) features include the details of all four whorls of a typical flower, viz. calyx, corolla, stamen, and pistil. It may also include the study of development (ontogeny) of the whorls. To understand the functionality of a flower one needs to study the functional features of all the four whorls with special emphasis on pollen, pistil and their interactions. Floral biology also includes study of morphology of pollen grains, stigma and style, estimation of pollen production, pollen viability, pollen fertility, stigma receptivity, and number of types of ovules. These provide an indication of the pollen: ovule ratio, ovule: seed ratio, type of pollination and breeding system of the species. Floral biology studies most importantly help in the identification of the sexual system of the species; it being hermaphrodite, monoecious or dioecious. Structural features of flower also include the spatial and temporal arrangement of its sex organs. The synchronization or asynchronization in the maturity of anther and stigmatic receptivity indicate the mating strategy operating in a species. Asynchronization in maturity is indicative of presence of dichogamy in a species while spatial separation marks the herkogamy. This information is essential for the technique of hybridization in crop breeding experiments.

- **Pollination Biology:** The transfer of pollen grains (male gametophyte) to conspecific stigma (female sporophyte) by the plant itself or through biotic or abiotic means is known as pollination. Abiotic pollination is through either wind or water; biotic pollination is carried out by pollinators such as insects, birds, bats, and others. These pollinators in turn are rewarded with pollen and nectar, fulfilling

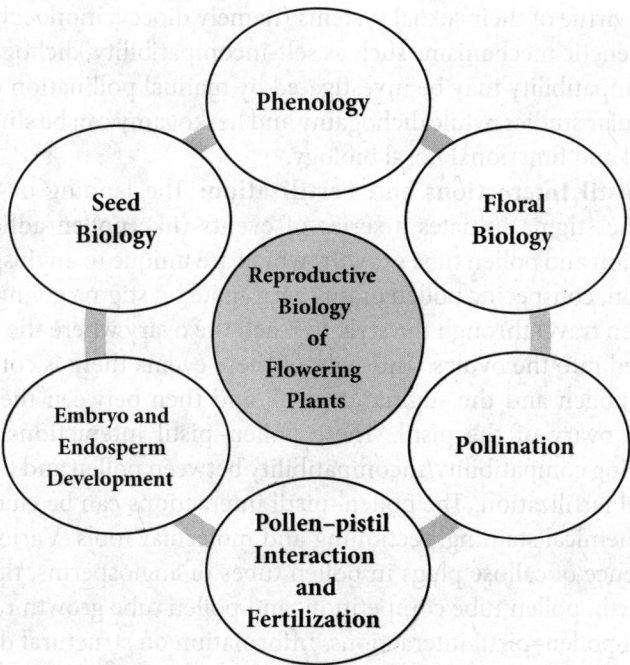

Figure 1.1 Various aspects of reproductive biology of flowering plants.

their nutritional requirements. Pollination biology is studied in the field by direct observations; with the help of photography and video recordings. Pollination biology needs thorough understanding of the floral biology of the species. The floral biology suites observed in a species constitute the pollination syndrome which offers clues of prospective pollinators of the species. Similarly, critical observation of the time of flower opening (anthesis) in a day is also important for studying the pollination mechanism as there is a correlation seen between the time of anthesis and the type of pollinator. For example, the flowers which exhibit anthesis during the day are usually pollinated by bees and other insects, while flowers opening in the evening or night are primarily pollinated by moths and bats.

- **Mating System:** Flowering plants display a remarkable diversity of mating patterns. Species in which ovules are predominantly fertilized by pollen from the same flower are known as autogamous, while allogamous species are those in which the ovules of a flower are fertilized by pollen from another conspecific flower present on the same plant or from another conspecific plant. Apart from the different mating systems, certain plants have the ability to develop seeds without fertilization (and even without pollination) through a phenomenon known as apomixis. The mating system of a species provides an insight about the gene flow and an estimation of the genetic makeup of the population. The mating system of a species can be deciphered through manual pollinations, bagging experiments, and by genetic marker studies (parent-progeny analysis). Many species avoid self-pollination which is achieved either by virtue of their sexual systems (namely dioecy, monoecy, etc.) or through specific genetic mechanisms such as self-incompatibility, dichogamy, herkogamy. Self-incompatibility may be investigated by manual pollination experiments, and by molecular studies, while dichogamy and herkogamy can be studied by exploring structural and functional floral biology.

- **Pollen–pistil Interactions and Fertilization:** The landing of pollen grains on conspecific stigma initiates a series of events (like pollen adhesion, hydration, germination and pollen tube growth) which are unique to angiosperms. Following pollination, conspecific pollen grains germinate on stigma giving out pollen tubes which then travel through the style to reach the ovary where the male gametes are discharged into the ovules. Throughout these events there is constant interaction between pollen and the stigma of pistil; and then between the pollen tube and style and ovary of the pistil. These pollen–pistil interactions are essential for determining compatibility/incompatibility between pollen and pistil and ensuring successful fertilization. The pollen–pistil interactions can be elucidated by means of histochemical staining, sectioning and molecular tools. Various other features like presence of callose plugs in pollen tubes of angiosperms, tip-oriented pollen tube growth, pollen tube competition, and pollen tube growth rate are important aspects of pollen–pistil interactions. Information on structural details of the pistil is crucial for understanding pollen–pistil interactions. This information has helped in understanding the evolution of angiosperms and their successful establishment on the earth.

- **Embryo-Endosperm Development:** Pollen tubes carry two male gametes, out of which one fuses with the egg cell and the other fuses with the polar nuclei. These two events together constitute 'double fertilization' which is a unique feature of angiosperms. The fusion product of a male gamete and egg cell is the zygote which develops into an embryo. The fusion of two polar nuclei with the second male gamete is called the triple fusion; and results in the formation of an endosperm. The study of development of embryos is known as embryology. The development of an embryo and its nutritive tissue, the endosperm, follows different developmental pathways in different angiosperms. Embryological studies are an important component of reproductive biology as these studies help to understand the development of a new sporophyte within ovule and ovary.

- **Seed formation, Dispersal and Recruitment:** The ultimate goal of reproduction in flowering plants is to produce seeds for perpetuation. Many of the seeds are of economic and commercial importance. Except in apomictic plants, seeds are formed from the fertilized ovules, representing new genotypes. The establishment of new genotypes is crucial for the maintenance of populations. The study of the development, dispersal and recruitment of seeds is known as seed biology which is an integral part of the reproductive biology of plants. Angiosperms exhibit a huge diversity in seed types and in their development mechanisms. The knowledge of seed dormancy, measures to break dormancy, seed viability, seed storage, dispersal mechanism and recruitment pattern is necessary for the conservation of a species.

1.2 Scope of Reproductive Biology of Angiosperms

Understanding plant reproduction and pollination of plants has been an area of interest for naturalists and scientists for centuries. During the course of the past several years considerable progress has been made in understanding the reproductive biology of angiosperms and its scope has widened with the integration of several disciplines such as ecology, population biology, genetics, physiology, molecular biology, biotechnology, and conservation biology. Depending on the problem at hand, reproductive biology can be integrated with other arenas for a stronger and more focused research. The integrated nature of the subject which involves field and laboratory based observations and hypothesis testing makes plant reproductive biology one of the most dynamic field of research. Some important research themes based on reproductive biological studies are summarized here:

- **Reproductive Biology and Conservation Biology:** Loss of biodiversity is one of the major challenges that has risen in the last century. Uncontrolled collection from the wild (plants or plant parts), deforestation, habitat fragmentation, climate change and anthropogenic interference have threatened biodiversity. These challenges have induced changes in the reproductive behavior of species with detrimental consequences on production of new individuals and hence sustanance of natural populations. These factors affect reproduction in various ways, viz. low/no availability of pollen grains (the situation becomes even more critical if the species is cross pollinated), loss of services offered by pollinators, reduced pollination (both cross or self) leading to decrease in fruit set, high inbreeding, diminished fitness and reduced regeneration. The changing temperature and abiotic factor regimes, adversely affect plant distribution and their reproduction. Due to rise in temperature, many species are showing altitudinal shifts which along with increase in concentration of CO_2 are affecting the phenology of species. Several plants are showing early flowering in a short duration which is becoming a threat to both the host plant and the pollinators relying on that particular plant for food or other requirements. Uncertain reproduction may be the major reason for the rarity of medicinal plants especially tree species in the wild. Thus, there is an inevitable need for planning conservation and restoration of biological diversity. One of the most important aspects of conservation and management of plants is the collection of baseline data on their reproductive biology. In the absence of such data, any conservation effort will remain ineffective. Thus, detailed information on the reproductive biology of plants is essential for developing effective strategies for their conservation and sustainable utilization. With this aim, elaborate studies on plant phenology, pollinators and plants distribution are gaining prominence.

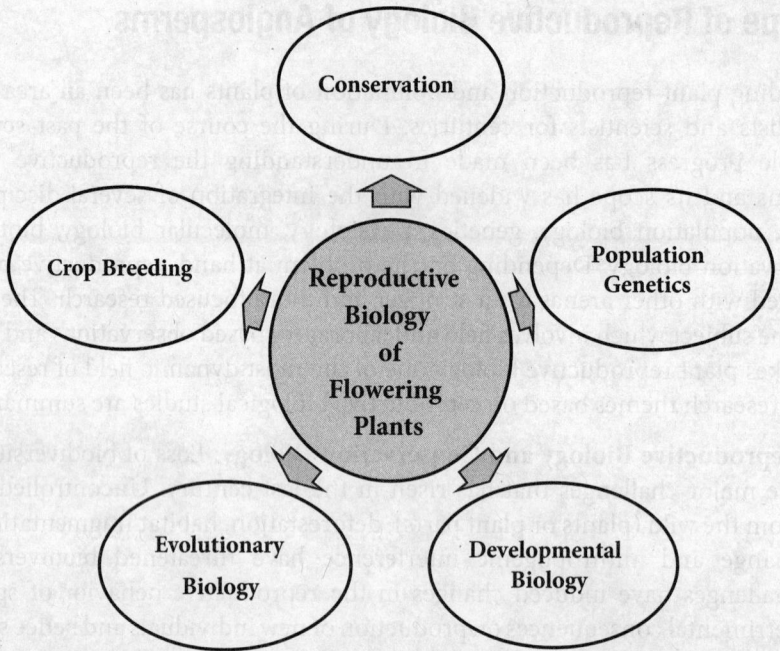

Figure 1.2 Various fields requiring data on reproductive biology of flowering plants.

- **Reproductive Biology and Developmental Biology:** Plant reproduction represents a highly coordinated and complex developmental process. In the past few decades molecular biology, genetics and biotechnological tools have been variously employed to elucidate underlying mechanisms, pathways and allied phenomena in the various aspects of plant reproduction. These studies have led to the discovery of several families of genes involved in flower formation (homeotic genes), development of anther and pollen, viability of pollen, embryo sac development, molecular genetics and the biochemical basis of pollen–pistil interactions, genetic control of self-incompatibility, embryo and endosperm patterning and apomixis. Understanding these developmental aspects of plant reproduction not only sheds light on how plants reproduce but can also be exploited in agriculture, especially the genes involved in male sterility and self-incompatibility. However, it is important to have basic knowledge of reproductive events for studying different aspects of developmental biology.

- **Reproductive Biology and Population Genetics:** Flowering plants display a diverse array of mating strategies and sexual systems. These reproductive systems have a profound influence on the patterns of genetic variation within and among plant populations. Principles of population genetics and phenotypic selection models are used to understand the adaptive significance of these mating strategies. Using molecular techniques, outcrossing rates, inbreeding coefficients, and paternity data, the success of mating system of a species can be evaluated. Gene flow is the transfer of genetic information within and between populations of a species. The genes in plants are exchanged in the form of pollen grains and seeds. The distance traveled by pollen grains (pollen flow) via biotic or abiotic means in a population or between populations has demographic consequences for a species. Therefore, it is imperative to study pollination biology of a species to have a better understanding of genetic structure of a population and its genetic diversity. Such information on pollen flow and consequent gene flow is also mandatory before the release of any transgenic crop species for commercial cultivation. It helps in planning the strategy to prevent unwanted introgression of transgenes into the wild and its ill-effects on pollinators. Studying reproductive biology at the population level can also provide important taxonomic insights into the delimitation of species even at the infraspecific level.

- **Reproductive Biology and Crop Breeding:** Genetic improvement of crops has been a constant endeavor of agriculturists for ensuring food security. Developing new varieties that are higher yielding, disease resistant, drought tolerant or better adapted to different environments and growing conditions, can be achieved through hybridization. This approach requires information on reproductive events for planning and carrying out breeding programs. To carry out manual pollinations, one must know the stage and duration of anther fertility and stigma viability, so that emasculation can be performed before the attainment of maturity of either stigma or anther. The studies on floral biology and embryo-endosperm development are also helpful for carrying out various *in vitro* protocols, viz. anther culture, pollen culture, embryo culture and such-like.

- **Reproductive Biology and Evolutionary Biology:** Research in plant evolutionary biology aims to understand the basis of adaptation and speciation in different plant systems. Today, reproductive systems of flowering plants are the focus of considerable research in plant evolutionary biology. This is primarily due to the ecological and evolutionary consequences of these reproductive adaptations which are often cited as the reason for the success and diversification of angiosperms. Therefore, the understanding of evolution of diverse sexual systems and different modes of pollination is crucial in understanding the general evolution of angiosperms. Application of molecular genetic approaches to understand the mechanisms responsible for changes in reproductive system among different lineages have added an altogether new dimension to reproductive biological studies. For example, genetic and molecular studies on dioecious model systems have shown significant variation in the inheritance pattern of sex, ranging from single locus to multiple loci to sex-specific chromosomes. Nearly 39 plant taxa have been examined for sex chromosomes among the angiosperms. Molecular genetic approaches can also help to decipher reproductive transitions which have evolved in a particular angiosperm lineage over the years.

Reproductive Biology of plants is a focal theme and understanding it is of utmost importance for managing the environmental challenges, for maintaining biodiversity and genetic resources, for crop improvement and understanding the ecology and evolution of plants.

Bibliography

Mangla, Y., Khanduri, P. and Tandon, R. (2016). Gender determination mechanisms in plants: foundation for unisexuality. *The International Journal of Plant Reproductive Biology* 8:103–114.

Mangla, Y., Tandon, R. Goel, S., and Raina, S. N. (2013). Structural organization of the gynoecium and pollen tube path in Himalayan sea buckthorn, *Hippophae rhamnoides* (Elaeagnaceae). *AoB PLANTS* 5: 1–11. plt015; doi:10.1093/aobpla/plt015.

Shivanna, K. R. and Tandon, R. (2014). *Reproductive Ecology of Flowering Plants: A Manual.* Springer-India.

Tandon, R., Shivanna, K. R. and Koul, M. (2020). *Plant Reproductive Ecology: Patterns and Processes.* Springer-India.

Mangla, Y. and Gupta, C. K. (2014). Love in the air-wind pollination: ecological and evolutionary considerations. In R. Kapoor, I. Kaur and M. Koul, eds., *Plant Reproductive Biology and Conservation* Delhi: I. K. International, pp. 234–45.

2

The Flower

In angiosperms diverse forms of flowers are accompanied by an array of mating strategies and sexual systems.

2.1 Introduction

Angiosperms possess a vast diversity of flowers which serve various purposes for the different groups of living beings, including humans. Due to their color, fragrance, and beauty flowers have always occupied a special place in human lives. Flowers are considered sacred across most cultures and have inspired much artistic expression. Apart from their aesthetic value, flowers possess myriad medicinal properties that further enhance their value to humans. Describing from a botanist's perspective though, the flower is a unit of reproduction in angiosperms. A flower may be defined as a modified determinate shoot system with four distinct whorls, viz. calyx, corolla, androecium and gynoecium arranged on a receptacle. Outer whorls, calyx and corolla are leaf-like structures which are not directly involved in reproduction. The two inner whorls, the androecium and the gynoecium harbor the reproductive organs of the flower and are the ones involved in reproduction. Flowering plants exhibit enormous diversity in size, shape, color, symmetry and the other morphological features (Fig. 2.1). This diversity in floral forms plays a huge role in ensuring pollinator services by different groups of pollinators. The diverse forms of flowers are accompanied by an array of mating strategies and sexual systems in angiosperms.

The timing of flowering in plants is critical for their reproductive success as both late and premature flowering can limit proper seed development. Plants also attempt to realize their reproductive potential by synchronizing their flowering to match pollinator availability. Floral induction is promoted by distinct environmental cues such as photoperiod, vernalization and endogenous regulators like phytohormones. These signalling cues are perceived in the leaves and the shoot apical meristem (SAM) for induction of flowering. Plants use genetic machinery to control all events starting from induction of flower to development of different whorls. Research in the last few decades has identified numerous genes which are involved in floral induction, floral meristem formation, and

Figure 2.1 Flower diversity in angiosperms. A. *Ranunculus* sp. Arrow indicates carpels. **B.** *Lavendula* sp. (Lamiacaeae) with bi-lipped corolla. **C.** *Potentilla* sp. (Rosaceae). **D.** *Innula racemosa* (Asteraceae), a capitulum. **E.** *Impatiens* sp. (Balsaminaceae), a zygomorphic flower. **F.** Cactus flower. **G.** *Manilkara zapota* (Spotaceae). **H.** *Fuchsia* sp. (Onagraceae), Note the colorful sepals. **I.** *Podophyllum hexandrum* (Berberidaceae). **J.** *Tropaeolum majus* (Tropaeolaceae). **K.** *Peristrophe* sp. **L.** *Ipomea* sp. (Convolvulaceae), funnel shaped flowers. **M.** *Bauhinia variegata* (Fabaceae). **N.** *Passiflora incarnata* (Passifloraceae). **O.** *Calotropis procera* (Apocynaceae). **P.** *Ixora* sp. (Rubiaceae). **Q.** *Oxalis* sp.(Oxalidaceae). **R.** *Hibiscus rosa-sinensis* (Malvaceae). **S.** *Primula* sp. (Primulaceae). **T.** *Alstroemeria* sp. (Alstroemerieae). **U.** *Yucca* sp. (Asparagaceae), bell shaped flowers. **V.** *Cynotis* (Commelinaceae), note the stamens with colorful hairs on filaments. **W.** *Arisaema* sp. (Araceae), also known as cobra plant, stigma extend from perianth tube like the tongue of a cobra. **See Color Plates (page 473).**

floral organ development. Genes which control floral organ development are called floral organ identity genes. These genes belong to the MADS box gene family and are also known as homeotic genes. The functioning of these genes is explained by the ABCDE model of flower development. This chapter gives an outline of the organization of a flower, sexual system seen in angiosperms and a summary of the components that play important role in the floral induction and floral organ development.

2.2 Organization of a Flower

Flowers on a plant bloom either singly or in groups called inflorescence. There are many types of inflorescence seen in angiosperms. The largest inflorescence among angiosperms is found in *Amorphophallus titanum* (Araceae) which is about 3 meters in height. Flowers come in all shapes, sizes, colors, forms, anatomical arrangement and with varied fragrances. They range in size from very small blossoms like *Wolffia* to giant blooms of *Rafflessia*. *Wolfffia microscopia,* has the smallest flower among angiosperm which is about 0.1mm across and *Rafflessia arnoldii,* has the largest flower, with diameters spanning up to a meter. Regardless of their diversity, all flowers have a common pattern of arrangement of the four whorls from outwards to inwards – calyx (comprised of sepals), corolla (comprised of petals), androecium (comprised of stamens), and gynoecium (comprised of carpels) (Fig. 2.2).

The two outer whorls i.e., calyx and corolla form the perianth of a flower (also called the **perigonium**) which is essentially a group of modified leaves. In some species perianth is undifferentiated into calyx and corolla and rather intergrade to form tepals which can be green (sepaloid, e.g., *Acer*) or colored (petaloid, e.g., *Lilium*). The tepals may be present in one or two whorls referred to as outer tepals and inner tepals (as seen in monocots, e.g., *Lilium, Amaryrllis,* orchids) (Fig. 2.3 A and 2.4 A). There are a few plants in which flowers lack perianth altogether like species of genera *Euphorbia, Podostemum, Polypleurum, Salix.*

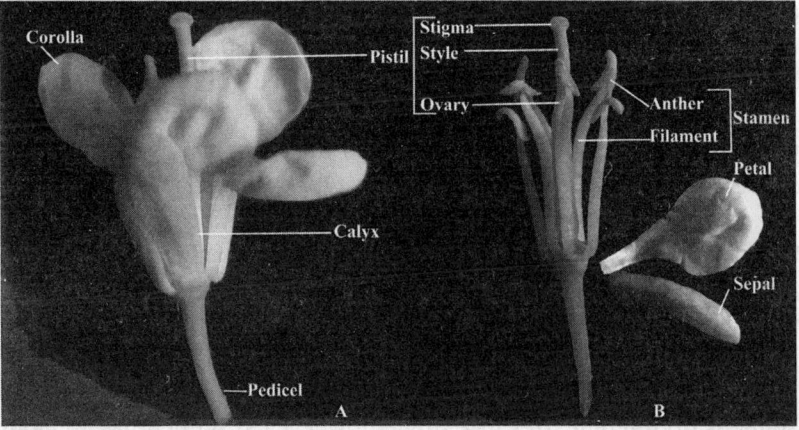

Figure 2.2 A. A typical flower (*Brassica* sp.). Distinct calyx (sepals) and corolla (petals) can be seen. **B.** Calyx and corolla are removed to show the androecium (stamens) and gynoecium (pistil). The floral parts are labelled. **See Color Plates (page 473).**

Figure 2.3 A. A flower of *Lilium longiflorum* showing petaloid tepals arranged in two whorls: outer 'o' and inner 'i'. st: stigma; an: anther. **B.** *Euphorbia* sp. Here the perianth is absent. Solitary female flower (♀) remains surrounded by several male flowers (♂) in a specialized inflorescence, cyathium (arrowhead). **See Color Plates (page 474).**

The calyx and corolla of flowers may provide peculiar shapes to flowers. Some common shapes of perianth are campanulate (bell-shaped), tubular (tube-like), papilionaceous (butterfly-like), ligulate (tongue-shaped), rotate (wheel-shaped), infundibular (funnel-shape), clavate (club-shaped), urceolate (urn-shaped) bilipped (two-lipped), and slaverform (trumpet-shaped) (Fig. 2.1 and 2.4). The calyx and corolla do not have a regulatory function in reproduction but aid in reproductive events like pollination. The calyx covers and protects other parts of the flower in the bud stage. Though mostly green in color, the calyx sometimes can be colorful too, in order to attract the pollinators as seen in *Fuschia* (Fig. 2.1 H), *Clerodendrum* sp. After the flowering, the sepals mostly fall off except in some plants where sepals persist in fruits (e.g., in Solanaceae) and may also become fleshy (e.g., in Nyctaginaceae). The corolla is the main attraction for pollinators in a flower and exhibit a fascinating diversity of shape, color, size and arrangement. The arrangement of petals in some families like Fabaceae and Orchidaceae is quite distinct as one of the petals is distinguished from other petals by either shape, markings or color (Fig. 2.4 A–B). Some flowers possess, markings called 'nectar guides' or 'guide marks' on petals which offer guidance to the pollinators (e.g., *Jacaranda mimosifolia*, orchids, *Digitalis purpurea*; Fig. 2.4 A, C). In some species, the corolla changes color as a visual signal for pollinators so that they can avoid visiting old flowers (e.g., *Pedicularis monbeigiana*, *Pulmonaria collina*; Oberrath & Gaese, 1999; Sun et al. 2005). Corolla color change also facilitates attraction of different sets of pollinators in species like *Combretum indicum* (syn. *Quisqualis indica*) where flowers change color from white to pink to red to shift from moth to bees to butterfly pollination (Yan et al. 2016).

The two inner whorls, the androecium and the gynoecium, are the ones which are directly involved in reproduction. The third whorl of a flower, the androecium comprises stamens. The number and size of stamens varies among taxa. Sometimes, a species may also have stamens of different sizes in a single flower. Each stamen has an elongated stalk or pedicel like structure called filament, which is topped by a lobed structure called the anther (Fig. 2.3 A). A mature anther typically consists of two compartments called thecae, with each theca containing two microsporangia in which the pollen grains develop. Thus, a typical anther is a tetrasporangiate structure with exceptions like Malvaceae, Cannaceae

Figure 2.4 A. A specialized flower of an orchid (*Phalaenopsis* sp.). The perianth is petaloid and arranged in two whorls of three each. One of the petals (in inner whorl) is modified and shows nectar guides on it called labellum, **ov:** ovary. **B.** An intact flower of *Crotolaria* sp. (side view) and a where petals are dissected out. Largest petal is the standard petal, two lateral petals appear like wings and two are fused to form a keel. **C.** Flower of *Digitalis purpurea* showing nectar guides (arrowhead). **See Color Plates (page 474).**

where anthers are monothecous and bisporangiate. Once pollen grains have matured, the anther dehisces and liberates them.

The inner most or the fourth whorl of a flower is the gynoecium which is made up of pistil/s. A typical pistil can be divided into three main parts: the basal ovary containing the ovules, the style: a conduit or path for pollen tubes to grow, and the terminal stigma on which pollen grains land and germinate. Each pistil is comprised of one or more carpels (each with its stigma, style and ovary). The number of carpels varies among species. In species with more than one carpel (pluricarpellate), carpels can be free (apocarpous) or can be completely fused (syncarpous). If the gynoecium is apocarpous, the number of carpels is the same as the number of pistils (e.g., *Ranunculus* sp., Fig. 2.1 A). However, in a syncarpous condition, a single pistil can be composed of number of fused carpels. Thus, pistil contains an ovary, one or more styles (may sometimes be absent), and one or more stigmas. The morphology of the stigma shows a lot of diversity in angiosperms, e.g., single capitate in *Brassica*, bilobed in *Solanum* sp., pentafid stigma in *Hibiscus rosa-sinensis* (Fig. 2.1 R; for details see Chapter 7).

Flowers of a large number of species possess all the four whorls and are considered complete. If any of the four whorls is missing, the flower is called an incomplete flower. Flowers bearing both the stamens and carpels are referred to as bisexual, perfect or monoclinous flowers. In several species (about 6–7%) of angiosperms; the flower may possess either of the two sexes i.e., stamen or pistil and are hence called unisexual or imperfect or diclinous flowers. The flowers possessing only stamens are called staminate or male flowers and the flowers possessing only pistils are referred to as pistillate or female flowers. Thus, flowers can have any of the three sexual forms: bisexual, male or female (Fig. 2.6).

Box 2.1

Bracts and Attraction

Bract is usually a small, green, leaf like structure present beneath a flower or inflorescence, e.g., in Asteraceae involucre bracts enclose the inflorescence. However, in some plants bracts are bigger and colorful and act as a source of attraction for pollinators, e.g., *Anthurium* sp. *Bougainvillea* sp.

Figure 2.5 A. *Blumea* sp. (Asteraceae). Arrow indicates the involucre bracts (arrow) enclosing the whole inflorescence **B–C.** *Anthurium* sp. and *Bougainvillea* sp. respectively; showing presence of colorful bracts (arrow). **See Color Plates (page 474).**

2.3 Sexual Diversity in Angiosperms

Depending on the arrangement or presence of staminate (male), pistillate (female) and/ or hermaphrodite flowers on the individual plants of a species or between plants in the population, an extensive sexual diversity may be displayed in flowering plants. Each species is characterized by a specific distribution of flower type or male and female reproductive structures on individual plants, and this is referred to as its sexual system. The diversity in sexual systems is also regarded as an adaptive diversification among flowering plants and associated with an impressive variety of mating strategies. The sexual systems in angiosperms can directly influence offspring quantity and quality, e.g., dioecy promotes cross-pollination and consequently enhances genetic variation (Richards 1997; Barrett 2010). An account of various sexual systems present among flowering plants with examples is provided below:

- **Hermaphroditism:** Individual plants bear bisexual flowers (Fig. 2.6 A–C). A great majority of the flowering plants families (72%) show presence of hermaphrodite flowers (Richards 1997; Renner 2014) and possess high incidences of selfing. Among some hermaphrodites, outcrossing is achieved either through morphological and phenological strategies such as herkogamy, heterostyly, flexistyly, enantiostyly, dichogamy, duodichogamy and heterodichogamy (Shivanna 2002; Barrett 2002; 2010) or through self-incompatibility mechanisms (Nasrullah & Nasrullah 1993). These strategies have been discussed in detail in Chapter 6 on '*Pollination*' and Chapter 8 on '*Self-incompatibility*.'

- **Monoecy:** Individual plants bear flowers of separate sexes, i.e., pistillate and staminate flowers (Fig. 2.6 D–F). Monoecy may prevent intra-flower self-

Figure 2.6 Major sexual systems in angiosperms. Hermaphrodite (A–C): **A.** Diagrammatic representation of a hermaphrodite plant bearing bisexual flowers. **B–C.** Hermaphrodite flowers of *Solanum* sp. and *Dianthus* sp.; respectively (an: anther; st: stigma). Monoecy (D–F): **D.** Diagrammatic representation of a monoecious plant. Here, unisexual flowers are present on the same plant. **E.** Unisexual flowers of monoecious *Luffa* sp. (i:male flower, an: anthers, ii: female flower with inferior ovary (ov), iii: close up of stigma (st) in female flower. **F.** *Ricinus communis*. Here male flowers (arrow) are present below the female flowers in the inflorescence. The exerted red stigma (st) of female flowers can be seen. Dioecy (G–I): **G.** Diagrammatic representation of dioecy in a species. Here, the male and female flowers are present on different plants. **H–I: H.** *Acer caesium*. Close up of a male flower with exerted anthers (an) (H) and pistilate flowers with exposed bifid stigma (st) and ovary (ov) (I). Arrowhead marks the male flower on male plant **(H)** and the female flower on female plant **(I)**. Here one may also note that perianth is not differentiated in to calyx and corolla (i.e., petaloid perianth). **See Color Plates (page 475)**.

pollination but not intra-individual self-pollination (Ainsworth 2000). It is reported in about 5% of angiosperms. Some common examples of families showing monoecy are Euphorbiaceae (*Phyllanthus, Jatropha, Ricinus*), Arecaceae (Coconut palm, Date palm), Moraceae (*Ficus* sp.), Poaceae (*Zea mays*), Cucurbitaceae (*Luffa* sp., *Cucurbita* sp., *Cucumis* sp.) (Fig. 2.6E). The arrangement of pistillate and staminate flowers on an individual plant varies among different species. For example, in cucurbits, the male and female flowers are randomly distributed. In *Ricinus communis*, the male flowers are arranged below the female flowers in an inflorescence (Fig. 2.6F) while in *Phyllanthus* sp. (Euphorbiaceae), staminate flowers are present towards the mother axis and pistillate flowers are present away from the axis. In *Ficus* spp., the inflorescence is a syconium, where the male flowers are present at the opening (ostiole) of the inflorescence and female flowers are near the base of inflorescence.

• **Dioecy**: Individual plants bear either male/staminate or female/pistillate flowers (Fig. 2.6 G–I). Dioecy is an extreme case of gender separation which prevents intra-individual self-pollination completely. The populations of dioecious species possess two types of plants i.e. plants bearing staminate flowers and plants bearing pistillate flowers. This is a strict mechanism which ensures cross pollination mainly facilitated by abiotic means (wind and water) or in some incidences by pollinators. Only 5–6% of all angiosperms are dioecious and are present in 43% of families (Renner 2014). Some examples of dioecious taxa are *Actidinia* sp. (Actinidiaceae), *Cannavis sativa* (Cannabaceae), *Morus alba* (Moraceae), *Salix, Populus* (Salicaceae), *Hippophae* (Elaeagnaceae), *Commiphora wightii* (Bursaraceae), *Acer oblongum, A. caesium* (Sapindaceae) (Fig. 2.6 H, I). Different pathways have been proposed for the evolution of dioecy among different dioecious taxa. Dioecy is a major evolutionary transition among flowering plants but many aspects of its evolution are yet to be understood.

2.4 Origin of Flower

The presence of flowers distinguishes the angiosperms from the gymnosperms (the only other seed-bearing group of plants). The word *Angiosperm* has been derived from the Greek word *Angeion* meaning vessel and *sperma* meaning seed, thus angiosperm is a plant group with a protected seed habit. Carpel is the structure enclosing and protecting the ovules (hence seed) in a flower because of which the evolution of angiosperms is also related to the origin of carpel. It has been suggested that the origin of the carpel and its subsequent morphological modifications may reflect the methods of acquittance of angiospermy. Let's recall the presence of megasporophylls (megaspore bearing leaf-like structures) in some pteridophytes (e.g., *Selaginella*) and also in gymnosperms (e.g., *Pinus, Cycas*). In gymnosperms, the ovules are naked, as the megasporophylls are not closed leaving the ovules exposed. While in angiosperms, the ovule bearing leaves referred to as carpels (equivalent to megasporophylls) are closed and cover the ovules. During the evolution of flowering plants,

Box 2.2

Intermediate Sexual States

Dioecy and Monoecy are considered two absolute sexual systems but there are many sexual intermediates that fall between these two extremes. These intermediate forms are distributed throughout angiosperm families suggesting independent evolutionary events (Ainsworth, 2000; Barrett 2002). For instance, Papaya exhibits a trioecious sexual system with three types of plants i.e. male, female, and plants with hermaphrodite flowers. This underscores the common saying by our ancestors that for fruiting one must plant at-least two papaya trees so that there is one female and one male tree. In Mango (*Mangifera indica*) also, female, male and hermaphrodite flowers may be found on single tree suggesting a polygamo-monoecious sexual system. Types of sexual systems (Geber et al. 1999) with examples are summarized below:

	Definition	Examples
Gynomonoecious	Individuals have female and bisexual flowers	*Atriplex, Cirsium (Asteraceae)*
Andromonoecious	Individuals have male and bisexual flowers	*Aralia, Gingidia, Lignocarpa, Olea europaea, Passiflora incarnata*
Polygamomonoecious	Individuals have male, female and bisexual flowers	Coconut palm, *Hippophae rhamnoides*
Gynodioecious	Individuals bear either female or bisexual flowers	*Plantago lanceolata, Daphne laureola, Leucopogan melaleucoides*
Androdioecious	Individuals bear either male or bisexual flowers	*Datisca glomerata, Ulmus minor*
Trioecious	Individual plants in a population bear either male, female or bisexual flowers	*Wurmbea dioica, Carica papaya*

the closure of carpel has occurred through different evolutionary-ontogenic routes. The carpel is thought to be derived from closure of megasporophylls that subtended ovules in the pre-angiosperm lineage. Following closure, some sites on the closed carpel provided seat to pollen grains for germination (as receptive surface) and delivering the pollen tubes. In the due course of evolution, the receptive surface of the carpel became limited and confined to a smaller region representing the present-day stigma in flowers. The presence of a confined stigma posed a constraint to the germination and growth of pollen grains. As a result, a need arose for a specialized carpel tissue that could provide retention and better transmission of the male gametes to the female gametophyte (embryo sac). This specialized carpel tissue is referred to as transmitting tissue or compitum, which represent the style in present day flowers. This tissue served as a site for pollen tube growth, nutrition, competition and selection. Here, it is noteworthy that in the entire sequence of carpel evolution, it was the ovary which evolved first, followed by stigma and later the style. The understanding of this sequence is the result of numerous ontogenetic and phylogenetic studies based on different groups of angiosperms (basal, eudicots). Several genera in basal and advanced groups of angiosperms can be arranged to represent the sequential events leading to the understanding of evolution of gynoecium (Endress 1994; Endress & Doyle 2009).

2.5 Induction of Flowering

Flower development is a highly coordinated multistep procedure involving floral induction, floral meristem formation and floral organ development. All the steps of flower development are under the strict control of a gene regulatory network that is composed of interacting genes and their protein products (Kinoshita & Richter 2020). The expression of all these genes is tightly regulated in a temporal and spatial manner to produce a functional flower essential for successful reproduction. Flower induction is a physiological process during which vegetative to reproductive transition of the shoot apical meristem takes place. Induction of flowering is promoted by different environmental cues in different species. While the photoperiod is crucial for flowering in rice, vernalization promotes flowering in *Arabidopsis*, and ambient temperature influences flowering in several orchids. Also, phytohormones (e.g., gibberellins) are known to enhance flowering in short day plants like *Cosmos* and *Pharbitis* (Wittwer & Bukowak 1962; Sun et al. 2014; Wang et al. 2019). After reaching the right developmental stage and perceiving the right environmental cues flowering-time genes are triggered. *FLOWERING LOCUS C (FLC)* is one of the best characterized genes in plants which is responsible for vernalization-dependent flowering time regulation and induction (e.g., in *Arabidopsis*, *Zea mays*). Another gene, *FLOWERING LOCUS T (FT)* regulates the photoperiod dependent transition of shoot apical meristem into floral meristem. Studies have highlighted the constant presence of these two genes in most examined angiosperms (Adeyemo et al. 2017; Kennedy & Geuten 2020). Apart from mediating transition of SAM to inflorescence meristem, flowering-time genes also activate the meristem identity genes. The floral meristem is established by the floral meristem identity genes, which include *LEAFY (LFY)*, *APETALA1 (AP1)* and *CAULIFLOWER (CAL)*, together with *UNUSUAL FLORAL ORGANS (UFO)*. The floral meristem elicits the genes regulating the floral organogenesis which then produce the four concentric whorls of floral organs (Wang et al. 2019). Details of floral organ identity genes are discussed in the next section.

2.6 ABCDE Model of Floral Organ Development

Research in last few decades has highlighted the development of a flower being a perfect orchestration between numerous genes. A number of genes with their functions have been identified which are involved in the formation of different whorls of a flower. These genes are called the floral organ identity genes, which work in a synchronized way to control floral organ development. These genes which specify regional and segmental identities of different floral organs are also known as homeotic genes. Their down-regulation or mis-expression can cause homeotic mutations within and outside the floral domains (mutants exhibit anomalies like the absence of one whorl or the presence of a whorl at the wrong position or the doubling of a whorl). Proteins encoded by floral organ identity genes possess a characteristic DNA binding domain, the MADS domain and thus belong to a MADS box family of transcription factors. Their synchronized functioning and regulation in the making of a flower was first comprehensively explained by Coen and Meyerowitz (1991) in form of the ABC model of flower development in the model species *Antirrhinum* and *Arabidopsis*. The ABC model explained three major classes of genes and transcription factors (class A, B and C) involved in the development of four floral organs – the sepal, the petal, the stamen and the carpel. A-class consists of *APETALA1* (*AP1*) and *APETALA2* (*AP2*) genes, B-class consists of *APETALA3* (*AP3*) and *PISTILLATA* (*PI*), genes, and C-class consists of a single gene *AGAMOUS* (*AG*). While A-class genes control development of sepals, B-class genes in combination with A-class genes regulate formation of petals. A coordinated expression of C-class genes and B-class genes is required for differentiation of stamens and the C-class gene alone is responsible for carpel development. Although, the ABC model elucidated the development of floral organs to a large extent, it was deemed insufficient to explain certain complications like the loss of B-function in *Antirrhinum* resulting in lack of the fourth floral whorl (i.e., gynoecium). This suggests that, over and above petals and stamens, B class genes are also necessary for the formation of carpel. Also, the ABC model failed to explain the formation of ovules. Thus, two other classes of genes, namely D and E were later added to the model and it was extended as the ABCDE model of floral organ development (Angenent & Colombo 1996; Pelaz et al. 2000; Theißen 2001). In *Arabidopsis*, class D genes are *SEEDSTICK* (*STK*), *SHATTERPROOF1* (*SHP1*) and *SHP2* and Class-E gene is *SEPALLATA* (*SEP*). According to the ABCDE model, class A + E genes specify development of sepals; class A+B+E work in conjunction to form petals; class B+C+E correspond to stamen differentiation; class C+E specify carpel development and class C+D+E together control ovule development (Fig. 2.7).

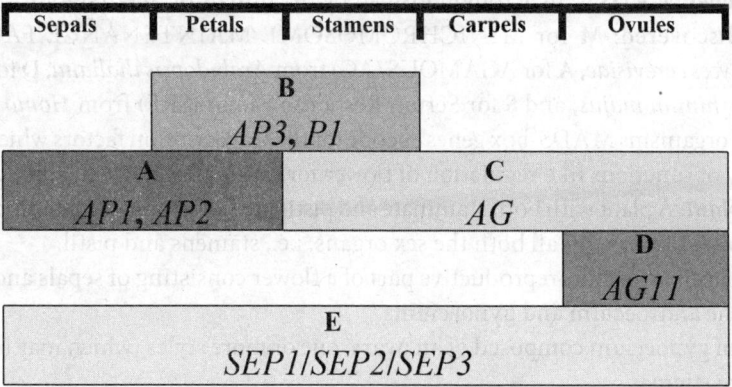

Figure 2.7 Diagrammatic representation of 'ABCDE model' for flower organ identity in *Arabidopsis*. (*Source*: Adapted and modified after Theißen et al. 2016.)

Box 2.3

Nectary

A nectary is regarded as a localized gland which synthesizes and secretes nectar. In plants, the nectary may be present either on the vegetative (petioles, stipules, leaf blades, cotyledons) or the reproductive parts (receptacle, sepal, petal, stamen, filament, anther, ovary, style, stigma, fruits). Based on their location, the nectaries which are present on floral parts (like anther, filament, ovary) are known as floral nectaries while rest as known as extrafloral nectaries. Floral nectaries play a crucial function in pollinator attraction by means of nectar production; as nectar serves as a reward. There is remarkable diversity in morphology and histology of nectaries among flowering plants. Likewise, the nectar secreted also varies in composition; differing in carbohydrates, alkaloids or other components. Families where floral nectaries are commonly present are Bignoniaceae (*Oroxylum indicum*), Malvaceae (nectaries associated with petals),, Solanaceae (*Datura*: ovary associated nectaries), Brassicaceae (*Arabidopsis thaliana*: stamen associated nectaries). Studies in *Arabidopsis* suggested that B, C, and E gene functions regulate nectary specification while in *Petunia* where nectaries are carpel associated; C-function gene is possibly required for nectary development.

Source: Morel et al. 2018; Phukela et al. 2020.

Glossary

Carpel: A unit of the pistil consisting of an ovary, a stigma, and usually a style.

Complete flower: Flowers that possess all the four floral whorls.

Dioecious Plants: Plant species with staminate and pistillate flowers on separate individual plants.

Hermaphrodite: A plant with only bisexual flowers.

Homeotic genes: A class of regulatory genes that control time and place of development along with the identity of tissues and organs.

Imperfect Flower: A flower possessing either of the two sex organs i.e., stamen or pistil.

Incomplete flower: A flower in which any of the four floral whorls is missing.

MADS Box genes: MADS is an acronym formed by the initials of the first four MADS-domain proteins discovered: **M** for MINICHROMOSOME MAINTENANCE FACTOR1 from *Saccharomyces cerevisiae*, **A** for AGAMOUS (AG) from *Arabidopsis thaliana*, **D** for DEFICIENS from *Antirrhinum majus*, and **S** for Serum Response Factor (SRF) from *Homo sapiens*. In all eukaryotic organisms MADS-box genes encode for the transcription factors which are involved in a variety of functions like regulation of flower formation, response to abiotic stress.

Monoecious plant: A plant with both staminate and pistillate flowers on the same individual plant.

Perfect flower: A flower with all both the sex organs, i.e., stamens and pistil.

Perianth: Perianth is the non-reproductive part of a flower consisting of sepals and petals which surround the androecium and gynoecium.

Pistil: A unit of gynoecium composed of an ovary, one or more styles (which may be absent), and one or more stigmas.

Tepals: Leaf-like structures representing the perianth when latter are not differentiated into sepals and petals.

Key Questions

Q2.1   Discuss the formation of a flower in *Arabidopsis* in light of the ABCDE model of floral organ development.

Q2.2   Write notes on:
- a. Sexual systems in angiosperms
- b. Structure and function of floral organs
- c. Induction of flowering
- d. Functions of calyx and corolla

Q2.3   Differentiate between:
- a. Dioecy and Monoecy
- b. Hermaphroditism and Unisexuality

Practicals

Exercise 2.1: To study important morphological features of randomly selected flowers

A flower is the reproductive unit of angiosperms. The study of reproductive biology of any flowering plant species starts with the detailed observations of its flower. While describing the morphology of a flower other than observing the overall shape and size of the flower, the morphological details of each floral whorl are also important.

Materials required

Flowers of randomly selected plants, measuring scale, hand lens, forceps, needles and stereomicroscope.

Procedure

1. Record whether the flowers are borne singly or in an inflorescence.
2. If flowers are in an inflorescence, describe the type of inflorescence.
3. Count the number of flowers in at-least 4–5 inflorescences and compute the average number of flowers produced per inflorescence.
4. Measure the size (length and width/diameter) of the flower and record its color and number of floral whorls present.
5. Record the number of units in each floral whorl i.e. number of sepals, petals, stamen and carpel.
6. Note whether the perianth is fused or not. Note the shape of the fused perianth and measure its length and width. If the perianth is not fused, measure the length of individual sepals and petals.
7. Record the position of the anthers and the stigma with respect to the corolla.
8. Dissect the stamens and measure the length of filament and also the length and width of anther.
9. Measure the length and diameter of the ovary, the length of the style and the size of the stigma (diameter/length).
10. Record the type of stigma and the style. Note whether the carpels are free or fused. To determine the number of carpels in a fused condition, cut the transverse section of the ovary. (*in general the number of carpels corresponds to the number of major vascular bundles*)
11. Tabulate your observations for each species (*suggestions about some species suitable for classroom studies are given in* Table 2.1).

Table 2.1 Parameters to study morphology of flowers of selected species

Species	Average number of flowers per inflorescence	Type of Inflorescence	Size of flower	Number of whorls	Perianth		Shape of corolla (flower)	Color	Other features (e.g., nectar guides)
					Distinguished	Not Distinguished			
Brassica									
Cucurbita sp.									
Hibiscus									
Hamelia									
Euphorbia milli									
Mangifera indica									
Solanum nigrum									

Table 2.2 Parameters to study sexuality of flowers of selected species

Species	Average number of flowers per inflorescence	Number of whorls in flower	Presence of sex organs		Possible sexual system
			Stamen	Pistil/Carpel	
Stellaria media					
Vinca rosea					
Malva sp.					
Hamelia patens					
Euphorbia milli					
Ricinus communis					
Morus alba					
Carica papaya					

Exercise 2.2: To observe the sexuality of randomly selected flowers

Sexuality of a species is determined by presence/absence of the sex organs i.e. stamens and pistil. Based on the presence of either male/female or both sex organs, flowers of a species can be unisexual or bisexual respectively. Adaptations in flowering plants are mostly directed to promote outcrossing. While unisexual flowers are strictly outcrossing bisexual flowers have adopted certain strategies to realize it. However, not all flowering plants with bisexual flowers are cross-pollinating.

Materials required

Flowers of randomly selected plants, hand lens, stereomicroscope, glass slides, needles, forceps.

Procedure

1. Select a number of flowers (minimum n=20–30) from each of the randomly selected plants (minimum n=5) of a species.
2. Observe each flower for its gender (bisexual/male/female) by observing the presence of stamens and pistil.
3. If the flowers are borne in inflorescence, record the gender (sex) of all the flowers. Also note the position of flowers of each sex in the inflorescence. Record your observation in a table (*suggestions of species suitable for classroom studies have been provided in* Table 2.2).

Bibliography

Adeyemo, O., S. Chavarriaga, P. Tohme, J. et al. (2017). Overexpression of *Arabidopsis FLOWERING LOCUS T (FT)* gene improves floral development in cassava (*Manihot esculenta*, Crantz). *PLoS ONE* 12: e0181460. https://doi.org/10.1371/journal. pone.0181460

Ainsworth, C. (2000). Boys and girls come out to play: the molecular biology of dioecious plants. *Annals of Botany* 86: 211–21.

Amasino, R. M., Cheung, A. Y., Dresselhaus, T. and Kuhlemeier, C. (2017). Focus on flowering and reproduction. *Plant physiology* 173: 1–4.

Angenent, G. C. and Colombo, L. (1996). Molecular control of ovule development. *Trends in Plant Science* 1:228–32.

Barrett, S. C. H. (2002). The evolution of plant sexual diversity. *Genetics* 3: 274–84.

Barrett, S. C. H. (2010). Understanding plant reproductive diversity. *Philosophical Transactions of the Royal Society: B-Biological Sciences* 365: 99–109.

Castelán-Muñoz, N., Herrera, J., Cajero-Sánchez, W., et al. (2019). MADS-box genes are key components of genetic regulatory networks involved in abiotic stress and plastic developmental responses in plants. *Frontiers in Plant Sciences* 10:853. doi: 10.3389/fpls.2019.00853

Coen, E. S. and Meyerowitz, E. M. (1991). The war of the whorls: genetic interactions controlling flower development. *Nature* 353: 31–37.

Endress, P. K. and Doyle, J. A. (2009). Reconstructing the ancestral angiosperm flower and its initial specializations. *American Journal of Botany* 96: 22–66.

Endress, P. K. (1994). *Diversity and Evolutionary Biology of Tropical Flowers*. Cambridge: Cambridge University Press.

Endress, P. K. (2011). Evolutionary diversification of the flowers in angiosperms. *American Journal of Botany* 98: 370–96.

Geber, M. A., Dawson, T. E. and Delph, L. F. (1999). *Gender and Sexual Dimorphism in Flowering Plants*. Heidelberg: Springer.

Kennedy, A. and Geuten K. (2020). The role of *FLOWERING LOCUS C* Relatives in cereals. *Frontiers in Plant Sciences* 11: 617340. doi: 10.3389/fpls.2020.617340

Kinoshita, A. and Richter, R. (2020). Genetic and molecular basis of floral induction in *Arabidopsis thaliana*. *Journal of Experimental Botany* 71: 2490–504.

Mabberley, D. J. (2008). *A Portable Dictionary of Plants, Their Classifications and Uses*. Seattle: University of Washington Botanic Gardens.

Morel, P., Heijmans, K., Ament, K., et al. (2018). The floral C-lineage genes trigger nectary development in *Petunia* and *Arabidopsis*. *The Plant Cell* 30: 2020–37.

Nasrallah, J. B. and Nasrallah, M. E. (1993). Pollen-stigma signaling in the sporophytic self-incompatibility response. *The Plant Cell* 5: 1325–35.

Oberrath, R. and Böhning-Gaese, K. (1999). Floral color change and the attraction of insect pollinators in lungwort (*Pulmonaria collina*). *Oecologia* 121: 383–91.

Pelaz, S., Ditta, G. S., Baumann, E., Wisman, E. and Yanofsky, M. F. (2000). B and C floral organ identity functions require SEPALLATA MADS-box genes. *Nature* 405: 200–3.

Phukela, B. and Adit, A. (2020). A snapshot of evolutionary history of floral nectaries across angiosperm lineages. In R. Tandon, K. R. Shivanna and M. Koul, eds., *Reproductive Ecology of Flowering Plants: Patterns and Processes*. Springer Nature Singapore Pte Ltd., pp. 105–29.

Renner, S. S. (2014). The relative and absolute frequencies of angiosperm sexual systems: dioecy, monoecy, gynodioecy, and an updated online database. *American Journal of Botany* 101: 1588–96.

Richards, A. J. (1997). *Plant Breeding Systems*. London: Chapman and Hall.

Rijpkema, A., Gerats, T. and Vandenbussche, M. (2006). Genetics of floral development in *Petunia*. *Advances in Botanical Research* 44: 237–78.

Shivanna, K. R. and Tandon, R. (2014). *Reproductive Ecology of Flowering Plants: A Manual*. Springer-India.

Shivanna, K. R. (2002). *Pollen Biology and Biotechnology*. New Delhi: Oxford & IBH Publishing Co. Pvt. Ltd.

Sun, C., Chen, D., Fang, J., et al. (2014). Understanding the genetic and epigenetic architecture in complex network of rice flowering pathways. *Protein Cell* 5: 889–98.

Sun, S. G., Liao, K., Xia, J. and Guo, Y. H. (2005). Floral color change in *Pedicularis monbeigiana* (Orobanchaceae). *Plant Systematics and Evolution* 255: 77–85.

Theißen, G. (2001). Development of floral organ identity: stories from the MADS house. *Current Opinion in Plant Biology 2001*, 4: 75–85.

Wang, S. L., Viswanath, K. K., Tong, C. G., et al. (2019). Floral induction and flower development of orchids. *Frontiers in Plant Science* 10: 1258.

Weigel, D. and Meyerowitz E. M. (1994). The ABCs of Floral Homeotic Genes. *Cell* 78: 203–9.

Wittwer, S. H. and Bukovac, M. J. (1962). *Some Gibberellin Effects on Flowering of Plants*. In R. Knapp, eds., *Eigenschaften und Wirkungen der Gibberelline*. Berlin, Heidelber: Springer, pp. 68–73.

Yan, J., Wang, G. and Sui Y. (2016). Pollinator responses to floral color change, nectar and scent promote reproductive fitness in *Quisqualis indica* (Combretaceae). *Scientific Reports* 6: 24408.

3

Brief Historical Account on Transformation of Classical Embryology to Integrated Reproductive Biology

The inception of interest in plant reproduction is as old as the inception of interest in biology. The science of sexual plant reproduction is more than 400 years old. During all these years there has been an accumulation of information which has greatly enriched our understanding of plant reproduction. Our current knowledge of plant reproduction is a result of continuous efforts of scientists world-wide that transformed it from an observational investigation to an important field of experimental science. This chapter provides a summary of important mile stones in the history of reproductive biology of flowering plants. To maintain conciseness, only significant contributors across the world including India have been mentioned in the chapter without undermining the importance of others whose ideas and concepts have shaped the science of plant reproduction today.

3.1 Early Discoveries

The long and venerable history of studies in plant reproduction dates back to seventeenth century with the ideas of European naturalists **Rudolf Jacob Camerarius** in Germany and **Nehemiah Grew** in England. Though, Grew first proposed the idea of sexual processes occurring in plants for generation of seeds, to Camerarius must go the principal credit for experimentally establishing the existence of plant sexuality. N. Grew, in an address to the Royal Society of London in 1676, had expressed the view that the stamens are the male organs of a flower and the pollen act as vegetable sperm. He is credited for

Rudolf Jacob Camerarius

documenting stamens as the male sex organ of plants in his book *The Anatomy of Plants* (1682). The experiments of Camemarius were primarily based on the removal of stamens and styles along with isolation of female plants in dioecious species. He provided evidence for the inevitability of both sex organs in seed formation. His publications *On the Sex of Plants* (1694) and *Botanical Works* (1697); are landmarks in the history of botany.

In the history of plant reproduction, **Adam Zalužanský** is a little-known botanist. He was a professor at the University of Prague and his book *Methodi herbariae libri tres* was published in 1592 and 1604. This book includes a chapter *De sexu plantarum in* which, almost a century before the work of Camerarius, sexuality of plants was suggested. Thus, it will not be incorrect if Adam Zalužanský is credited with the conceiving the idea of sexuality in plants. Interestingly, the correspondence of a Dutch merchant, **Willem Bosman**, indicates recognition of the sexuality of plants among the papaya growers of West Africa much before it was known to the scientific world (Baker 1979). In 1700, while visiting the Guinea Coast of West Africa, Bosman wrote to a correspondent in Holland '*There grow multitudes of papaya-trees all along the Coast; and these are of two sorts, viz. the male and female or at least they are here so called, on account that those named males bear no fruit but are continually full of blossoms, consisting of long white flowers; the females also bears the same blossom, though not so long, nor so numerous. Some have observed that the females produce their fruits in greatest abundance when the males grow near them.*'

3.2 Era of Exploration

In the eighteenth century noted contributions in the field of plant reproduction were mainly made by European naturalists. In 1729, **Carolus Linnaeus** published his first paper on the sexuality of plants which gained considerable attention and acceptance. Following which, in 1735 he proposed a 'sexual system' of classification of seed plants in his book *Systema Naturae*. This system of classification laid emphasis on the importance of reproductive characters and classified seed plants purely on the basis of their sex organs. In the mid-eighteenth century, **Joseph Kölreuter** (1761–1766) published a series of papers on his observations of pollination by insects. Through his work he established that transfer of pollen from stamen to stigma is essential for seed set in plants. Kölreuter recognized the importance of insects as

Joseph Kölreuter

agents of pollen transfer in plant fertilization. However, he believed that insect-mediated pollen transfer brought about self-pollination only. Kölreuter is also believed to have made the first systematic study of plant hybrids. He cultivated plants with the purpose of studying their fertilization and development. He performed various experiments (particularly with the tobacco plants), which included artificial fertilization and production of fertile hybrids between plants of different species. This was followed by the work of a German theologist and naturalist **Christian Konrad Sprengel**, which was the first attempt to uncover the reason for diversity of flowers. His book *Das entdeckte Geheimniss der Natur im Bau und in der Befruchtung der Blumen* made an important contribution to our understanding of the role of insects in plant reproduction. He studied relationships between flowers and their pollinating insects and concluded that floral architecture is adapted to facilitate their pollinators. This work which was unfortunately not appreciated until almost 50 years later by the likes of Charles Darwin is believed to have initiated the discipline of floral ecology (Vogel 1996). His work was also important for its recognition of cross-pollination for the first time as Sprengel concluded that 'nature did not intend any flower to be fertilized by its own pollen'. Around same time, in 1799, **Thomas Knight** published his results on the crosses he made between varieties of garden pea and concluded that cross fertilization leads to superior progenies. This was an evidence for the adaptive significance of cross pollination in plants.

3.3 Years that Laid the Foundation

Nineteenth century was a great period for research in the field of plant reproduction when some very fundamental discoveries were made. Despite Kölreuter and Sprengel's work, the existence of cross fertilization in seed plants was still being questioned in the beginning of the nineteenth century. The contention was put to rest with notable observations and experimental contributions by **Carl Friedrich von Gärtner**, **William Herbert, Fritz Müller** and **Hermann Müller, George Henslow** and certainly by **Charles Darwin**. Darwin can be considered the motivator for generating wide interest in plant-pollinators interactions and the reproductive biology of angiosperms. Though Darwin is mostly remembered for his work with tortoises and finches, he was immensely interested in plants and spent much of his time studying botany. Darwin published three significant works (books) on the subject: *On the*

Charles Darwin

Various Contrivances by Which British and Foreign Orchids are Fertilised by Insects, and on the Good Effects of Inter-crossing (1862), *The Effects of Cross and Self-fertilisation in the Vegetable Kingdom* (1876), and *The Different Forms of Flowers on Plants of the Same Species* (1877). These books laid intellectual foundations for evolutionary aspects of reproductive biology in plants which remain valid even today. He tested his hypothesis with rigorous pollination experiments which led him to establish for the first time that diversity of floral forms evolved to facilitate cross-pollination. He proved that cross fertilization increases offspring performance and also gave insights on the evolution and function of sexual polymorphisms. In 1862, while working on orchids from Madagascar he came across an orchid with beautiful and star-shaped flowers called *Angraecum sesquipedale* that possessed an exceptionally long spur with nectary (up to 30 cm). In his book on orchid pollination, Darwin suggested that this extreme feature (deep situated nectary) may have evolved alongside a moth with an exceptionally long tongue to pollinate it. While Darwin was hypothesising about an unknown species of moth purely based on the size of the flower, little did he know that there actually existed a moth with such a long tongue in the Congo region of Africa in the 1830s (*Xanthopan morganii*). More than 20 years after Darwin's death in 1907, a subspecies of the gigantic Congo moth was identified in Madagascar and named as *X. morganii praedicta,* after Darwin's prediction. However, whether this moth pollinated *A. sesquipedale* remained elusive till 1992, when nearly a century later, it was observed that *X. morganii praedicta* actually pollinated this African orchid. Such was the acumen and precision of Darwin's predictions.

G B Amici

During this time, there were other scientists who were working to unravel the mysteries of various plant reproductive processes and made some remarkable discoveries. **Giovanni Battista Amici** was an Italian astronomer whose interests were also in biological observations, especially botany. In forty years of microscopic observations, he made several contributions in plant biology, the notable ones being on fertilization processes in seed plants leading to the discovery of pollen tubes. In 1822, while observing hairs on the stigma of *Portulaca oleracea*, he noticed some granules lodged on them. The granules were actually pollen grains. It was serendipitous that while making his observations he saw a pollen grain attached to the hair suddenly split open and sent out a kind of tube or 'gut' (as he called it) which entered the tissues of the stigma (Maheshwari 1950). However, Amici could not trace the pollen tubes for long and eventually lost sight of them but he was convinced that the pollen tube continues to lengthen out until it reaches the ovary. After several years of efforts, finally in 1846, Amici presented a clear exposition of the overall process of fertilization of a plant, beginning with pollination till the full development of the embryo. Amici also elucidated the existence of the 'germinative vesicle' or 'embryo sac' in the ovule before fertilization which results in an embryo. Amici's discovery of pollen tubes was also established by a young French botanist **Adolphe Brongniart** (1827), who examined a large number of pollinated pistils in order to understand the interaction between the pollen and the stigma. He found that exit of pollen tubes from the pollen grains and their penetration into the stigma and style was a generalized phenomenon. Brongniart was a very prolific botanist who later worked extensively on fossils and came to be known as the *father of paleobotany*.

Though Amici was the first to observe pollen tubes, the stages of pre- and post-fertilization and that of ovule development were primarily studied by **Wilhelm Hofmeister** (1849). He showed that the egg cell and/or the synergids that comprised the egg apparatus, were present in the ovule before the arrival of the pollen tubes. Either the egg or synergids could be fertilized by the pollen tubes and eventually undergo embryogenesis. He traced the formation of embryo from egg up to its maturity in seed. Hofmeister also observed the formation of microspore tetrads.

Although research in apomixis began in the twentieth century, the initial discovery of apomixis in higher plants occurred much before in 1839 and is attributed to **John Smith**. He observed a solitary female plant of an Australian species *Alchornea ilicifolia* (syn. *Cœlebogyne ilicifolia*) that produced seeds at Royal Botanic Gardens, Kew in the absence of male plants (Bicknell & Koultnow 2004). However, the term 'apomixis' to account for this type of reproduction of plants which did not involve male plants was later introduced by **Hans Winkler** in 1908.

W. Hofmeister

Eduard Strasburger was a German botanist who is regarded as founder of modern plant cell biology. His most notable contribution in the field of plant reproduction is the observation of fusion of the male sperm nucleus with the female egg nucleus during fertilization (1894). He proved that the pollen tube does not stay intact, but its apical tip disintegrates upon contact with the embryo sac and that one of its 'nuclei fuses with the nucleus of the egg' (Maheshwari 1950; Cresti & Linskens 1999). This union of male and female nuclei was observed in *Monotropa* and has been referred to as syngamy. Strasburger also reported the guidance of the tube by synergids and described the pollen grains in a number of species of angiosperms and

E. Strasburger

gymnosperms. He showed that out of the two cells in a ripe spore (the pollen grain), the smaller one is the generative cell which further divides to give rise to the two male nuclei (Chamberlain 1897). The division mostly takes place in the pollen tube, but in some cases, may take place in the spore itself such that the mature spore may contain three nuclei.

The nineteenth century concluded on a great note with the discovery of double fertilization in 1898 by **Sergius Nawaschin**. The breakthrough occurred when Nawaschin (1898; 1899) showed that in ovules of *Lilium martagon* and *Fritillaria tenella* (Liliaceae), both male gametes from the pollen tube penetrated the embryo sac. While one of them

S. Nawaschin

fused with the nucleus of the egg cell, the other fused with the polar nuclei (at that time known as the definitive nucleus) floating in the central cell, initiating a second fertilization event (Raghavan 2003). The phenomenon of 'double fertilization' observed by Nawaschin was later confirmed in *L. martagon* and *L. pyrenaicum* by **Léon Guignard** (1899) in France. In 1903, **Coulter and Chamberlain**, in their classical book *Morphology of Angiosperms*, mentioned that 16 families of angiosperms, encompassing about 40 genera and over 60 species have a second fertilization event.

3.4 Broadening Horizons

In the first half of the twentieth century, a great deal of attention was given to the study of plant embryology. The advancement of microscopy tools and micro techniques stimulated research in plant reproduction and initiated its spread to laboratories around the world. Several scientists in India, France, USA and the former Soviet Union made notable contributions towards understanding embryogenesis in most of the families of angiosperms (Willemse 2008). Most notable among them was **Panchanan Maheshwari**, whose vision and novel contributions made him the pioneer of plant embryology in India. His research group investigated over 100 families of angiosperms. He is credited with building a school of integrated plant embryology at the Department of Botany, University of Delhi, which was internationally acclaimed. Under his guidance, the Department informally and popularly came

P. Maheshwari

to be known as *Delhi School of Embryology*. Initial studies from this school largely utilized traditional methods of wax-embedded sectioning to study developmental events associated with embryological processes (Shivanna & Tandon 2020). Nevertheless, he always laid emphasis on the importance of integration of techniques such as histochemistry, electron microscopy and tissue culture for further advances in embryology. He is also well known for introducing the techniques of plant tissue culture for the first time in the country during 1960s which widened the scope of embryological studies (Shivanna & Tandon 2020). His book *An Introduction to the Embryology of Angiosperms,* published in 1950, is a classic in the field and one of the most cited biological textbooks ever. In this book, Maheshwari classified embryology into descriptive, phylogenetic and experimental forms. He established the International Society of Plant Morphologists in 1951, and started the international journal-Phytomorphology. Maheshwari and co-workers (Kanta et al. 1962) were also the forerunners in attempting to eliminate the barriers of stigma and style to overcome self-incompatibility by intra ovarian pollination (*see Chapter 8 for details*).

Another well-known researcher working during this time was **Bengaluru Gundappa Lakshminarayana Swamy** who was an internationally acknowledged botanist and also a gifted Kannada writer. Working on basal angiosperms, Swamy made important contributions on the origin of angisoperms in Irving Widmar Bailey's laboratory between 1948 and 1953. He was an adept taxonomist, anatomist and embryologist and this multi-faceted experience culminated into co-authoring a textbook entitled *From Flower to Fruit – Embryology of Flowering Plants*. He contributed immensely to the embryology of family Orchidaceae. B. G. L. Swamy along with Parameswaran in 1959, prepared a detailed analysis and classification of helobial endosperm and reported its restriction to monocots.

Post 1950's, plant reproductive biology entered a new era of ultrastructural studies and experimentation. This period included use of electron microscopy and witnessed the rise of experimentations to answer questions in plant reproduction. Some of the major contributors during this period were:

William August Jensen studied the cell differentiation associated with plant embryos and embryo sac development with the combination of techniques of histochemistry and electron microscopy. Jensen also explained the organization of female gametophyte (1965–1975) and generated valuable insights into the ultrastructure and composition of the egg and central cell of *Gossypium hirsutum*. He also authored *Botanical Histochemistry* (1962), a text that continues to be popular among embryologists.

Jack Heslop-Harrison is credited with a host of scientific discoveries and for generating fresh perspectives in the field of plant reproductive biology by using a variety of techniques ranging from genetical, cytological, anatomical, physiological, to biochemical (Gunning 2000). The main focus of his research was mechanism of flowering, male sterility, pollen development, pollen–pistil interactions, and self-incompatibility. He elucidated the micro-morphology and morphogenesis of the pollen wall using electron microscopy and also explained the existence of two self-incompatibility mechanisms: sporophytic and gametophytic. He deduced the presence of pollen wall enzymes and proteins and envisaged link of pollen wall proteins in incompatibility reactions. Heslop-Harrison also studied the biology of the pollen tube and its interaction with the female parts of the flower. His books *New Concepts in Flowering Plant Taxonomy* (1953) and *Sexuality of Angiosperms* (1972) remain key texts on the subject.

Ravinder Nath Kapil was a specialist of plant embryology and morphology who made embryological investigations into a number of little-known and interesting families of flowering plants like Orchidaceae, Euphorbiaceae, Leguminaceae, Scorphulariaceae, specifically in relation to their phylogeny. His research deciphered several facts about the development of ovules, endosperm and the haustoria in plant reproduction. He demonstrated the co-occurrence of mono, bi, and tetrasporic types of embryo sacs in the same species (*Delosperma cooperi*). He used embryological features in the delimitation and understanding of taxonomic relationships of various taxa and families.

Narinder Nath Bhandari worked on the embryology of primitive flowering taxa Ranunculaceae, Magnoliaceae, Winteraceae, and Viscaceae (in 1971). Using histochemical techniques, he investigated the nutrition of the embryo. In association with Professor Barbara Haccius and her group at the Institut für Spezielle Botanik der Universität, in Germany, he carried out morphogenetic studies on induction of androgenic haploids in *Nicotiana tabacum*. He also authored a classical book *Staining Techniques – A Manual* (1997); which is a true reflection of his expertise and skill in microtechniques.

Brij Mohan Johri was the first to report the entry of pollen grains in the style and ovary of *Butomopsis lanceolata* employing suction technology (similar to gymnosperms). He also reported the presence of the largest synergid haustorium and a 3-nucleate branched haustoria in antipodals of *Quinchamalium* (Santalaceae). He worked on several developmental processes related to sexual reproduction in many angiosperm families and elucidated the haustorial role of pollen tubes. Johri also examined the feasibility of endosperm culture, and reported the various growth responses of mature endosperm in different media and conditions. This was a significant contribution as prior to this, endosperm was considered to be a dead tissue. Johri was also successful in culturing the endosperm and embryo of various semi-parasites belonging to families Santalaceae and

B. M. Johri

Loranthaceae without any physical contact with the host tissue or the addition of extracts of parent tissue to the nutrient medium. He authored (and co-authored) two books: *The Angiosperm Pollen: Structure and Function* (1989), and *Comparative Embryology of Angiosperms* (1992). B. M. Johri also edited the popular *Experimental Embryology of Vascular Plants* (1982) and *Embryology of Angiosperms* (1984).

Haploid plants are of special interest to the geneticist and plant breeders for making homozygous diploid plants by chromosome doubling. Considering the immense theoretical and applied value of haploids, different methods were tried by scientists worldwide to develop haploids. **Katayama** and **Nei** (1964) suggested a brilliant approach for haploid formation through pollen culture but did not succeed in their experiments. The breakthrough came in 1964 when **Sipra Guha** and **Satish Chandra Maheshwari** reported induction and differentiation of embryos from the pollen grains of *Datura*. This was the first report of production of haploid embryoids and seedlings from the pollen grains of plant, a technique with great implications in the field of plant breeding and agriculture.

S. C. Maheshwari

David D. Cass and Scott D. Russell put forward the concept of the male germ unit in *Plumbago zeylanica*, where the sperm cells and vegetative nucleus function as a single male unit (also known as Male Germ Unit, MGU) in pollen (Russell & Cass 1981; Russell 1984). In their studies, Russell and Cass provided the details of unequal distribution of plastids and mitochondria in sperm cells. They also elucidated how two sperm cells are directly linked together by a common extracellular matrix with one of the sperms being associated with the vegetative nucleus through a cytoplasmic projection.

Robert Bruce Knox was an innovative scientist who applied new techniques of immunochemistry and histochemistry to advance the knowledge on plant reproduction. He was first to discover cell recognition mechanisms in the reproductive systems of flowering plants and explained how plants recognize self-pollen grains from cross-pollen grains (Kenrick & Lediges 2002). His study of pollen grains led to the discovery of various hydrolytic enzymes and proteins in pollen cell wall, also considered human allergens. Knox along with **McConchie and Jobson** (1985; 1987) provided computer-assisted three-dimensional reconstruction of the ultrastructure of sperm cells and male germ unit of *Brassica campestris* and *Zea mays*.

Holenarasipur Yoganarasimham Mohan Ram, fondly referred to as HYM, was a trained embryologist. He was considered a doyen of Indian botanists because of his wide interest and knowledge of plants. His major research contributions were: reasons for changes in flower color (pre to post pollination), sex expression in flowering plants, and *in vitro* culture of bamboo and aquatic angiosperms. He has to his credit, the first report of *in vitro* culture of an insectivorous plant *Utricularia* (Mohan Ram & Dutta 1966). HYM also initiated work on tissue culture of banana varieties using pulp (the triploid endosperm) in nutrient media containing different plant growth regulators.

H. Y. Mohan Ram

He and his group were the pioneers in establishing the fact that sex expression can be changed or reversed with the applications of growth regulators (Mohan Ram & Sett 1982). While working on *Lantana camara*, he proposed that color changes could play a role in conserving pollinator energy and also in the production of nectar by flowers (Mohan Ram & Mathur, 1984; Mathur & Mohan Ram 1986). He also initiated and carried out research on an embryologically intriguing family of aquatic angiosperms – the Podostemaceae, most of whose Indian members are endemic to the Western Ghats. Various aspects of reproduction in commercially important plant species, viz. *Commiphora wightii*, *Acacia senegal*, *Boswellia*, *Elaeis guinnensis* (oil palm) were also unraveled in his laboratory.

Kundaranahalli Ramalingaiah Shivanna is a dynamic and enthusiastic plant reproductive biologist working on broader aspects of reproductive biology including pollination and pollen biotechnology of flowering plants. His comprehensive paper with Dr. Y. Heslop-Harrison (in 1977), classifying the stigmas of over 1000 species (growing at the Royal Botanic

Gardens, Kew) based on their structure and showing that the presence of extra-cellular components on the surface of the stigma is a general feature of angiosperms, irrespective of their structure, was a significant contribution. It stimulated extensive studies throughout the world on the stigma and pollen–pistil interactions. Some of the other major contributions from Shivanna's research group include: (a) providing experimental evidence to demonstrate the role of plasma membrane in pollen viability, (b) highlighting the importance of pollen vigor in assessing the quality of pollen, (c) demonstrating that self-incompatibility responses in some heteromorphic species operates across species limits, (d) isolating the male gametes from 2-celled pollen systems for the first time, (e) demonstrating that, unlike many other pollination stimuli, self-incompatibility recognition and rejection are confined to the style and are not

K. R. Shivanna

transmitted to the ovules, (f) formulation of an hypothesis that the egg in angiosperms has lost the ability to recognize incompatible male gametes and (g) investigations on the reproductive biology of a number of endangered and economically important species such as *Commiphora, Boswellia, Sterculia*, oil palm and several endemic species of the Western Ghats. Biotechnological studies of his group included (a) standardization of a new technique of ovule pollination *in vitro* and its use to overcome sexual incompatibility and (b) production of over 50 interspecific and intergeneric hybrids between wild and cultivated species of *Brassica* (mustard), to tap genes imparting resistance to diseases and abiotic stresses from wild species. Currently, he is associated with the Ashoka Trust for Research in Ecology and the Environment (ATREE), Bangalore as an Indian National Science Academy (INSA) Honorary Scientist. He has authored and edited several books on the subject and some of them are: *The Angiosperm Pollen: Structure and Function* (1985, Shivanna KR, Johri BM), *Pollen Biology: A Laboratory Manual* (1992, Shivanna KR, Rangaswamy NS), *Pollen Biology and Biotechnology* (2002, Shivanna KR), *Reproductive Ecology of Flowering Plants: A Manual* (2014, Shivanna KR, Tandon R), *Reproductive Ecology of Flowering Plants: Patterns and Processes* (2020, Tandon R, Shivanna KR, Koul).

Sant Saran Bhojwani made significant advances in the field of experimental embryology. He gained a reputation in the field of plant tissue culture through his work on pollen tetrad culture, pollen culture (for haploid plants) and endosperm culture (for raising triploid plants) in various commercially important species. As a young student working in Prof. B. M. Johri's laboratory, he achieved the feat of differentiating normal shoot buds from the endosperm of *Exocarpus cupressiformis*. This was to be of considerable practical importance in plant genetics and improvement (1965). In 1972, he reported isolation of microspore protoplasts using helicase enzyme, for the first time ever, in laboratory of Professor E. C. Cocking at the University of Nottingham. Bhojwani and his team worked on a range of basic and applied aspects of *in vitro* plant morphogenesis. His book *The Embryology of*

Angiosperms which he co-authored with **Professor S. P. Bhatnagar**, was first published in 1974 (now in its 6th edition) is a seminal contribution that has helped students in grasping the basics of the embryology of angiosperms.

Nanjangud Sreekantaiah Rangaswamy made significant contributions in the field of experimental embryology, and morphogenesis. Rangaswamy and his group (Tupy & Rangaswamy 1973) highlighted that rRNA and DNA contents increase in the style of plants following compatible pollinations. One of the pioneers in morphogenetic studies of nucellus culture, mature endosperm cultures and test-tube fertilization; he demonstrated monopolar seedling development from 'unorganized' embryos. Rangaswamy reported a 'built-in' chemical mechanism in the developing seeds of cotton and cucurbits which represses nucellar embryony. He also showed that gossypol and cucurbitacin inhibit somatic embryogenesis and induce androgenic embryos in mulberry and *Brassica* sp. from microspores. His work estalished that the simultaneous type of microsporogenesis is a feature of only the tetrahedral microspore tetrads. He also emphasized on screening the pollen grains alone for salt stress instead of the whole plants in angiosperms.

N. S. Rangaswamy

3.5 Integration of Ecological, Evolutionary and Genetical Approaches to the Study of Plant Reproductive Systems

In the last few decades, reproductive biological studies have seen an increase in the use of molecular tools for understanding underlying genetics, molecular and biochemical pathways and mechanisms. At the same time, integration of plant population biology, ecology and evolution has broadened the dimensions of plant reproductive biology. Today, this field is an integral part of evolutionary ecology, quantitative genetics, comparative biology, phylogenetics and population genetics with high impact research as reflected by the profuse research articles appearing in influential scientific journals. A number of botanists have explored these aspects of plant reproductive systems and the contributions of a few eminent ones follow:

Joseph H. Williams, (presently associated with Department of Ecology and Evolutionary Biology, University of Tennessee, Knoxville, USA) works to reconstruct historical sequences of developmental changes leading to the reproductive process in ancient flowering plants. The prime focus of his research is to decipher the pollen–pistil interactions and fertilization events in basal angiosperms, and also the evolution of reproductive traits in flowering plants. His studies on basal angiosperms have indicated that pollen tube growth rate innovations are an important aspect of early angiosperm reproduction. **William E. Friedman** is the Arnold Professor of Organismic and Evolutionary Biology, at Department of Organismic and Evolutionary Biology, Harvard University, Boston, as well as Director of the Arnold Arboretum and Faculty Fellow of the Arnold Arboretum. Friedman was the first scientist to investigate double fertilization and embryo development in *Ephedra*. Significantly, he suggested that homologies in reproduction exist between several non-flowering plants (gymnosperms) and flowering plants. His research focuses on developmental, phylogenetic and evolutionary biology. He follows an integrated approach using morphology, anatomy and reproductive biology to study various developmental processes. He also works on origins and diversification of angiosperms, and in particular, the establishment of double fertilization and endosperm as defining biological features of angiosperms.

Integration of ecological and evolutionary approaches to the study of plant reproductive systems has been championed by **Spencer C. H. Barrett**, Professor Emeritus, at the Department of Ecology and Evolutionary Biology, University of California, Berkley. He has provided novel insights on plant sexual diversity and mating systems using both field and laboratory approaches. His work has been unique in finding answers to diverse questions on the ecology and evolution of plant reproductive systems. His group has deciphered several ecological and genetic causes and consequences of gender differentiation as well as pollination mechanisms (wind pollination and ambophily). Barrett has also carried out research on experimental and comparative studies of mating-system evolution. Such analysis has greatly enriched our understanding of evolution of angiosperms. **Peter K. Endress**, Professor Emeritus at the Institute of Systematic Botany, University of Zurich, Switzerland, works towards deciphering the evolution of angiosperms. Through his work he has determined the floral development, evolution of flowers and carpel using morphological, anatomical and phylogenic approaches. His contributions to the understanding of evolution and structural details of various routes to angiospermy are notable. He has authored several books which are indispensable for those working on floral evolution, most notable being, *Diversity and Evolutionary Biology of Tropical Flowers* (1994).

Paula J. Rudall also works on evolutionary aspects of angiosperms at Royal Botanic Garden, Kew, and is part of the Angiosperm Phylogeny Group. Her work emphasizes upon the importance of reproductive features such as pollen aperture in the evolution of eudicots. She has also analyzed microsporogenesis across angiosperms with an evolutionary perspective. **Amots Dafni**, is Professor Emeritus at University of Haifa, Israel, with primary research specialization in pollination ecology and pollinator interactions. The ecological context in which flowers interact with their pollen vectors has long been the focus of his research. His other areas of interest include medicinal botany and ethno-botany. He has authored and edited many books on the subject which are comprehensive including the methods and techniques in pollination biology from both botanical and zoological perspectives. His popular books are: *Pollination Ecology: A Practical Approach* (1994), *Pollen and Pollination* (2000), *Practical Pollination Biology* (2005).

Box 3.1

Some Indian Researchers Currently Working in the Field of Plant Reproductive Biology

Plant reproductive biology in India has been an active field of research for a very long time. Currently, a number of researchers across India are actively contributing to scientific pursuit of reproductive biology. Namrata Sharma, Professor at University of Jammu, works on pollination ecology and conservation biology of plants of North-Western Himalayas. Her colleague, Professor Veenu Kaul also works on the reproductive biology of plants of the region with special interest in the cytological variability and characterization of sex chromosomes in dioecious systems. Professor Rajesh Tandon, University of Delhi, has been conducting research in plant reproduction, developmental biology and conservation biology for several years. His research work spans across India from Ladakh to Kerala with focus on the reproductive ecology of many plant species including threatened and endemic plants. He is also a co-author of a practical manual on *Reproductive Ecology of Flowering Plants* and a reference book *Reproductive Ecology of Flowering Plants: Patterns and Processes.* Professor Vishnu Bhat at University of Delhi works on the genetic and molecular mechanism governing apomixis. His team works on the characterization of genes associated with apomixis and use genetic transformation to suppress putative candidate apomictic genes in understanding the process of apomixis. Krishna Kumar Dwivedi, Sr. Scientist at Indian Grassland and Fodder Research Institute, Jhansi, also works on epigenetic regulation of apomixis, development of molecular markers for apomictic and sexual lines and cytoplasmic male sterility. Soubhadra Devy, a senior fellow at the Suri Sehgal Centre for Biodiversity Conservation, Ashoka Trust for Research in Ecology and the Environment (ATREE), is engaged in the study of canopy ecology, conservation biology with main focus on pollination ecology of wild and cultivated plants and on the sustenance of plant-pollinator/animal interactions in human dominated landscapes. In recent years she has been working on the plants of Darjeeling and Sikkim in North-Eastern Himalayas. Along with her colleagues she has established the Agastyamalai Community Based Conservation Centre at Tirunelveli, Tamil Nadu. Aluri Jacob Solomon Raju is a Professor of Environmental Sciences at Andhra University. He has worked extensively on pollination biology of several plants of the Eastern Ghats of India in the last three decades. His work on butterfly pollination and mangroves of Eastern Ghats of India is notable. There are several researchers who have bridged the gap of studying plants and animals in isolation. They may not be botanists by training but have contributed significantly towards understanding of the plant-animal interactions. Professor Renee Borges, Indian Institute of Science, Bangalore, is one such scientist who works on the evolutionary ecology of species interaction, mainly on figs and fig wasps, ant–plant mutualisms, fungus-growing termites and pollination systems in the Western Ghats. Palatty Allesh Sinu, Department of Animal Science, University of Kerala, specializes in ecological studies of tropical terrestrial systems especially plant–animal interaction studies. His current research is based at the Western Ghats Biodiversity hotspot and Sub-Himalayan Plains of Northeast India. The Centre for Pollination Studies (now renamed as Centre for Agro-ecology and Pollination Studies), based in the Department of Zoology, University of Calcutta, primarily focuses on pollination in agro ecosystems.

Looking at the history and progress made in the field of plant reproductive biology, one can see that plant reproduction has always captured the attention of researchers. With the advancement in theory and experimentation directed at testing models, study of plant reproductive systems will continue to attract students and unveil the exciting findings in the decades to come.

Bibliography

Baker, H. G. (1979). Anthecology: Old Testament, New Testament, Apocrypha. *New Zealand Journal of Botany* 17: 431–40.

Barrett, S. C. H. (2010). Darwin's legacy: the forms, function and sexual diversity of flowers. *Philosophical Transactions of the Royal Society B: Biological Sciences* 365: 351–68.

Bhandari, N. N. (1971). Embryology of the Magnoliales and comments on their relationships. *Journal of the Arnold Arboretum* 52: 285–304.

Bhojwani, S. S. and Cocking, E. (1972). Isolation of protoplasts from pollen tetrads. *Nature New Biology* 239: 29–30.

Bicknell, R. A. and Koltunow, A. M. G. (2004). Understanding apomixis: recent advances and remaining conundrums. *The Plant Cell* 16: S228–S245.

Chamberlain, C. J. (1897). The Pollen Grain. *Botanical Gazette* 23: 423–30.

Cresti, M. and Linskens, H. F. (1999). The discovery of sexual reproduction in higher plants. *Acta Biologica Cracoviensia Series Botanica* 41: 19–29.

Dickinson, H. G. (2000). Pollen–stigma interactions: an international conference dedicated to the memory of Jack Heslop-Harrison. *Annals of Botany* 85 (Supplement A): 1–3.

Discovery of Sexuality in Plants. (1933). *Nature* 131: 392. https://doi.org/10.1038/131392b0

Gunning, B. E. S. (2000). John Heslop-Harrison. 10 February 1920–7 May 1998: Elected F.R.S. 1970. *Biographical Memoirs of Fellows of the Royal Society* 46: 197–217.

Harley, S. M. (2010). Charles Darwin's botanical investigations. *The American Biology Teacher* 72: 77–81.

Heslop-Harrison, Y. and Shivanna, K. R. (1977). The receptive surface of the angiosperm stigmas. *Annals of Botany* 41: 1233–58.

Horner, H. T. (2014). In Memorium: William A. Jensen, 1927–2014. *Plant Science Bulletin* 60: 201–2.

Johri, B. M. and Bhojwani, S. S. (1965). Growth response of mature endosperm in cultures. *Nature* 298: 1345–47.

Kanta, K. Rangaswamy, N. S. and Maheshwari, P. (1962). Test tube fertilization in flowering plants. *Nature* 194: 1214–17.

Kapil, R. N. and Prakash, N. (1966) Co-existence of mono-, bi- and tetrasporic embryo sacs in *Delosperma cooperi* Hook. f. (Aizoaceae). *Beiträge zur Biologie der Pflanzen* 42: 381–92.

Katayama, Y. and Nei, M. (1964). Studies on the haploidy in higher plants. *Report of Laboratory of Plant Breeding, Faculty of Agriculture University of Miyazaki, Japan* 2: 1–78.

Kenrick, J. and Ladiges, P. Y. (2002). Robert Bruce Knox 1938–1997. *Historical Records of Australian Science* 14: 67–88.

Maheshwari, P. (1950). *Introduction to the Embryology of Angiosperms.* McGraw-Hill Book Company.

Maheshwari, R. Shivanna, K. R. Doreswamy, R. et al. (2014). A doyen of Indian botanists: H. Y. Mohan Ram. *Current Science* 106: 305–9.

Mathur, G. and Mohan Ram, H. Y. (1986). Floral biology and pollination of *Lantana camara*. *Phytomorphology* 36: 79–100.

Mc Conchie, C. A., Jabson, S. and Knox, R. B. (1985) Computer-assisted reconstruction of the male germ unit in pollen of *Brassica campestris*. *Protoplasma* 127: 57–63.

McConchie, C. A., Hough, T., and Knox, R. B. (1987). Ultrastructural organization of the sperm cells of *Zea mays*. *Protoplasma* 139: 9–19.

Mohan Ram, H. Y. and Mathur, G. (1984). Flower color changes in *Lantana camara*. *Journal of Experimental Botany* 35: 1656–62.

Mohan Ram, H. Y. and Sett, R. (1982). Modification of growth and sex expression in *Cannabis sativa* by Aminoethoxyvinylglycine and Ethephon. *Zeitschrift Für Pflanzenphysiologie* 105: 165–72.

Mohan Ram, H. Y. and Dutta S. (1966). In vitro culture of *Utricularia*. *Current Science* 35: 48–50.

Natesh, S. (2018). Tributes to Professors N. N. Bhandari and H. Y. Mohan Ram. *Phytomorphology* 68: 55–64.

Raghavan, V. (2003). Some reflections on double fertilization, from its discovery to the present. *New Phytologist* 159: 565–83

Russel, S. D. (1984). Ultrastructure of the sperm of *Plumbago zeylanica* II. Quantitative cytology and three-dimensional organization. *Planta* 162: 385–91.

Russell, S. D. and Cass, D. D. (1981). Ultrastructure of the sperms of *Plumbago zeylanica*. l. Cytology and association with the vegetative nucleus. *Protoplasma* 107: 85–107.

Shivanna, K. R. and Tandon, R. (2020). Developmental biology of dispersed pollen grains. *International Journal of Developmental Biology* 64: 7–19.

Shivanna, K. R. Mohan Ram H. Y. (2005). Contributions of Panchanan Maheshwari's school to angiosperm embryology through an integrative approach. *Current Science* 89: 1820–34.

Shivanna, K. R. and Mohan Ram, H. Y. (2018). *Current Science* 115: 168–71.

Tupy, J., and Rangaswamy, N. S. (1973). The investigation of the effect of pollination on ribosomal RNA, transfer RNA and DNA contents in styles of *Nicotiana alata*. *Biologia Plantarum* 15: 95–101.

Vogel, S. (1996). Christian Konrad Sprengel's theory of the flower: The cradle of floral ecology. In D. G. Lloyd and S. C. H. Barrett, eds., *Floral Biology: Studies on Floral Evolution in Animal Pollinated Plants*. Chapman & Hall, New York, pp. 44–62.

Willemse, M. T. M. (2008). History and prospects of plant sexual reproduction congresses, the IASPRR and sexual plant reproduction. *Sexual Plant Reproduction* 21: 89–97.

4

The Anther and Male Gametophyte

Pollen grain, the 'male gametophyte' houses genetic material destined for union with female gamete.

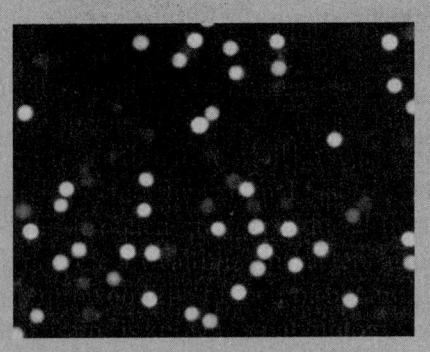

4.1 Introduction

Life cycle of an angiosperm is characterized by alternation of generation between a diploid sporophyte and a haploid gametophyte. Unlike lower plants, gametophytic generation in angiosperms is much shorter and dependent on sporophytic generation for its development. Gametophyte develops from the cells of a sporophyte in preparation for reproduction. The gametophytic cells undergo meiotic division and produce haploid gametes within the specialized structures of a flower. While the male gametophyte develops within the anther, the female gametophyte develops within the ovule. Pollen grain is the male gametophyte in flowering plants and contains the two male gametes (also called the sperm cells). Pollen grains are also involved in the formation of pollen tubes to facilitate the movement of sperm cells for fertilization with female gametes.

The male reproductive organ in flowering plants is the stamen. It consists of two morphologically distinct parts, the anther and the filament (Fig. 4.1 A). Filament is an entirely sporophytic structure which anchors the stamen to the flower. It also contains vascular tissue for transporting water and nutrients. The anther on the other hand contains both sporophytic and gametophytic tissues that are responsible for producing and releasing pollen grains. Anther development is a perfectly timed and orchestrated event which follows different pathways in different groups of angiosperms. Development of pollen grains (male gametophyte) takes place within the anther and is divided into two phases. It begins with the meiosis in the microspore mother cells to produce four haploid microspores, each of which later develops into a pollen grain and the process is called as **microsporogenesis**. This is followed by a second phase of pollen development where the formation of two sperm cells takes place and the process is known as **microgametogenesis**.

Pollen development includes participation of various sporophytic cells of the anther and the associated molecules. Pollen grains vary immensely in size, shape and surface characteristics among different plant species. At maturity, the pollen grains are surrounded by an elaborate cell wall which consists of a thin inner wall known as the intine, and an outer thicker wall called the exine. The shape and the external features of the exine are highly variable, and often used to distinguish pollen grains produced by different species. The morphological features of pollen provide a wealth of characters and sometimes can even help in the identification of different taxon. The branch of botany dedicated to scientific study of living and fossilised pollen grains is known as **palynology**.

Normal development and timely dissemination of pollen grains is one of the prerequisites for the success of sexual reproduction among angiosperms. Thus, understanding anther and pollen development is essential to study plant reproduction. This chapter deals with development of anther, anther structure, pollen development, pollen structure and its diversity in angiosperms.

4.2 Anther Structure and Development

4.2.1 Structure

Anthers are the pollen containing units of the stamen. The anther typically consists of two compartments called thecae, with each theca containing two microsporangia enclosing the sporogenous cells which undergo meiosis and develop into pollen grains. The two adjacent thecae are separated by a connective tissue to which the filament of a stamen is attached. The microsporangia within a theca are separated by a group of cells called the septum. Each microsporangium is surrounded by the four layers of anther wall, namely, epidermis, endothecium, middle layers and tapetum. Each anther layer/tissue carries out its own specialized tasks (Table 4.1).

A fully differentiated anther has several highly specialized cells and tissues, which are either sporophytic or gametophytic. The diploid sporophytic tissue includes the anther wall layers and gametophytic tissue includes the haploid sporogenous tissue or the microspores that fill the microsporangia. Cross-section of a typical anther shows the four microsporangia (tetrasporangiate) with developing pollen grains and appears four lobed (Fig. 4.1 B-C). In a few plants such as *Hibiscus* (Malvaceae), *Moringa* (Moringaceae) and *Wolffia* (Araceae), the anthers are monothecous and bisporangiate. *Arceuthobium* is a unique plant with only one microsporangium per anther.

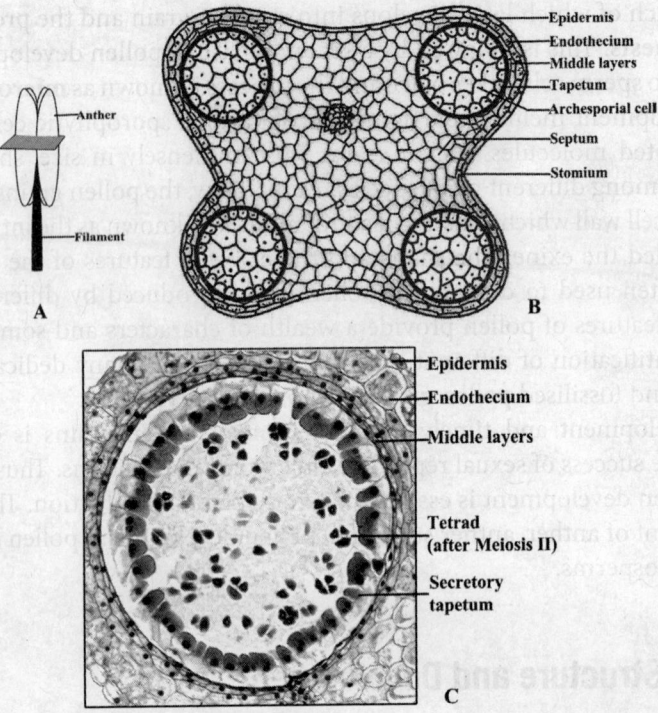

Figure 4.1 **A.** Diagram of a stamen. **B.** Semi-diagrammatic sketch of the transverse section of a bithecous, tetralocular/tetrasporangiate anther with sporogenous tissue and wall layers. Plane of section is transverse as marked in A. **C.** An anther locule showing different wall layers in lily. (*Source:* IASPRR). **See Color Plates (page 476).**

Table 4.1: Summary of cell types and their functions in anther

Cell type	Function
Connective	Joins the anther theca together; connects anther to the filament and extends support to the whole anther
Epidermis	Prevents water loss; facilitates gaseous exchange; anther dehiscence
Stomium	Anther dehiscence
Endothecium	Support and anther dehiscence
Middle layers	Anther dehiscence
Tapetum	Pollen development; pollen wall formation; pollen nutrition
Sporogenous cells	Formation of male gametophyte (pollen)
Vascular bundle	Connection and nutrient supply from mother plant

4.2.2 Development

The development of anther and pollen is a critical step in plant reproduction. It is essential for successful fertilization and seed formation. Anther development can be divided into two general phases (Goldberg et al. 1993). During phase I, the morphology of the anther is established and microspore mother cells (MMCs) undergo meiosis. During phase II, pollen grains differentiate and the anther enlarges and subsequently dehisces to release the pollen grains.

4.2.2.1 *Phase I*

Phase I marks the initiation of anther primordium and establishment of anther morphology and its cellular components. It begins with stamen initiation and development, which is is governed by a combination of class B and C genes described under the ABCDE model for determination of floral organ identity (Goldberg et al. 1993; Kelliher et al. 2014). The role of phytohormones and epigenetic regulation of stamen development has also been documented (refer to Box 4.1 for details). Stamen primordium is round or oval in shape. In dicots (e.g., *Arabidopsis*) it comprises three floral meristematic layers of cells: L1 (outer), L2 and L3 (inner). In monocots, either all three layers are present, or, only two are present, L1 and L2, with L2 being bi layered (e.g., rice, maize; Linde & Walbot 2019). A short while after stamen primordium initiation, the anther and filament compartments differentiate. During the development of anther, divisions in the primary layers of stamen primordium result in the formation of anther lobes, sporophytic cell layers (anther wall layers) and an inner most layer of sporogenous cells. In dicots, derivatives of L1 and L2 meristem zones contribute to the formation of the anther lobes, the wall layers and the sporogenous tissue. The outermost layer or the L1 forms the epidermis. It only divides in anticlinal plane in order to cope up with the volume of tissue differentiating and developing inside. It never divides in periclinal plane except for the formation of a stomium at the time of anther dehiscence. L2 cells develop into archesporial cells in each anther lobe. The L3 meristem zone is involved in the differentiation of vasculature within the filament and the connective tissue. The connective layer differentiates into a small group of cells separating the two pollen sacs in each half of the anther (Fig. 4.2).

Archesporial cells derived from L2 are radially elongated and densely cytoplasmic with a large and prominent nucleus. Initial divisions in these cells are periclinal. This results in formation of Primary Parietal cells (PP) on the outside towards the epidermis and Primary Sporogenous cells (PS) towards the inside. By this time, the two bilaterally symmetrical theca have been established within the anther and the anther assumes a four-lobed appearance. The parietal and sporogenous tissue are present towards the four corners and a vascular strand runs in the centre establishing a connection with the filament. Soon, the PS undergo a number of divisions to generate the meiocytes or the Microspore Mother Cells (MMCs). On the other hand, the PP cells further divide periclinally and generate Inner Secondary Parietal cells (ISP) and Outer Secondary Parietal cells (OSP) (together known as secondary parietal layer). In many angiosperms, OSP and ISP cells do not divide any further and eventually develop into the endothecium and the tapetum, respectively. However, in other angiosperms another layer/s may develop known as middle layer/s which is located between the endothecium and tapetum. It may either originate from the periclinal divisions in OSP or ISP cells or from both (Kelliher et al. 2014; Xue et al. 2021).

To summarize: L1 gives rise to the epidermis and the stomium; L2 gives rise to the microspore mother cells, endothecium, middle-wall layers and tapetum; and L3 gives rise to the connective tissue and the vascular bundle (Fig. 4.2). In monocots (e.g., maize), the L2-derived cells generate the endothecium and the secondary parietal cells, which in turn generate the middle and the tapetum layers (Kelliher & Walbot 2011).

Typically, the individual tissues or the cell types of an anther are derived from a single primary layer as described above. However, sometimes, tapetum may originate from both the L2 and the L3 layers. Tapetal cells present along the inner portion of the pollen sacs (towards connective tissue), develop from the L3 layer, and those that line the outer portion of the pollen sacs develop from the L2 layer. In such cases tapetum is said to be dual origin.

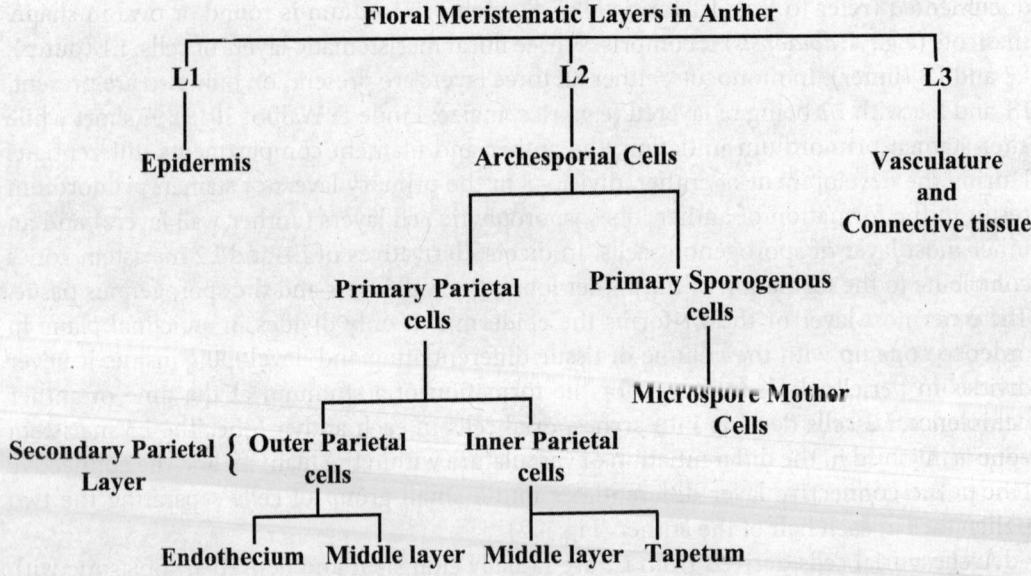

Figure 4.2 Schematic diagram for anther wall ontogeny.

The anther wall of a species is composed of a specific number of wall layers. However, number of wall layers may differ among species depending upon the number of divisions the Secondary Parietal Cells (SPCs) undergo. On the basis of the number of divisions in SPCs, four types of anther development have been described by Davis (1996): Basic (type I), dicotyledonous (type II), monocotyledonous (type III) and reduced (type IV). These anther development types are mostly family specific. However, some families can have more than one type. For example Commelinaceae exhibits both type I and type III.

- **Type I:** In the basic type, both the layers, i.e., OSP and ISP, divide periclinally and differentiate to form the four layers of anther wall excluding the epidermis i.e., the endothecium, two or three middle layers, and the tapetum as seen in *Juglans regia* (Juglandaceae).

- **Type II:** In the dicotyledonous type, only the OSP layer divides periclinally to give rise to the endothecium and a single middle layer. The ISP does not divide further and differentiates into the tapetum. Thus, only three layers are formed; these being the endothecium, a single middle layer and the tapetum. *Smilax davidiana* (Smilacaceae) exhibits Type II development.

- **Type III:** In the monocotyledonous type of anther development, the OSP layer does not divide any further and directly forms the endothecium. The ISP layer divides again giving rise to the tapetal layer and a middle layer. Here too, like in Type II, only three layers are formed (the endothecium, a middle layer and the tapetum) but the origin of middle layer is different from Type II. Example: *Guzmania* (Bromeliaceae).

- **Type IV:** This is the reduced type of anther development where the primary parietal cells undergo periclinal divisions to form two-layered SPCs, which directly transform into an endothecium and a tapetum. The middle layers are absent. Such a type of development is seen in members of Podostemaceae like *Polypleurum wallichii*.

4.2.2.2 *Phase II*

At the end of phase I, formation of the anther wall layers towards the periphery and the differentiation of sporogenous tissue in the centre of each locule (microsporangium) is complete. In the phase II, pollen grains develop inside each microsporangium. This phase involves two processes: microsporogenesis and microgametogenesis. During pollen development, the anther too undergoes enlargement to accommodate the developing pollen grains. Details of pollen development are provided in Section 4.5. Once the development of the pollen grains is accomplished, the anther splits or dehisces to release the pollen grains. This phenomenon of liberating the pollen grains is known as **anther dehiscence**. Anther dehiscence is the last step in the development of an anther and is temporally coordinated with the differentiation of pollen grains. Many of the processes of pollen development occur concurrently, and there are significant variations in relative timing between species. However, a number of distinct stages can be recognized, e.g., a total of 15 stages have been described in *Arabidopsis thaliana* (Gomez et al. 2015) and wheat (Browne et al. 2018) (Fig. 4.3).

4.3 Anther Wall Layers

As mentioned earlier, the anther wall consists of four distinct sporophytic layers from outside to inside. These are the epidermis, the endothecium, the middle layers and the tapetum. Generally, at pollen maturity both the middle layers and the tapetum are degenerated leaving only an epidermis and endothecium at the time of anther dehiscence.

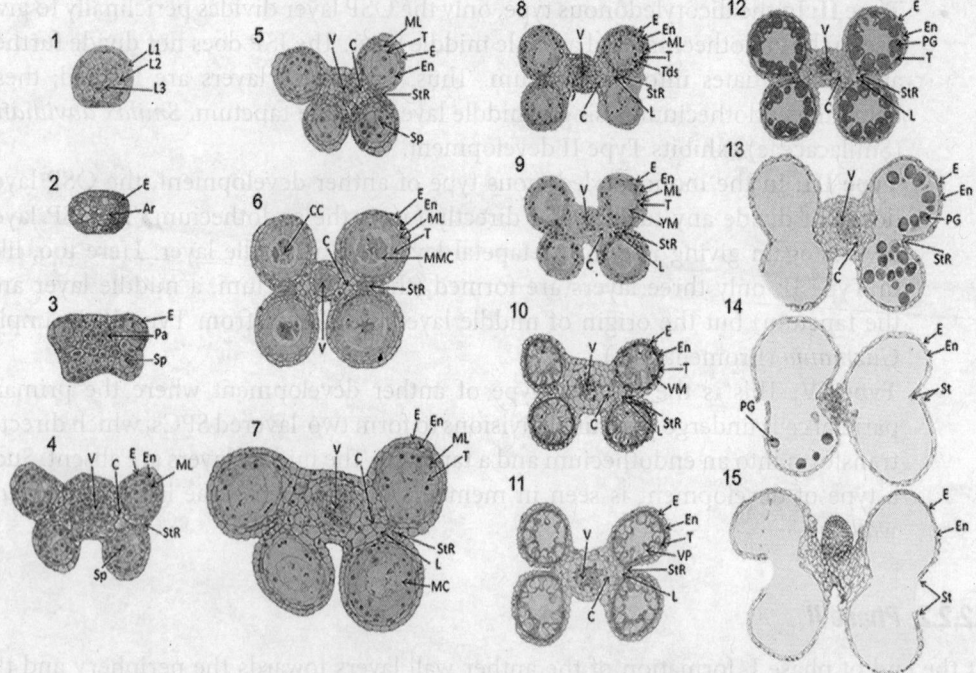

Figure 4.3 Development of anther and pollen grains in wheat exhibit 15 distinct stages. **Stages 1–5:** Early stages of anther development involving the formation of anther shape, the cellular differentiation of the four locules and different wall layers. **Stage 6:** Microspore mother cells developed from sporogenous tissue enveloped in callose. **Stages 7–8:** Microsporogenesis. **Stage 9:** The young microspores are released from the tetrads and the tapetum is at its largest and most active stage, with more mitochondria than during the earlier stages. **Stage 10:** Degradation of the middle layer and subsequently, the tapetum. During this time the microspores are growing and accumulating sugars and other nutrients, which are released via Ubisch bodies into the locular cavity. It is also called the vacuolate microspore stage **Stage 11:** The bi-nucleated microspores **Stage 12:** The microspores contain a vegetative nucleus and two nuclei in the generative cell. The tapetal layer is almost fully degraded and the pollen grains are largely starch filled at this stage. **Stage 13:** The anther becomes bilocular at this stage with the degeneration of the septum between the upper and lower locules. **Stage 14:** The stomium degrades, allowing dehiscence to occur. Filament elongation occurs and the anther eventually protrudes from the floret. **Stage 15:** It marks the completion of the developmental program of the anther, with occurrence of dehiscence and pollen being released. The anther senesces; eventually detaching from the filament. E: epidermis, Ar: archesporial cells, Pa: parietal tissue, V: vasculature, C: connective, En: endothecium, ML: middle layers, stR: stomium region, Sp: sporogenous tissue, T: Tapetum, MMC: microspore mother cells, CC: central callose, L: lacunae, Tds: tetrads, YM: young microspores, VM: vacuolated microspores, VP: vacuolated pollen, PG: pollen grain, St: stomium. (*Source:* Browne et al. 2018, published under CC BY 4 License.)

4.3.1 Epidermis

The epidermis is the outermost wall layer of the anther comprising of columnar cells. This layer initially encloses the anther primordium and later archesporial cell and its products. As the anther develops, the epidermal cells undergo repeated anticlinal divisions increasing

Box 4.1

Phytohormones Involved in Stamen Development

It is a well-established fact that phytohormones regulate flowering. However, the growing evidences from mutant studies have unraveled that phytohormones like jasmonic acid, auxin, and gibberellins are also indispensable for stamen development. Gibberellins regulate early stamen development while jasmonic acid and auxin are essential for later stages of stamen development. Auxin acts indirectly through jasmonic acid and plays a role in anther dehiscence, pollen maturation and filament elongation (Song et al. 2013). Many of the genes and proteins involved in biosynthesis or regulatory function of phytohormones also exert an effect on stamen development as shown by analysis of their mutants. Some of them are summarized below:

Gene/protein	Mutant	Effect
YUC genes (encodes flavin monooxygenases, essential for auxin biosynthesis)	double mutants *yuc2 yuc6*	Abnormal filament elongation, anther dehiscence, and pollen maturation
	yuc1 yuc4	No stamen or fewer stamens
P-glycoprotein (PGP1), PIN-FORMED 1 (PIN1), PIN3, and *PIN7* (required for auxin transportation)	*pin1, pin3, pin7*	No stamen or fewer stamens
ent-CDP synthase (CPS) for conversion of geranylgeranyl diphosphate (GGDP) to *ent*-CDP, (essential for biosynthesis of Gibberellins)	GA-deficient mutant *ga1-3, GA 20-oxidase (GA20ox) and GA 3-oxidase (GA3ox)*	Male-sterile due to un-elongated filament, and arrested anther development
Gibberellin Insensitive Dwarf 1 (GID1a/b/c) (gibberellins receptors)	*gid1a gid1b gid1c* triple mutant	Mutant exhibits failure of filament elongation and arrested anther development
Fatty acid desaturase (FAD), phospholipase A1 (PLA), 13-lipoxygenase (LOX), allene oxide synthase (AOS), allene oxide cyclase (AOC), OPC-8:0 CoALigase (OPCL), (JA-biosynthetic enzymes)	Mutation in any gene	Failure of filament elongation, delayed anther dehiscence, and non-viable pollens
MYB21, MYB24, and MYB57 (MYB transcription factors Related to cellular proliferation) {MYB named after the gene of the avian myeloblastosis virus}	*coi1-1, myb21 myb24, myb21, myb24, myb57*	Male-sterility due to failure of filament elongation, delayed anther dehiscence, and nonviable pollens
AUXIN RESPONSE FACTOR3 (ARF3) and ARF4, PHABULOSA 3 (PHB3)		For developing stamen symmetry

Source: Based on Song et al. 2013.

the girth of epidermis to cope up with the rapidly enlarging anther. The epidermis persists till the time of anther dehiscence (Fig. 4.1 B-C). Typically, the epidermis of anther is single layered but there are some exceptions such as *Lagerstroemia indica* where the epidermis is differentiated into upper and lower epidermis (Rezanezad 2008). Also, epidermis of anther may be cuticularized as in maize or may contain stomata as in *Lilium hybrida* and *Aloe ciliaris* or even have epidermal hairs as in *Chelone glabra*. Being the outermost wall layer, the main function of epidermis is to protect the anther. The epidermis is also known to assist in gaseous diffusion, prevention of moisture loss and in the dehiscence of the anther lobes by differentiating stomium. Recent mutant analysis in rice and maize indicates a role of epidermis in pollen fertility too (Fan et al. 2016).

4.3.2 Endothecium

Endothecium is a layer present next to the epidermis and is derived from the descendants of the L2 primary layer. It is formed in the protuberant parts of the anther. Endothecium is predominantly single layered among flowering plants. The cells of endothecium are radially elongated, uninucleate, vacuolated and hygroscopic (Figs. 4.1 B-C and 4.4). The endothecium attains maximum development just before the dehiscence of anther. During that time, the walls of endothecial cells develop characteristic secondary wall thickenings on the inner tangential wall; alternatively, the thickening may be restricted to more or less center of the wall. The outer tangential wall remains thin. The hygroscopic nature of endothecial cells combined with the differential expansion of the outer and inner tangential walls promotes anther dehiscence. Thus, by the virtue of its specialized nature, the endothecium together with the stomium helps in the anther dehiscence.

The secondary wall thickenings of endothecial cells, also known as endothecial thickenings exhibit a variety of patterns, e.g., plate like, U-shaped, helical, reticulate. These thickenings mostly contain α-cellulose and a small amount of lignin. However, in *Borago officinalis*, *Phaseolus vulgaris* (Whatley 1982), and orchids (Freudenstein 1991), the wall thickenings in endothecium are chiefly made of lignin. Many mutant analysis studies have identified several genes and proteins which regulate the development of wall thickenings

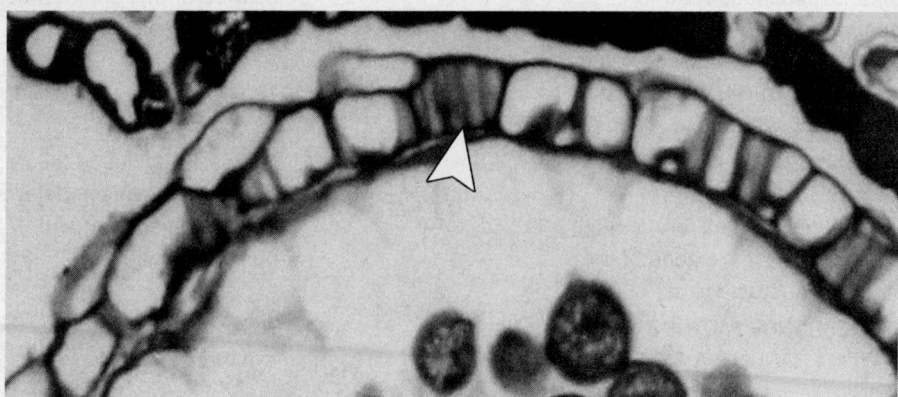

Figure 4.4 Part of mature anther showing endothecial wall thickenings (arrowhead).

in the endothecium, e.g., IRX (IRREGULARXYLEM), CA2 (carbonic anhydrase 2), AHP4 (Arabidopsis histidine-containing phosphor transfer factor 4), SAF1 (secondary wall thickening-associated F-box1) and CBSX2 (cystathionine β-synthase domain-containing protein) (Wang et al. 2015). Any disruption in the thickening of cell walls of endothecium leads to male sterility because of arrested anther dehiscence and pollen release.

Box 4.2

Genes Involved in Anther Development

The development of stamen and differentiation of anther wall layers is under the complex network of genes, transcription factors and proteins. It is well known that stamen development is under the regulation of the B and C class transcription factors/homeotic genes (Goldberg et al. 1993). However, even after stamen specification, the B and C class genes along with SEPALLATA genes continue to express during stamen development (Bowman et al. 1991; Scott et al. 2004). Thus, it was proposed that these classes of transcriptional factors may be involved in activating/regulating many of the genes which are involved in anther wall layer specification and their functioning. Recent investigations have also highlighted the role of microRNAs in anther development. The abundant 21-nucleotide (nt) and 24-nt phased, secondary small interfering RNAs (phasiRNAs) in the anther of grasses (*these small RNAs are produced by the phased action of a DICER enzyme*) are two such classes of phasiRNA. These two are temporally as well as spatially separated in their existence, i.e., 21-nt phasiRNA are in found during early/premeiotic anther development while 24-nt phasi RNA is seen in meiotic stages. Spatially, 21-nt class amasses in the immature layers and the 24-nt class can be seen in the tapetum. It is suggested that 21-nt class regulates differentiation of epidermis while mutant (*ocl4*), defective in tapetum lacks 24-nt type phasiRNA. With an ever-increasing repository of work on anther and its development, a number mutants, genes, proteins and transcription factors have been identified. Some of them and their effects/functions are summarized below.

Mutants/Genes	Plant	Effect on anther development and differentiation
ameiotic1	Maize	Archesporial /Pollen mother cells defects
callose, somatic, and microspore defect1 (csmd1)	Maize	Pre-meiotic somatic defects and post-meiotic gametophytic defects
serk1 serk2 double mutant (leucine-rich-repeat receptor-like kinase)	Arabidopsis	Failure of secondary parietal cells periclinal divisions
msp1 (LRR receptor-like kinase)	Rice	Failure of somatic cell specification and over proliferation of archesporial cell
mpk3, mpk6 (Mitogen activated protein kinase)	Arabidopsis	
outer cell layer4 (ocl4)	Maize	Epidermis defects, Endothecium fails to differentiate
male sterile23 (ms23), ms9, ms11, ms13, ms14	Maize	Early tapetum defects

myb33 myb65 (GAMYB-like transcription factor), *Gibberellic acid MYB*	*Arabidopsis*	Failure of tapetum differentiation
udt1 (*UNDEVELOPED TAPETUM 1*) bHLH (basic helix-loop-helix) transcription factor	*Arabidopsis*	Failure of tapetum differentiation, meiotic arrest
tdf1 (defective in tapetal development and function 1) and *ams (aborted microspore)*	*Arabidopsis*	Defective tapetum and aborted microspores
roxy1 roxy2 (GRX encoding genes) {Glutaredoxin i.e. GRX proteins utilize glutathione to reduce disulfide bonds or to remove covalently attached glutathione in target proteins}	*Arabidopsis*	Defective male sterile anthers
microsporeless 1 (mil1)	Rice	
male sterile converted anther 1 (msca1)	Maize	
tip2 (TDR INTERACTING PROTEIN2)	Rice	Failure of all three layers of anther wall
MEIOSIS ARRESTED AT LEPTOTENE 1 (MEL1)	*Arabidopsis*, Rice	Transition from floral organ specification to anther development
SUP (SUPERMAN)	*Arabidopsis*	Controls proliferation of stamen cell
C2H2, AP2/ERF, bZIP, WRKY and *MYB*	*Apium graveolens*	Anther development
NOZZLE/SPOROCYTELESS (SPL) (Transcription factor)	*Arabidopsis*	Necessary for archesporial cell differentiation and their meiotic entry
MSCA1 (Male Sterile Converted Anther1) (a glutaredoxin)	Maize	Affect archesporial cell fate acquisition
OsTDL1a (Oryza sativa TAPETUM DETERMINANT1–LIKE 1a), TAPETUM DETERMINANT1 (TPD1), EXTRA SPOROGENOUS CELLS (EXS)/EXCESS MICROSPOROCYTES1 (EMS1)	Rice, *Arabidopsis*	Determines number and differentiation of archesporial cells

MULTIPLE ARCHESPORIAL CELLS 1 (MAC1)	*Arabidopsis*, Rice	Stimulates the single layer of L2 cells to differentiate as primary parietal cells and divide periclinally to establish endothecium and the secondary parietal cell
EXCESS MICROSPOROCYTES1 (EMS1)	*Arabidopsis*	Regulates stamen differentiation, also associated with origin of middle layer from outer secondary parietal cells
NST1 and NST2 (NAC Secondary Wall Thickening Promoting Factor)	*Arabidopsis*	Regulates endothecium wall thickening
TCP24 (a transcription factor)	*Arabidopsis*	Regulates endothecium wall thickening, overexpression in mutants disrupts anther dehiscence
GRF1, GRF2, GRF3, and GRF5 (GROWTH-REGULATING FACTOR)	*Arabidopsis*	Tapetum degeneration and pollen wall formation
TAZ1 (TAPETUM DEVELOPMENT ZINC FINGER PROTEIN1)	*Petunia*	Mutation leads to abnormal development and premature degeneration of the tapetum, resulting in microspore infertility
AtMYB103	*Arabidopsis*	Development of tapetum
ETERNAL TAPETUM1 (EAT1)	*Arabidopsis*	Regulate programmed cell death in the tapetum
DYSFUNCTION TAPETUM1 (DYT1)	*Arabidopsis*	suppresses periclinal division in tapetum
UNDEVELOPED TAPETUM 1 (UDT1)	Maize	
TAPETUM DEGENERATION RETARDATION (TDR)	Rice	Necessary for pollen fertility and their development
21-nt PHAS loci	Maize	Mutations in two of the many 21-nt PHAS loci confer day length- and temperature-dependent male sterility

4.3.3 Middle Layers

Layers of thin-walled cells situated between the tapetum and endothecium are known as middle layers. The number of middle layers may vary from one to four among different species. E.g., one in *Arabidopsis*, two in *Juglans* sp., three to four in *Magnolia stellata* and *Lilium* sp. (Fig. 4.1 C). However, there are many species and families where middle layers are absent, e.g., *Polypleurum walichii*, *Hippophae rhamnoides*, Lemnaceae, Najadaceae. The cells of the middle layer/s are usually ephemeral. They become flattened and get crushed by early meiosis in the pollen mother cell. However, these layers may persist in some taxa like *Ranunculus* and *Lilium*. Sometimes the middle layer adjacent to the endothecium may even develop fibrous thickenings and behave as the endothecium. These layers can also act as storage centres for starch and other reserves which are later mobilized during the development of the pollen (Bhojwani et al. 2015). Studies in *Arabidopsis* have shown that middle layers play a key role in correct and coordinated pollen maturation, anther dehiscence and filament elongation (Cecchetti et al. 2016). Mutants which lack the middle layers show non-viable pollen grains, altered tapetum development and indehiscent anthers (Mizuno et al. 2007). In the dioecious plant *Actinidia deliciosa* (kiwifruit), the middle layer possibly regulates the production of male-sterile and male-fertile flowers by altering the timing of its own cell death (Falasca et al. 2013). The middle layer also shows secretory activity.

4.3.4 Tapetum

The tapetum is the innermost and the most conspicuous layer of the anther wall. Though the tapetum is usually derived from the secondary parietal layer, it may have a dual origin in a few species like *Alectra thomsoni*, *Tecoma stans*, *Excentrodendron hsienmu*, and *Nicotiana*. In *A. thomsonii*, the inner tapetum is derived from the cells of the connective (L3 primary layer), and the outer tapetum is derived from the secondary parietal layer (L2 primary layer). Often the two layers are dimorphic with tapetal cells derived from the parietal layer being isodiametric and small; and those derived from cells of the connective being larger and radially elongated.

Tapetal cells are metabolically active and show the presence of dense cytoplasm and prominent nuclei. They also exhibit transfer cell like activity that aids in water and nutrient supply to microspore mother cells and in synthesis and secretion of molecules associated with the pollen wall formation. Tapetum cells may also be binucleate-multinucleate (Fig. 4.1 C) which is believed to support the increased transcriptional load required for the synthesis of pollen wall constituents (Yang et al. 2003). Tapetum also serves as a sink of nutrients from mother plant. It receives nutrient supply from the mother plant via the vascular bundle of the filament. The cells in the outer layers of anther wall are connected via plasmodesmata but the tapetal layer is symplastically isolated. Thus, delivery of sugars into the tapetum is apoplastic which involves expression of the cell wall invertase gene in the tapetum. Carbohydrate mobilization to the tapetum and its genetic control is believed to play an important role in guaranteeing pollen development under stress conditions. The tapetum in spermatophytes is broadly grouped into secretory and amoeboid (invasive) types, differing primarily in the extent of their intrusion into the locule during microspore development (Pacini 2010).

- *Amoeboid Tapetum (or periplasmodial tapetum or invasive tapetum):* This type of tapetum is characterized by the early breakdown of the inner and radial walls of the cells. The protoplast of neighboring tapetal cells fuse, enlarge and gradually invade into the anther locule. Such tapetum can be seen in the family Annonaceae, Lauraceae, Asteraceae and genera *Crocus, Beta, Alisma, Arum, Butomus, Tradescantia,* and *Tinantia.* The breakdown of the cells of amoeboid tapetum is programmed cell death that leads to the reorganization of cellular content. Such a breakdown of the tapetum is required for release of materials for pollen development. At the structural level, the programmed cell death in tapetum is achieved by sequential elimination of the cellular structures. For example, in both *Lobivia rauschii* and *Tillandsia albida*, programmed cell death includes cytoplasmic shrinkage, oligonucleosomal cleavage of DNA, vacuole rupture, and swelling of the endoplasmic reticulum (Papini et al. 1999).

In species like *Tradescantia* and *Hypoxis*, the tapetal cells at the pollen mother cell stage develop plasmodesmata in their radial walls and connect their cytoplasm. This is followed by the degeneration of both radial and tangential cell walls. The cells then undergo fusion facilitated by local appearance of microtubular arrays. Conspicuous membrane sacs also migrate and accumulate at the plasma membranes near the fusion sites resulting in the formation of a common membrane covering tapetal cytoplasm which is also known as 'perispore membrane'. During meiosis in the MMCs, the tapetal cells which are now in the form of a multinucleate periplasmodium, form long extensions and start intruding into the locule. By the time pollen mother cells complete their meiosis or are near the tetrad stage, the multinucleate tapetal periplasmodium covers individual tetrad within a vacuole. The liberation of microspores from the tetrad is facilitated by the enzyme 'callase' which is produced by the periplasmodium. All the nutrients (lipids, proteins, starch) required for the development of pollen grains as well as precursors for sporopollenin (pollen wall material) are also provided by the periplasmodium. Plasmodial mass is consumed by developing pollen grains or gradually degenerates by the young pollen stage or late free microspore stage. The remnants of plasmodium can be seen in anther locule (Furness & Rudall 1998).

In certain species such as *Costus* and *Globba*, tapetal cells do not fuse by means of plasmodesmata. Their walls breakdown gradually and discharge their cytoplasm into the anther locule. In such incidences, the perispore membrane is absent. Here, the tapetum may be multilayered and with multiple nuclei. Some workers call this type of tapetum as invasive tapetum (*Sensu*; Tiwari & Gunning 1986; Furness & Rudall 1998).

- *Secretory Tapetum:* The secretory or glandular tapetum is the most common type of tapetum among flowering plants. Cells of secretory tapetum maintain their position and are secretory in nature. They fill the anther locule with their secretions which are later used for pollen development, e.g., *Arabidopsis*, Poaceae. However, cells of secretory tapetum undergo post-meiotic degeneration via programmed cell death (refer to Fig. 4.18). The timing of programmed cell death of secretory tapetal cells varies among species. For instance, in *Brachypodium* and rice, the degeneration commences at the tetrad stage and is complete by the bicellular pollen stage (Zhang et al. 2011). However, in wheat, the breakdown of tapetum cells appears to begin during the vacuolated microspore stage. The degeneration of the tapetum is required for release of pollen wall materials onto the developing pollen grains and is also important for normal dehiscence of the anther (Pacini 2010).

One key characteristic of secretory tapetum is the production of spheroid electron-dense structures called orbicules or Ubisch bodies. Ubisch bodies are small, granular sporopollenin containing structures. Rosanoff (1865) was the first to discover Ubisch bodies (synonym: orbicules) and Rowley (1962) coined the term Ubisch bodies to honor **Gerta von Ubish**, a German biologist who pioneered the study on Ubisch bodies (Verstraete et al. 2014).

The general structure and functioning of the secretory tapetum has been observed in *Helleborus foetidus* (Echlin & Godwin 1969), *Allium cepa* (Risueno et al. 1969) and in several members of Poaceae (Maheshwari 1950). When the sporogenous tissue is undifferentiated, the cells of secretory tapetum are mostly multinucleate, with large nucleoli, few mitochondria, plastids and numerous ribosomes in their cytoplasm. Their cytoplasm is also comprised of some electron dense, osmiophilic granular, grey bodies which are called **pro-Ubisch bodies**. However, dictyosomes and smooth endoplasmic reticulum are uncommon at this stage. As the microspore mother cells differentiate, the tapetal cells show further aggregation of ribosomes and pro-ubisch bodies. The thickening of the tapetal cell wall towards the anther locule becomes more irregular. At the tetrad stage there is further increase in size of nuclei, number of pro-ubisch bodies and wall thickenings. The pro-Ubisch bodies aggregate more towards the tapetal wall facing the locule. By the time the pollen is released from the tetrad, plasmodesmata develop on the radial walls of tapetal cells. At this stage, two remarkable characteristics of the tapetal cytoplasm can be observed. The first is the appearance of pro-Ubisch bodies in the tapetal cytoplasm surrounded by a zone of ribosomes; which radiate from the pro-Ubisch bodies like the spokes of a wheel. The limiting membrane of the pro-Ubisch body is usually discontinuous at those places which correspond to the insertion of the rays of radiating ribosomes. The second change observed is the association of a large number of vesicles with the dictyosomes. These vesicles are found aggregated at the periphery of the dictyosomes and are inconspicuous in earlier stages of development. At this stage, the tapetal cell membrane becomes highly convoluted and soon the pro-ubisch bodies pass through these convoluted regions of the membrane by the process of membrane fusion and extrusion. These pro-Ubisch bodies accumulate in the space between the tapetal membrane and the tapetal cell wall. Here, the pro-Ubisch bodies get coated with the sporopollenin, and now called as **Ubisch bodies**. The wall facing anther locule gradually starts to disintegrate and the Ubisch bodies get liberated through spaces or channels into the anther locule. Then, they get deposited over the pollen grains and contribute to the formation of exine layer of the pollen wall. By the time the pollen wall formation is complete, the dissolution of radial tapetal walls begins and the tapetum undergoes programmed cell death. However, in wheat (*Triticum*) the Ubisch bodies are released in the locule directly by secretion and there is no channel created for their release. In Brassicaceae, ultrastructural studies reveal the involvement of specialized tapetum organelles, **elaioplasts or tapetosomes** in formation of exine wall. These elaioplasts are small droplets derived from the endoplasmic reticulum and are coated with oleosin.

Box 4.3

Differences between Amoeboid and Glandular Tapetum

Amoeboid Tapetum	Glandular Tapetum
• The tapetum breaks down early, and the contents of the cell (protoplasm) extrude into the sporogenous cavity between the young pollen grains to provide nourishment.	• The tapetal cells remain intact as they supply nutrients in the form of secretions to the developing pollen grains.
• Tapetal cells fuse among themselves to form a tapetal periplasmodium, hence also called periplasmodial tapetum. Periplasmodial mass moves in to the anther locule and engulfs the MMCs/pollen tetrad.	• Cells do not fuse with each other and cytoplasm does not move into the locule. Tapetal cells remain in their place till they eventually disintegrate.
• Hydrolytic enzymes secreted by the tapetal dictyosomes are responsible for the dissolution of the tapetum walls.	• No secretion of hydrolytic enzymes is involved.
• Ubisch bodies are generally not produced. Amoeboid mass itself produces the precursors for pollen wall material.	• Orbicules/Ubisch bodies are secreted which carry sporopollenin and help in exine formation.
• Plant species with a periplasmodial tapetum have a reduced volume of locular fluid as the microspores directly get engulfed in the tapetum cytoplasm providing nutrition. Hence, there is no requirement for abundant locular fluid.	• Plant species with such a tapetum have a larger volume of locular fluid due to secretory activity of the cells.
• Abundant plasmodesmata connecting the tapetal cells are seen. The number of ribosomes and dictyosomes in the cell cytoplasm is relatively reduced.	• There is a very high number of ribosomes and endoplasmic reticulum in tapetal cell cytoplasm.
• It is seen in *Alisma, Butomus, Tradescantia, Typha*.	• Such tapetum can be observed in *Helleborus*, Wheat, Rice, *Brassica, Arabidopsis*.

Functions of Tapetum: Tapetal cells are metaboloically very active and are generally polyploid and/or multinucleate. Active tapetum cells are essential for the development and maturation of pollen grains as they provide nutrition to the developing pollen grains and the precursors for pollen wall material. The tapetum also forms an interface between the sporophyte and the male gametophyte and therefore controls reproductive development. Tapetal cells also give rise to a temporary cell wall of callose which separates the microspores in the tetrad. They also secrete the enzyme 'callase' which is essential for the release of pollen grains from the microspore tetrad by dissolution of callose. Any mutation affecting callose deposition and its dissolution hampers microspore development. Also, premature or delayed degeneration of the tapetum leads to sterility of pollen grains (refer to Box 4.2 for details).

Tapetum is actively involved in the biosynthesis and secretion of sporopollenin for the exine wall. It also contributes the proteins (sporophytic origin) to the exine which assist in pollen–pistil interactions. It is also suggested that the normal development of tapetum is indispensable for the completion of the male meiosis. Recent studies in the field of development biology have indicated that a disrupted tapetum development causes alterations in the expression of a large number of genes involved in male meiosis (Lei & Liu 2020). Programmed cell death responses of tapetum generate nutrients for the locule fluid which feed developing pollen grains. The main nutritional activity of the tapetum occurs during the microspore stage, with degeneration normally occurring near the end of the uninucleate microspore stage.

Box 4.4

Orbicules or Ubisch Bodies

Orbicules or Ubisch bodies, are sporopollenin carrying particles lining the inner tangential or the radial tapetal cell walls. These were discovered more than a century ago by Rosanoff (1865) and are common in anthers of basal angiosperms and 85% of the monocots except Orchidaceae, Commelinales and Zingiberales. Within eudicots, Asteraceae, Brassicaceae and the majority of Lamiaceae members lack orbicules. The formation of orbicules/ubisch bodies is associated with secretory tapetum. However, there are several exceptions where plants with invasive/ amoeboid tapetum also produce orbicules, e.g., Amaranthaceae (*Beta vulgaris*), Apocynaceae (*Vinca rosea*), Asteraceae (*Cosmos bipinnatus*), Fabaceae (*Acacia conferta, Acacia iteaphylla, Acacia subalata*) (for details see Verstraete et al 2014, Furness & Rudall 1998).

Orbicules develop simultaneously with the growing pollen exine and are composed of sporopollenin. Pro-orbicules are the progenitors of orbicules, former are produced by the endoplasmic reticulum of secretory tapetum cells. Orbicules are usually smaller than 1 μm, but larger orbicules with a diameter up to 15 μm too are reported in *Quararibea* (Malvaceae), which is an exception. The smaller orbicules may fuse into larger compound aggregates in the later stages of their development. In some instances, such aggregates constitute platelets extending far across the tapetal cell surface and cross cell boundaries. Functions of orbicules have been debatable. Mostly, orbicules are believed to be involved in sporoderm formation. The idea that orbicules transport sporopollenin between the tapetum and the developing microspores was first put forward by Maheshwari (1950). The striking similarity between the pollen exine and the surface ornamentation of orbicules provides some clues. The walls of the orbicules apparently consist of sporopollenin, the main constituent of the mature microspore exine. However, Heslop-Harrison (1968) has refuted the role of orbicules in exine formation and regard orbicules as the only by-products of tapetum degeneration. Since orbicules offer potential micromorphological character to the pollen grain of a species, they are regarded as taxonomically important.

4.4. Anther Dehiscence

Anther dehiscence is the last step of anther development. Once pollen grain development is accomplished, the anther splits or dehisces to release the pollen grains. Dehiscence of anther occurs through a group of modified epidermal cells called stomium. Based on the position of the stomium, the major types of anther dehiscence in flowering plants are the following (Simpson 2010) (Fig. 4.5):

i) **Longitudinal:** Dehiscence suture/opening is parallel to the longitudinal axis of theca, e.g., in lily, *Brassica*. This is the most common type of dehiscence among flowering plants.

ii) **Transverse:** Dehiscence at the right angles to theca, e.g., *Ocimum sanctum*

iii) **Poricidal:** Dehiscence by a pore at one end of the theca, e.g., in Solanaceae, Ericaceae, grasses, *Casssia* sp.

iv) **Valvular:** Dehiscence through a pore covered by a flap of tissue appearing like a valve, e.g., in Lauraceae.

The research targeting anther dehiscence in numerous species (Lily, *Arabidopsis*, rice, maize, members of Solanaceae) has shown that the basic process of anther dehiscence is conserved among flowering plants, though the position of the final opening of the anther might vary (Wilson et al. 2011). In general, anther dehiscence involves a switch from cellular differentiation to degeneration combined with structural and physiological changes in the anther. It involves three major cell types: the septum, the stomium and the endothecium (Fig. 4.6).

Distinct events mark anther dehiscence which includes: degeneration of tapetum and middle cell layer immediately after meiosis, selective deposition of secondary thickening in endothecium, enzymatic digestion of cell walls at the septum between the two locules, pollen swelling, differential endothecial and epidermal cell expansion and finally dehydration (Scott et al. 2004; Ma 2005). Longitudnal anther dehiscence as observed in *Gasteria verrucosa* (Keijzer 1987) starts when the generative cell of the pollen grain detaches itself from the intine (refer to Section 4.5). At this stage, the cells of the epidermis and the endothecium start to expand both radially and tangentially followed by deposition of secondary thickenings on the inner tangential walls of the endothecium. Notably, the cells of stomium do not undergo expansion like the neighboring epidermal cells. Soon the cells of septum start dissociating due to the activity of hydrolytic enzymes such that locules of theca get merged and the anther becomes bilocular. As the pollen grains develop further in the locules, there is an outward pressure exerted from the inside of the locule forcing the anther to increase in size. The tangential swelling of the epidermis and the endothecium help in enlarging the circumference of the anther wall. However, the 'spring-like' bands of secondary thickenings on the inner tangential wall of the endothecium resist this expansion of the anther wall. These opposing mechanical forces of outward expansion and inward restriction bend the locule wall inwards breaking the small epidermal cells of stomium. Hence, resulting in the opening of anther locules with their wall bent inwards. This is followed by dehydration of the epidermal and endothecial cells due to their exposure to the outside environment. In addition to the epidermis and the endothecium, the connective and the filament also dehydrate. Loss of water from the wall layers causes the locule wall to bend outwards, consequently releasing pollen grains (Fig. 4.6).

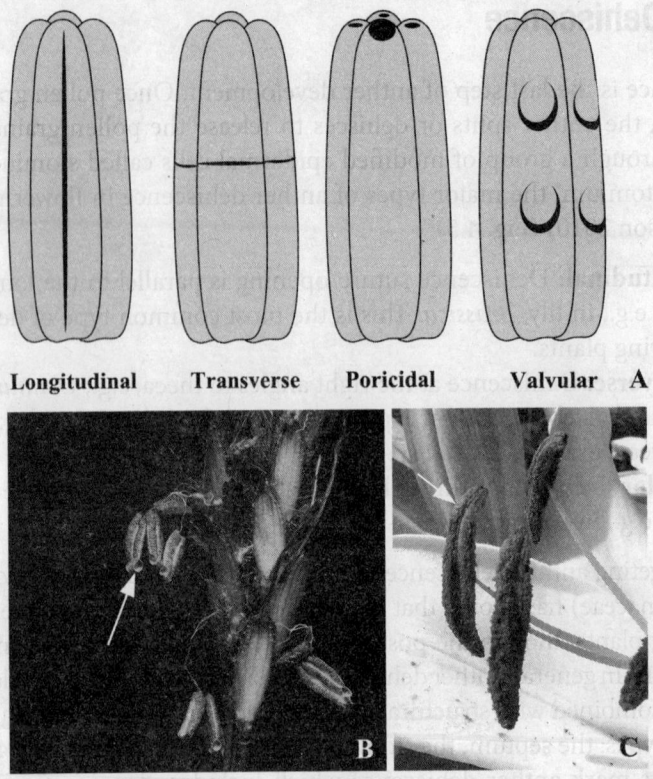

Figure 4.5 **A.** Major types of anther dehiscence in flowering plants. **B.** Poricidal anther (arrow) in grass. **C.** Dehiscence by a longitudinal slit (arrow) in lily. **See Color Plates (page 476).**

Dehydration in the final stages of dehiscence is an essential process that provides the final force for anther opening. It has been suggested that dehydration occurs, either as a consequence of evaporation via stomata on the adaxial side of the anthers or due to active removal of water (Keijzer 1987). Evidence for the latter is seen in tomato where conversion of starch to sucrose selectively occurs in the anther connective tissue. This results in an increased osmotic potential with the effect of removal of water from the anther. Also, in *Arabidopsis*, there are high concentrations of the H^+-ion sucrose transporter around the connective tissue of anthers. These transporters increase osmotic potential and induce dehydration of the surrounding regions in the anther (Stadler et al. 1999). Likewise, in *Petunia NECTARY1* (*NEC1*) and *NEC2* genes responsible for starch to sugar regulation in the filament and stomium have been characterized. The altered balance of starch to sugar regulates water potential in the filament and the stomium (Ge et al. 2001).

Dehiscence requires activation of many genes, including those that encode hydrolytic enzymes such as RNases, proteases, and cellulases (Wilson et al. 2011). Phytohormones like jasmonic acid and auxin have also been shown to provide a critical signal for dehiscence (Scott et al. 2004; Cecchetti et al. 2016). There are numerous mutant studies which show that mutations in genes encoding for endothecial thickenings, degeneration of septum and stomium lead to failure of anther dehiscence and prevalence of male sterility (refer to Box 4.5 for details).

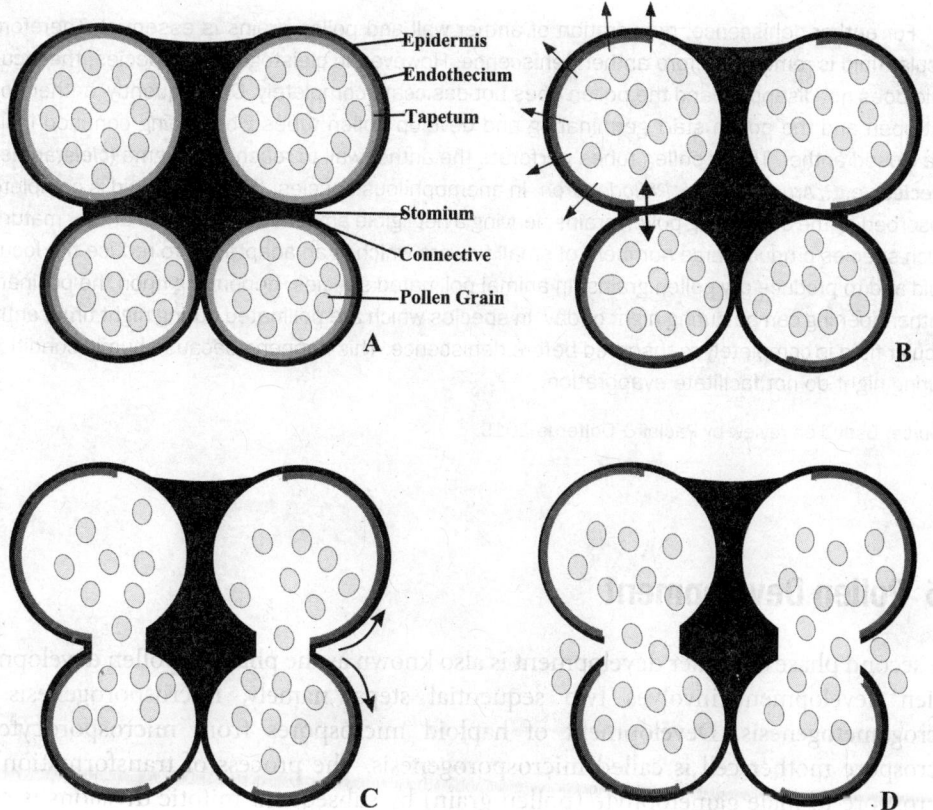

Figure 4.6 Anther dehiscence. A. Mature anther with pollen grains. **B.** Anther after breakdown of tapetum. The anther locule gets filled with a fluid contributed by the degeneration of tapetum and vascular supply. This increase in volume exerts an outward pressure on anther wall (outward arrows, only shown in one locule). **C.** The disintegration of septum joins the locules and, further adds to the pressure on the stomium. The endothelial wall thickenings and epidermis restrict the expansion of anther locules. Dehydration of anther at final stages of dehiscence and shrinkage of epidermal cells due to evaporation leads to increased tension (curved arrows) at stomium region and eventually favors its opening (**D**).

Box 4.5

Anther Locular Fluid and Dehiscence

The anther locule remains filled with fluid contributed by both tapetum and the vascular supply through the filament. The volume of the locular fluid varies with the anther morphology (aseptate or septate), its size, number and the type of pollen dispersal unit. Aseptate anthers usually have more locular fluid as compared to the septate anthers. The anthers comprising pollen as single units (monads) have more locular fluid than those where pollen are dispersed as aggregates (polyads). Species which produce a large number of small flowers also have less locular fluid, e.g., grasses.

For anther dehiscence, dehydration of anther wall and pollen grains is essential. Therefore, locular fluid is removed before anther dehiscence. However, in cleistogamous species, the locular fluid does not disappear and the pollen does not desiccate completely. Consequently, anther does not open and the pollen start germinating and develop pollen tubes while being confined inside the closed anther. These pollen tubes perforate the anther wall to reach the stigma (*cleistantheric species*), e.g., *Arum italicum, Philodendron*. In anemophilous species, the locular fluid is completely absorbed by the developing pollen grains, leaving a negligible amount at the time of anther maturity. Such species produce large numbers of small flowers which is an adaptation to reduce the locular fluid and to produce dry pollen grains. In animal pollinated species, depending upon the pollinator, anther opening can be during night or day. In species which are pollinated during night time, anther locular fluid is completely reabsorbed before dehiscence. This happens because humid conditions during night do not facilitate evaporation.

Source: Based on review by Pacini & Dolferus 2019.

4.5 Pollen Development

The second phase of anther development is also known as the phase of pollen development. Pollen development involves two sequential steps, namely, microsporogenesis and microgametogenesis. Development of haploid microspores from microsporocytes or microspore mother cell is called microsporogenesis. The process of transformation of a microspore to male gametophyte (pollen grain) by subsequent mitotic divisions is called microgametogenesis.

4.5.1 Microsporogenesis

Phase II of anther development is initiated with the differentiation of the pollen mother cells (PMCs) or microspore mother cells (MMCs) or microsporocytes from the sporogenous tissue. The sporogenous cells may directly function as PMCs or may divide in several planes to give rise to the PMCs. The PMCs then undergo meiosis to give rise to haploid microspores (Fig. 4.7). Significant changes occur in the PMCs during the prophase of meiosis I. In the pre-leptotene stage, the PMCs possess a dense cytoplasm with a prominent nucleus and a cellulosic cell wall. At this stage the PMCs are rich in organelles and remain interconnected with each other and also with tapetal cells through plasmodesmata. With the progression of meiosis, the protoplasts of PMCs withdraw considerably from the cell wall creating a space between the plasma membrane and the cell wall. Soon, each meiocyte synthesizes a temporary callose wall (a polymer of β-1,3-glucan) in the space between the primary cell wall and the plasma membrane. In most angiosperms, the callose is synthesized by callose synthases secreted by the meiocytes (Scott et al. 2004). Development of this callose wall breaks the cytoplasmic connections existing between the tapetum and the PMCs. On the other hand, the plasmodesmata between the meiocytes enlarge to form 'cytomictic

Box 4.6

Regulation of Anther Dehiscence

Anther dehiscence is a multistep process involving several cell types of anther wall. Several key regulators of anther dehiscence have been identified over the years. Several genes have been characterized in *Arabidopsis* for the enzymatic breakdown of the septum. These include *ARABIDOPSIS DEHISCENCE ZONE POLYGALACTURONASE1* (*ADPG1*) and *ARABIDOPSIS DEHISCENCE ZONE POLYGALACTURONASE2* (*ADPG2*). These two genes along with *QUARTET2* gene are required for anther dehiscence. *ADPG1* and *ADPG2* encode polygalacturonase (PG), which helps in the degradation of the cell wall material (pectin in middle lamella). These genes work under the influence of jasmonic acid, ethylene and abscisic acid for their anther dehiscence related polygalacturonase activity. A similar PG gene, *PS-2* has been described in tomato which expresses synchronously during different stages of anther dehiscence; its mutation leading to failure of anther dehiscence (reviewed in Ma et al. 2005; Wilson et al. 2011).

Likewise, there are several studies which report non-dehiscence mutants and also suggest an essential role of endothecial thickenings in the dehiscence of anthers. In the *ms35* mutant of *Arabidopsis*, the endothecium cells fail to accumulate thickenings and the anther does not dehisce even after pollen maturation and stomium degeneration. In another non-dehiscent mutant, although the pollen appears normal, the endothecium degenerates resulting in failure of stomium breakage. In *Arabidopsis myb26* mutant, anther development initially appears to be normal. However, after meiosis the tapetum and middle cell layers degrade and the endothecium fails to expand. Interestingly, pollen development and subsequent septum degradation occur normally but, as the anther dehydrates, the endothecium cells collapse, resulting in failure of retraction of the anther walls and no release or anther dehiscence occurs. There are also examples of precocious endothecium breakdown in *Arabidopsis* anthers, as a consequence of overexpression of the plantacyanin gene. Overexpression of the plantacyanin gene brings about premature programmed cell death of the endothecium due to high levels of Reactive Oxygen Species (ROS). Increased expression of the plantacyanin gene is also seen in the *receptor-like protein kinase2* (*rpk2*) mutant, which displays a lack of the middle cell layer, tapetal hypertrophy of the anthers, and defects in endothecial thickening. Jasmonic acid biosynthetic enzyme mutants i.e., *fatty acid desaturation* (*fad*), *opr3* (mutation in 12-oxophytodienoic acid reductase), *delayed-dehiscence1* (*dde1*) and *dde2*, and *defects in anther dehiscence 1 (dad1)*; also show non-dehiscence of anthers.

Source: Ma et al. 2005; Wilson et al. 2011; Nelson et al. 2012.

channels' to promote synchrony within the microsporocyte mass (Heslop-Harrison 1968). These channels can be up to 0.5 mm in diameter. Thus, each PMC gets enveloped by a callose wall except at cytomictic channels which connect the PMCs to each other. This stage is also called the central callose stage or the syncytium stage. The cytomictic channels may also sometimes lead to migration of chromatin and chromosomes between proximate meiocytes, a phenomenon known as 'cytomixis'. It is an anomaly which can lead to pollen sterility, e.g., *Clematis* (Kumar et al. 2010). By the end of prophase I the cytomictic channels get blocked and synchrony is gradually lost.

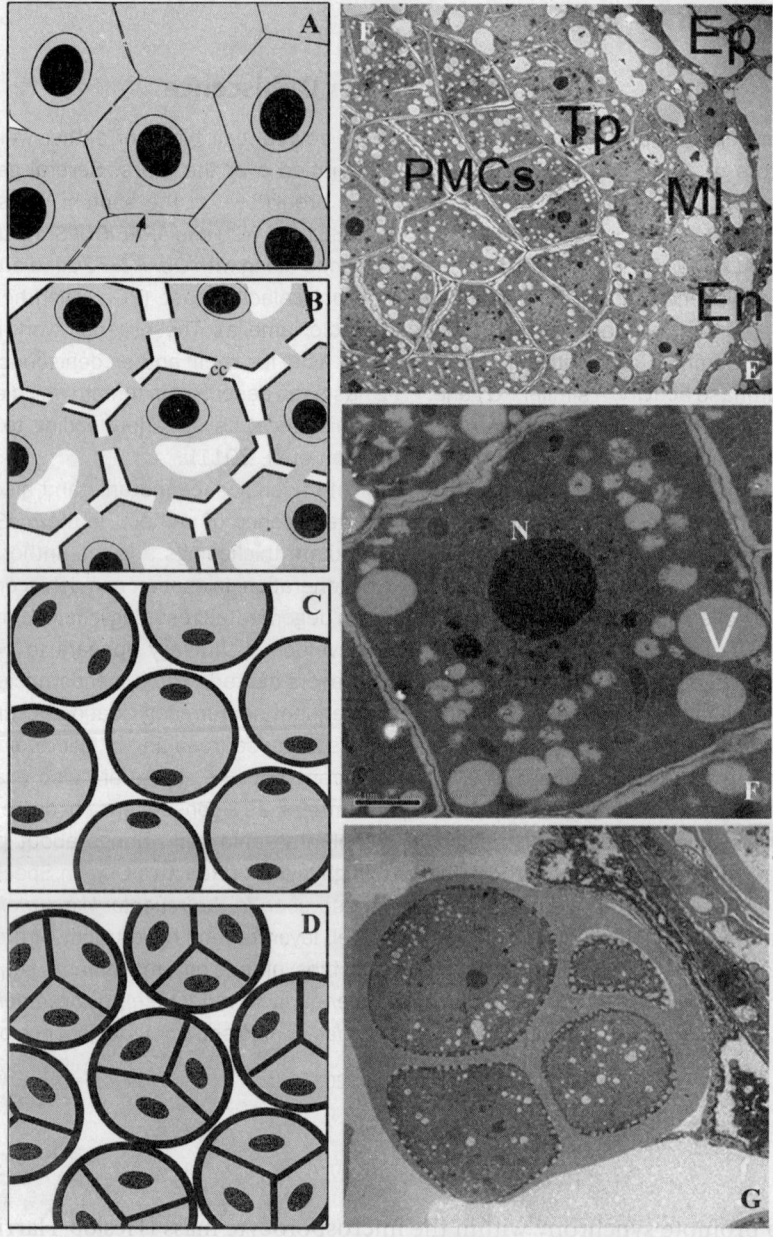

Figure 4.7 A–D Diagrammatic representation of microsporogenesis. A. Pollen mother cells (PMCs) or microspore mother cells (MMCs) with prominent nuclei connected with each other through plasmodesmata (arrow). **B.** Appearance of callose wall around each MMC along with the formation of the cytomictic channels (cc). **C.** Meiosis I. **D.** A microspore tetrad after meiosis, where four microspores remain encased in a common callose wall. **E–F. Transmission electron micrographs. E.** An anther locule with wall layers and compactly arranged MMCs. **F.** A MMC with a prominent nucleus (N) and a few vacuoles (V). **G.** Tetrad of micropsores after meiosis II of megasporogenesis. Note the microspores are enclosed in a common callose wall. Ep: epidermis, Ml: middle layers, En: endothecium, Tp: tapetum. (*Source*: E–G: Li et al. 2016, published under CC BY License.)

Like other meiotic cells, the principal events of meiosis like the chromosome pairing, recombination, and segregation; are common to PMCs (Fig. 4.8). Thus, at the end of meiosis there is formation of a tetrad of haploid cells enclosed in a thick callose wall, and each one of these haploid cells is called the 'microspore'. The microspores in the tetrad may be arranged differently. The arrangement is species specific and can be tetrahedral, isobilateral, decussate, linear or t-shaped. However, even in a single species different arrangements of tetrads can be observed, e.g., in *Aristolochia elegans* five types of arrangements, namely, tetrahedral, isobilateral, decussate, linear, and t-shape are seen. In *Pseuduvaria trimera*, tetrahedral, tetragonal, rhomboidal, decussate, and T-shaped can be even seen in one anther locule (Li & Xu 2018). Similarly, isobilateral, decussate, and linear tetrads are observed in *Pancratium maritimum* (Tütüncü 2017).

At the end of meiosis, the external and intersporal callose walls of the tetrad dissolve to release the individual microspores as a result of callase activity (β-1,3-glucanase). You may recall that the callase enzyme is secreted by the tapetum. Appropriate timing of callase secretion is critical in microsporogenesis as its failure or delay may cause male sterility. Callase synthesis is shown to be controlled by transcription factor AtMYB80 (of MYB-family). Other than pollen maturation, this factor is also linked to the development of tapetum and to its programmed cell death (Hafidh et al. 2016). The release of the microspores from the tetrad marks the end of microsporogenesis.

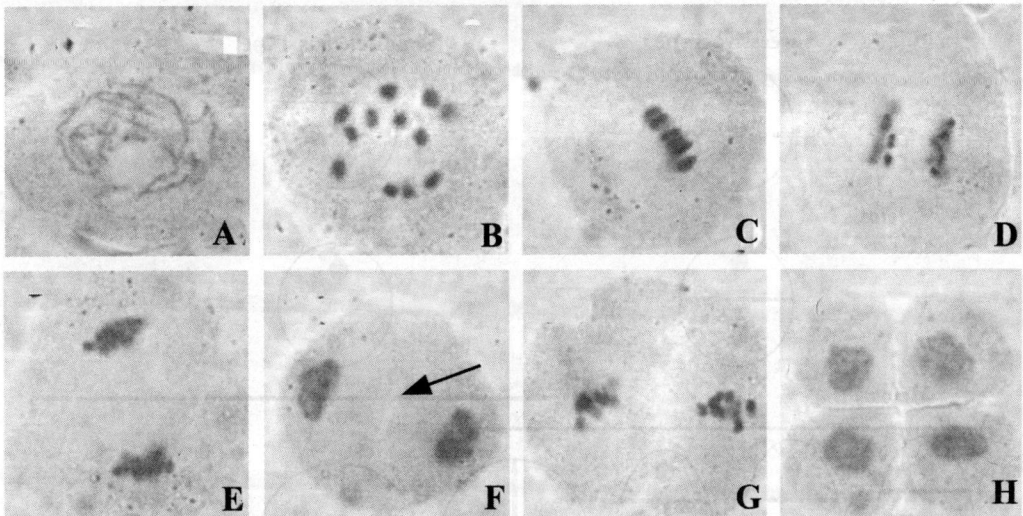

Figure 4.8 Meiosis of microspore mother cells in rice. A–B. Prophase **C.** Metaphase I **D.** Anaphase I. **E.** Telophase I **F.** Formation of a centrifugal cell plate (arrow, successive cytokinesis) after meiosis I. **G.** Dyad after completion of cytokinesis. **H.** A tetrad after meiosis II (tetragonal type). (*Source:* Qin et al. 2015 published under CC BY License.)

4.5.1.1 *Cytokinesis during Microsporogenesis*

Cytokinesis during microsporogenesis takes place through the formation of intersporal walls composed of callose. The cytokinesis may occur immediately after meiosis I or may occur after completion of meiosis II. Based on the timing of occurrence, two types of cytokinesis are observed during male meiosis; successive and simultaneous. If cytokinesis occurs after meiosis I and meiosis II, then it is of the successive-type and if it occurs only after the completion of Meiosis II, then it is of the simultaneous-type. Successive cytokinesis can thus result in a transitory two-celled stage after meiosis I and subsequently in a tetrad after meiosis II. At the dyad stage, the two cells are embedded within the pollen mother cell wall, and separated by a callose wall. However, simultaneous microsporogenesis results in tetrad formation only after the completion of meiosis II (Fig. 4.9).

In general, the successive-type of cytokinesis is typically observed in monocotyledonous species whereas the simultaneous-type is characteristic of dicotyledons (with exceptions like Ranunculales, Proteales, Malpighiales and Malvales). This variation in cytokinesis also results in various arrangements of microspores in a tetrad (Fig. 4.9). Recent studies have drawn a correlation between the type of pollen cytokinesis and shape of the tetrad. In general, tetrads obtained through successive cytokinesis can be tetragonal, decussate, T-shaped, Z-shaped and linear, whereas tetrads resulting from simultaneous cytokinesis can be tetrahedral, rhomboidal, tetragonal and decussate (Albert et al. 2011; Storme & Geelen 2013) (Fig. 4.9).

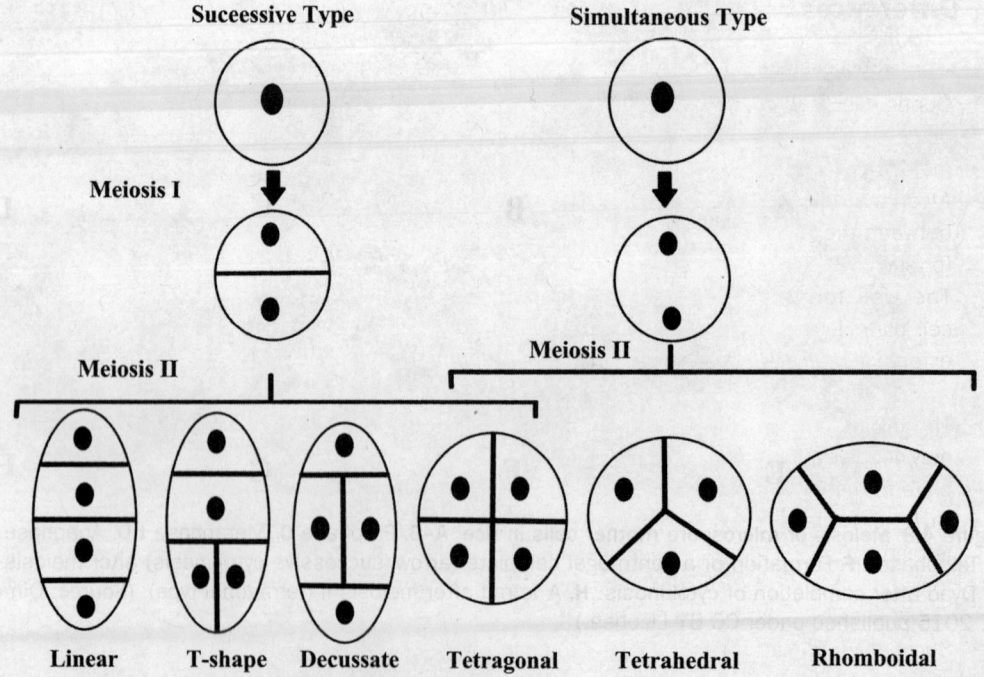

Figure 4.9 Diagrammatic representation of types of cytokinesis in male meiosis in flowering plants resulting in different types of tetrad arrangement.

Role of phragmoplast and microtubules in cytokinesis: Cell wall formation during successive microsporogenesis occurs centrifugally, similar to that in mitosis. It involves the establishment of a centrifugally expanding microtubular structure called the phragmoplast. The position of this cell plate is determined by the spindle structure in meiosis I. At anaphase I, three types of the spindle microtubules are seen. First type of microtubules pulls the chromosomes towards the two poles, and they are called the centromere microtubules. The second type of microtubules are connected to the two poles and are called the pole microtubules. A third type of microtubules, called the phragmoplast microtubules appear in the centre of the dividing meiocyte during anaphase I. This third type of phragmoplast microtubules may be derived from depolymerized centromere microtubules as seen in *Allium cepa* or in wheat (Shamina et al. 2007). Later, phragmoplasts expand centrifugally and form the cell plate by guiding vesicle-mediated deposition of cell wall components. Contrastingly, in the simultaneous-type of cytokinesis, cell plates are formed centripetally. There is inward-oriented furrowing of the callosic parental wall to partition the meiotic cytoplasm. The centripetal cell plate formation is mediated by phragmoplast-like microtubular structures called radial microtubule arrays which demarcate the position of a new cell wall (Shamina et al. 2007; Zhang et al. 2011; Storme & Geelen 2013).

Box 4.7

Differences between Simultaneous and Successive Wall Formation

Successive Wall formation	Simultaneous Wall formation
• During microsporogenesis, the cytokinesis or cell wall formation occurs after each meiotic division.	• During microsporogenesis, the cytokinesis or the cell wall formation occurs after both the meiotic divisions.
• After meiosis I, wall formation occurs between the two nuclei and a dyad cell is formed	• After meiosis I, no wall formation or a dyad stage occurs. The two nuclei remain in a common cytoplasm.
• The wall formation is centrifugal with cell plate forming in the centre and then extending laterally.	• Wall formation is centripetal in nature with furrows growing towards the centre of the cell to eventually meet and divide the cell into four parts.
• The division in the two dyad cells may or may not be synchronous because of the cell plate formation between the dyads.	• The two haploid nuclei always undergo the second meiotic division in synchrony as they are lying in the common cytoplasm.
• Such cytokinesis generally results in tetragonal (isobilateral), linear and sometimes to T-shaped tetrads.	• Such cytokinesis generally results in more variable tetrad morphology giving rise to tetragonal, rhomboidal and tetrahedral tetrads.
• Successive meiotic cytokinesis is present in monocots like *Allium* and Wheat.	• Simultaneous meiotic cytokinesis is present in dicots like *Nicotiana*.

4.5.1.2 *Role of Callose during Microsporogenesis*

The development of a callose wall around meiocytes during microsporogenesis is a common feature in flowering plants with a few exceptions. There are several biological roles that have been attributed to callose during microsporogenesis. The callose wall isolates the meiocytes from other sporophytic tissues and prevents them from dehydration under water stress conditions (Li et al. 2010). Callose has also been proposed to act as a barrier, or molecular filter transmiting only those signals that are indispensable for meiosis into the meiocytes (Heslop-Harrison 1968; Rodriguez-Garcia & Majewska-Sawka 2011). Callose is also believed to function as a temporary wall isolating the products of meiosis from each other to prevent cell cohesion and fusion. Once it is achieved, microspores are released upon callose dissolution (Waterkeyn 1962). Callose also serves as a template for the formation of the species-specific exine-sculpting patterns seen on mature pollen grains (Waterkeyn & Bienfait 1970). However, there are exceptions like *Pandanus odoratissimus* (Periasamy & Amalathas 1991) where microsporogenesis is not disrupted even though the species lacks callose naturally.

4.5.2 Microgametogenesis

After the completion of meiosis in MMCs the resultant microspores get liberated from common callose wall. The free microspore stage (uninucleate stage) extends until mitosis, after which the microspore becomes a microgametophyte or the pollen grain. The mitosis within the microspore results in a vegetative cell and a generative cell or two-celled male gametophyte. The generative cell undergoes another mitotic division to give rise to the two sperm cells (or the male gametes). Thus, microgametogenesis comprises events that lead to the development of a mature pollen grain from a unicellular microspore. The general process of microgametogenesis is described below and in Fig. 4.10.

By the end of meiosis, the anther locule gets filled with the tapetum derived secretions (in case of secretory tapetum) or with disintegration products of the tapetum (as in amoeboid tapetum). Upon release from the tetrad, the free microspores increase rapidly in volume due to absorption of the nutrients like the proteins, lipids, carbohydrates and secondary metabolites along with the water present in the anther locule. At this (uninucleate) stage, the microspore cytoplasm shows a centrally localized nucleus, numerous plastids (without starch grains), mitochondria, dictyosomes and a number of small vacuoles scattered in the cytoplasm. Except in orchids, these vacuoles coalesce to form a large vacuole which occupies most of the cytoplasm (sometimes even more than half of the volume of the microspore) and further adds to the volume of the micropsore. Vacuolation is accompanied by the displacement of the centrally placed nucleus and the cytoplasm to the periphery (Fig. 4.10 B–C).

The first mitotic division (mitosis I) in microspore is asymmetric, producing a smaller generative cell and a larger vegetative cell. This asymmetry in division is attributed to asymmetrical microtubular arrangement. At the onset of mitosis, the peripherally placed nucleus loses its membrane. The chromosomes get attached to the mitotic spindles which are placed asymmetrically in the cytoplasm. The spindle itself is asymmetrical with two unequal halves. The peripheral pole of spindle is typically wide and is in contact with the plasma membrane of the microspore; while the interior pole of the spindle is more or less pointed and is towards the centre of the microspore. During cytokinesis, phragmoplasts forming the cell plate expand centrifugally but in a unique curved fashion. Hence, a curved cell plate separates a small, lens-shaped (or convex) generative cell (GC) and a large vegetative cell (VC), with each cell containing a haploid nucleus (Liu et al. 2011) (Fig. 4.10 D–F).

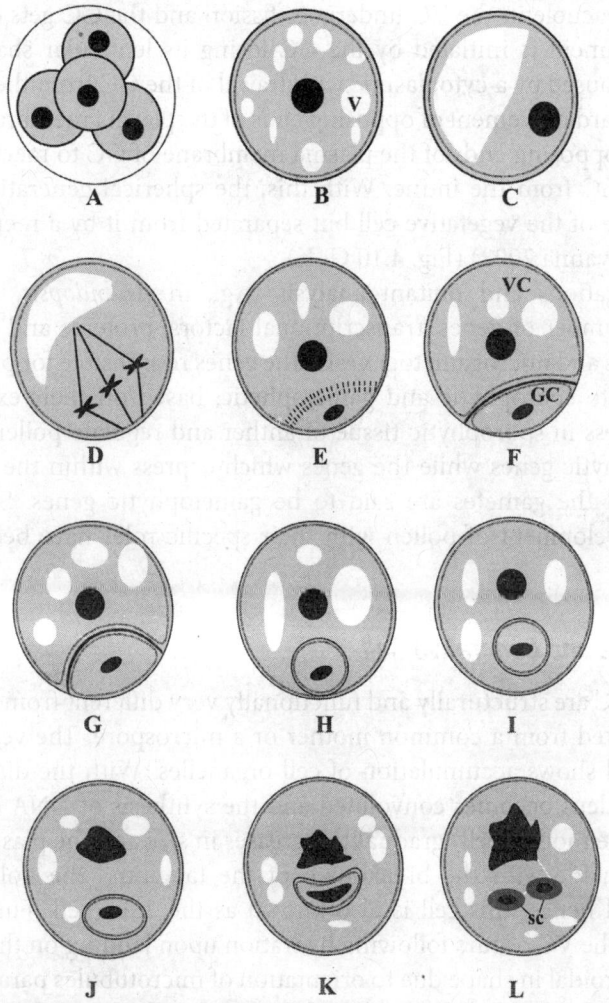

Figure 4.10 Diagrammatic representation of process of microgametogenesis. A. Microspore tetrad after meiosis. **B.** A uninucleate micropsore after release from tetrad. Note the presence of numerous small vacuoles (v) in cytoplasm and a centrally placed nucleus. **C.** The vacuoles merge to form a large vacuole that pushes the nucleus towards the periphery. **D.** Occurrence of an asymmetrical spindle for mitosis. **E.** A curved cell plate expanding centrifugally to give rise to lens shape generative cell. **F.** A microspore with a large vegetative cell **(VC)** and a smaller generative cell **(GC)**. **G–I.** Stages during the detachment of a generative cell. **J.** The VC nucleus becomes lobed. **K.** The generative cell elongates and partly covers the lobed VC nucleus. This stage typically represents bicelled stage of pollen. **L.** In some species the GC divides and gives rise to two sperm cells within the pollen before its dispersal. Such pollen is called three-celled pollen.

While the vegetative cell constitutes the bulk of the young pollen grain with most of the plastids and mitochondria, the generative cell inherits a very small amount of the microspore cytoplasm (with few or no cell organelles). The cell plate between VC and GC consists of callose. However, the callose wall does not envelope the whole generative cell and is strictly confined to the hemispherical cell plate between the VC and the GC. On the outer side, the generative cell remains attached to the inner/intine layer of the pollen wall. Immediately after mitosis, the vacuole in the VC undergoes fission and the GC gets detached from the intine. The detachment is initiated by the GC losing its lenticular shape and becoming spherical. This is caused by a cytoplasmic withdrawal of the GC from the contact sites with intine causing inward movement of opposing ends of the plasma membrane of the GC. This finally causes the opposing ends of the plasma membrane of GC to meet and fuse, thereby pinching off the GC from the intine. With this, the spherical generative cell moves and floats in the centre of the vegetative cell but separated from it by a membrane only (cell-within-a-cell) (Shivanna 2002) (Fig. 4.10 G–K).

Recent investigations and mutant analysis (e.g., in *Arabidopsis*, rice, maize) have characterized a number of genes, transcriptional factors, proteins and kinases regulating microsporogenesis and microgametogenesis. The genes responsible for pollen development are categorized into sporophytic and gametophytic, based on their expression site. The genes which express in sporophytic tissue of anther and regulate pollen development are called the sporophytic genes while the genes which express within the microspore (male gametophyte) and the gametes are said to be gametophytic genes. Some of the genes regulating the development of pollen with their specific roles have been summarized in Box 4.9.

4.5.2.1 *Vegetative and Generative Cell*

The VC and the GC are structurally and functionally very different from each other though both have originated from a common mother or a microspore. The vegetative cell grows post mitosis I and shows accumulation of cell organelles. With the disappearance of the nucleolus, the nucleus becomes convoluted and the synthesis of DNA stops. The vacuole (inherited from the mother cell) gradually decreases in size and the plastids develop starch grains which coincide with the breakdown of the tapetum. The pollen tube develops from this cell and hence, this cell is also known as the 'tube cell'. Further polarization/differentiation in the VC occurs following hydration upon landing on the stigma.

The GC is spheroidal in shape due to orientation of microtubules parallel to its long axis. It possesses very little cytoplasm, none or a few mitochondria, a nucleus with a prominent nucleolus and abundant endoplasmic reticulum. Notably, its cytoplasm has a very low number of plastids and some species like *Medicago* sp. and *Plumbago zeylanica* completely lack plastids. The organelles within the GC remain distributed on the periphery of the nucleus, forming a dense annulus. Subsequently, the GC undergoes a symmetric second mitotic division to produce the two sperm cells (or the male gametes). After division, the sperm cells acquire maturity which is characterized by a significant increase in their cytoplasmic content, number of mitochondria and plastids (Russell & Strout 2005). The GC and its derivatives i.e., the sperm cells are never 'free' of the vegetative cell in which they are embedded till the time they are released for fertilization.

4.5.2.2 Bi- and Tri-cellular Pollen Grains

The timing of the second mitotic division resulting in two sperm cells varies among species (Fig. 4.10 J–L). In some, it takes place before the dehiscence of anthers and in others it may happen after anther dehiscence. Thus, the number of cells and nuclei present in the pollen grain at the time of anther dehiscence is variable among species. About 70% of flowering plants shed their pollen grains at a bicellular stage. Such pollen grains with a VC and a GC are called bicellular pollen. The GC divides to form two sperm cells after pollen tube formation, as seen in Orchidaceae and Nymphaeaceae. In many angiosperm taxa, however, the second mitotic division (pollen mitosis II) takes place before anthesis such that pollen at anthesis is tricellular with one VC and two sperm cells, e.g., Asteraceae, Lamiaceae, Brassicaceae and Poaceae. Very few species like *Annona cherimola* release both bi- and tricellular pollen grains. When this occurs, the ratio between bi-cellular and tri-cellular pollen grains in an anther is dependent on environmental factors such as temperature and relative humidity during the last phases of maturation.

Box 4.8

Differences between Vegetative Cell and Generative Cell

Vegetative Cell	Generative Cell
• Vegetative cell is larger in size.	• Generative cell is smaller in size.
• Vegetative cell does not undergo further division. Vegetative cell produces a pollen tube.	• Generative cell undergoes mitosis and produces two sperm cells.
• Nucleus is convoluted and the DNA content remains the same.	• The nucleus is spheroidal and the synthesis of DNA increases the DNA content.
• Rich in plastids and mitochondria.	• Plastids and mitochondria are absent or very low in number.
• Metabolically active.	• Metabolically inactive/dormant except during mitosis.
• Accumulates the starch grains.	• Does not accumulate starch grains.

Box 4.9

Summary of Genes/Proteins Involved in Microsporogenesis and Microgametogenesis

Gene/Proteins	Plant	Role
ASY1, HOP2	*Arabidopsis,* Rice	Regulates meiotic homolog pairing during microsporogenesis
TUBG1 and *TUBG2* (Tubulin Genes)	*Arabidopsis*	Required for spindle and phragmoplast organization in microspores
HINKEL (AtNACK1) and *TETRASPORE (AtNACK2)*	*Arabidopsis*	Cell plate expansion and cytokinesis defects during microsporogenesis
ArabidopsisMS1 (AMS1), AtMYB103	*Arabidopsis*	Expresses in tapetum and behave genetically as sporophytic genes required for normal pollen development at the onset of pollen development
NEDD1 (Neural precursor cell Expressed, Developmentally Down-regulated 1)	*Arabidopsis*	Mutants exhibit defects in microtubular organization in developing microgametophyte
HOP1(HORMA-domain protein), AHP2 (Arabidopsis thaliana histidine phosphotransfer proteins)	*Arabidopsis,* Rice	Essential for sperm cell assembly
MOR1/GEM1, TMBP200	Tobacco	Mutation caused pollen mitosis I could be extrapolated to mitosis in somatic cells
Gemini Pollen 1 (Gem1)	*Arabidopsis*	Mutants produced single-celled pollen, exhibit impaired pollen cytokinesis
Cyclin-Dependent Kinase (CDKA;1), *F-BOX-Like 17 (FBL17)* gene	*Arabidopsis*	Transiently expressed in the male germ cell. Disruption leads to retarded S-phase, and delayed germ cell entry in mitosis; until the pollen grain germination. FBL17 mutants also produced pollen with only single sperm. cdka;1 mutant germ cells, fertilized the egg cell or the central cell
TIO (Two in One), Ser/ Thr protein kinase FUSED	*Arabidopsis*	Localizes in phragmoplast. Essential role in cell plate expansion during microgametogenesis. Mutant pollen shows complete asymmetric nuclear division, and binucleate pollen. No sperm cell formation.

DUO1, DUO3 (Encodes a novel conserved R2R3 MYB protein)	*Arabidopsis*	Specifically expresses in the male germline, mutants develop only single sperm cell as mitosis II fails. The duo1 germ cells are also incompletely differentiated and do not result in fertilization
Retinoblastoma-related (RBR) protein	*Arabidopsis*	Loss of RBR results in hyperproliferation of the vegetative cell and to a lesser extent the generative cell, leading to pollen with four sperm cells (*Note it is a phenotype opposite to that observed in cdka;1 mutant*)
GCS1, GEX1, GEX2 and GEX3	Arabidopsis, Maize, Lily	Required for germline specification
GUM (Germ Unit Malformed) and MUD (MGU displaced)	*Arabidopsis*	Abnormal organization or position of the MGU
CELLULOSE SYNTHASE-LIKE D4 (CSLD4)	*Arabidopsis*	Expresses in vegetative cells and required for proper pollen tube growth
HAPLESS 2 (HAP2), also known as GENERATIVE CELL-SPECIFIC 1 (GCS1)	*Arabidopsis*	Expresses in sperm, required for efficient pollen tube guidance to ovules
AtGPAT1 (encodes a membrane bound glycerol-3-phosphate acyltransferase)	*Arabidopsis*	Exhibits abnormal tapetum morphology and defects in the formation of the pollen wall
FLP1 (Flowering Promoting Factor 1)	*Arabidopsis*	Mutant produces pollen grains with defective exine that has irregular sporopollenin patterns

Source: Based on Liu et al. 2011, Twell 2011 and references therein.

4.6 Male Germ Unit (MGU)

Russell & Cass (1981) using electron microscopy in 3-celled *Plumbago zeylanica* demonstrated that the two sperm cells are directly linked to each other. Also, one sperm remains consistently associated with the vegetative nucleus through sperm cell extensions (Fig. 4.10 L). This intimate association that exists between the two sperms, and between one sperm and the vegetative cell nucleus is known as the Male Germ Unit (MGU), a term coined by Dumas et al. (1985). In 2-celled systems, association between vegetative nucleus and sperm cells is seen in the pollen tube. Thus, the existence of MGU is a universal feature of flowering plants, though the timing of establishment of union may vary. Some examples of the species with 3-celled pollen where MGU has been studied are *Beta vulgaris, Brassica, Spinacia oleracea*. The species with 2-celled pollen and MGU are *Gossypium, Euphorbia, Nicotiana, Petunia, Acacia*, and *Aloe cilliaris*. Family Poaceae which has 3-celled pollen is an exception, as MGU is absent in the family. In grasses, the connection between vegetative nucleus and the sperm cells is temporary and exists for a very short duration. As soon as sperm cells enter into the pollen tube, the vegetative nucleus gets detached from them, although the sperms remain connected during the pollen tube growth. However, in maize the cytoplasmic connections are present between vegetative nucleus and sperm cells even during their passage into the pollen tube (Shivanna 2002; McCue et al. 2011).

4.6.1 Structure

The structure of MGU was deduced from transmission electron microscopic studies (Russell & Cass 1981) and computer assisted 3-D models (Russell 1984; McConchie et al. 1985; Mogensen 1992) as observed in *Plumbago zeylanica* (Fig. 4.11). In *P. zeylanica*, the sperms are encompassed by the VC plasma membrane. Out of the two, one sperm cell remains consistently associated with the vegetative nucleus and is designated as S_{vn} or sperm cell 1. S_{vn} has a 30μm long slender projection of about <1μm width called 'cytoplasmic projection' that embraces the vegetative nucleus over a large surface area and is partially surrounded by lobes of the vegetative nucleus. In other species (viz. *Nicotiana, Hippeastrum vitatum, Brassica, Petunia hybrida*) also the sperm cells are not only in direct physical association with the vegetative nucleus but are also partially surrounded by it (Mogensen 1992; Lalanne & Twell 2002). The second sperm cell which is not in direct contact with the vegetative nucleus is denoted as S_{ua} or Sperm 2. The S_{ua} maintains a direct connection with the other sperm cell (S_{vn}) by a common transverse cell wall containing plasmodesmata. The other side is free and floats in the cytoplasm of vegetative cell. It lacks projection or if present such a projection may be very short (Fig. 4.11).

The size of the cytoplasmic extension of S_{vn} is variable among species and within a species at different developmental stages, e.g., cytoplasmic projections usually appear short in the ungerminated pollen grain and elongated in the pollen tube. Studies have shown that the cytoplasmic projections between the vegetative nucleus and the generative cell are present before mitosis II or pollen germination (in case of 2-celled pollen). These projections are thought to possess a microtubule-based cytoskeletal framework that may provide structural support to the cytoplasmic projection. It maintains the shape of sperm cells as well. It has also been conjectured that cytoplasmic projection may exhibit RNA and/or protein transport, facilitating the communication between the somatic vegetative cell nucleus and the sperm cells (McCue et al. 2011).

4.6.2 Functions

- The MGU ensures sperm-to-sperm connections during pollen tube transport. This close proximity of sperms is significant for the two fertilization events that initiate embryo and endosperm development to occur simultaneously.
- In certain incidences due to the entry of two or more pollen tubes in to the embryo sac, the egg cell and polar nuclei of an embryo sac may get fertilized by sperms cells belonging to different pollen tubes. Such cases of double fertilization are called heterofertilization which leads to formation of endosperm and embryo which lead genetically different. Close association of sperm cells in MGU prevents heterofertilization.

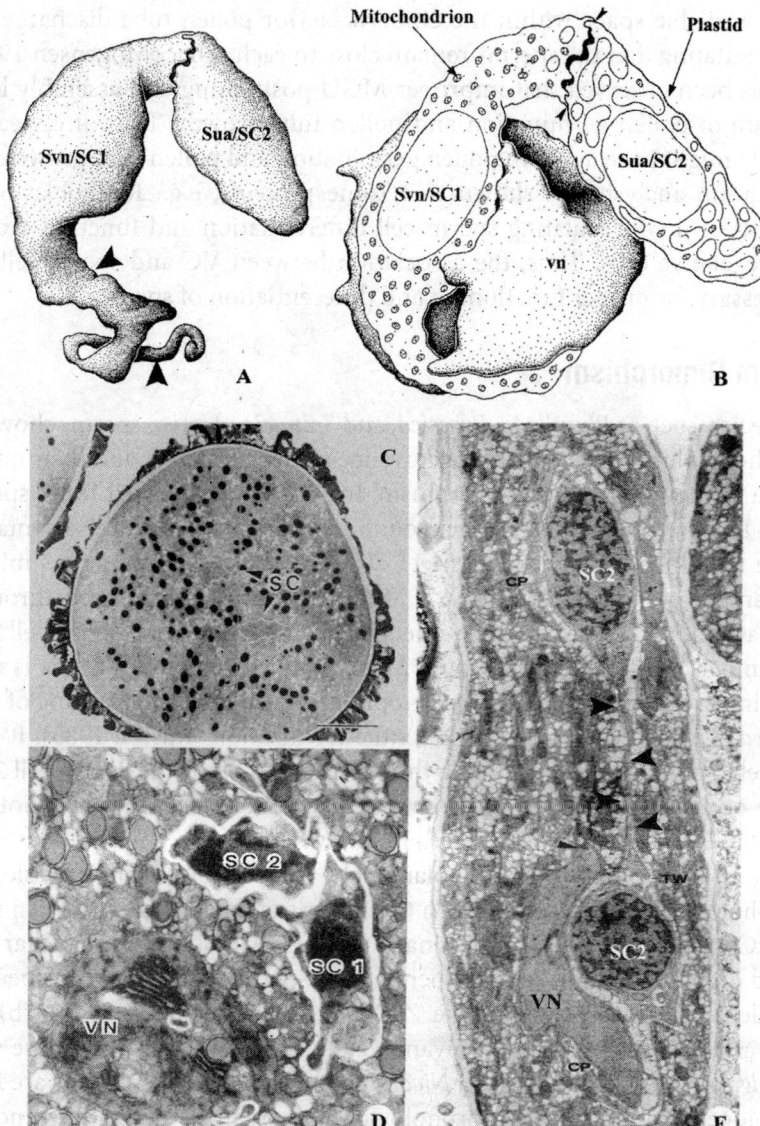

Figure 4.11 Male germ Unit. A–B. Computer assisted reconstruction of MGU in *Plumbago zeylanica* exhibiting pronounced sperm dimorphism. The larger sperm (S_{vn}/SC1) is associated with vegetative nucleus and has a long projection (arrowheads). The smaller sperm (S_{ua}/SC2) is connected to SC1 via common wall. There is a marked difference in the number of mitochondaia and plastids between the sperm cells (see text for details). C–E. Transmission electron micrographs showing: C–D. *Brassica campestris*. C. A mature pollen grain with two sperm cells (sc) The two sperm cells (SC1 and SC2) are present in the cytoplasm of vegetative cell. D. The male germ unit. Note the tail of sperm cell1 (SC1) is present in the embayment of the vegetative nucleus (VN). The RER is also associated with VN. E. Longitudinal section of the pollen tube of *Nicotiana tabacum* showing MGU. SC1: Sperm cell 1 which is closely associated with VN. Sperm cell 2 (SC2) is connected with SC1 by tail like extension (arrow heads). The anterior of each sperm exhibits presence of cytoplasmic projections (cp). (*Source*: A–B: Russell 1984, C–D: after McConchiee at al. 1985; E: after Yu et al. 1989, *reproduced with permission*.)

- The available space within the embryo sac for pollen tube discharge is limited, necessitating that the sperms remain close to each other (Mogensen 1992).
- It has been observed that improper MGU positioning and assembly leads to the failure of pollen germination and pollen tube growth. Thus, it appears that the MGU might have a role in pollen germination and pollen tube growth.
- Mutation analysis has shown that some proteins, e.g., Retinoblastoma-related (RBR) protein, regulating sperm cell differentiation and function are expressed in vegetative cells. Thus, the association between VC and sperm cells might be necessary for proper functioning and differentiation of sperm cells.

4.6.3 Sperm Dimorphism

In several species, such as *Plumbago*, *Brassica*, and *Spinacia*, the two sperms show variability in terms of their volume (size), shape, occurrence of cell organelles and their number. This phenomenon is known as 'sperm dimorphism'. In *P. zeylanica* (Russell 1984) sperm cell S_{vn} or sperm cell 1, is characterized by a larger volume and surface area and also contains a larger nucleus than the other sperm cell (Fig. 4.11 D-E). S_{vn}, possesses numerous mitochondria (upto 311) and few or no plastids (0-2). Mitochondria are distributed throughout the sperm body as well in the cytoplasmic projection. In contrast, S_{ua} or sperm cell 2 possesses a smaller number of mitochondria (upto 52), while the number of plastids is more (upto 46); these being evenly distributed in the cytoplasm. Ultrastructural tracking of sperm cells in *P. zeylanica* provide evidences of preferential fertilization (Russell 1985). It was shown that Sperm cell 1 preferentially fuses with the polar nuclei, while the Sperm cell 2 invariably fertilizes the egg cell. This observation shows that double fertilization is not a random process and MGU is a 'polarized fertilization unit' where one sperm is pre-destined to fuse with the egg and the other with the polar nuclei (Heslop-Harrison & Heslop-Harrison 1984). Morphological differences between the sperms of a pair have also been reported in *Brassica* (McConchie et al. 1987a) and spinach (Wilms 1986), but the organellar differences are restricted only to the number of mitochondria. Sperm dimorphism has been reported to variable degrees in other species like *Zea mays* (McConchie et al. 1987b); *Gladiolus gandauensis* and *Rhododendron* sp. (Shivanna et al., 1988). However, in some species like *Hordeum vulgare*, *Petunia hybrida*, and *Nicotiana tabacum,* the sperm cells are isomorphic (as cited in Mogensen 1992). Thus, morphological and organelle-based differences between the sperms of a pair can vary from species to species.

4.7 Pollen Wall: Structure, Synthesis and Features

One of the key adaptations for colonization of land plants is the development of a durable pollen wall (also known as sporoderm). The major component of the pollen wall is the highly resistant biopolymer sporopollenin. It is a highly cross-linked polymer of hydroxylated fatty acids, aliphatic compounds, and, possibly phenolics conjugated by ether and ester bonds (Wang & Dobrista 2018).

The composition, features and synthesis of pollen wall are unique compared to other plant cell walls. The morphology of pollen wall exhibits species-specific diversity. The pollen wall also varies chemically due to deposition of various pollen coat substances. This diversity in physical and chemical composition is functionally important for the plants to distinguish its own pollen from that of other plants. Pollen wall diversity also serves a taxonomical function, forming the basis of 'Palynology'.

4.7.1 Pollen Wall Structure

The pollen wall of flowering plants has multiple layers which are laid down in regulated manner during pollen development. The pollen wall chiefly comprises of two layers; the inner layer 'intine' and the outer layer 'exine'. These two layers are different from each other in their structure and features. The intine surrounds the microspore cytoplasm and is composed of cellulose and pectin which make it less elastic. The exine is made of sporopollenin, providing elasticity to the pollen wall and resistance against numerous environmental factors. Structurally, exine is highly ornamented and shows vast diversity. An important component of exine is the predetermined spaces or gaps in it that allow the emergence of pollen tubes. These gaps are known as the germinal apertures (GA).

Despite the enormous diversity in patterns exhibited by exine; the general architecture is quite similar across species. The sub-layers of exine have been given different names by different scientists and two major sets of terminologies proposed by Erdtman (1952) and Fægri (1956) are frequently used. A comparison between the two sets of terminology has been depicted diagrammatically in Fig. 4.12. According to Erdtman, mature exine can be subdivided into two sub-layers i.e., sexine (the outer sculptured layer) and nexine (the simpler inner sub-layer located just above the intine). The sexine often consists of several structures: the radially elongated rods known as 'baculae or columellae' and the roof-like structure formed by the fusion of heads of the baculae known as 'tectum'. Other elements like spikes, spines, bumps, globules develop over the tectum giving the exine its unique ornamentation. These elements are also referred to as the supratectal elements. The inner sublayer, the nexine may be subdivided into nexine 1 and nexine II. According to Fægri (1956), exine can be divided into two sublayers ektexine and endexine. While the ektexine comprises of tectum, columellae and foot layers (nexine 1), the endexine is single layered (nexine 2). Not all species have all the sub-layers or structures of the exine. Pollen grains of many species neither develop the supratectal elements nor the tectum or the nexine substructures. For example, aquatic plants or plants living and pollinating in extremely wet environments, exhibit reduced or no exine deposition.

4.7.2 Pollen Wall Synthesis

The pollen wall synthesis overlaps with the pollen development. After the completion of cytokinesis (leading to the tetrad stage), wall deposition begins within the microspore and is dependent on the activity of the tapetum (Fig. 4.13 and 4.14).

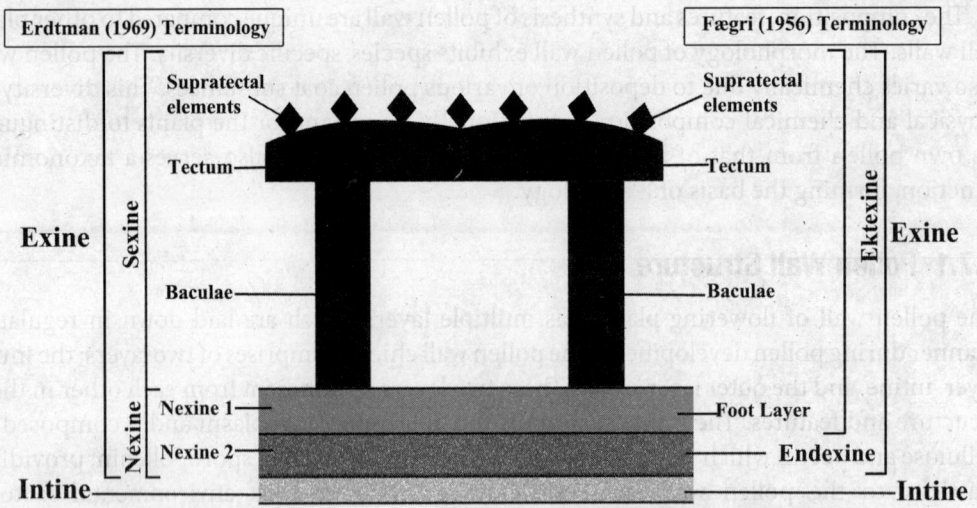

Figure 4.12 Diagrammatic sketch of pollen wall layers and comparison of widely accepted terminologies.

Exine: At the end of meiosis, each microspore in a tetrad is surrounded by a callose wall. Soon, a microfibrillar material starts accumulating between the plasma membrane and the callose wall of each microspore. This low electron density microfibrillar material is largely composed of cellulose and known as primexine. It is secreted chiefly by the microspore cytoplasm and also by the tapetum (Blackmore et al. 2007; Wang & Dobrista 2018). Primexine is an ephemeral structure and functions as an elaborate glycocalyx matrix. Its orientation and arrangement guides the patterned deposition of sporopollenin precursors which results in the ornamentation of exine (Rowley & Dahl 1977). Therefore, primexine may be regarded as scaffold/stencil/blueprint/template for exine patterns.

There are several models of development that have been proposed to explain the conversion of primexine into exine (Fig. 4.13). The timing of the formation and the shape of supratectal elements is variable among different species. For example, spines in *Hibiscus, Catananche caerulea, Solanum virginianum* are deposited after the release of microspores from the tetrad. On the other hand, the protectum assumes the echinate pattern by the tetrad stage in some Asteraceae, Nymphaceae (*Nuphar*) (Blackmore et al. 2007; Wang & Dobrista 2018). However, certain aspects of exine development are universal:

1. Sporopollenin precursors are primarily synthesized and secreted by the microspore itself till they are in tetrad encased in common callose wall. Once the dissolution of the tetrad takes place, secretion of the sporopollenin precursors is taken over by the tapetum (packed in orbicules or Ubisch bodies). These precursors of sporopollenin are incorporated at specific positions on free microspores. This may be noted that the deposition of sporopollenin precursor derived by tapetum does not provide any pattern to exine but only increases its thickness.

2. The areas that are destined to become apertures lack any deposition of primexine. These areas accumulate plates of endoplasmic reticulum which are oriented parallel to the plasma membrane and prevent the secretion of primexine (Heslop-Harrison 1968).

- **Primexine mediated exine development:** According to transmission electron microscope studies (Heslop-Harrison, 1963, 1968), column like structures known as probaculae (singular: probacula) develop within primexine with the gradual deposition of sporopollenin leading to formation of early sexine. Further deposition and accumulation of sporopollenin extends the length of probaculae resulting in formation of baculae and the tectum. Once the pollen is released from the tetrad, more sporopollenin accumulates, leading to further development and enlargement of baculae and tectum. With lateral growth, the baculae join at the base and form another layer of pollen wall called the nexine 1. Scanning electron microscope (SEM) studies, however, show that primexine initially appears to be uniform, but progressively, some regions (from the outside) become less solid and eventually disperse, leaving only the solid elements corresponding to the tectum and the columellae (Blackmore & Barnes 1985; 1987; 1988). Sporopollenin first accumulates at the outermost part of the primexine corresponding to the tectum. It is later deposited between the intratectal elements like the columellae, and finally is deposited over the foot layer or nexine I. This model is yet to resolve the ontogeny of nexine 2, although a few studies suggest that it stems from the inside of nexine 1 after the dissolution of the callose wall (Blackmore et al. 2007). It is also suggested that the spaces between the tectum and nexine I get filled with the tryphine and hydrophobic materials derived from the degeneration of tapetum that constitute the second layer of nexine- the nexine 2 (Ariizumi & Toriyama 2011). Recently, presence of arabinogalactan proteins (AGPs), which are generally rich in hydroxyproline have also been seen in the nexine (Jia et al. 2015).

- **Plasma membrane undulation model:** This model for exine development is observed in pollen grains of *Brassica, Lilium, Caesalpinia, Arabidopsis* (Takahshi 1991; Zhou et al. 2015). According to this model, the plasma membrane of microspores in tetrads, begins to undulate to form lumina and protuberances. These invaginations acquire a distinctive pattern in mature exine (e.g., reticulate in Lily). There is patterned deposition of primexine in the depressions and lumina of the plasma membrane. The raised areas are the points at which probaculae are formed in a more or less a hexagonal pattern which is followed by formation of protectum. By the time callose wall disappears, a smooth protectum is clearly distinct. The nexine layer is typically the last one to develop and is in place before the release of pollen from the tetrad (e.g., in *Arabidopsis*). By the time microspores undergo mitotic division II, the exine pattern may be completed. The order of the events just described may differ between taxa, e.g., in some species like *Bougainvillea spectabilis* and *Lilium longiflorum* formation of protectum takes place before the probaculae formation and the nexine layer is typically the last to develop (Wang & Dobrista 2018).

- **Tensegrity model:** Another model for exine development was proposed by Southworth & Jernstedt (1995) for the generation of echinous (spine like) microreticulate pattern in Compositae pollen. Known as 'Tensegrity Model', it proposed that exine patterns are generated by the physical properties of the callose wall, the primexine and the osmotic pressure, and cytoskeletal tension generating in the microspore. According to this model, callose is involved along with primexine in the pollen surface patterning.

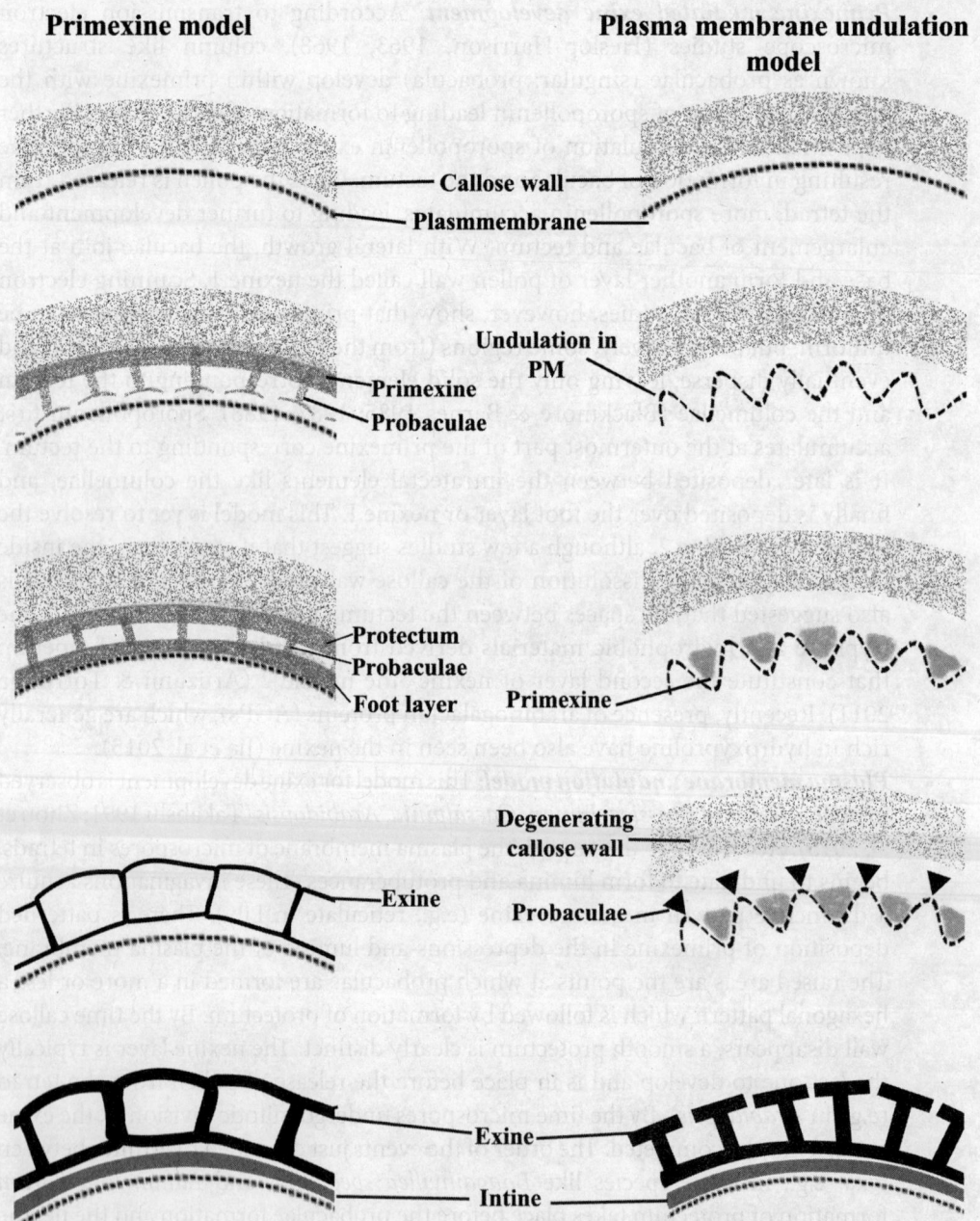

Figure 4.13 Comparative schematic representation of exine synthesis as proposed by two different models.

Box 4.10

Genes Involved in Exine Formation

A number of mutants have been identified that affect the exine formation. Such studies have led to characterization of a number of genes which are involved in exine formation during pollen development. For example, In *dex1* mutants, the normal undulating pattern of the plasma membrane is not seen. This causes primexine to deposit randomly outside the plasma membrane, resulting in an exine surface that lacks a normal reticulate pattern (Paxson-Sowders et al. 1997; 2001). Another pollen pattern mutant, *faceless pollen-1* in *Arabidopsis* produces pollen with an almost smooth surface without a prominent reticulate pattern (Ariizumi et al. 2003). Some other genes are: *DEFECTIVE IN EXINE FORMATION1 (DEX1), MALE STERILE1 (MS1)/HACKLY MICROSPORE (HKM), NO PRIMEXINE AND PLASMA MEMBRANE UNDULATION1 (NPU1), FACELESS POLLEN-1 (FC-1), RUPTURED POLLEN GRAIN 1 (RPG1), NO EXINEFORMATION1 (NEF1), SPONGY2 (SPG2),* and *UNEVEN PATTERN OF EXINE1 (UPEX1)*. All these genes are found to express in the tapetum establishing the significance of tapetum in the formation of the primexine. Also, genes controlling synthesis of precursors of sporopollenin (at microspore stage) have also been found to express in tapetal cells, e.g., *CYP703A2, CYP704B2, DEFECTIVE POLLENWALL2*.

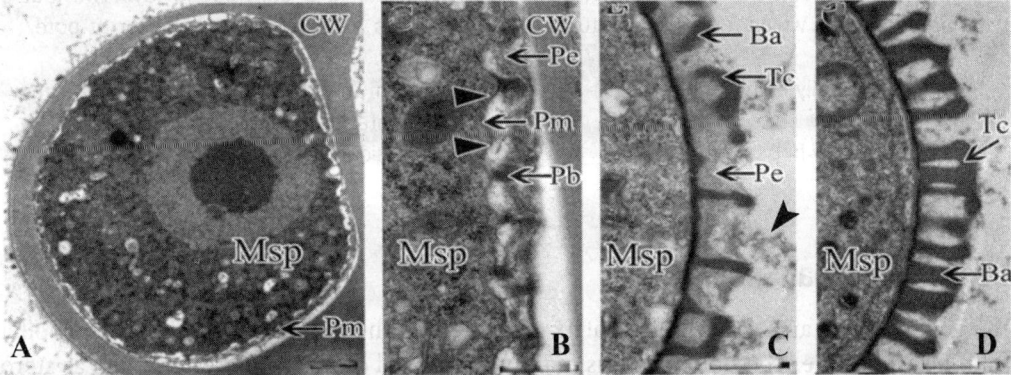

Figure 4.14 A. A microspore microspore (**Msp**) in tetrad surrounded by a thick callose wall (**cw**). **B.** The undulation in plasma membrane of microspore (**Pm**, arrow heads). Primexine (**Pe**) is deposited between the plasma membrane and the callose wall. The growing probaculae (**Pb**) can be seen growing on the peaks of plasma membrane undulation. Notably, the presence of callose wall indicates the presence of microspores in tetrad. **C–D.** Microspore after being released from tetrad. There is absence of callose wall and also, can be seen the developing baculae (**Ba**), tectum (**Tc**) and the sporopollenin (arrow head). (*Source:* Chang et al. 2012, published under CC BY License.)

Intine: The electron-transparent intine layer is composed of cellulose and pectin. It forms immediately after differentiation of the endexine. Intine synthesis starts before the first mitotic division and is always completed by the time the vegetative and generative cell are formed. Accumulation of intine is greatest at the vacuolated microspore stage, where it begins below the apertures. Soon it thickens rapidly under the aperture and extends around the perimeter of the pollen grain. Like other somatic cell walls, the intine matrix substances originate from the golgi apparatus and leave the cytoplasm via exocytosis. Recently, an ABC (ATP-binding cassette) transporter has been identified in rice (*OsABCG3*) which is localized in pollen plasma membrane and helps in the transport of intine components to the pollen wall (Chang et al. 2018). Another recent study suggests the synthesis of intine to be under the control of various factors such as the flavanols present at the vacuolated stage as seen in rice (Zhang et al. 2020).

Box 4.11

Differences between Exine and Intine

Exine	Intine
• It is the outer wall layer of a pollen grain.	• It is the inner wall layer of a pollen grain.
• It is multi-layered and further divided into sublayers: ektexine and endexine	• It is single layered.
• It is composed of sporopollenin.	• It consists primarily of cellulose, pectins and hemicellulose.
• The exine is acetolysis resistant.	• Intine gets disintegrated with acetolysis.
• The exine usually has one or more thin areas or gaps which form the apertures.	• The intine is a continuous layer without any gaps except that it is thicker at pore/aperture.
• Exine is usually highly sculptured.	• It is an inner smooth layer with no ornamentation or sculpturing.
• Taxonomically important.	• Taxonomically not important.

4.7.3 Pollen Coat Substances

The pollen coat is an extracellular matrix deposited on the outermost surface of the pollen grain i.e., the exine. The pollen coat is mainly composed of non-polar esters such as sterol esters and saturated acyl groups. The substances for the pollen coat are derived from the anther tapetum released after the breakdown of the tapetal cells. These substances eventually fill the spaces and cavities of the highly sculpted exine.

- **Pollenkitt:** The sticky substances present over the pollen surface (and between interstices of the exine) are collectively known as the pollenkitt substances. The term pollenkitt was first used by Knoll (1930) (as reported in Pacini & Hesse 2005). Pollenkitt is a hydrophobic mixture of materials, composed mainly of saturated and unsaturated lipids, carotenoids, flavonoids, proteins and carbohydrates. Usually, the anther locule gets filled with pollenkitt substances released from the tapetum, which can either be reabsorbed by pollen grains or get deposited on the surface of the pollen. While the resorption of pollenkitt substances is more common in plants with anemophilous pollination, deposition of pollenkitt substances over pollen grains is a characteristic feature of zoophilous plants making the pollen sticky. Some of the functions assigned to pollenkitt substances, are listed below (Pacini & Hesse 2005):

1. The pollenkitt substances keep the pollen grains in aggregates thus promoting their collective dispersal and pollen presentation for zoophily.
2. The pollenkitt substances enable adhesion of pollen grains to the bodies of insects or other dispersal agents and also serve as reward for pollinators.
3. The pollenkitt substances prevent the pollen grains from desiccation after they are dispersed.
4. They also protect the pollen grains from the ill effects of UV-radiation.
5. Proteins held in pollenkitt are responsible for pollen–pistil interactions such as pollen recognition, adhesion and pollen hydration.
6. Sporophytic incompatibility is manifested by proteins held in exine as well as the pollenkitt substances.

- **Tryphine:** It is a type of pollen coat substance restricted to the family Brassicaceae. The word tryphine has originated from the Greek words *tryphos* and *trypheros*, meaning fragment and soft, respectively. Tryphine is a thick, highly viscous coating composed of a proteinaceous fibro-granular and a lipidic component. The initial description of tryphine formation was given by Dickinson & Lewis (1973) in *Raphanus*. Tryphine is tapetally synthesized; the proteinaceous part is secreted by the endoplasmic reticulum and the lipidic component is derived, mainly, from degraded elaioplasts. A peculiar feature of tryphine is that it never shows association with orbicules. Tryphine is involved in manifestation of self-incompatability reactions.

- **Pollenkitt-like structures:** Viscin threads and elastoviscin are the special pollenkitt-like substances restricted to only a few genera and families. Viscin threads are non-viscous, non-sticky thread like structures that help in aggregation of pollen grains. Like pollenkitt, viscin threads help in zoophilous pollination. However, unlike pollenkitt, visicin threads originate from sporopollenin, remain attached to exine and are acetolysis resistant. Viscin threads are seen in Onagraceae, Ericaceae (*Rhododendron* sp.) and Caesalpiniaceae. Elastoviscin is a viscous, sticky and gum-like substance of high viscosity which connects pollen grains in Orchidaceae and Asclepiadaceae. Elastoviscin is derived from degeneration of tapetal cell cytoplasm and its main components are spherosomes. As compared to viscin threads, the elastoviscin is not resistant to acetolysis (Hesse 1981; Schil & Wolter 1986; Pacini & Hesse 2005).

4.8 Characteristics of Pollen

The study of morphological characteristics of pollen grains is known as Palynology. The pollen wall structure, especially the exine, is one of the most important characters used (for details refer to Exercise 4.8) for palynological studies. Apart from the pollen wall, the other characteristics used to describe pollen are the pollen shape, size, apertures, and dispersal units.

Box 4.12

Pollen Presentation

The method of presenting the pollen to the dispersing agents is known as pollen presentation. In the case of cleistogamy, pollen is not presented because flowers do not open. Primary pollen presentation indicates the manner in which the pollen is presented by the anther. For primary presentation, anthers may release pollen in batches or partition them into groups using pollenkitt, tryphine or viscin threads to provide aid to the dispersing agent. The significance of these methods mainly lies in the targeted deposition of pollen for successful mating. Secondary pollen presentation is the display of pollen on floral parts other than the anther. This means that when pollen is mature and the anther opens, pollen adheres to other parts of the flower such as non-fertile components of androecium (staminodes, e.g., in *Zingiber densissimum*), gynoecium (e.g., in *Gardenia, Pavetta, Canna*) or perianth/floral bracts (Barman et al. 2020). The part of the flower onto which the pollen is loaded for the presentation is known as the pollen presenter. Presentation of pollen on a surface other than the primary pollen presenter may protect pollen from desiccation or exploitation, help in cross- or self-pollination, and increase the efficiency of pollen delivery (Yeo 1993). The secondary pollen presentation may also serve to extend the male phase of a flower through the protection and regulated release of pollen grains (Howell et al. 1993). There is great variation in the specific mechanism of transfer of pollen to secondary pollen presenter. However, nearly all secondary pollen presenters receive pollen from anthers dehiscing introrsely (Howell et al. 1993).

4.8.1 Pollen Dispersal Unit

Pollen dispersal unit (PDU) refers to the number of pollen grains united together at the time of their release from the anther. The term 'pollen dispersal unit' was introduced by Pacini (1997) to indicate the different ways in which the pollen is presented to dispersal agents (Pacini & Hesse 2005). The microspores in a tetrad usually get separated from each other prior to their release from anther. Such single, separated pollen grains are known as monads, the most common type of dispersal unit seen in angiosperms. In some incidences, however, the pollen grains remain united at the time of anther dehiscence. Such type of pollen grains are known as compound pollen grains. Union of two or more pollen grains might be because of pollen grains sharing a common wall or presence of some sticky material like viscin threads or elastoviscin. Based on the number of pollen grains present in a dispersal unit, various types of dispersal units are recognized which are (Fig. 4.15) listed below:

- **Monad:** It is a dispersal unit consisting of a single pollen grain. Evolutionarily it is the most primitive type, widely seen in both monocots and dicots. Monads can be observed in *Hibiscus, Tradescantia, Vinca, Betula humilis, Aster*, maize, rice. (Fig. 4.15 A and 4.23)
- **Dyad:** Very rarely pollen grains are released from anther in pairs and each pair is known as a dyad, a characteristic feature of the sub-family Podostemoideae (of family Podostemaceae). Dyads are seen in *Polypleurum stylosum, Griffithella hookeriana* (both members of Podostemaceae) and also in *Scheuchzeria* (a member of family Scheuchzeriaceae). Here, due to a simple fusion of the tectal layers the two pollen grains are fused (Volkova et al. 2016) (Fig. 4.15 A–B).

- **Tetrad:** It is a unit of four pollen grains where the microspores do not separate after meiosis and the four pollen grains are dispersed as a unit. Examples are: *Epidendrum centropetalum* (Orchidaceae), *Xanthosoma ceronii*, (Arecaceae), *Fagus* (Fagaceae), *Typha angustifolia* (Typhaceae), and *Rhododendron* (Ericaceae). Tetrads can be further classified on the basis of arrangement of the pollen grains. These could be a tetrahedral tetrad (e.g., *Rhododendron* sp.), T-shaped tetrad (e.g., *Polyanthes* sp.) rhomboidal tetrad (e.g., *Camptocarpus* sp.), linear tetrad (*Mimosa pudica*), tetragonal tetrad (e.g., *Philydrum* sp.) and a decussate tetrad (e.g., *Magnolia grandiflora*). *Chlorospatha* sp. is an interesting example as the tetrad here remains covered by a continuous exine (Fig. 4.15A, C).
- **Polyad:** A dispersal unit of more than four pollen grains (in multiples of 4) is called a polyad. Polyads develop due to mitotic divisions in PMCs before meiosis resulting in prepolyads. The number of cells in a prepolyad depends on the number of times a PMC undergoes division. Later, meiosis occurs in PMCs in the usual way resulting in polyads. Therefore, the number of pollen grains in a polyad may vary. There are 8 pollen grains in a polyad of *Calliandra tergemina* and *Acacia*, 32 in *Albizia saman* (Fig. 4.15 A, D).
- **Massula:** The massula (plural: massulae) is an aggregation of irregular numbers of pollen grains by fusion. The number of pollen grains in a masssula is always less than the locular content of a theca. Mostly found in Orchidaceae, (Fig. 4.15 E) such dispersal units may be held together by viscous materials like pollenkitt and tryphine. The term massula is often erroneously used for the dispersal units in members of Fabaceae-Mimosoideae which produce more than four pollen grains (Halbritter et al. 2018).
- **Pollinium:** It is a compact and coherent mass of pollen, usually consisting of massulae, united together and attached to a slender, mealy or elastic extension known as caudicle. In other words, all the pollen grains derived from archesporial tissue of a theca aggregate as one dispersal unit. Commonly seen in Orchidaceae and subfamily Asclepiadoideae of Apocynaceae; where several massulae of an anther locule may form one pollinium. Massulae within a pollinium are variable and differ in shape, size, and numbers of pollen grains. Sometimes each locule of an anther may be subdivided by a septum resulting in more than two pollinia in an anther (Halbritter et al. 2018).
- **Pollinarium:** It is a collective term used for all the pollinia of an anther sticking together with the help of secretions or tissues like viscidium, caudicle, stipe, or sometimes even the anther cap. A pollinarium becomes attached to an insect during pollination as seen in *Spiranthes spiralis* (Orchidaceae). While the pollinarium is characterized by the presence of separate appendages for each pollinium in Asclepiadoideae, pollen sacs are attached to a common appendage in members of Orchidaceae (Fig. 4.15 F-G).
- **Pseudomonad** or **Cryptotetrad:** These are a characteristic feature of the family Cyperaceae, where the meiotic products do not separate and form a unique permanent tetrad. Post meiosis, only one nucleus remains functional and the

other three nuclei degenerate. Thus, a mature pollen grain appears monad but is homologous to a tetrad as traces of degenerated microspores may be observed. E.g., *Carex atrata*, *Scirpus sylvaticus* (Fig. 4.16).

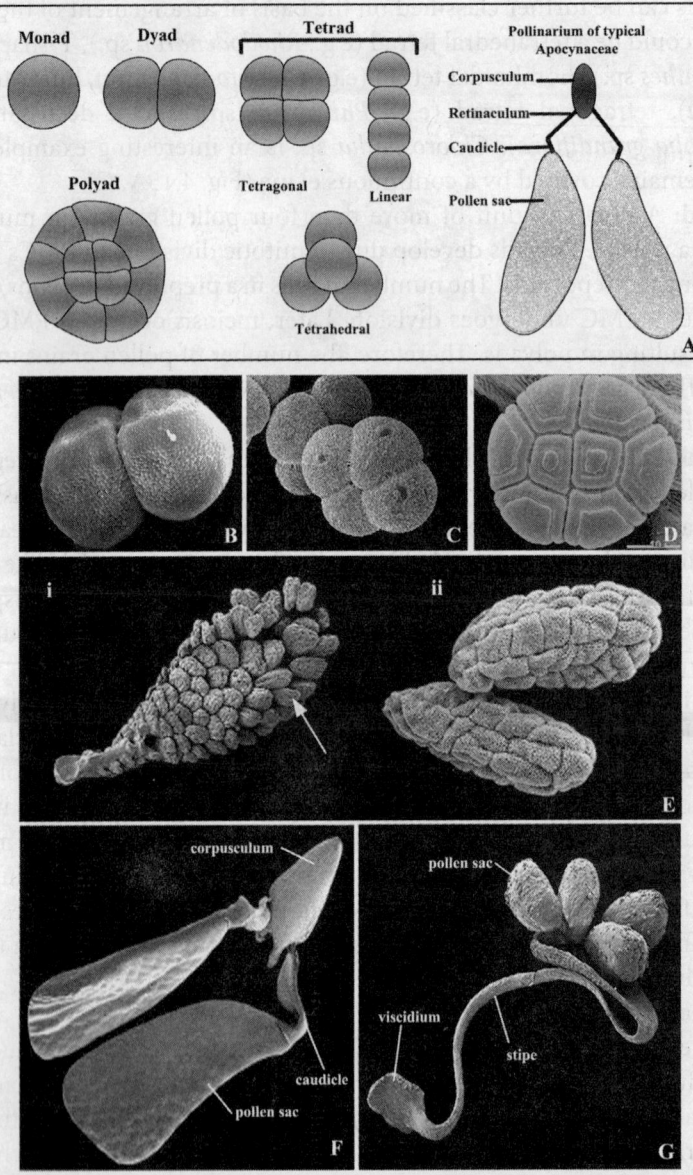

Figure 4.15 Pollen Dispersal Units (PDUs). A. Diagrammatic sketch of PDUs. **B–G.** Scanning electron micrographs of pollen dispersal units **B.** Dyad (*Zeylanidium olivaceum*, Podostemaceae) **C.** Tetrad (*Typha* sp.) **D.** Polyad (*Acacia* sp.) **E. i.** Massulae in pollinarium (*Gennaria diphylla*, Orchidaceae). Arrow marks a massula. **ii.** Two massulae, where numerous pollen grains are aggregated. **F.** A typical pollinarium of Apocynaceae (*Asclepias syriaca*). **G.** Pollinarium (*Ornithocephalus myrticola*, Orchidaceae). (*Source*: C and G: paldat.org, *reproduced with permission.*)

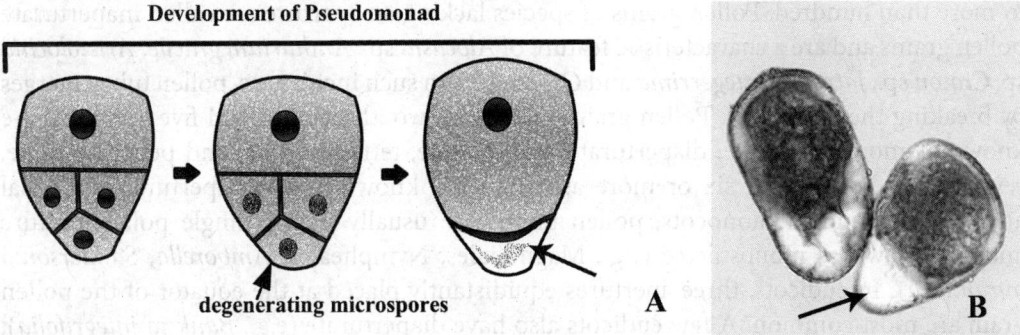

Figure 4.16 A. Pseudomonad development. **B.** Scanning electron micrograph of a pseudomonad in *Carex* sp. Arrows mark the remnants of degenerated microspores. (*Source B*: Halbritter et al. 2018.)

Box 4.13

Pollen Developmental Arrest

Following anther dehiscence pollen grains need to survive through hostile environments. To survive environmental exposure during their travel; pollen grains exhibit a state where both primary metabolism and development are arrested; and this is known as pollen development arrest. This arrest in pollen development is strongly associated with acquisition of desiccation tolerance to extend pollen viability during travel. Period of development arrest can vary and may last from a few days to a few weeks (e.g., herbaceous annual, perennial plants) and up to several months in some woody gymnosperms and angiosperm plants. The accumulation of soluble carbohydrates such as sucrose plays a role in membrane stabilization and protection against osmotic stress during desiccation.

Source: Pacini & Dolferus 2019.

4.8.2 Pollen Aperture

An aperture is a region of the pollen wall where the exine is either thin or completely absent. Pollen wall at aperture differs significantly from the rest of the grain in its morphology and anatomy. Aperture serves as a site of water uptake and eventually pollen germination, pollen tube emergence and regulation of pollen hydration (refer to Box 4.14 for details). In most flowering plants, the apertures are typically lined by thickened intine only. The position, number and morphology of aperture result in a pattern characteristic for each species. Though these features are highly conserved for pollen grains of a plant species, they vary widely across flowering plants (Furness & Rudall 2004; http://www.paldat.org). Terms used to describe the position and morphology of pollen aperture are further discussed in Section 4.8.3 and Fig. 4.17.

Aperture can be positioned at poles, at the equator or may even be distributed throughout the surface of a pollen grain. The number of apertures in angiosperms can vary from zero to more than hundred. Pollen grains of species lacking an aperture are called inaperturate pollen grains and are a characteristic feature of *Alocasia* sp., *Anthurium gracile*, *Aristolochia* sp. *Croton* sp., *Jatropha integerrima* and *Orobanche*. In such incidences, pollen tube emerges by breaking the exine wall. Pollen grains with one, two, three, four and five apertures are known as monoaperturate, diaperturate, triapeturate, tetraperturate, and pentaperturate, respectively. Pollen with six or more apertures are known as pantoaperturate. In basal angiosperms, and in monocots, pollen grains are usually with a single polar aperture and are known as monosulcate (e.g., Magnoliales, Nympheales, *Amborella*, *Sandersonia aurantiaca*). In eudicots, three apertures equidistantly placed at the equator of the pollen grain are most common. A few eudicots also have diaperturate (e.g., *Banksia integrifolia*), tetraperturate (e.g., *Pulmonaria officinalis*), pentaaperturate (e.g., *Viola arvensis*) and pantoaperturate pollen (e.g., *Cerastium tomentosum*).

Domains or regions which would subsequently establish as pollen apertures are demarcated in the microspore plasma membrane at the tetrad stage of microspores. These membrane domains do not receive deposition of primexine and are also deficient in plasma membrane undulation. This eventually leads to these domains receiving little or no exine deposition. An investigation of mutant inaperturate pollen grains of *Arabidopsis* has led to identification of a protein INP1 (Inaperturate Pollen1), which anchors the microspore plasma membrane at the sites of aperture formation (Zhou & Dobrista 2019). Recently, a lectin receptor-like kinase in *Oryza sativa*, OsDAF1, has also been recognized for aperture formation. OsDAF1 remains evenly distributed in microspore mother cells, but gets confined to the distal pre-aperture site at the tetrad stage (Zhang et al. 2020). These studies suggest aperture formation is a highly regularized process governed by molecular mechanistic associations.

Box 4.14

Harmomegathy

Upon release from anther, the pollen grains are subjected to a drier environment and this results in their dehydration. This further leads to changes in the osmotic pressure and volume of the pollen cytoplasm. In order to survive this dehydration, pollen grains possess an ability to fold their wall onto themselves to prevent further desiccation, and this feature is known as Harmomegathy. It is the characteristic folding of pollen grain wall to accommodate the decrease in cellular volume due to water loss. Upon reaching a compatible stigma, pollen once again experience a humid environment and rehydrate and harmomegathy is reversed (the pollen wall unfolds). Apertures are known to play a primary role in harmomegathy as they are more elastic than the remainder of the pollen wall. This elasticity is attributed to thicker pectin rich intine at the pores/furrows because of which the pollen cytoplasm remains hydrated during travel. However, occurrence of pollen coat substances reduces the harmomegathic effects.

Source: Volkova et al. 2013; Halbritter et al. 2018.

4.8.3 Pollen Polarity

Pollen grains possess two different poles: a proximal pole (the area next to the centre of the tetrad) and a distal pole (the area towards the outside of microspore in a tetrad and opposite to the proximal pole). The polarity of pollen can be examined only at the tetrad stage. A straight line passing through the poles and through the centre of the tetrad is called the *polar axis*. The equatorial plane of a pollen includes its equatorial diameter and a line passing through the equatorial plane is called the *equatorial axis*. The equatorial axis is perpendicular to the polar axis and divides the microspore/pollen into a proximal and a distal half, comparable to the northern and southern hemisphere of the Earth. The two poles of the pollen grain may have same or different features. When the pollen has identical proximal and distal poles, it is said to be isopolar. The equatorial axis divides the pollen grain into two equal halves in 'isopolar' pollen. In contrast, when the proximal and distal halves of a pollen differ either in shape, ornamentation or apertures (e.g., one with aperture and one without aperture), it is called a 'hetropolar' pollen. Pollen grains in dicot species usually possess one polar and one equatorial view (Fig. 4.17). In monocots, the pollen grains can be observed in four different views; these being distal polar, proximal polar, and two different equatorial views (Simpson 2010; Halbritter et al. 2018). This is primarily because of the distal position of the apertures in most monocots.

The polarity of the pollen also influences the terminology used for apertures. A circular aperture in a pollen grain is commonly called a pore and the pollen is referred to as 'porus' (length/width ratio <2:1; plural: *pori*). If a porus aperture is present distally, it is called an 'ulcus' and if the pori are present all over the pollen grain surface, the aperture type is categorized as 'pantoporate'. Similarly, an elongated aperture is called a 'colpus' (length/width ratio >2:1, plural: colpi). If the colpus occurs distally, it is called a 'sulcus'. An aperture which combines the features of both porus and colpus is termed as 'colporus' (plural: *colpori*), where the aperture shape is like a colpus but has a circular pore in the centre (as in Leguminosae). Colpi and pori may be present simultaneously in some taxa like *Melastoma sanguineum* and *Phacelia campanularia*. This simultaneous occurrence of colpi and pori is referred to as heteroaperturate.

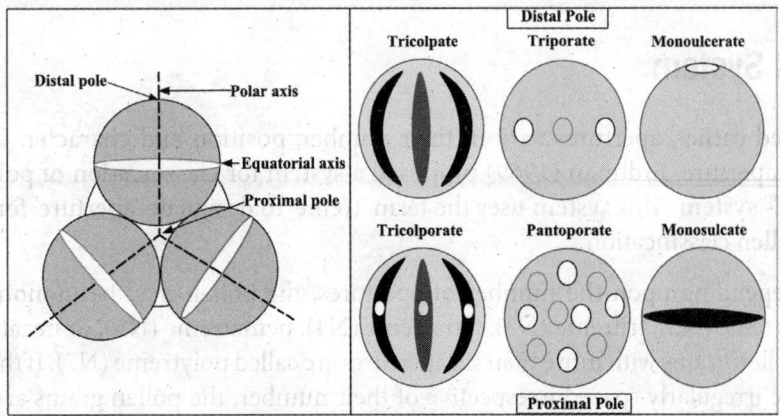

Figure 4.17 A. Pollen tetrad exhibiting polarity. **B.** Types of apertures. (*Source*: Based on Halbritter et al. 2018.)

4.8.4 Pollen Symmetry

On the basis of plane of symmetry, the pollen grains may be symmetric or asymmetric. The asymmetric pollen grains have no plane of symmetry and are rare in occurrence. Symmetric grains may have a single plane of symmetry (bilateral pollen) or may be radially symmetrical (radiosymmetric). A radiosymmteric pollen will produce identical halves when cut in any plane. A radially symmetrical, isopolar pollen will have one horizontal and two or more vertical planes of symmetry. On the other hand, radially symmetrical heteropolar pollen lack a horizontal plane of symmetry.

4.8.5 Pollen Shape

The shape of the pollen grains varies from species to species. It may be circular, spheroidal, elliptical, triangular, rectangular, quadrangular or any other geometrical shape (Fig. 4.23). Erdtman (1952) categorized eight classes for pollen grains based on their shape considering the ratio of polar axis (PA) and equatorial diameter (ED). When viewed equatorially, the ratio between the PA and ED multiplied by 100 gives an indication of the shape of the pollen (Table 4. 2).

Table 4.2 Pollen shape types (after Erdtman 1952)

Shape	(PA/ED) X 100
Oblate	50-75
Sub-oblate	75-88
Oblate-spheroidal	88-99
Spheroidal	100
Prolate-spheroidal	101-114
Sub-prolate	114-133
Prolate	134-200
Per-prolate	>200

4.9 NPC System

As discussed earlier, apertures vary in their number, position and character. Using these features of aperture, Erdtman (1969) proposed a system for classification of pollen known as the **NPC- system**. This system uses the term 'treme' to designate 'aperture' for preparing keys for pollen classification.

- Depending upon the number of apertures, the pollen may be monotreme (N1), ditreme (N2), tritreme (N3), tetratreme (N4), pentatreme (N5), or hexatreme (N6). Pollen grains with more than six apertures are called polytreme (N7). If the apertures are irregularly spaced irrespective of their number, the pollen grains are described as anomotreme (N8). Pollen grains without apertures are called atreme (N0).

- According to the position of aperture, there are 7 groups (P0–P6). The pollen grains are designated as 'catatreme' (P1) when the aperture is on the proximal face and are designated 'anatreme' (P3) when the aperture is on the distal face. The pollen is designated as 'Anacatatreme' (P2) when the apertures appear both in proximal and distal faces. If the centre of the aperture is located on the equator, the grains are referred to as 'zonotreme' (P4). 'Dizonotreme' (P5) are like zonotreme, but with two rows of apertures on the equatorial region. Pollen grains in which the apertures are more or less uniformly distributed all over the surface are called 'pantotreme' (P6). Pollen grains having uncertain position of aperture are designated as P0.
- Based on the characters of the apertures of the pollen grains, seven groups are recognized (C0–C6). If the character of the aperture is not known (character unknown), it is designated as C0. Pollen having an aperture like thin area or leptoma is designated as C1. Pollen with one leptoma is called monolept. Pollen with three slit like colpi are called trichotomocolpate and belong to category C2. The remaining character classes C3, C4, C5 and C6 include colpate (with colpus), porate (with circular aperture or pore), colporate (with both colpi and pore), pororate (with pore) pollen grains, respectively.

Based on above apertural features, NPC classification designates each pollen type with a three-digit number. The first digit denotes the number of aperture(s), the second digit denotes the position of the aperture and the third digit denotes the characters of the aperture. For example, if number 133 is assigned to a pollen grain, it denotes a monosulcate pollen grain. This system, however, does not work well with heteroaperturate or inaperturate (formula 000) pollen, or even with aggregated pollen dispersal units. The NPC-system also ignores other pollen characters including shape and ornamentation which are considered indispensable for a complete description of a pollen grain.

4.10 Nemec Phenomenon

Occurrence of embryo sac like structures in pollen grains is known as the Nemec Phenomenon, named after Nemec who observed them for the first time in 1898. Nemec, observed eight nuclei arranged in Polygonum-type embryo sac organization in the pollen grains of petaloid anthers of *Hyacinthus orientalis*. He hypothesized that the eight nuclei were a result of three successive mitotic divisions in the vegetative nucleus of a pollen grain.

Later, De Mol (1923) and Stow (1930, 1934) also observed 'Nemec phenomenon' in the anthers of other varieties of *Hyacinthus orientalis* which were subjected to special conditions. Stow traced the development of the pollen embryo sacs in detail and showed that initially the microspores increase in size to form large sac-like bodies followed by three successive divisions of the nucleus. Out of the eight daughter nuclei, three position themselves at the end where the exine was still intact, three at the opposite end, and two in the middle. The six nuclei at the two poles organized into cells, while the remaining two in the centre fused. Since the three cells at the exine end were found to remain healthy for a much longer time, Stow regarded them as the egg and synergids. Thus, he proposed that, it is not the divisions of the vegetative or the generative nucleus which give rise to the pollen-embryo sacs but the divisions of the microspore nucleus itself. Also, pollen-embryo sacs were always accompanied by a large number of dead pollen grains. Stow suggested that the former secrete a 'necrohormone' which causes an abnormal growth of the surviving pollen grains. He also observed that when the pollen-embryo sacs were placed on an agar medium, together with some normal pollen grains of another variety, the pollen tubes formed from the latter coiled around the pollen embryo sacs.

Subsequently such structures were also reported in *Ornithogalum nutans* and *Leptomeria billardierii*, and in a few anthers of *Heuchera micrantha* (Saxifragaceae) by Geitler (1941), Ram (1959) and Vijayaraghavan and Ratnaparkhi (1977) respectively. It was shown that pollen-embryo sacs can be formed by division of microspore mother cell, microspore nucleus, vegetative nucleus or the generative nucleus. The Nemec phenomenon is seen as an expression of the dominance of female potency over the male in microspore and proves the microspore to be a labile cell (Vijayaraghavan & Ratnaparkhi, 1977).

Box 4.15

Non-reproductive Roles of Pollen

Other than performing the crucial role of delivering sperm nuclei to the female reproductive parts, pollen grains also play other important roles. Pollen grains are known to be a storehouse of energy and a concentrated vitamin rich food which is eaten by insects and also stored as a reserve. Besides being vital as diet for insects, pollen has become an important component of animal feed in many countries. It serves as a source of proteins and a natural substitute for antibiotics for raising poultry and livestock. Pollen has also been used traditionally by humans as a food supplement as they have high content of carbohydrates, proteins, lipids, vitamins (with the prevalence of group-B vitamins), carotenoids (such as lutein, and β-carotene), minerals and polyphenols. Due to multi-nutrient properties, pollen grains are acknowledged as the 'only perfectly complete food', and 'the world's best food product' and can be found in different forms in the market such as granules, capsules, tablets, pellets, and powders.

Pollen grains are also one of the most common triggers of seasonal allergies. Pollinosis, also known as seasonal allergic rhinitis, pollen allergy or hay fever, is a major health problem for humans all over the world. It is characterized by conjunctivitis and/or bronchial asthma as a result of sensitization to pollen components. Most allergenic pollens are produced from abundantly distributed wind-pollinated plants. Plants of the Poaceae are the main source of pollen allergens. Some species which cause allergic reactions with their pollen are *Lolium perenne* (rye-grass), *Poa pratensis* (Kentucky bluegrass), *Anthoxanthum odoratum* (sweet vernal grass), *Cynodon dactylon* (Bermuda grass). Reactivity to pollen allergens occurs due to cross-reactivity of IgE antibodies to pollen proteins. Allergenic proteins are usually located in the pollen protoplast and are released during the rehydration process. These pollen allergens are putatively involved in stress responses and metabolic processes such as cell wall metabolism during pollen development. Unfortunately, pollen allergens are difficult to avoid because of their extremely small size and high concentration in the air.

Source: Chen et al. 2016; Kostic et al. 2020; Shivanna 2020.

4.11 Pollen Development and Metabolism

• **Role of vacuoles:** Vacuoles of several types are observed in pollen grains at various stages of development. Vacuolization can occur once or twice (depending on species) during pollen development. Pollen mother cells, like undifferentiated meristematic cells, are originally devoid of vacuoles but small roundish vacuoles start to develop at telophase II. Vacuoles appear again at the young microspore stage and are essential for asymmetric mitotic division I. This cycle of vacuolization is always followed by storage of starch in amyloplasts, leading to disappearance of vacuoles and formation of new cytoplasm. Functions of vacuoles vary according to the developmental stage of the pollen. Autophagic vacuoles degrade organelles in the microspore after meiosis, which can be regarded as 'a cytoplasm clean-up' following the transition from the diploid sporophytic to the haploid gametophytic state (Pacini et al. 2011). Vacuoles also play a role in increasing the volume of pollen grain with the formation of new cytoplasmic components such as mitochondria and amyloplasts. Vacuolization is also associated with the storage of pectins during intine synthesis. Mature pollen usually lack vacuoles and reduced vacuolization at maturity may be associated with the reduction in pollen size during presentation and dispersal. However, in species producing pollinia small vacuoles with watery content may be present.

• **Amylolysis and amylogenesis during pollen development:** Accumulation and mobilization of sugars is common during pollen development. There is a close correlation between cycles of amylolysis (starch mobilization), amylogenesis (starch accumulation) and the metabolic demands at various stages of pollen development. These cycles also help in growth and differentiation of different layers of the anther wall like fibrous thickenings of the endothecium, formation of primexine and intine, and in some cases, also the exine (Horner & Pearson 1978; Hedhley et al. 2016).

Depending on the species, there are one or two phases of starch accumulation in amyloplasts during pollen grain development; the first being at microspore mother cell stage and the other at bicellular pollen stage. Plastids are completely starch-free at the onset of meiosis. Starch accumulation and the differentiation of pro-plastids into amyloplasts starts after telophase. Plastids divide once again in the vegetative cell of pollen before starch engorgement. In most species, starch stored in the amyloplasts of the vegetative cell is hydrolysed before anther opening and pollen dispersal. Thus, mature pollen can be starchy or starchless, depending on the presence or absence of starch grains in the vegetative cell amyloplasts. For example, the pollen grains of *Mercurialis annua* are starchless, the pollen grains of *Oenothera* have abundant starch content.

During amylolysis, starch molecules from the amyloplasts are converted to protective sucrose and other oligosaccharides; to provide energy for cell metabolism and callose synthesis. These also function as an osmolyte in protecting pollen membranes under stress (Firon et al. 2012; Pressman et al. 2012).

- **Adjustment of osmotic pressure and water balance in pollen:** Water balance of pollen grains is crucial for maintaining their viability (Heslop-Harrison & Heslop-Harrison 1985; 1992). Carbohydrates play a significant role in maintaining the internal turgor pressure of pollen, thus regulating its water balance with the surrounding environment (Pacini et al. 2006). Starch is totally or partially converted to pectins, glucose, fructose, sucrose, and to some unknown polysaccharides before the dehiscence of the anther, depending on the species. Presence of sucrose in the final stage of pollen maturation imparts desiccation tolerance and longer viability to the pollen (Hoekstra et al. 2001). On the other hand, species with low percentage of sucrose are often partially hydrated and lose water and vaibility quickly (Nepi et al. 2001).

Box 4.16

Water Status and Types of Pollen

Pollen grains are classified into two types on the basis of water content at the time of their dispersal, namely, orthodox and recalcitrant. Orthodox pollen is the one dispersed in partially desiccated form while the recalcitrant pollen are those which are dispersed in partially hydrated form. Note that either of the 2-celled or 3-celled pollen can be orthodox or recalcitrant. Also, both orthodox and recalcitrant pollen may or may not have starch at maturity. Generally, it may be observed that self-pollinating plants produce hydrated/recalcitrant pollen grains which do not exhibit developmental arrest. However, their viability period is very short as observed in rice and wheat (Pacini & Dolferus 2016; 2019). Some of the differences in orthodox and recalcitrant pollen grains are summarized below.

Orthodox pollen	Recalcitrant pollen
• Desiccation tolerant	• Desiccation sensitive
• Water at dispersal is less than 30%.	• Water at dispersal is more than 30%
• Size: 30–100 μm	• Size: 15–30/70–150 μm
• 1–6 furrows and pores	• 0–12 (or more) pores, No furrows
• Example: *Olea europaea, Lycopersicum peruvianum, Nelumbo nucifera, Canna indica.*	• Example: *Cucumis melo, Plantago* sp., *Ruellia prostrata, Spinacia oleracea,* Poaceae.

Source: Based on Pacini & Dolferus 2019.

Box 4.17

Cytoplasmic Male Sterility

Inability to produce or to release viable or functional pollen grains is known as male sterility. Male sterility results from failure of development of functional stamens, microspores or sperm cells. Male sterile plants are unable to produce functional pollen, however, the female gametes function normally. Male sterile lines are known to occur in a large number of species and are highly sought in crop breeding programs. Hybrid seed production involves crosses between two genetically different parents. The main problem encountered in producing hybrid seeds is self-pollination of parental lines which would produce seeds that are not hybrids. To avoid self-pollination, emasculation (removal of anthers) is the usual practice but it is time consuming and very tedious. Thus, using a male sterile parent can avoid enormous manual work of emasculation and pollination and speed up the hybridization program.

There are various types of male sterility seen in flowering plants. One of them is Cytoplasmic Male Sterility (CMS), which is a maternally inherited trait and has been reported in self-pollinating crop species including maize, rice, cotton, and a number of vegetable crops. CMS arises due to a mutation in the mitochondrial genome (mtDNA). The mutation can arise either spontaneously in the mtDNA (autoplasmy) or can be expressed following cytoplasmic substitutions due to nuclear-mitochondrial incompatibility (alloplasmy) (Prakash et al. 2009). Nuclear genotype of a male sterile line is identical to that of the normal pollen of the species. Cytoplasmic male sterility phenotypes include degenerate anthers, aborted pollen, carpelloid and petaloid stamens (Chase 2007). During hybridization, CMS lines are used as a female parent in the crosses. The progenies obtained would always be male sterile since the cytoplasm comes primarily from female gamete only.

In many cases, it has been found that male fertility can be restored by nuclear-encoded fertility restorer (*Rf*) gene(s). These fertility restoration nuclear genes suppress the cytoplasmic dysfunction caused by mitochondrial genes. Till date, nine *Rf* (*restorer of fertility*) genes have been isolated in seven plant species e.g., *Rfo* (*Rfk1*) in radish and *Brassica* (Chen & Liu 2014). Most of the identified *Rf* genes encode PPR (pentatricopeptide repeat) proteins. For example, in the CMS-Boro II system in rice, mitochondrial or f79 encodes a cytotoxic peptide responsible for CMS, whereas two PPR proteins encoded by the *Rf-1* locus in the Boro II system block the production of this cytotoxic peptide (Wang et al. 2006).

Stable CMS lines of crops can be produced by introducing a sterility inducing cytoplasm from the wild relatives through interspecific-intergeneric hybridizations, or by mediating the mitochondrial rearrangement via somatic fusions, or through somatic hybridization via protoplast fusion (Prakash et al. 2009). A sterility inducing cytoplasm identified in a wild population of *Raphanus sativus* (known as Ogura CMS) was introduced into *B. napus* and *B. oleracea* (Bannerot et al. 1974). Similarly, CMS-WA (wild abortive) rice was developed in *indica* rice cultivars from a male-sterile plant found in a natural population of the wild rice *Oryza rufipogon*. CMS-Boro II rice arose from a wide cross based on the cytoplasm of Chinsurah Boro II (*O. sativa* sub-sp. *indica*) and the nucleus of Taichung 65 (sub-sp. *japonica*). However, the well-known male-sterile Texas cytoplasm in maize arose spontaneously in a breeding line (as mentioned in Eckardt 2006).

Glossary

Anther: Pollen grain containing unit of a stamen.

Aperture/germ pore: Area of the pollen wall where exine deposition is generally thin or absent and intine is thick. It serves as a site for pollen tube emergence.

Callose: It is a polysaccharide made up of β-1,3-glucan units.

Caudicle: A slender, elongate stalk like extension of the pollinium to which the masses of pollen are attached.

Exine: External elastic layer of pollen cell wall composed of sporopollenin.

Generative cell: The small lenticular cell formed after Mitosis I in the microspore which later divides to form two sperm cells.

Harmomegathy: The ability of pollen wall to fold in order to accommodate changes in the pollen volume due to water loss. It helps in preventing desiccation of pollen grains.

Heterofertilization: It is the incidence of fertilization of the egg cell and polar nuclei of an embryo sac by sperms cells belonging to different pollen tubes.

Intine: Inner continuous pecto-cellulosic layer of pollen wall.

Locular fluid: The fluid secreted by the tapetum which fills the anther locule and serves to nurture pollen.

Locule: Central cavity in the anther where pollen grains develop. In cross section, a typical anther shows four locules.

Male germ unit: It is a close association formed between the vegetative cell and the two sperm cells so that they behave as a single unit.

Microgametogenesis: Development of pollen grains (male gametophyte) from microspores via two mitotic divisions which lead to the formation of one vegetative cell and two male gametes.

Microsporogenesis: Process of formation of microspores through meiosis in microspore mother cell.

Nemec phenomenon: The presence of embryo sac-like organization in the pollen grains. It was first described by Nemec in 1898 in the petaloid anthers of *Hyacinthus orientalis*.

Orbicules: Acellular, sporopollenin containing spheroidal bodies originating from tapetum. Also known as Ubisch bodies.

Palynology: The study of morphological features of pollen grains is known as palynology.

Perispore membrane: Membrane enclosing the plasmodial mass after the disintegration of amoeboid tapetum as in *Tradescantia, Hypoxis*.

Phragmoplast: A complex assembly of microtubules, actin filaments and associated molecules involved in the formation of cell plate during cytokinesis.

Pollen developmental arrest: State of physiological and metabolic arrest in pollen grains to reduce water content before dispersal.

Pollen viability: Ability of pollen grains to deliver sperms for fertilization.

Pollenkitt: Glue like hydrophobic mixture of saturated and unsaturated lipids, carotenoids, flavonoids, proteins and carbohydrates which make pollen grains sticky.

Primexine: An ephemeral glycoclayx-like fibrillar polysaccharide deposited between the microspore plasma membrane and the callose wall.

Sperm dimorphism: When sperms derived from the same generative cell are different from each other with respect to volume, size and composition of cell organelles, the phenomenon is called sperm dimorphism as observed in *Plumbago zeylanica*.

Sporopollenin: Chemically and biologically resistant elastic polymer forming the exine layer of pollen wall. It is a mixture of carotene and carotenoid esters.

Stamen: The male reproductive organ of flowering plants consisting of an anther and a filament.

Stipe: A band or strap of columnar tissue connecting pollinia to the viscidium.

Tapetosomes: Organelles containing endoplasmic reticulum derived vesicles and oleosin-coated lipid droplets, found in the tapetal cells during pollen maturation. These are observed in Brassicaceae.

Tryphine: A heterogenous mixture of hydrophilic and hydrophobic substances that serves as pollen coat in family Brassicaceae.

Vegetative cell: The larger of the two cells formed after mitosis I in micropsore which is involved in the formation of pollen tube.

Viscidium: A viscid, sticky part of the rostellum dispersed with the pollinia as a unit which serves to attach pollinia to the pollinators.

Key Questions

Q4.1 Fill in the blanks:
a. ....has monothecous anthers.
b. Orbicules are spheroidal bodies originating from............................
c. ....is an example of plant species exhibiting sperm dimorphism.
d. The concept of MGU was based on observations of ....
e. ....has tapetum of dual origin.
f. Amoeboid tapetum is characterized by disintegration and dispositioning before.............
g. ....is an example of 2-celled pollen grain.
h. In .... pollen grain, the presence of MGU can be seen only during pollen tube growth.
i. .... family lacks presence of MGU.
j. *Brassica* shows presence of special cell organelles in their tapetum known as ....
k. Poricidal anther dehiscence can be seen in family......................
l. In Lily, anther dehiscence takes place by ....line of dehiscence.

Q4.2 Differentiate between:
a. Exine and Intine
b. Endothecium and Endothelium
c. Microsporogenesis and Microgametogenesis
d. Secretory and Amoeboid tapetum
e. Young and Mature anther
f. 2-celled and 3-celled pollen grains
g. Orthodox and Recalcitrant pollen grains

Q4.3 Write a note on:
a. Ubisch bodies or Orbicules
b. Microsporogenesis
c. Ontogeny of exine
d. Pollen wall sculpturing
e. Sperm dimorphism
f. Primexine
g. Anther dehiscence
h. Role of callose in pollen development

Q4.4 Discuss the anther wall layer ontogeny with the help of suitable diagrams.

Q4.5 Elaborate the concept and importance of pollen dispersal units among flowering plants with suitable examples and diagrams.

Q4.6 What is a male germ unit? Briefly describe its structure and functional significance.

Q4.7 Tapetum is a master regulator of pollen development. Justify the statement with suitable examples.

Q4.8 Define pollen polarity and its applications.

Q4.9 Explain the concept of NPC system of pollen classification.

Q4.10 Define microgametogenesis. Give an account of the process with suitable diagrams.

Q4.11   What are pollen coat substances? Enlist their roles.

Q4.12   Draw well labelled diagrams of:
   a.   Bithecous, tetralocular anther with MMCs
   b.   Anther locule with secretory tapetum
   c.   Anther locule with amoeboid tapetum
   d.   MGU
   e.   Anther locule with wall layers and microspore tetrad
   f.   Microgametogenesis
   g.   Mature dehiscing anther

Q4.13   Select the odd one out and justify the answer:
   a.   Callose, microsporogenesis, anther wall, microgametogenesis
   b.   Endothecium, epidermis, exine, septum
   c.   Vegetative cell, sperm cell, sporogenous cell, pollen
   d.   Exine, pollenkitt, pollination, sperm cell
   e.   Epidermis, endothecium, tapetum, exine
   f.   2-celled pollen, 3-celled pollen, MGU, microsporogenesis

Practicals

Exercise 4.1: To study anther wall layers and microsporogenesis

The stages of anther development correspond to the stages of pollen development (microsporogenesis and microgametogenesis). The anther is considered immature when the anther layers (at least the endothecium, archesporial cells) are not fully differentiated. It is referred to as young just for the convenience of identification of its initial developmental stage i.e., from differentiation of archesporial cells till microgametogenesis. Anther with fully developed pollen grains and ready to dehisce is called a mature anther. However, it is very difficult to draw a clear boundary between a young and a mature anther and such terminology is not of much scientific significance. Recent studies have rigorously classified the anther and pollen development to number of stages to better understand the ontogeny and differentiation of different cell layers and also the pollen. For example, in *Arabidopsis* and wheat, anther and pollen development have been divided into 15 distinct stages each with characteristic feature/s (See Fig. 4.3). Referring to the account of anther development given earlier in this chapter (Figure 4.3), one can easily analyze the different wall layers along with the sporogenous tissue in a transverse section of an anther. Some of the key features of major stages of anther and pollen development in lily (*Lilium* sp.) have been listed below.

Transverse section of a young anther (Fig. 4.18 A)

- The anther is bithecous, bilobed and tetralocular.
- The anther lobes are joined by connective tissue with a vascular bundle.
- Each locule is separated by 2–3 layers of cells called septum.
- All the wall layers (epidermis, middle layers, endothecium and tapetum) are distinguishable.

- The locules are filled with microspore mother cells.
- Filament is also visible; its position suggests the versatile nature of the anther.

T.S. of a young anther (a locule, Fig. 4.18 B–D) showing microsporogenesis

- The anther locule is filled with numerous enlarged microspore mother cells which are at meiosis I stage of division (B) and in dyad stage (C). The microspore tetrads can also be seen in figure D.
- All the wall layers of the anther (epidermis, middle layers, endothecium and tapetum) are distinctly visible.
- Epidermis is single layered with barrel-shaped, uninucleate cells with large vacuoles.
- Endothecium is single layered and present only on the protuberant parts of the anther (Fig. 4.18 D).
- 2–3 middle layers are present below the endothecium. The cells of middle layers are flattened and uninucleate.
- Tapetum is single layered and of secretory type. It surrounds the locule completely.
- Tapetal cells are elongated and multinucleate (2–3 nuclei per cell) with nuclei towards the locular side.
- Tapetal cells also comprises of a large vacuole present towards periphery and adjacent to middle layers (Fig. 4.18 D).
- A densely staining cytoplasm is located towards the anther locule in the tapetal cells.

T.S. of a young anther (a locule, Fig. 4.18 D–E) showing secretory tapetum and its degeneration

- The anther shows a secretory type of tapetum which is single layered and surrounds the locule completely.
- The secretion from the tapetal cells have filled the locule.
- Tapetal cells are elongated, multinucleate (2–3 nuclei per cell) with nuclei located towards the locular side.
- Tapetal cells also contain a large vacuole present towards periphery/adjacent to middle layers (Fig. 4.18 D).
- A densely staining cytoplasm can be seen towards the anther locule.
- The tapetal cells are seen at their original position till the end of meiosis I which is a key feature of the secretory tapetum.
- Gradual degeneration of tapetum is initiated at the late microspore tetrad stage (compare Fig. 4.18 D with E). The middle layers also show disintegration.

Exercise 4.2: To study a mature anther and its dehiscence

A mature anther at dehiscence shows peculiar features that are listed below and shown in Fig. 4.19.

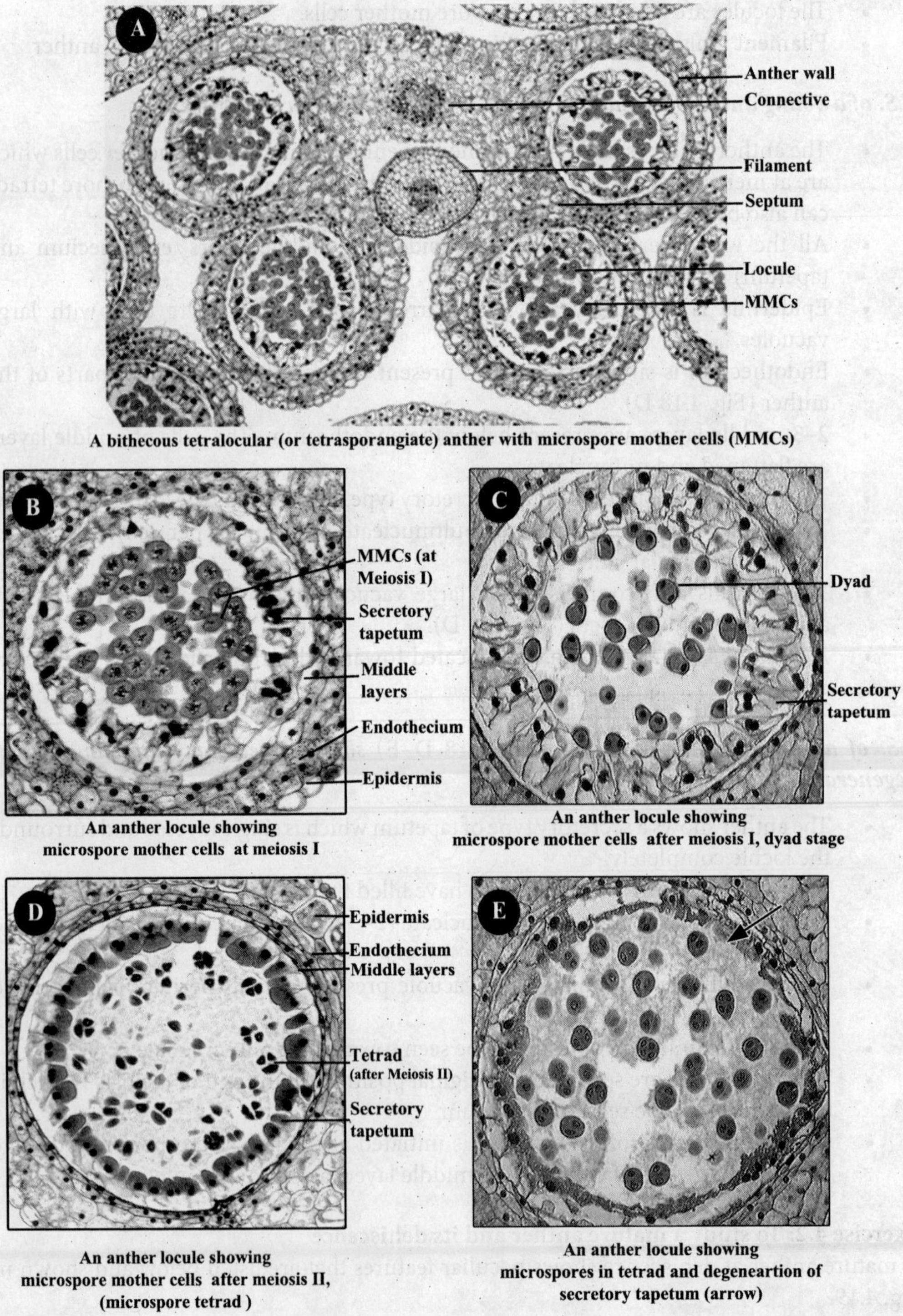

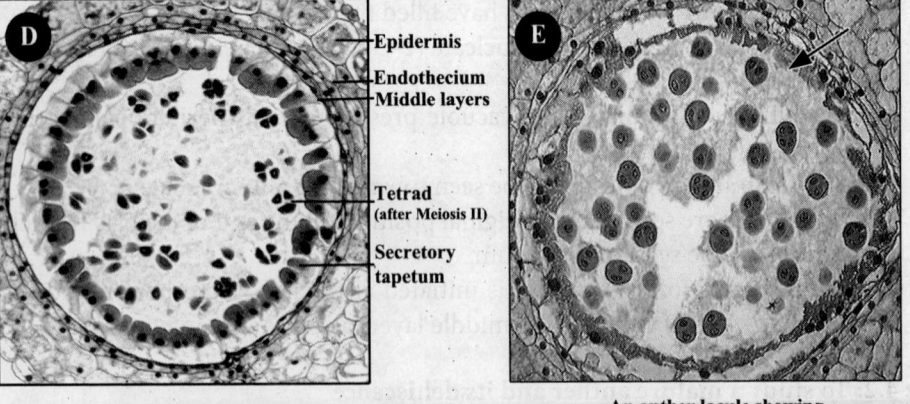

A
Anther wall
Connective
Filament
Septum
Locule
MMCs
A bithecous tetralocular (or tetrasporangiate) anther with microspore mother cells (MMCs)
B
MMCs (at Meiosis I)
Secretory tapetum
Middle layers
Endothecium
Epidermis
An anther locule showing microspore mother cells at meiosis I
C
Dyad
Secretory tapetum
An anther locule showing microspore mother cells after meiosis I, dyad stage
D
Epidermis
Endothecium
Middle layers
Tetrad (after Meiosis II)
Secretory tapetum
An anther locule showing microspore mother cells after meiosis II, (microspore tetrad)
E
An anther locule showing microspores in tetrad and degeneartion of secretory tapetum (arrow)

Figure 4.18 Anther wall layers and pollen development in lily. See Color Plates (page 477). (*Source*: Courtesy of University of Wisconsin, Stevens biology laboratory and IASSPR.)

Transverse section of a mature anther

- At maturity, the middle layers and the tapetum (irrespective of its type) disintegrate.
- Endothecial cells show deposition of secondary thickenings in their radial walls.
- The remnants of disintegrated tapetum can be seen.
- The septum is intact.
- Young microspores are liberated from the tetrads and the different layers of the pollen wall (intine and exine) are well-developed.
- Pollen grain cytoplasm is dense with a single nucleus (uninucleate stage).

Transverse section of an anther at dehiscence

- As the anther matures, the middle layers and the tapetum (irrespective of its type) disintegrate.
- Endothecium shows deposition of secondary thickening in the radial walls of its cells.
- The remnants of disintegrated tapetum can be seen.
- The septum is dissolved by enzymatic hydrolysis thus merging the locules of a theca. The anther takes on a bilocular appearance.
- Mature pollen grains are at a 2-celled stage.
- Stomium is the weakest point at the merger of two locules.

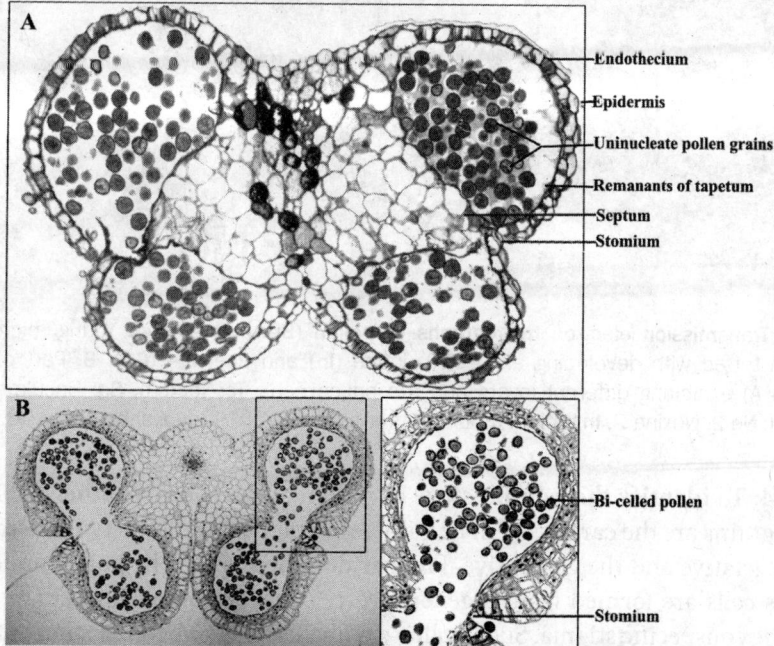

Figure 4.19 Transverse section of anthers. A. A mature anther in *Hippophae rhamnoides* where the septum has started to degenerate. **B.** Mature anther of lily at dehiscence. **See Color Plates (page 478).** (*Source:* Courtesy of Stevens biology Lab, University of Wisconsin.)

Exercise 4.3: To study the ultrastructure of pollen wall layers using transmission electron micrograph

A mature pollen wall is multi-layered and shows stratification. Transmission electron microscopy has helped enormously in the understanding of its ultrastructure and ornamentation. The details of pollen wall structure and development have been given in Section 4.5. Some features of the pollen wall are listed below.

- Rod shaped projections make the baculum layer of exine.
- The tops of baculae columns are fused and form a distinct tectum layer of ektexine.
- The tryphine (pollen coat substances) gets accumulated in the intersections and spaces of baculae.
- Endexine is visible as two very thin, but distinct layers namely nexine I and nexine II.
- A thin intine is present between the plasma membrane of the pollen cytoplasm and the nexine layers.

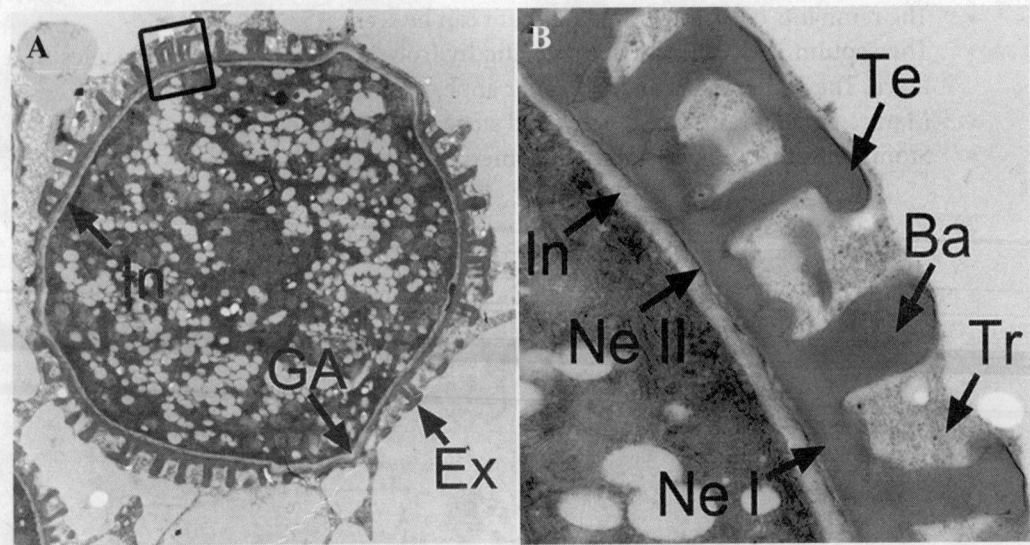

Figure 4.20 Transmission electron micrographs of pollen (Brassica sp.). A. Young microspore after release from tetrad with developing exine (**Ex**), intine (**In**) and aperture (**GA**). B. Part of pollen wall (as boxed in A) exhibiting different layers and the pollen coat. **Te:** Tectum, **Ba:** Bacula, **Tr:** Tryphine, **Ne I:** Nexine I, **Ne 2:** Nexine 2, **In:** Intine. (*Source*: After Lin et al. 2019, published under CC BY License.)

Exercise 4.4: To identify the cellular stage of pollen at the time of anther dehiscence

The pollen grains are the carriers of male gametes in plants. At maturity they comprise two cells: the vegetative and the generative. Further development of the pollen grains wherein two sperms cells are formed from a generative cell may accomplish after the landing of pollen on the conspecific stigma. Such pollen grains which are liberated from anther, before the sperm formation are called 2-celled pollen grains. In certain other species, the pollen grains are dispersed from the anther only after the formation of two sperm cells. Such pollen grains have two sperm cells and a vegetative cell before anther dehiscence, and are hence referred to as 3-celled pollen grains. The cellular stage at which pollen gets dispersed from the anther is linked with the viability, fertility and longevity of pollen. In general, 3-celled pollen exhibit shorter viability and fertility, as compared to 2-celled pollen which remain viable and fertile for a longer duration. The strategies for *in vitro* storage of both the types of pollen grains are different and the information about the stage of pollen at the time of dehiscence is valuable.

Protocols for nuclear staining that may be employed for observing pollen at the time of anther dehiscence utilize fluorophores like Feulgen, DAPI (4',6-diamidino-2-phenylindole), EtBr, propidium iodide (commonly used to detect the pollen nuclei). However, the staining procedure is different for pollen grains with pollenkitt substances. A pre-treatment step needs to be done for the species with sticky pollen grains.

Materials Required
Flowers, HCl (0.5N), Tween-20 (or Triton X or any other liquid detergent), Eppendorf's tubes (or Micro Centrifuge Tube 1.5 or 2 ml)/glass vials, pipettes, droppers.

Procedure
1. Collect the pollen grains by teasing/crushing mature anthers (4–5) in a micro centrifuge tube using distilled water. Take out the pollen grains as a suspended mass in water and remove the rest of the debris of the anther.
2. Add a small drop of liquid detergent in tube and mix well with water (this step will help in removing the lipids and proteins present in pollenkitt substances).
3. Centrifuge the tube (~1000–2000 rpm) for 2–3 minutes. Pollen mass will separate as a pellet.
4. Throw the supernatant and suspend the pellet in water. Now repeat step 3 (an additional step of washing may be followed to remove detergent).
    Steps 1-4 are also considered as pre-treatment steps and can be omitted where the pollen grains are dry or lack sticky substances.
5. Finally suspend the pellet in 1 ml 0.5N HCl and incubate the tube at 37°C for about 20-30 minutes.
6. Centrifuge the tube (~1000–2000 rpm) for 2–3 minutes. Pollen mass will get separated as a pellet. Resuspend the pellet in water and centrifuge. Discard the supernatant.
7. Add 0.1–0.5 ml of staining solution (depending upon the pellet size) and suspend the pollen grain pellet in it. Incubate the tube at room temperature (working solutions of dyes PI: 1 mg/10ml in distilled water, EtBr: 1 µg/10µl in distilled water) or at 4°C (in case of Schiff's reagent, DAPI: 5 ng/µl in 50% glycerol) for about 20–30 minutes.
8. Observe the pollen grains by spreading them over the glass slide and laying down the coverslip.
9. When observed under a UV-light, the pollen nuclei emit fluorescence. The excitation and emission spectrum of some of the DNA fluorophores is given in the Table 4.3.

Table 4.3

DNA Fluorophore	Excitation/Emission	
Ethidium Bromide	510 nm/ 605 nm	Emits orange fluorescent light
DAPI (4',6-diamidino-2-phenylindole)	364–405 nm/470 nm	Emits blue fluorescent light
Propidium Iodide	535 nm/617 nm	Emits red-orange fluorescent light
Hoechst 33258	350 nm/461 nm	Emits blue-cyan fluorescent light
Feulgen (Basic Fuchsin)	540 nm/630 nm	Emits green fluorescent light
Acridine Orange	502 nm/526 nm	Emits orange fluorescent light

Results: Depending upon the species, 2 or 3 nuclei emit fluorescence under UV-illumination in a fluorescence microscope. (Fig. 4.21)

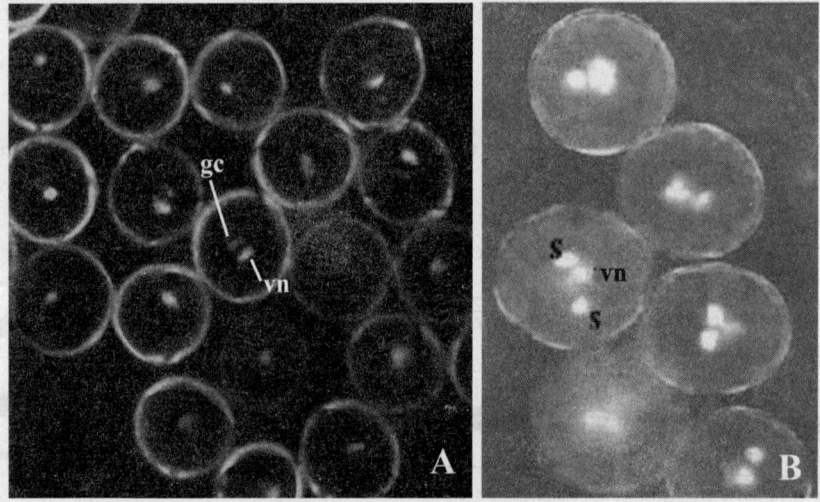

Figure 4.21 A. Two-celled pollen grains of *Hippophae rhamnoides* (stained with propidium iodide). **B.** Three-celled pollen grains of *Indotristicha ramosissima* (stained with EtBr). **gc:** generative cell; **vn:** vegetative nucleus; **s:** sperm cell. **See Color Plates (page 478).**

Exercise 4.5: To study the ultrastructure of male germ unit

The association between the vegetative cell and the sperm cells is known as male germ unit (MGU). The MGU has a functional significance. The vegetative nucleus helps the gametes to travel in a guided manner inside the pollen tube. In most of the 2-celled pollen grains, the MGU forms inside the pollen tube after the germination of pollen, where as in 3-celled pollen, MGU formation occurs before anther dehiscence.

MGU in Brassica
(McConchie et al. 1985, refer to Fig. 4.11 C–D)

- *Brassica* is an example of three-celled pollen where MGU is established before the dehiscence of anther. Two sperm cells can be seen in a mature pollen grain.

- In Fig. 4.11 D, the vegetative nucleus is convoluted. The cytoplasm in the embayments contains endoplasmic reticulum which is associated with the nuclear membrane of the vegetative nucleus.
- The sperm cells are floating in the cytoplasm of the vegetative cell. Each sperm cell is encased within its own plasma membrane.
- One of the sperm cells (sperm cell 1) extends a tail to penetrate into the embayments of convoluted vegetative nucleus through a cytoplasmic connection. These cytoplasmic connections between the vegetative nucleus and the sperm cell 1 (SC1) comprise electron lucent material.
- Sperm cell 2 (SC2) is connected to SC1 and also has a tail but does not penetrate the vegetative nucleus.
- The size/volume of SC1 is larger than SC2. Thus, sperm dimorphism can be observed.
- The SC1 has a greater number of mitochondria (10–34) than SC2 (4–11). Both sperm cells lack plastids (Dumas et al. 1987).

MGU in Nicotiana (refer to Fig. 4.11 C)

- *Nicotiana* is an example of two-celled pollen where MGU is established after pollination and hydration. Thus, MGU can be seen in the pollen tube. *All other features of MGU are the same as in Brassica.*

Exercise 4.6: To test the pollen grain fertility and viability

The pollen viability and fertility are the parameters used to assess the quality of pollen grains. Pollen viability is defined as the ability of the pollen grains to complete its function of delivering the male gametes into the embryo sac. On the other hand, pollen fertility is just the ability of the pollen grain to germinate. After being shed from the anthers, the pollen grains are exposed to the prevailing environmental conditions. Due to environmental exposure, the pollen grains might lose their ability to germinate and deliver the sperm cells. The viability and fertility of pollen grains decline gradually over a period of time. The period of pollen viability and fertility is variable among species. Note that while a fertile pollen may not be a viable pollen; a viable pollen is always a fertile one.

Based on the duration of viability, pollen longevity can be decided. Pollen grains in angiosperms are classified into three groups: (i) pollen grains with short life span (short-lived pollen), (ii) pollen grains with mid-life span and (iii) pollen with long life span (long-lived pollen). The short-lived pollen grains lose their viability within a few minutes-hours after anther dehiscence. Such pollen grains are found in Poaceae (sorghum, wheat), Commelinaceae, and Alistamaceae. In some species, this short span of pollen viability is associated with their being shed at the 3-celled stage. Long-lived pollen grains are present in members of families like Rosaceae, Fabaceae, Saxifragaceae, where pollen remain viable for about 6 months. It is important to know the duration of pollen viability and fertility before performing or using stored pollen grains for manual pollination. The identification of floral developmental stages with high pollen viability is mandatory for pollen collection during crop-breeding exercises. We suggest use of flowers of *Catharanthus roseus*, *Impatiens* sp., *Crotolaria* sp., *Zephyranthes* sp., *Tradescantia*, *Brassica*, and *Bombax ceiba* for class-room exercises.

I. Test for fertility: Acetocarmine staining

Materials Required
Flowers, Acetocarmine stain, glass slides, coverslips, needle, Improvised Humidity Chamber (IHC), compound microscope

Procedure

1. Dissect some undehisced mature anthers of a flower.
2. Keep anthers over clean glass slide and incubate them in IHC for 15–20 minutes.
3. Generally, the anther dehisces under humidity, if not then gently squeeze hydrated anthers to liberate the pollen grains on a glass slide.
4. Add a drop of Acetocarmine solution on to the slide with pollen grains.
5. Mix pollen grains with acetocarmine using a clean needle.
6. Place a coverslip and observe under microscope.

Results
Fertile pollen grains would stain dark red-brown (Fig. 4.22 A).

Precautions

1. Before performing the test, pollen must be hydrated.
2. Do not add extra stain.

II. Pollen Viability Tests

There are several tests available to test pollen viability. The simplest test for pollen viability is the germinability test. However, even viable pollen grains might not germinate in a given germination medium as pollen grains from different species respond differently in different germination media. Hence, one has to standardize the concentration of various nutrients in the medium. Recent studies have highlighted the addition of PEG, amines to enhance the percentage of pollen germination under *in vitro* conditions. Thus, to assess the viability of pollen, other tests may also be performed along with germination experiments.

A. *In Vitro* Germination Test (refer to Chapter 7, Exercise 7.3)

B. Tetrazolium Test (or TTC test)

Principle: This test is based on the reduction of a soluble colorless tetrazolium salt to reddish insoluble formazan in the presence of dehydrogenases. For the test, 2,3,5 triphenyl tetrazolium chloride (TTC) salt is commonly used. When the pollen grains are incubated with TTC solution, if they are viable, tetrazolium will get converted into formazan and pollen grains turn red brown. This is one of the frequently used tests to estimate pollen viability which usually gives satisfactory results. However, the major limitation of this test is the false coloration of pollen grains due to oxidation of TTC in presence of atmospheric oxygen and light. To avoid this one must observe the pollen grains present in the centre of the coverslip and leave the pollen grains towards the margin of the coverslip. Also ensure that the coverslip is quickly lowered over the pollen grains to minimize their contact with air. Incubation of pollen grains with the TTC solution must be done in dark for 20–30 minutes.

Materials Required
TTC solution (freshly prepared), glass dropper, glass slides, coverslips, needle, Improvised Humidity Chamber (IHC), compound microscope

Procedure

1. Dissect undehisced mature anthers from a flower.
2. Keep anthers over clean glass slide and incubate them in IHC for 15–20 minutes.
3. Generally, the anther dehisces under humidity, if not, then gently squeeze hydrated anthers to liberate the pollen grains on a glass slide.
4. Add a drop of TTC solution to the slide and suspend small amount of pollen grains in the solution with help of a needle.
5. Gently place a coverslip and transfer the preparation to IHC for incubation in dark for 30-40 minutes. Observe under microscope.

Result
Viable pollen grains would appear red (Fig. 4.22 B).

Precautions

1. Only use hydrated pollen grains for the test.
2. Incubate the slides with TTC solution in dark.
3. Consider the pollen grains present in the centre of the coverslip/coverglass.
4. Sometimes TTC salts cause itching and irritation on skin, and so must be handled carefully.

Preparation of TTC Solution: Prepare 2% solution of Triphenyl tetrazolium chloride in 15% sucrose.

C. Fluorochromatic Reaction (FCR) Test

Principle: The FCR test is also referred as fluorescein diacetate test (FDA test). This test for pollen viability was developed by Heslop-Harrison & Heslop-Harrison in 1979. This is a commonly employed test in research laboratories to assess the viability of pollen grains. The FCR test assesses the viability of pollen on the basis of two features of a pollen grain; first being, the integrity of the plasma membrane of the vegetative cell and the second being the presence of active esterases in the pollen cytoplasm. Fluorescein diacetate is a non-polar, non-fluorescent dye which passes freely inside the pollen cytoplasm. The hydrolysis of acetate molecules of FDA by active esterases present in the pollen cytoplasm results in free fluorescein, which is polar and fluorescent. Being polar; fluorescein cannot pass through the plasma membrane of the pollen grain and accumulates inside the pollen cytoplasm. Such pollen when observed under a fluorescence microscope appear fluorescent bright green. The membrane integrity is lost in non-viable pollen grains and hence, appears dull, and does not fluoresce.

Materials Required
Flowers, FDA solution (freshly prepared), glass dropper, glass slides, coverslips, needle, Improvised Humidity Chamber (IHC), Fluorescence microscope

Procedure:

1. Dissect undehisced mature anthers from a flower.
2. Keep anthers over clean glass slide and incubate them in IHC for 15–20 minutes.
3. Generally, the anthers dehisce under humid conditions, if not, then gently squeeze these hydrated anthers to liberate the pollen grains on a glass slide.
4. Add a drop of FDA solution to the slide.
5. Suspend a small amount of pollen grains in solution with help of a needle.
6. Gently place a cover glass and incubate the preparation in IHC for 5–10 minutes.
7. Observe under fluorescence microscope (Ex filter: 450–490; Blue range).

Result
Viable pollen grains fluoresce bright green (Fig. 4.22 C).

Precautions

1. Only use hydrated pollen grains for the test.
2. Incubate the slides in dark and ensure not to expose the slides to light.

Preparation of FDA Solution: Prepare stock solution of Fluorescein diacetate (2–5 mg/ml) in acetone. To 2–5 ml of sucrose solution (10–15%) add a few drops of FDA stock solution until the resulting mixture shows persistent turbidity.

Preparation of IHC: Take a petri dish, place a blotting sheet in it, sprinkle water enough to wet the sheet but not overflow it, cover with another petri dish.

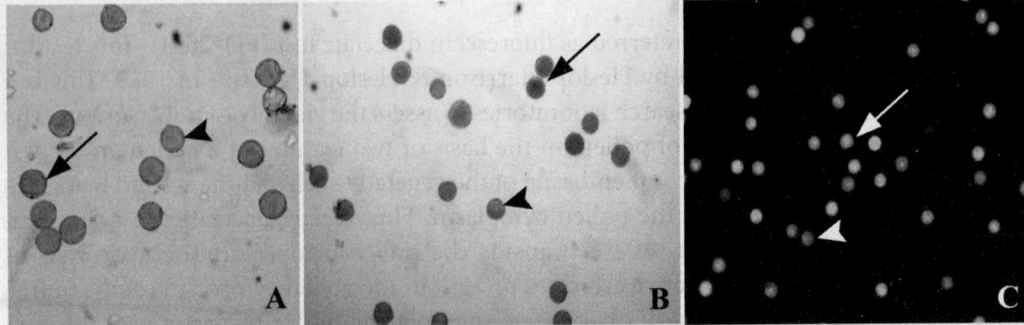

Figure 4.22 **A.** Acetocarmine test. The fertile pollen are deep red (arrow) while the sterile pollen are faintly red (arrowhead). **B.** TTC test. **C.** FCR test. In B and C, the arrow indicates viable pollen grain while arrowhead indicates non-viable pollen grain. **See Color Plates (page 478).**

Exercise 4.7: To estimate the production of pollen grains in a flower

The production of pollen per flower in angiosperms is highly variable. Studies have suggested the amount of pollen production in a species is directly linked to its mode of pollination and breeding system. Pollen production in a flower can be estimated by several ways. Two commonly used methods of direct estimation and pollen suspension method have been described below.

Materials Required

Flowers, Acetocarmine, Tween-20 (or Triton X or any other liquid detergent like Teepol), 20% alcohol, Eppendorf's tubes (or Micro Centrifuge Tube 1.5 or 2 ml) or glass vials, micropipettes and their tips, droppers, Haemocytometer, or glass slides with markings (of 1 x 1 cm^2 or equal to coverslip), teflon pestle and glass rod.

Procedure

Method 1

Direct estimation: This method is suitable for species where pollen production is low.

1. Collect the flowers with mature anthers and count the total number of anthers in the flower.
2. Take one anther and squash it in a small drop of acetocarmine.
3. Remove the debris and place the coverslip over the anther squash.
4. Count the number of pollen grains present in area covered by coverslip under a stereozoom microscope.
5. Multiply the observed pollen number with number of anthers to estimate pollen production in a flower.

Method 2

Pollen suspension method: This method is suitable for species where pollen production is very high.

1. Collect the flowers with mature anthers and count the total number of anthers in the flower.
2. Gently dissect out an anther and put in a micro centrifuge tube or in a glass vial.
3. Add 1ml of pollen suspension medium (20 ml ethanol + 80 ml of water + 5–10 drops of Tween20 or teepol) into the centrifuge tube (*alcohol and detergent helps in removing the pollenkitt substances*).
4. Squash the anther with help of a teflon pestle or glass rod within the micro centrifuge tube.
5. Remove the anther debris. Centrifuge the tube (~1000–2000 rpm) for 1 minute. Pollen mass will get suspended as a pellet.
6. Throw the supernatant and suspend the pellet in pollen suspension medium (1 ml).
7. Place 100–500 µl of pollen suspension on a glass slide and cover it with coverslip. Then count the number of pollen grains. One can also use a haemocytometer to make this observation.

8. Calculations: First calculate the pollen grains present in 1ml of suspension, e.g., 10 X number of pollen grains counted in 100μl suspension

Or,

2 X number of pollen grains counted in 500 μl suspension

9. The final estimation can be done using the given formula:

Total pollen in a flower = number of anthers X number of pollen grains in 1 ml of suspension

Precautions

1. Anthers must not be at the dehiscence stage.
2. Counting must be done with sufficient number of sample flowers.
3. Pollen with pollenkitt substances generally gets clumped. To avoid clumping an appropriate detergent may be used to wash off such substances.
4. High volumes of pollen suspension must not be taken as it will escape from the coverslip and the estimated number of pollen would be erroneous.
5. Scan the complete area covered by coverslip while counting the pollen grains. It is advisable to use coverslips of size 1 cm² and low volumes of suspension.

Exercise 4.8: To study the morphological features (shape, exine pattern and aperture) of pollen grains of different species

The study of morphological features of pollen is known as 'Palynology'. Pollen grains exhibit a huge diversity in morphology within the flowering plants. Pollen features are an important taxonomic tool for species identification and their delimitation. For example, *Salix and Populus* of Salicaceae can be distinguished on the basis of pollen characters. *Nelumbo* has been separated from Nymphaeaceae into a distinct family Nelumboaceae based on tricolpate pollen of *Nelumbo* as compared to monosulcate condition seen in Nymphaeaceae. The details of important pollen features, viz. pollen polarity, apertures have been provided in Section 4.8.3. Here the pollen shape and exine patterns are provided.

Pollen shape refers to the P/E-ratio which is the ratio of the length of the polar axis (P) to the equatorial diameter (E) (Table 4.2). Pollen shape also refers to the 3-dimensional form of a pollen grain in relation to the P/E ratio. Pollen grains can be variously shaped like spheroid, cup, boat, cube, tetrahedral, triangular, pentagonal, hexagonal, and so on. The pollen grains in which the polar axis is equal to the equatorial diameter are called the spheroidal or isodiametric pollen. They are categorized as 'Prolate' when their polar axis is longer than the equatorial diameter. The pollen grains with polar axis shorter than the equatorial diameter are called 'Oblate' (Halbritter et al. 2018).

The exine pattern of a pollen grain is decided by shape, size, density and angle of orbicules deposition. The exine patterns are highly variable among flowering plants but quite conserved within a family. The terminology for exine pattern is very extensive and elaborate. A list of terms has been provided for reference here (based on Halbritter et al. 2018).

Exine: Tectate (roof like)

- **Psilate:** Pollen wall with a smooth surface, e.g., Elaeagnaceae (*Hippophae rhamnoides*), Zingiberaceae (*Hedychium gardnerianum*), Boraginaceae

(*Lithospermum officinale*), Apocynaceae (*Vinca minor*), Fabaceae (*Anthyllis vulneraria*).

- **Perforate:** Exine with small holes or depressions (>1 μm in diameter), e.g., Boraginaceae (*Pulmonaria officinalis*), Lecythidaceae (*Napoleonaea imperialis*), Polygonaceae (*Rumex acetosa*).
- **Scabrate:** Exine with minute sculpture elements (less than 1μm in diameter, undefined shape). Pattern may be irregular as in *Populus alba, Betula*.
- **Reticulate:** Sculpturing elements are present in 'network/mesh-like pattern'. Ridges are arranged in a network with lumina of size 1 μm or more in diameter. The breadth of ridge is denoted by term 'Muri'. Such exine pattern can be seen in Brassicaceae (*Cardamine pratensis, Arabidopsis*), Lamiaceae (*Physostegia virginiana, Ajuga genevensis*), Acanthaceae (*Beloperone guttata*), Bromeliaceae (*Aechmea azurea*), Cucurbitaceae (*Ecballium elaterium, Luffa cylindrica*), Salicaceae (*Salix daphnoides*), Fabaceae (*Trifolium rubens*).
- **Foveolate:** Surface having roundish lumina (holes or depressions) of 1μm (or more) in diameter. Usually, the distance between two adjacent lumina is larger than their diameter. Examples include: *Lavandula angustifolia*, Bromeliaceae, Gesneriaceae (*Glossoloma ichthyoderma*).
- **Striate:** Exine with elongated ornamentation of elements. The length of ornamentation elements is at least 2 times of the width; running more or less parallel. Muri may present. Surface may look like a fingerprint, e.g., Lythraceae (*Cuphea llavea*), Rosaceae (*Malus sylvestris, Crataegus laevigata*), Rutaceae (*Ruta graveolens*). When the elongated sculpturing elements are irregularly arranged, the exine is termed as 'Rugulae' or 'Regulate'.

Exine: Intectate/atectate

- **Granulate:** Any sculptural element less than 1μm in diameter (shape may vary).
- **Verrucate:** 'Wart-like' ornamentation elements; >1 μm tall and typically broader and shorter in height, e.g., Aristolochiaceae (*Aristolochia arborea*), Papaveraceae (*Corydalis cava*), Fabaceae (*Calliandra tergemina*), Plantaginaceae (*Plantago media*).
- **Baculate:** Rod-shaped ornamentation elements (bacula, never pointed), longer than wide, >1 μm high rod-like, free standing elements. Present in Santalaceae (*Viscum album*), Nymphaeaceae (*Nymphaea alba*), Rutaceae (*Erythrochiton brasiliensis*).
- **Gemmate:** Sculpturing elements (gemma) higher than 1μm; approximately of same width as height; constricted at their base; 'balloon-like'. Acanthaceae (*Stenandrium dulce*), Proteaceae (*Hakea kippistiana*).
- **Clavate:** Club-shaped sculpturing elements (clavae), or rods with knob heads, appearing 'lollipop-like' (pila); height greater than 1 μm; diameter of clavae or pila is smaller than its height; thicker at apex than at base. Commonly observed in *Jatropha multifida* and *Geranium* sp.
- **Echinate:** Pointed sculpturing elements (echini) 1μm or greater in height. Present in Asteraceae, *Nuphar lutea, Malva neglecta, Hibiscus* sp.
- **Lophae/lophate:** Massive exine ridges, lacunae: depressed areas surrounded by lophae. Characteristic of Cactaceae (*Opuntia basilaris*), Amaranthaceae (*Pfaffia tuberosa*), and Asteraceae (*Gazania* sp., *Hieracium hoppeanum*).

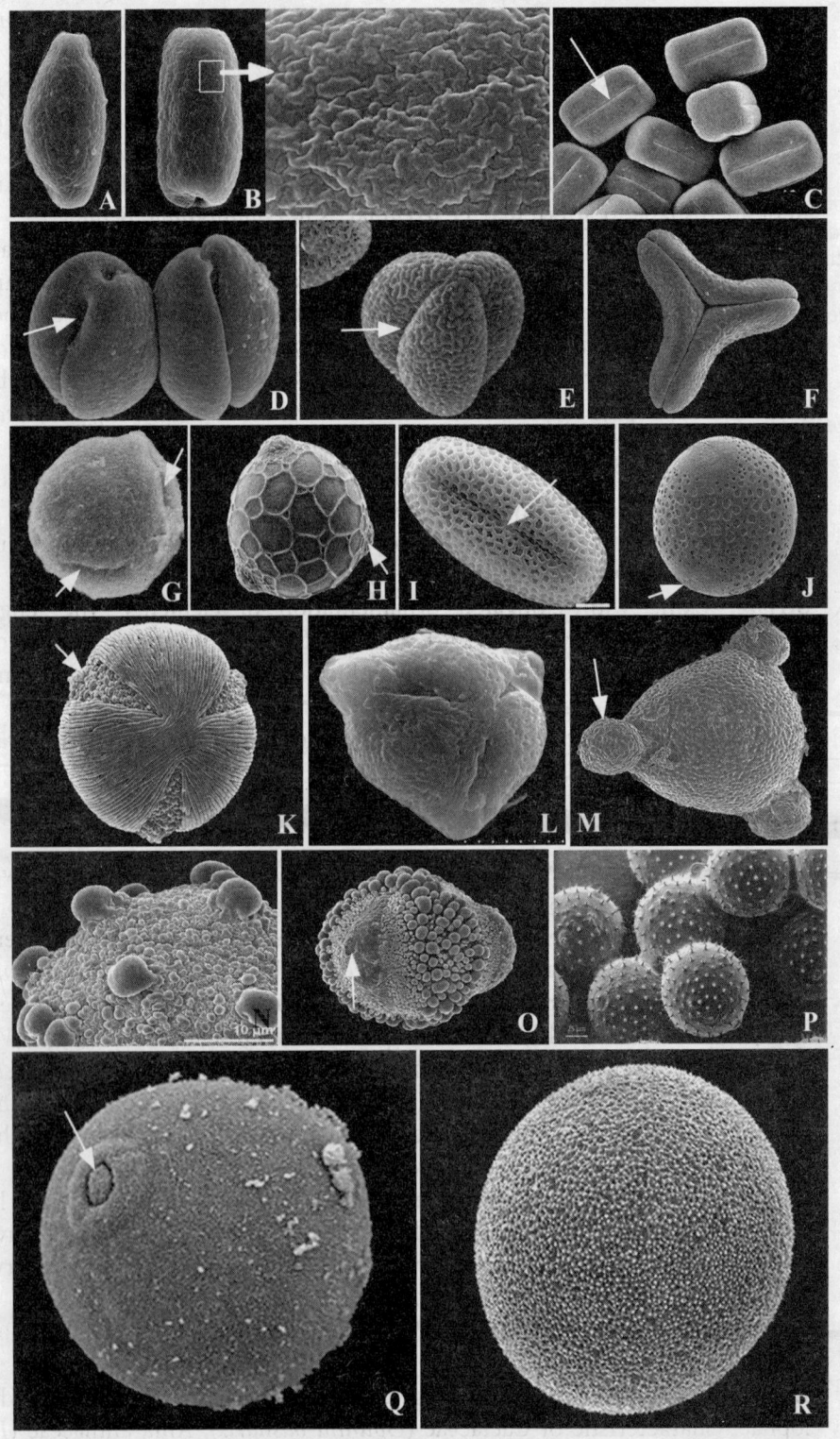

Figure 4.23 Scanning electron micrographs of pollen grains of many species showing different shapes, types of apertures and exine pattern. **A.** Ellipsoid pollen grain in *Ficus maclellandii*. **B.** Cylindrical pollen grain in *Ficus ischnopda* with rugulate granulate exine ornamentation. **C.** Tetraangular, tetracolpate pollen of *Sideritis romana* (Lamiaceae). Arrow marks a colpus. **D.** Bicolporate pollen of *Nymphaea lotus* with verrucated ornamentation with microgranules. **E.** Tricolporate pollen of *Nelumbo nucifera* with regulated ornamentation without microgranules. Arrow marks a colpus. **F.** Triangular, syncolpate pollen (in polar view) of *Loranthus europaeus*. The syncolpate condition arises due to fusion of colpi. Here three colpi have fused. **G.** Tricolporate, spherical pollen with smooth exine surface (*Hippophae rhamoides*). Two colpi (arrows) are visible in this plane. **H.** Prolate spheroidal pollen grain of *Vigna* sp., with reticulate exine pattern. **I.** Oval shape pollen of *Arabidopsis thaliana* exhibiting tricolporate condition (colpus is marked with arrow) and reticulae exine. **J.** Tricolpate pollen with foveolate exine of *Glossoloma ichthyoderma*. (colpus is marked with arrow). **K.** Pollen grain (in polar view) of *Neoalsomitra sarcophylla* with striae/striate exine ornamentation. **L.** Circular-triangular, isopolar pollen (polar view) of *Spiraea tomentosa* exhibiting tricolporate condition. The exine sculpturing is striate-foveolate. **M.** Mature pollen grain of *Vangueria infausta:* suboblate to spheroidal, isopolar, radially symmetrical and triporate. A protuberant globular oncus (intinous structure occurring beneath the apertures) projects from each aperture (arrow). Exine is microreticulate (i.e., with very small diameter of lumina). **N.** Part of pollen gemmate exine pattern in *Stenandrium dulce,* showing gemmae of different sizes **O.** Pollen (in equatorial view) with clavate exine sculpturing in *Ilex aquifolium*. Arrow marks the germ pore. **P.** Echinate exine sculpturing in pollen of *Cucurbita pepo*. **Q.** Pollen of *Triticum aestivum* showing aperture (arrow) surrounded by annulus; a collar like structure. **R.** Inaperturate pollen grain of *Populus alba*. (*Source:* A–B: Wang et al. 2014; C, F, H, J, N, O, R: Halbritter et al. 2018; D–E: Tungmunnithum et al. 2020; I: Chang et al. 2012; P: Pacini & Dolferus 2019; Q: Singh et al. 2017; L: Wrońska-Pilarek et al. 2019; M: Tilney et al. 2014, published under CC BY License.)

Bibliography

Albert, B. Raquin, C. Prigent, M. et al. (2011). Successive microsporogenesis affects pollen aperture pattern in the tam mutant of *Arabidopsis thaliana*. *Annals of Botany* 107: 1421–6.

Albert, B. Ressayre, A. and Nadot, S. (2011). Correlation between pollen aperture pattern and callose deposition in late tetrad stage in three species producing atypical pollen grains. *American Journal of Botany* 98: 189–96.

Ariizumi, T. and Toriyama, K. (2011). Genetic regulation of sporopollenin synthesis and pollen exine development. *Annual Review in Plant Biology* 62: 437–60.

Barman, C. Singh, V. K. and Kakkar, M. (2020). Secondary pollen presentation in flowering. In R. Tandon, K. R. Shivanna and M. Koul, eds., *Reproductive Ecology of Flowering Plants: Patterns and Processes*. Springer Nature Singapore, pp. 197–214.

Bhojwani, S. S., Bhatnagar, S. P. and Dantu, P. K. (2015). *The Embryology of Angiosperms*. New Delhi: Vikas Publishing House Pvt. Ltd.

Blackmore, S. and Barnes, S. H. (1985). *Cosmos* pollen ontogeny: a scanning electron microscope study. *Protoplasma* 126: 91–9.

Blackmore, S. and Barnes, S. H. (1987). Pollen wall morphogenesis in *Tragopogon porrifolius* (Compositae: Lactuceae) and its taxonomic significance. *Review of Palaeobotany and Palynology* 52: 233–46.

Blackmore, S. and Barnes, S. H. (1988). Pollen ontogeny in *Catananche caerulea* L. (Compositae: Lactuceae) I. Premeiotic phase to establishment of tetrads. *Annals of Botany* 62: 605–14.

Blackmore, S., Wortley, A. H. Skvarla, J. J. and Rowley J. R. (2007). Pollen wall development in flowering plants. *New Phytologist* 174: 483–98.

Bonner, L. and Dickinson, H. (1990). Anther dehiscence in *Lycopersicon esculentum* II. *New Phytologist* 115: 367–75.

Bowman, J. L., Drews, G. N. and Meyerowitz, E. M. (1991). Expression of the *Arabidopsis* floral homeotic gene AGAMOUS is restricted to specific cell types late in flower development. *Plant Cell* 3: 749–58.

Browne, R. G., Iacuone, S. Li, S. F. et al. (2018). Anther morphological development and stage determination in *Triticum aestivum*. *Frontiers in Plant Science* 9. doi:10.3389/fpls.2018.00228.

Cecchetti, V., Altamura, M. M., Serino, G. et al. (2007). ROX1, a gene induced by rolB, is involved in procambial cell proliferation and xylem differentiation in tobacco stamen. *The Plant Journal* 49: 27–37.

Cecchetti, V., Celebrin, D., Napoli, N. et al. (2016). An auxin maximum in the middle layer controls stamen development and pollen maturation in Arabidopsis. *New Phytologist* 213: 1194–207.

Chaban, I. A., Kononenko, N.V., Gulevich, A. A. et al. (2020). Morphological features of the anther development in tomato plants with non-specific male sterility. *Biology* 9: 32.

Chang, H. S., Zhang, C., Chang, Y. H. et al. (2012). *NO PRIMEXINE AND PLASMA MEMBRANE UNDULATION* Is essential for primexine deposition and plasma membrane undulation during microsporogenesis in *Arabidopsis*. *Plant Physiology* 158: 264–72.

Chang, Z., Jin, M., Yan, W. et al. (2018). The ATP-binding cassette (ABC) transporter OsABCG3 is essential for pollen development in rice. *Rice* 11: 58. doi.org/10.1186/s12284-018-0248-8.

Charzinska, M., Murgia, M., Milanesi, C. and Cresti, M. (1989). Origin of sperm cell association in the 'male germ unit' of Brassica pollen. *Protoplasma* 149: 1–4

Chase, C. D. (2007). Cytoplasmic male sterility: a window to the world of plant mitochondrial-nuclear interactions. *Trends in Genetics* 23: 81–90.

Chen, L. T. and Liu, Y. G. (2014). Male sterility and fertility restoration in crops. *Annual Review of Plant Biology* 65: 579–606.

Chen, M. et al. (2016). Origin and Functional prediction of pollen allergens in plants. *Plant Physiology* 172: 341–57.

Chichiriccò, G. (1999). Developmental stages of the pollen wall and tapetum in some *Crocus* species. *Grana* 38: 31–41.

Clément, C. and Pacini, E. (2001). Anther plastids in angiosperms. *Botanical Reviews* 67: 54–73.

CQ, A. O. (2013). Anther wall development, placentation, sporogenesis and gametogenesis in *Smilax davidiana* A.DC: a contribution to the embryology of *Smilax*. *South African Journal of Botany* 88: 459–65.

Dickinson, H. G. and Lewis F. R. S. (1973). The formation of the tryphine coating the pollen grains of *Raphanus*, and its properties relating to the self-incompatibility system. *Proceedings of Royal Society of London B* 184: 149–65.

Dumas, C, Knox, R. B. and Gaude T. (1984). Emerging physiological concepts in fertilization. *Plant Physiology* 15: 17–20.

Dumas, C, Knox, R. B. and Gaude T. (1984). Pollen–pistil interactions: new concepts from electron microscopy and cytochemistry. *International Review in Cytology* 90: 239–72.

Dumas, C., Knox, R. B. and Gaude T. (1985). The spatial association of the sperm cells and vegetative nucleus in the pollen grain of *Brassica*. *Protoplasma* 124: 168–74.

Echlin, P. and Godwin, H. (1969). The ultrastructure and ontogeny of pollen in *Helleborus foetidus* L.: the formation of the pollen grain wall. *Journal of Cell Science* 5: 459–77.

Eckardt, N. A. (2006). Cytoplasmic Male Sterility and Fertility Restoration. *The Plant Cell* 18: 515–17.

El-Ghazaly, G. (1999). Tapetum and orbicules (Ubisch bodies): development, morphology and role of pollen grains and tapetal orbicules in allergenicity. In: M. Cresti, GC, and Moscatelli A., eds., *Fertilization in Higher Plants*. Berlin: Springer, pp. 157–73.

Erdtman, G. (1952). Pollen morphology and plant taxonomy. *Geologisca Foreningens.* Stockholm Forhandlingar. 74: 526–7.

Erdtman, G. (1969). *Handbook of Palynology – An Introduction to the Study of Pollen Grains and Spores.* Copenhagen: Munksgaard.

Fagri, K. (1956). Recent trends in palynology. *The Botanical Review* 22: 639–64.

Falasca, G., D'Angeli, S., Biasi, R. and Fattorini, L. (2013). Tapetum and middle layer control male fertility in *Actinidia deliciosa. Annals of Botany* 112: 1045–55.

Fan, Y., Yang, J., Mathioni, S. M. et al. (2016). PMS1T, producing phased small-interfering RNAs regulates photoperiod-sensitive male sterility in rice. *Proceedings of National Academy of Sciences* U.S.A. 113: 15144–9.

Firon, N., Nepi, M. and Pacini, E. (2012). Water status and associated processes mark critical stages in pollen development and functioning. *Annals of Botany* 109: 1201–14.

Freudenstein, J. (1991). A systematic study of endothecial thickenings in the orchidaceae. *American Journal of Botany* 78: 766–81.

Furness, C. A. and Rudall P. J. (2004). Pollen aperture evolution – a crucial factor for eudicot success? *Trends in Plant Science* 9: 154–8.

Furness, C. A. and Rudall, P. J. (1998). Tapetum and systematics in monocotyledons. *Botanical Review* 64: 201–39.

Garcia, C. (2002). Anther wall formation in Solanaceae species. *Annals of Botany* 90: 701–6

Ge, Y. X., Angenent, G. C., Dahlhaus, E. et al. (2001). Partial silencing of the NEC1 gene results in early opening of anthers in *Petunia hybrida. Molecular Genetics and Genomics* 265: 414–23.

Goldberg, R. B., Beals, T. P. and Sanders, P. M. (1993). Anther development: basic principles and practical applications. *Plant Cell* 5: 1217–29.

Gomez, J. F., Talle, B. and Wilson, Z. A. (2015). Anther and pollen development: A conserved developmental pathway. *Journal of Integrated Plant Biology* 57: 876–91.

Hafidh, S., Fíla, J. and Honys, D. (2016). Male gametophyte development and function in angiosperms: a general concept. *Plant Reproduction* 29: 31–51.

Halbritter, H., Ulrich, S., Grímsson, F. et al. (2018). *Illustrated Pollen Terminology.* Springer OPEN. Switzerland. doi:10.1007/978-3-319-71365-6.

Hedhly, A., Vogler, H., Schmid, M. W. et al. (2016). Starch turnover and metabolism during flower and early embryo development. *Plant Physiology* 172: 2388–402.

Hernández-Pinzón I., Ross, J. H. E., Barnes, K. A., et al. (1999). Composition and role of tapetal lipid bodies in the biogenesis of the pollen coat of *Brassica napus. Planta* 208: 588–98.

Heslop-Harrison, J. and Heslop-Harrison, Y. (1984). The disposition of gamete and vegetative-cell nuclei in the extending pollen tubes of a grass species, *Alopecurus pratensis* L. *Acta Botanica Neerlandica* 33: 131–4.

Heslop-Harrison, J. and Heslop-Harrison, Y. (1985). Germination of stress-tolerant *Eucalyptus pollen. Journal of Cell Science* 73: 135–57.

Heslop-Harrison, J. and Heslop-Harrison, Y. (1992). Cyclical transformation of the actin cytoskeleton of hyacinth pollen subjected to recurrent vapour-phase hydration and dehydration. *Biological Cell* 75: 245–52.

Heslop-Harrison, J. (1963). An ultrastructural study of pollen wall ontogeny in *Silene pendula. Grana Palynologica* 4: 7–24.

Heslop-Harrison, J. (1968). Pollen wall development. *Science* 161: 230–7.

Hesse, M., Vogel, S. and Halbritter H. (2000). Thread-forming structures in angiosperm anthers: their diverse role in pollination ecology. *Pollen and pollination.* 281–92.

Hesse, M. (1981). Pollenkitt and viscin threads: their role in cementing pollen grains. *Grana* 20: 145–52.

Hoekstra, F. A., Golovina, E. A. and Buitink, J. (2001). Mechanisms of plant desiccation tolerance. *Trends in Plant Science* 6: 431–8.

Horner, H. T. and Pearson, C. B. (1978). Pollen wall aperture development in *Helianthus annus* (Compositae: Heliantheae). *American Journal of Botany* 65: 293–309.

Howell, G. J., Slater, A.T. and Knox, R. B. (1993). Secondary pollen presentation in angiosperms and its biological significance. *Australian Journal of Botany* 41: 417–38.

Huysmans, S., El-Ghazaly, G. and Smets, E. (1998). Orbicules in angiosperms: morphology, function, distribution, and relation with tapetum types. *Botanical Reviews* 64: 240–72.

Jia, Q. S., Zhu, J., Xu, X. F. et al. (2015). *Arabidopsis* AT-hook protein TEK positively regulates the expression of arabinogalactan proteins for nexine formation. *Molecular Plant* 8: 251–60.

Keijzer, C. (1987). The processes of anther dehiscence and pollen dispersal. I. The opening mechanism of longitudinally dehiscing anthers. *New Phytologist* 105: 487–9.

Kelliher, T., Egger, R. L., Zhang, H. and Walbot, V. (2014). Unresolved issues in pre-meiotic anther development. *Frontiers in Plant Sciences* 5: 347. doi: 10.3389/fpls.2014.00347.

Kelliher, T. and Walbot, V. (2011). Emergence and patterning of the five cell types of the *Zea mays* anther locule. *Development Biology* 350: 32–49.

Kliwer, I. and Dresselhaus, T. (2010). Establishment of the male germline and sperm cell movement during pollen germination and tube growth in maize. *Plant Signaling and Behavior* 5: 885–9.

Knoll, F. (1930). Über Pollenkitt und Bestäubungsart. Z. *Botany* 23: 609–75.

Kostić, A. Ž. et al. (2020). The application of pollen as a functional food and feed ingredient: the present and perspectives. *Biomolecules* 10: 84.

Kumar, P., Singhal, V. K, Kaur, D. and Kaur S. (2010). Cytomixis and associated meiotic abnormalities affecting pollen fertility in *Clematis orientalis*. *Biologia Plantarum* 54: 181–4.

Lalanne, E. and Twell, D. (2002). Genetic control of male germ unit organization in *Arabidopsis*. *Plant Physiology* 129: 865–75.

Lei, X. and Liu B. (2020). Tapetum-dependent male meiosis progression in plants: increasing evidence emerges. *Frontiers in Plant Sciences* 10: 1667. doi: 10.3389/fpls.2019.01667.

Li, B. and Xu, F. (2019). Formation pattern in five types of pollen tetrad in *Pseuduvaria trimera* (Annonaceae). *Protoplasma* 256: 53–68.

Li, T., Gong, C. and Wang, T. (2010). RA68 is required for postmeiotic pollen development in *Oryza sativa*. *Plant Molecular Biology* 72: 265–77. doi: 10.1007/s11103-009-9566-y.

Lin. S., Miao, Y., Su, S. et al. (2019). Comprehensive analysis of Ogura cytoplasmic male sterility-related genes in turnip (*Brassica rapa ssp. rapifera*) using RNA sequencing analysis and bioinformatics. *PLoS ONE* 14(6): e0218029. doi.org/10.1371/journal.pone.0218029.

Linde, K. V. D. and Walbot, V. (2019). Pre-meiotic anther development. *Current Topics in Development Biology* 131: 239–56.

Liu, B., Ho, C. M. K. and Lee Y. R. J. (2011). Microtubule reorganization during mitosis and cytokinesis: lessons learned from developing microgametophytes in *Arabidopsis* thaliana. *Frontiers in Plant Sciences* 2: 27. doi: 10.3389/fpls.2011.00027.

Luza Juvenal, G. and Polito, V. S. (1988). Microsporogenesis and anther differentiation in juglans regia l.: a developmental basis for heterodichogamy in walnut. *Botanical Gazette* 149: 30–6.

Ma, H. (2005). Molecular genetic analyses of microsporogenesis and microgametogenesis in flowering plants. *Annual Review of Plant Biology* 56: 393–434.

Maheshwari, P. (1950). *An Introduction to the Embryology of Angiosperms*. New York. McGraw-Hill.

Mascarenhas, J. P. (1989). The male gametophyte of flowering plants. *The Plant Cell* 1: 657–64.

McConchie, C. A., Hough, T. and Knox, R. B. (1987b). Ultrastructural analysis of the sperm cells of mature pollen of maize, *Zea mays*. *Protoplasma* 139: 9–19.

McConchie, C. A., Jobson, S. and Knox, R. B. (1985). Computer-assisted reconstruction of the male germ unit in pollen of *Brassica campestris. Protoplasma* 127: 57–63.

McConchie, C. A., Russell, S. D. Dumas, C. et al (1987a). Quantitative cytology of the sperm cells of *Brassica campestris* and *B. oleracea. Planta* 170: 446–52.

McCormick, S. (2004). Control of male gametophyte development. *The Plant Cell* (online) 16 (suppl1), S142–S153.

McCue, A. D., Cresti, M., Feijo, J. A. and Slotkin, R. K. (2011). Cytoplasmic connection of sperm cells to the pollen vegetative cell nucleus: potential roles of the male germ unit revisited. *Journal of Experimental Botany* 62: 1621–31.

Mizuno, S., Osakabe, Y., Maruyama, K. et al. (2007). Receptor-like protein kinase 2 (RPK 2) is a novel factor controlling anther development in *Arabidopsis thaliana. The Plant Journal* 50: 751–66.

Mogensen, H. L. (1992). The male germ unit: concept, composition and signification. *International Review of Cytology* 140: 129–47.

Murgia, M., Charzynska, M., Rougier, M. et al. (1991). Secretory tapetum of *Brassica oleracea* L.: polarity and ultrastructural features. *Sexual Plant Reproduction* 4: 28–35.

Nepi, M., Franchi, G. G. and Pacini, E. (2001). Pollen hydration status at dispersal: cytophysiological features and strategies. *Protoplasma* 216: 171–80.

Oriani, A. and Scatena, V. L. (2015). Anther wall development, microsporogenesis, and microgametogenesis in *Abolboda* and *Orectanthe*: contributions to the embryology of Xyridaceae (Poales). *International Journal of Plant Sciences* 176: 324–2.

Pacini, E. and Dolferus, R. (2016). The trials and tribulations of the plant male gametophyte. In A. Shanker, ed., *Understanding Reproductive Stage Stress Tolerance, Abiotic and Biotic Stress in Plants– Recent Advances and Future Perspectives.* London: Intech, pp. 703–54.

Pacini, E. and Dolferus, R. (2019). Pollen developmental arrest: maintaining pollen fertility in a world with a changing climate. *Frontiers in Plant Sciences* 10: 679. doi: 10.3389/fpls.2019.00679.

Pacini, E. and Franchi, G. G. (1999). Pollen grain sporoderm and types of dispersal units. *Acta Societatis Botanicorum Poloniae* 68: 299–305.

Pacini, E. and Franchi, G. G. (2020). Pollen biodiversity – why are pollen grains different despite having the same function? A review. *Botanical Journal of the Linnean Society* 193: 141–64.

Pacini, E., Guarnieri, M. and Nepi, M. (2006). Pollen carbohydrates and water content during development, presentation, and dispersal: a short review. *Protoplasma* 228: 73–7.

Pacini, E. and Hesse, M. (2005). Pollenkitt – its composition, forms and functions. *Flora* 200: 399–415.

Pacini, E. (2010). Relationships between tapetum, loculus and pollen during development. *International Journal of Plant Sciences* 171: 1–11.

Pacini, E. (1997). Tapetum character states: analytical keys for tapetum types and activity. *Canadian Journal of Botany* 75: 1448–59.

Papini, A., Mosti, S. and Brighigna, L. (1999). Programmed-cell death events during tapetum development of angiosperms. *Protoplasma* 207: 213–21.

Paxson-Sowders, D. M., Owen, H. A. and Makaroff, C. A. (1997). A comparative ultrastructural analysis of exine pattern development in wild-type *Arabidopsis* and a mutant defective in pattern formation. *Protoplasma* 198: 53–65.

Periasamy, K. and Amalathas, J. (1991). Absence of callose and tetrad in the microsporogenesis of *Pandanus odoratissimus* with well-formed pollen exine. *Annals of Botany* 67: 29–33.

Pressman, E., Shaked, R., Shen, S. et al. (2012). Variations in carbohydrate content and sucrose-metabolizing enzymes in tomato *(Solanum lycopersicum* L.) stamen parts during pollen maturation. *American Journal of Plant Science* 3: 252–60.

Qin, P., Wang, Y., Li, Y. et al. (2013). Analysis of cytoplasmic effects and fine-mapping of a genic male sterile line in rice. *PLoS ONE* 8(4): e61719. doi:10.1371/journal.pone.0061719.

Ram, M. (1959). Occurrence of embryo sac-like structures in the microsporangia of *Leptomeria billardierii* R. Br. *Nature* 184: 914–15.

Rezanezad, F. (2008). The structure and ultrastructure of anther epidermis and pollen in *Lagerstroemia indica* L. (Lythraceae) in response to air pollution. *Turkish Journal of Botany* 32: 35–42.

Risueño, M. C., Giménez-Martín, G., López-Sáez, J. F. et al. (1969). Origin and development of sporopollenin bodies. *Protoplasma* 67: 361–74.

Rodriguez-Garcia, M. I. and Majewska-Sawka, A. (2011). Is the special callose wall of microsporocytes an impermeable barrier? *Journal of Experimental Botany* 12: 1659–63.

Rowley, J. R. (1962). Nonhomogeneous sporopollenin in microcspores of *Poa annua* L. *Grana Palynologica* 3: 3–19.

Rowley, J. R. and Dahl, A. O. (1977). Pollen development in *Artemisa vulgaris* with special reference to glycoclayx material (1). *Pollen et Spores* 14: 169–284.

Russell, S. D. (1984). Ultrastructure of the sperm of Plumbago zeylanica. II. Quantitative cytology and three-dimensional organization. *Planta* 162: 385–91.

Russell, S. D. (1984). Ultrastructure of the sperm of *Plumbago zeylanica*. *Planta* 162: 385–391.

Russell, S. D. (1985). Preferential fertilization in *Plumbago*: Ultrastructural evidence for gamete-level recognition in an angiosperm. *Proceedings of National Academy of Sciences USA* 82: 6129–32.

Russell, S. D. and Cass, D. D. (1981). Ultrastructure of the sperms of *Plumbago zeylanica*. l. Cytology and association with the vegetative nucleus. *Protoplasma* 107: 85–107.

Russell, S. D. and Strout, G. W. (2005). Microgametogenesis in *Plumbago zeylanica* (Plumbaginaceae). 2. Quantitative cell and organelle dynamics of the male reproductive cell lineage. *Sexual Plant Reproduction* 18: 113–30.

Sajo, G. M., Furness, C. A., Prychid, C. J. and Rudall, P. J. (2005). Microsporogenesis and anther development in Bromeliaceae. *Grana* 44: 65–74.

Schill, R. and Wolter, M. (1986). On the presence of elastoviscin in all subfamilies of the Orchidaceae and the homology to pollenkitt. *Nordic Journal of Botany* 6: 321–4.

Scott, R. J., Spielman, M. and Dickinson, H. G. (2004). Stamen structure and function. *The Plant Cell* 16 (suppl 1) S46–S60

Shamina, N. V., Gordeeva, E. I., Kovaleva, N. M. et al. (2007). Formation and function of phragmoplast during successive cytokinesis stages in higher plant meiosis. *Cell Biol International* 31: 626–35.

Sharma, A., Singh, M. B. and Bhalla, P. L. (2015). Ultrastructure of microsporogenesis and microgametogenesis in *Brachypodium distachyon*. *Protoplasma* 252: 1575–86.

Shi, X., Han, X. and Lu, T. (2015). Callose synthesis during reproductive development in monocotyledonous and dicotyledonous plants. *Plant Signaling and Behavior* 11: e1062196.

Shivanna, K. R. (2002). *Pollen Biology and Biotechnology*. Oxford and IBH Publishing House Co. Pvt. Ltd. India.

Shivanna, K. R. (2020). Nonreproductive roles of pollen grains. *Journal of Indian Botanical Society* 100: 45–58.

Singh, M., Kumar, M., Thilges, K. et al. (2017). MS26/CYP704B is required for anther and pollen wall development in bread wheat (*Triticum aestivum* L.) and combining mutations in all three homeologs causes male sterility. *PLoS ONE* 12: e0177632. doi.org/10.1371/ journal.pone.0177632.

Singh, M. B. and Bhalla, P. L. (2007). Control of male germ-cell development in flowering plants. *Bio Essays* 29: 1124–32.

Song, S., Qi, T., Huang, H., and Xie, D. (2013). Regulation of stamen development by coordinated actions of jasmonate, auxin, and gibberellin in *Arabidopsis*. *Molecular Plant* 6: 1065–73.

Southworth, D. and Jernstedt, J. A. (1995). Pollen exine development precedes microtubule rearrangement in *Vigna unguiculata* (Fabaceae): a model for pollen wall patterning. *Protoplasma* 187: 79–87.

Stadler, R., Truernit, E., Gahrtz, M. and Sauer, N. (1999). The AtSUC1 sucrose carrier may represent the osmotic driving force for anther dehiscence and pollen tube growth in *Arabidopsis*. *The Plant Journal* 19: 269–78.

Storme, de N. and Geelen D. (2013). Cytokinesis in plant male meiosis. *Plant Signaling and Behavior* 8: e23394. doi.org/10.4161/psb.23394.

Stow, I. (1930). Experimental studies on the formation of the embryo sac-like giant pollen grains in the anthers of *Hyacinthus orientalis*. *Cytologia* 1: 417–39.

Stow, I. (1934). On the female tendencies of the embryo sac-like giant pollen grains of *Hyacinthus orientalis*. *Cytologia* 5: 88–108.

Takahashi, M. (1991). Exine pattern formation by plasma membrane in *Bougainvillea spectabilis* Willd. (Nyctaginaceae). *American Journal of Botany* 78: 1063–69.

Tilney, P. M., van Wyk, A. E. and van der Merwe, C. F. (2014). The epidermal cell structure of the secondary pollen presenter in *Vangueria infausta* (Rubiaceae: vanguerieae) suggests a functional association with protruding onci in pollen grains. *PLoS ONE* 9(5): e96405. doi:10.1371/journal.pone.0096405.

Tiwari, S. C. and Gunning, B. E. S. (1986). Cytoskeleton, cell surface and the development of invasive plasmodial tapetum in *Tradescantia virginiana* L. *Protoplasma* 133: 89–99.

Tiwari, S. C. and Gunning, B. E. S. (1986). Development of tapetum and microspores in Canna L.: an example of an invasive but non-synctial tapetum. *Annals of botany* 57: 557–63.

Tungmunnithum, D., Renouard, S., Drouet, S. et al. (2020). A critical cross-species comparison of pollen from Nelumbo nucifera Gaertn. vs. *Nymphaea lotus* L. for authentication of Thai medicinal herbal tea. *Plants* 9(7): 921. //doi.org/10.3390/plants9070921.

Tütüncü Konyar, S. (2017). Ultrastructural aspects of pollen ontogeny in an endangered plant species, *Pancratium maritimum* L. (Amaryllidaceae). *Protoplasma* 254: 881–900.

Ünal, M., Filiz Vardar, F. and Aytürk, O. (2013). Callose in plant sexual reproduction. In Marina Silva-Opps, ed., *Current Progress in Biological Research*. Intech open.

Verstraete, B., Moon, H. K., Smets, E. et al. (2014). Orbicules in flowering plants: a phylogenetic perspective on their form and function. *Botanical Reviews* 80: 107–34.

Vijayaraghavan, M. R. and Ratnaparkhi, S. (1977). Pollen-embryo sacs in *Heuchera micrantha* Dougl. *Caryologia* 30: 105–19.

Volkova, O. A., Margarita,. V. R., Sokoloff, D. D. and Elena, E. S. (2016). A developmental study of pollen dyads and notes on floral development in *Scheuchzeria* (Alismatales: Scheuchzeriaceae). *Botanical Journal of the Linnean Society* 182: 791–810.

Volkova, O. A., Severova, E. E. and Polevova, S. V. (2013). Structural basis of harmomegathy: evidence from Boraginaceae pollen. *Plant Systematics and Evolution* 299: 1769–79.

Walbot, V. and Egger,. RL. (2016). Pre-meiotic anther development: cell fate specification and differentiation. *Annual Reviews in Plant Biology* 67: 365–95.

Wang, G., Chen, J., Li, Z-B., Zhang, F-P. and Yang, D-R. (2014). Has pollination mode shaped the evolution of *Ficus pollen*? *PLoS ONE* 9(1): e86231. doi: 10.1371/journal.pone.008623.

Wang, H., Mao, Y., Yang, J. and He, Y. (2015). TCP24 modulates secondary cell wall thickening and anther endothecium development. *Frontiers in Plant Sciences* 6: 436. doi: 10.3389/fpls.2015.00436.

Wang, R. and Dobritsa, A. A. (2018). Exine and aperture patterns on the pollen surface: their formation and roles in plant reproduction. *Annual Plant Reviews* 1: 1–40.

Wang, Z., Zou, Y., Li, X. et al. (2006). Cytoplasmic male sterility of rice with Boro II cytoplasm is caused by a cytotoxic peptide and is restored by two related PPR motif genes via distinct modes of mRNA silencing. *The Plant Cell* 18: 676–87.

Waterkeyn, L. and Bienfail, A. (1970). On a possible function of the callosic special wall in *Ipomea purpurea* (L.) *Roth. Grana*10:13–20. doi: 10.1080/00173137009429852.

Waterkeyn, L. (1962). Les parois microsporocytaires de nature callosique chez Helleborus et Tradescantia. *Cellule.* 62: 225–55.

Whatley, J. M. (1982). Fine structure of the endothecium and developing xylem in *Phaseolus Vulgaris*. *New Phytologist* 91: 561–70.

Wilson, Z. A., Song, J., Taylor, B. and Yang, C. (2011). The final split: the regulation of anther dehiscence. *Journal of Experimental Botany* 62: 1633–49.

Wrońska-Pilarek, D., Wiatrowska, B. and Bocianowski, J. (2019). Pollen morphology and variability of invasive *Spiraea tomentosa* L. (Rosaceae) from populations in Poland. *PLoS ONE* 14(8): e0218276. https://doi.org/10.1371/journal.pone.0218276.

Xue, J. S., Yao, C., Xu, Q. L. et al. (2021). Development of the middle layer in the anther of *Arabidopsis*. *Frontiers in Plant Sciences* 12: 634114.

Yamamoto, Y., Nishimura, M., Hara-Nishimura, I. et al. (2003). Behavior of vacuoles during microspore and pollen development in *Arabidopsis thaliana*. *Plant and Cell Physiology* 44: 1192–201.

Yang, S., Xie, L. F., Mao, H. Z. et al. (2003). Tapetum determinant1 is required for cell specialization in the *Arabidopsis* anther. *Plant Cell* 15: 2792–804.

Yu, H. S., Hu, S. Y. and Zhu, C. (1989). Ultrastructure of sperm cells and the male germ unit in pollen tubes of Nicotiana tabacum. *Protoplasma* 152: 29–36.

Zhang, D., Luo, X. Zhu, L. (2011). Cytological analysis and genetic control of rice anther development. *Journal of Genetics and Genomics* 38: 379–90.

Zhang, D. B. and Yang, L. (2014). Specification of tapetum and microsporocyte cells within the anther. *Current Opinion in Plant Biology* 17C: 49–55.

Zhang, X., Zhao, G., Tan, Q. et al. (2020). Rice pollen aperture formation is regulated by the interplay between OsINP1 and OsDAF1. *Nature Plants* 6: 394–403.

Zhang, Y. C., He, R. R., Lian, J. P. et al. (2020). OsmiR528 regulates rice-pollen intine formation by targeting an anthocyanin to influence flavonoid metabolism. *Proceedings of the National Academy of Sciences* 117: 727–32.

Zhang, Y. N., Wei, D. M., Song, Y. Y. et al. (2011). Microtubule organization during successive microsporogenesis in Allium cepa and simultaneous cytokinesis in *Nicotiana tabacum. Biologia Plantarum* 55: 752.

Zhao, B., Shi, H., Wang, W. et al. (2016). Secretory COPII Protein SEC31B is required for pollen wall development. *Plant Physiology* 172: 1625–42.

Zhou. Q., Zhu, J. and Cui, Y. L.. et al. (2015). Ultrastructure analysis reveals sporopollenin deposition and nexine formation at early stage of pollen wall development in *Arabidopsis. Science Bulletin* 60: 273–6.

Zhou, Y. and Dobritsa, A. A. (2019). Formation of aperture sites on the pollen surface as a model for development of distinct cellular domains. *Plant Science* 288: 110222.

5

The Ovule and Female Gametophyte

Ovules are the progenitors of seeds.

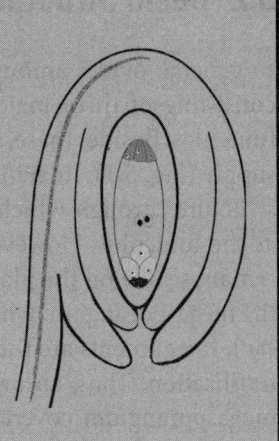

5.1 Introduction

Sexual reproduction involves delivery of sperm cells, via the pollen tube, to the egg cell present in the embryo sac, where fertilization occurs and the new sporophyte is formed (Dumas & Mogensen 1993). While the formation of male gametophyte (pollen grains) takes place within the anther, the female gametophyte (embryo sac or megagametophyte) develops within the ovule. Thus, the ovule can be defined as a specialized sporophytic structure within which development of female gametophyte or mega-gametophyte takes place. Ovule is the site for delimitation of megasporocyte, production of a functional megaspore (megasporogenesis) and eventually formation of embryo sac (megagametogenesis). The embryo sac harbors the female gamete or the egg, which subsequently gets fertilized by the male gamete to form an embryo. In angiosperms, apart from the female gametophyte and egg cell development, important reproductive events such as pollen tube attraction and guidance, double fertilization, and embryo and endosperm development all occur within the ovule. The ultimate result of all these events is the formation of seed and therefore, the ovule is also considered the developmental precursor or progenitor of the seed.

Among angiosperms, different modes of female gametophyte ontogeny are seen, leading to different types of female gametophytes. The cells of female gametophyte are very peculiar in their ultrastructure and with the help of electron microscopy, great details of these cells are known. In various species, besides the typical parts of ovule, there are many specialized structures associated with the ovules which aid in pollen tube guidance and facilitate fertilization. All these aspects of structure and development of the angiosperm ovule, female gametophyte and their types have been discussed in the present chapter. The chapter also includes exceptions to these developmental patterns and details of extra ovular structures.

5.2 Basic Structure of Ovule

In general, ovules among angiosperms are fundamentally similar in their basic structure, consisting of three major tissues: a nucellus, protective coat(s) or integument(s), and a funiculus. Besides these, a typical ovule also consists of a micropyle, a chalaza and its vascular supply (Fig. 5.1). In angiosperms, the ovules remain enclosed in the ovary, and a stalk like structure through which ovules remain attached to the ovary wall or placenta is known as the funiculus. However, not all the ovules may have a funiculus as certain ovules may remain sessile on the placenta (e.g., in *Acharia*, Achariaceae). The nucellus is equivalent to the megasporangium which produces and encloses a haploid megagametophyte (or embryo sac). Integuments are the tough protective layers of ovule that develop into seed coats after fertilization. Thus, an ovule can be morphologically visualized as a structure harboring a megasporangium covered by one or two integuments attached to the ovary wall with a funiculus. Nonetheless, angiosperms display an extreme diversity in ovules by the way of arrangement and presence of these tissues, some of which will be discussed in succeeding paragraphs.

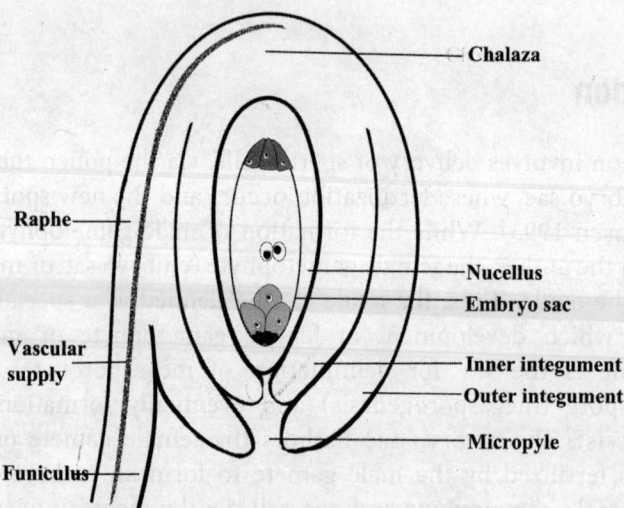

Figure 5.1 Diagrammatic sketch of a typical ovule in angiosperms. See Color Plates (page 479).

Within a mature ovule three regions are distinguishable: the distal part which comprises nucellus covered by the integuments, the median part, named chalaza, from which the two integuments originate and the proximal region where the funiculus develops (Fig. 5.1). The funiculus and the chalaza are intercalary structures and thus less demarcated than the nucellus and the integuments. Esau (1977) has defined chalaza as the region extending from the base of the integuments to the point of attachment of the funiculus. A single vascular bundle extends from the placenta through the funiculus to the chalaza (Endress 2011). In most of the angiosperms, ovules are curved and as this curvature occurs, a part of the funiculus gets fused with the main body of the ovule forming a ridge which is known as raphe.

Most commonly, the ovules of angiosperms are associated with two integuments and hence, are bitegmic, in contrast to gymnosperms where the ovules possess a single integument and are unitegmic (Gasser et al. 1998). However, certain angiosperm families like Ranunculaceae (*Anemone* sp., *Oxygraphis glacialis*, *Helleborus* sp.), Menispermaceae, many members of Solanaceae, Sabiaceae (*Meliosma* sp.), and Ceratophyllaceae (*Ceratophyllum* sp.) exhibit the presence of unitegmic ovules. During the course of evolution, unitegmic ovules evolved from the bitegmic ones either as a result of fusion of the two integuments as seen in Myrtaceae; or, due to reduction and loss of the outer integument as seen in basal angiosperms like Ceratophyllaceae (Igersheim & Endress 1998). In some angiosperms, highly reduced ovules with no integuments are also found. Such ovules which lack the presence of integuments are called ategmic ovules. Some members of Oleaceae and mycotrophic Gentianaceae have highly reduced ovules without differentiation into nucellus and integument (ategmic) (Bouman et al. 2002).

The nucellar tissue is almost completely surrounded by the two integuments, except for a narrow opening at the apical end called the micropyle. Structurally, the micropyle is the point where the integuments terminate and form a narrow canal through which pollen tubes can enter the nucellus. The portion of the micropyle formed by the outer integument is called the exostome and the one formed by the inner integument is called the endostome. In Resedaceae, the micropylar canal is not straight and instead follows a zigzag pattern as the exostome and endostome are not aligned in a straight line. In Podostemaceae, micropyle is organized by outer integument alone, as the inner integument ceases to grow after it reaches a certain height. In some extreme cases, such as in *Cassytha, Hernandia, Quisqualis*; a micropyle is not formed at the time of ovule maturity. Instead, the parts of the nucellus directly touch the funiculus or the obturator or ovary wall such that a narrow gate is formed above the nucellus (Endress 2011).

5.3 Development of Ovule

Pistil, the female reproductive organ, is composed of a single carpel or a number of carpels that are often fused. The basal portion of the carpel is the ovary which houses the ovules. The lateral margins of the carpels contain a meristematic tissue called the carpel margin meristem (CMM). During ovule development, CMM first gives rise to the placenta, the septum and the transmitting track. Later, the ovules develop from the placenta of the ovary wall in the form of a mound called ovule primordium (Fig. 5.2 A). This ovule primordium originates from the sub-epidermal tissue of the placenta through periclinal divisions. After initiation, the primordium elongates and close to its apex the two integument primordia appear almost simultaneously as mounds (Fig. 5.2 B, C). The position at which the integuments are initiated marks the prospective chalaza. Usually, the inner integument is dermal in origin, whereas the outer integument is typically derived from both dermal and subdermal layers (Bouman et al. 2002). The part of the ovule primordium which is above the level of the inner integument primordium is the site for nucellus. Thus, delimitation of nucellus is dependent on the initiation of integuments and if for some reason the integuments are not initiated, then the ovule remains morphologically undifferentiated (Endress 2006). This suggests that integument initiation is a crucial event in ovule morphogenesis and the demarcation of nucellus, chalaza and funiculus are dependent on it (Schneitz et al. 1997).

Both periclinal and anticlinal divisions occur in the integuments leading to increase in their size; such that, integuments first cover the nucellus and soon overgrow it (Fig. 5.2 D–E). Further, the ovules undergo curvature during development so that the micropyle comes to lie near the placenta and provides easy access to the pollen tubes (Fig. 5.2 F). Notably, the degree of curvature varies among angiosperms and is an important developmental feature in classifying different types of ovules (see Section 5.4 for details). Curved ovules have a raphe, a conspicuous area through which the vascular bundle runs from the funiculus to the chalaza. Raphe develops due to the extension of the ovule on one side beyond the funiculus and is considered merely a developmental by product of ovule curvature (Endress 2011).

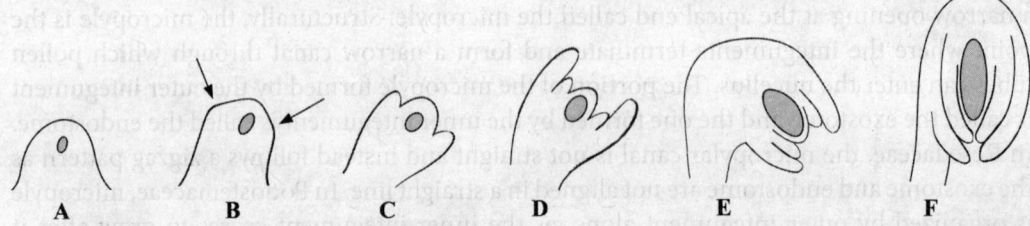

Figure 5.2 Development of an anatropous ovule. A. Ovule primordium. **B.** Integuments appear as narrow domes (arrows show outer integument). **C.** Outer integument differentiates initially and ovule curvature begins with divisions in the funicular region. **D–E.** Growth of integuments and curvature of the ovule continues further. The integuments first cover the nucellus and later outgrow it. **F.** Mature anatropous ovule shows nearly 180° curvature since its inception. (*Source*: Based on Ceccato et al. 2013; Endress 2011.)

5.4 Ovule Diversity

5.4.1 Types of Ovule on the Basis of Degree of Curvature

Ovules in most angiosperms undergo curvature but the degree of ovule curvature is variable among different taxa. The extent of curvature depends on the difference in the growth of the integuments and funiculus in different ovules. Curvature helps in securing a position for micropyle close to the placenta such that the pollen tubes can easily enter the ovules. Degree of curvature is an important basis for classification of ovules. Six different types of ovules are recognized primarily on the basis of curvature, and on the placement of morphological axis of the ovule with respect to placenta (Fig. 5.3):

- **Orthotropous or Atropous (ortho: straight, tropous: turn):** Such ovules lack curvature and the micropyle lies opposite to the placenta at an angle of 180° from it (Fig. 5.3 A). Consequently, the body of the ovule is erect or straight and the chalaza, funiculus and micropyle lie in a straight line. Orthotropous ovules

Box 5.1

Ovule Development: Genes and Proteins Involved

Ovule initiation and development involves a number of genes. It begins with expression of ovule identity genes in the floral meristem followed by expression of genes involved in regulation of ovule primordia initiation and ovule growth. Some genes that control ovule development are also involved in the growth of other floral organs. A majority of these genes encode transcription factors and some are mitochondrial proteins or chromatin remodeling factors indicating that ovule development is under the control of a complex genetic and molecular network. Some genes and proteins regulating the development of ovule are summarized below:

Gene	Encoded Protein	Function
SHATTERPROOF (SHP1 and SHP2)	MADS-box gene family of transcription factors	Ovule identity in floral meristem
SEEDSTICK (STK)	MADS-box gene family of transcription factors	Ovule identity in floral meristem
LEUNIG (LUG)	Glutamine-rich protein with seven WD repeats	Carpel fusion
SPATULA (SPT)	Helix–loop–helix(bHLH) transcription factor	Carpel fusion
REVOLUTA (REV)	Member of the class III Homeodomain-Leucine Zipper (HD-ZIP III) family	CMM and placenta development
AINTEGUMENTA (ANT)	AP2 domains containing transcription factor	CMM development and control of ovule primordia growth, integument formation
CUP-SHAPED COTYLEDON (CUC1 and CUC2)	Transcription factor of the NAC transcription factor family	Controls the number of ovule primordia that develop from the placenta
HUELLENLOS (HLL)	Mitochondrial ribosomal protein	Control of integument formation and ovule growth
SHORT INTEGUMENTS 2 (SIN2)	Mitochondrial DAR GTPase	Integument development
INNER NO OUTER (INO)	YABBY-domain transcription factors	Formation of outer integument
TOUSLED (TSL)	Putative protein kinase	Formation of inner integument
SUPERMAN (SUP)	Transcriptional repressor with a C2H2 zinc-finger DNA-binding domain	Asymmetric growth of outer integument

are characterized by the presence of vasculature that starts from the base of the funiculus and goes up to the chalazal-nucellar region as a straight line. Such ovules are found in Amborellaceae, Chloranthaceae, Ceratophyllaceae, Polygonaceae, Piperaceae and Acanthaceae.

- **Anatropous (ana: backward or up, tropous: turn):** The body of the ovule becomes completely inverted because of the extensive growth of the funiculus such that the micropyle comes to lie very close to it. Here, the integuments, nucellus and chalaza are parallel to the long axis of the funiculus (Fig. 5.3 B). In anatropous ovules, vascular strands traverse from the base of the funiculus to the chalazal-nucellar region and take positions opposite the micropyle because of 180° curvature (Simpson 2010). This is the most common type of ovule in angiosperms, seen in 82% of the families (Davis 1966). Nucellus is straight (unbent) both in anatropous and orthotropous ovules.

- **Hemitropous (hemi: half, tropous: turn):** The body of the ovule is placed transversely at right angles to the funicle and parallel to the placenta. The micropyle and chalaza lie in one straight line (Fig. 5.3 C) as seen in *Ranunculus*. A recent study (Shamrov 2018) described two subtypes within hemitropous ovules, viz. *hemi-anatropous* and *hemi-orthotropous*.

  (a) **hemi-anatropous:** This type of ovule is in between hemitropous and anatropous types of ovule. Most of the elements such as, integuments, nucellus and chalaza are placed as in an anatropous ovule but the main axis of the ovule is at an angle of 45° from the funiculus. This is due to the turning of the ovule up to 135° towards the placenta, instead of 180°, as seen in an anatropous ovule. Such ovules can be seen in members of Juncaginaceae.

  (b) **hemi-orthotropous:** Here, the vascular strand passes through the funiculus towards the chalaza in a straight line. However, the micropyle, the integuments and nucellus are at right angles to the funiculus. Such ovules are found in families Malphigiaceae and Primulaceae.

- **Campylotropous (kampylos: curved):** Such ovules are characterized by a curved morphological axis such that the alignment between the chalaza and micropyle is lost. As a result, the body of the ovule appears curved or bent and micropyle and chalaza do not lie in a straight line (Fig. 5.3 D), e.g., in Leguminosae. Curvature in the case of campylotropous ovules is less than the curvature in the case of anatropous ovules but more than that of hemitropous ovules. A campylotropous ovule also differs from an anatropous ovule in having a nucellus which is bent along the lower side, as viewed in the mid-sagittal section (*section along the plane of symmetry*) (Simpson 2010; Endress 2011). Based on the axis of vasculature, campylotropous ovules can be either *ortho-campylotropous* or *ana-campylotropous*. Like orthotropous ovules, *ortho-campylotropous* ovules possess a vasculature which is straight, leading from the base of the funiculus to the middle of the nucellus, e.g., *Capsella bursa-pastoris* (Brassicaceae). However, the nucellar body is bent along the lower side. In *ana-campylotropous* ovules, the conducting vascular strand traverses from the base of the funiculus and curves towards the chalazal region (Fig. 5.3 D). Such ovules are present in Chenopodiaceae and Capparaceae.

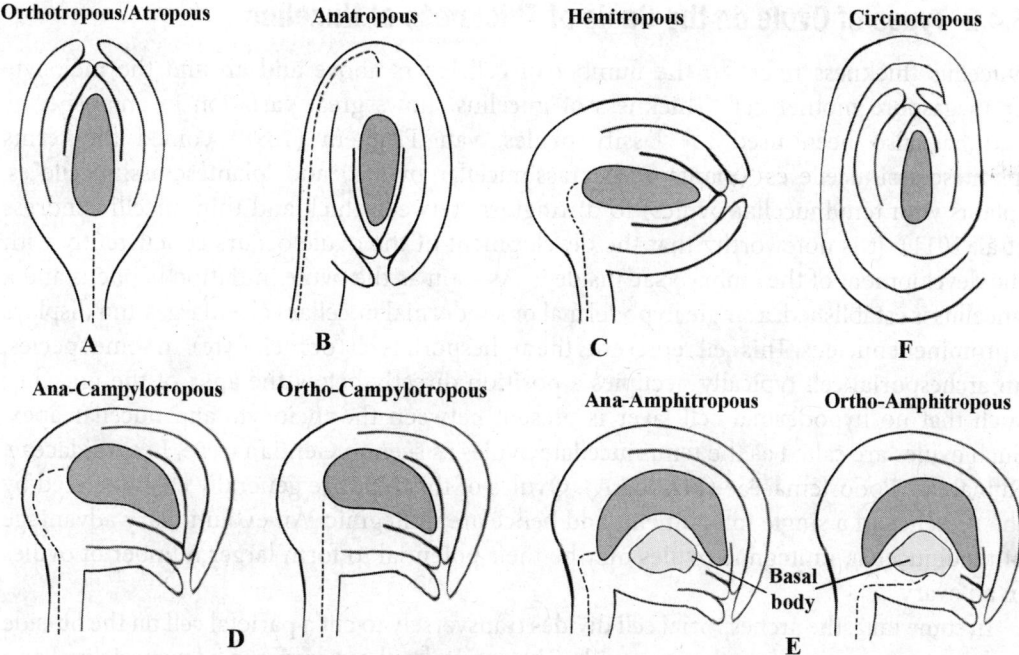

Figure 5.3 Types of ovules based on degree of curvature (and respective position of micropyle and chalaza). Broken line shows the vascular supply. (*Source*: Adapted and modified from Simpson 2010; Maheshwari 1950.)

- **Amphitropous (amphi: on both sides, tropous: turn):** The ovule is characterised by the curvature extending to the nucellus, thereby making the nucellus as well as embryo sac horse-shoe shaped as seen in Crossosomataceae, Papaveraceae and Alistamaceae. Amphitropous ovules, can also be of *ortho-* and *ana-* types. Ana-amphitropous and ortho-amphitropous ovules are similar to the ana-campylotropous and ortho-campylotropous type respectively except that the nucellus is bent along both the lower and upper sides and have differentiated cells at the angle of the bend called a 'basal body' (Simpson 2010; Endress 2011). (Fig. 5.3 E).
- **Circinotropous (circino: circular, tropous: turn):** This type of ovule is also characterized by the presence of nucellus and funiculus which are in the same line initially. But, due to rapid growth of the funiculus on one side, the ovule becomes anatropous. The curvature continues further and the micropyle again points upwards as in orthotropous ovules (Fig. 5.3 F). Thus, the funiculus grows and curves 360° to assume the shape of a circle. Such ovules are characteristic of Cactaceae.

5.4.2 Types of Ovule on the Basis of Thickness of Nucellus

Nucellus thickness refers to the number of cell layers above and around the meiocyte or megaspore mother cell. Thickness of nucellus shows great variation in angiosperms and has also been used to classify ovules. van Tieghem (1898) coined the terms 'plantescrassinucelle´es' (plants with crassinucellar ovules) and 'plantescrassinucelle´es' (plants with tenuinucellar ovules) to distinguish between thick and thin nucelli (Endress et al. 2011). It is noteworthy that the development of the ovule occurs concurrently with the development of the embryo sac inside it. As soon as the ovule initiation happens and a nucellus is established; a single hypodermal or subdermal nucellar cell enlarges and displays a prominent nucleus. This cell represents the archesporial cell (or meiocyte). In some species, an archesporial cell typically occupies a position directly below the apex of the nucellus; such that no hypodermal cell layer is present between the meiocyte and nucellar apex. Such ovules are called as the tenuinucellate ovules as seen in Gentianaceae, Lecythidaceae, Rubiaceae, Podostemaceae (Fig. 5.4 A). Ovules of this type are generally characterized by the presence of a single integument, and hence are unitegmic. An evolutionary advantage of tenuinucellar, unitegmic ovules may be their potential to form larger number of ovules in an ovary.

In some taxa, the archesporial cell divides transversely to cut a parietal cell on the outside and a sporogenous cell on the inner side. The parietal cell can undergo a few periclinal and anticlinal divisions to give a cushion to the sporogenous cell, such that it lies embedded in the nucellus. Ovules of this type which have one or more hypodermal cell layers between the meiocyte and nucellar apex are called crassinucellate ovules (Fig. 5.4 B); examples being the families Euphorbiaceae, Trigoniaceae, Zingiberaceae and Elaeagnaceae. Usually, bitegmic ovules are crassinucellate in nature.

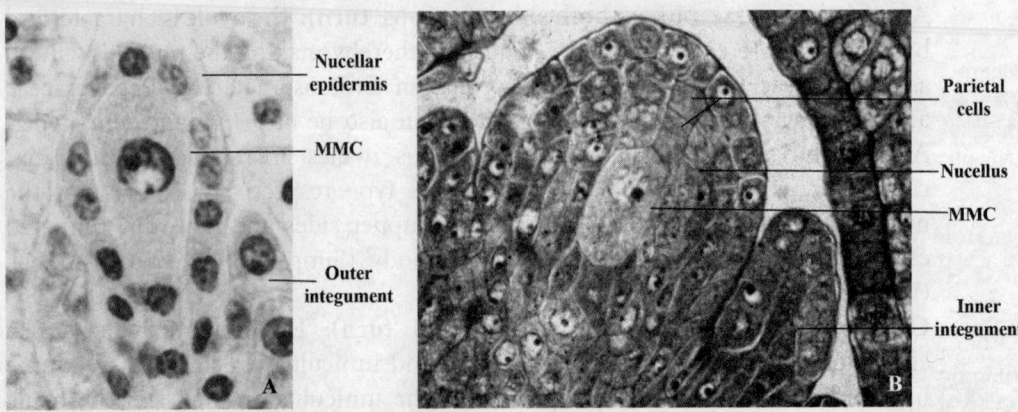

Figure 5.4 A. Tenuinucellate ovule in *Zeylanidium lichenoides*, note megaspore mother cell (MMC) with a single layer of nucellar epidermis. **B.** Crassinucellate ovule in *Hippophae rhamnoides*, note the layers of parietal cells between nucellar epidermis and MMC.

Based on the diversity seen among the flowering plants, the classical distinction between crassinucellate and tenuinucellate ovules has been further refined into six categories by Endress (2011) (Fig. 5.5):

a. Crassinucellate: With more than one hypodermal cell layer between the meiocyte and nucellar apex (Fig. 5.5 A)
b. Weakly crassinucellate: With just one hypodermal cell layer between the meiocyte and nucellar apex (e.g., *Dichelostemma*, Asparagaceae) (Fig. 5.5 B)
c. Pseudocrassinucellate: The meiocyte becomes sub-hypodermal due to periclinal cell divisions in the epidermis of the nucellar apex (e.g., *Sagittaria*, Alismataceae). The term *'pseudo-crassinucellate'* was coined by Davis (1966) in *Nigella damascena*, where the sporogenous cell becomes sub-hypodermal either because of divisions in the parietal cell or also sometimes due to periclinal division in the nucellar epidermis (Fig. 5.5 C).
d. Tenuinucellate: Without any hypodermal tissue in the nucellus (Fig. 5.5 E)
e. Incompletely tenuinucellate: Without a hypodermal cell layer between the meiocyte and nucellar apex, but with some hypodermal tissue at the flanks of nucellus and/or below the meiocyte (e.g., *Nemophila*, Boraginaceae) (Fig. 5.5 D)
f. Reduced tenuinucellate: Tenuinucellate but with meiocyte partly extending below the nucellus, thus with a partial inferior position (e.g., *Phyllis*, Rubiaceae) (Fig. 5.5 F)

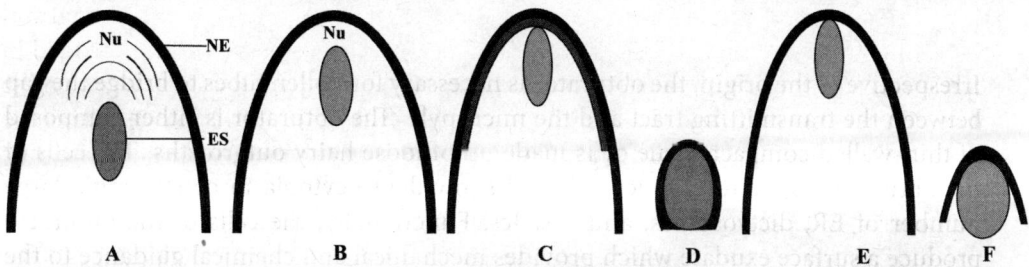

Figure 5.5 Types of ovules on the basis of thickness of nucellus. Embryo sac (ES) is shown in grey oval. **A.** Crassinucellate. **B.** Weakly crassinucellate. **C.** Pseudocrassinucellate. (Divisions in nucellar epidermis) **D.** Tenuinucellate. **E.** Incompletely tenuinucellate. **F.** Reduced tenuinucellate. (**NE:** nucellar epidermis, **Nu:** nucellus, modified after Endress 2011.)

5.5 Structures Associated with Ovules

Other than the primary morphological features of an ovule, a few special auxiliary structures may be present in ceratin taxa of angiosperms. Some of them are discussed below:

• **Obturator:** It is an outgrowth of any ovular tissue near the micropyle presumed to guide the pollen tube to the micropyle (Fig. 5.6 A). Obturator is widespread among flowering plants, reported in nearly 70 families of monocots and dicots (Losada & Herrero 2017). Developmental origin of obturator is variable among angiosperms. Mostly, it develops from the carpel flanks above the placenta (e.g., in Cuscutaceae, Euphorbiaceae), sometimes it may also develop from funiculus of ovule (*Aegle marmelos, Crinum*), or portion of style (Thymeleaceae, Rosaceae) near the micropyle.

Box 5.2

Differences between Crassinucellate and Tenuinucellate Ovules

Crassinucellate ovule	Tenuinucellate ovule
• The archaesporial cell divides transversely cutting a parietal cell on the outside and a sporogenous cell on the inner side.	• The archaesporial cell directly functions as the megaspore mother cell.
• Due to further divisions in the parietal cell, the sporogenous cell is embedded in the nucellus and is sub hypodermal.	• The sporogenous cell is hypodermal.
• The nucellus is more than one cell layer thick around the meiocyte.	• A single-layered nucellar tissue covers the meiocyte.
• Usually, seen in bitegmic ovules.	• Usually, seen in unitegmic ovules.
• Crassinucellate ovules are considered the primitive type of ovules.	• Tenuinucellate ovules are considered an advanced condition in angiosperms.
• Ranunculaceae, Haloragaceae, Euphorbiaceae, Trigoniaceae, Gingiberaceae are the common examples.	• Gentianales, Lamiales, Burmanniaceae, Hydatellaceae, Aracaceae, Solanaceae are the common examples.

Irrespective of the origin, the obturator is necessary for pollen tubes to bridge the gap between the transmitting tract and the micropyle. The obturator is either composed of thin-walled compact tissue or is made up of loose hairy outgrowths. The cells of the obturator at the ultrastructural level show dense cytoplasm containing a large number of ER, dictyosomes, and vesicles. Functionally, the cells of the obturator produce a surface exudate which provides mechanical and chemical guidance to the growing pollen tube. However, the timing and nature of this secretion is variable. For example, in *Actinidia deliciosa* copious exudates are present on the obturator right from the time of flower opening (González et al. 1996). In *Prunus persica*, obturator enters a secretory phase only after the arrival of pollen tubes (Arbeloa & Herrero 1987). The co-occurrence of pollen tube arrival and secretion from the obturator has also been reported in *Rhododendron* (Palser et al. 1989), and *Phellodendron* (Zhou et al. 2004). The obturator shrinks and disappears once the fertilization is accomplished.

- **Hypostase:** The hypostase was first described by van Tieghem (1893). It is a well-defined group of thick-walled cells situated between the bases of the two integuments and top end of the vasculature entering from the raphe. The cells are connected more or less directly with the inner integument, but it is not an integumentary structure. The cells of hypostase are thick-walled due to lignification, are heavily cutinized and are poor in cytoplasmic content. Hypostase typically but not always extends into the nucellus almost to the base of the megagametophyte. It has been suggested that hypostase prevents the downward growth of the embryo sac into the nucellus; secrets some hormones and enzymes for the growth of embryo sac; help supply water and nutrients to embryo sac from the vascular strand; and stabilize water balance of resting seeds over long periods of dormancy (Tiwari 1983). It is also thought to produce certain enzymes or hormones or even play a protective role in mature seeds. Hypostase in ovules can be seen in families like Onagraceae, Bignoniaceae (Fig. 5.6 B), Elaeagnaceae, and Liliaceae.

- **Epistase:** A secondary group of cells similar in characteristics to the hypostase is often found between the integuments at the apex of the nucellus in maturing seeds of some species. These nucellar cells are known as the epistase. The inner tangential walls of cells of epistase are lignified and thickened. In *Clarkia elegans* and *Costus malortieanus,* there is a well-developed epistase consisting of a single layer of vertically elongated cells. A very conspicuous epistase is also seen in *Arachis hypogaea,* where cells has very large starch grains and thick mucopolysaccharides rich cell walls. It differentiates during the later stages of ovule development from the nucellar epidermis. In *Costus,* the epistase forms a cap-like structure of cutinized cells during advance stages of embryo development. The epistase in *Cabomba caroliniana* (Cabombaceae), *Nymphaea gardneriana* and *Victoria cruziana* (Nymphaeaceae) differentiates from nucellar epidermis. In these species, the cells of epistase possess transfer cell like activity, a metabolically active cytoplasm and shows thickening on their inner tangential walls. This results in secretory nature of epistase to provide chemotropic guidance to the pollen tube for its growth (Zini et al. 2016).

- **Endothelium:** In nearly 65% of the angiospermous species, most of the nucellus degenerates before the embryo sac reaches maturity (Takhtadzhian 1991). As a consequence, the embryo sac comes in direct contact with the inner integument. In such cases, the innermost cell layer of the inner integument may differentiate into a unique cell layer termed the endothelium (Fig. 5.7 A and B). It is separated from the embryo sac by two layers of cuticle. The thickness and continuity of cuticle is variable not only among taxa but also at different places in the same species (Kapil & Tiwari 1978). Endothelium have been identified among the members of 65 families, mostly dicotyledons. Endothelial cells exhibit great diversity in their morphology, cytology, differentiation, and behavior. Sometimes endothelial cells develop wall projections which are thought to be involved in secretion and even contain multivesicular bodies. Large quantities of proteins, carbohydrates, ascorbic acid and some enzymes such as amylases, proteases are also known to occur in endothelial cells (Kapil & Tiwari 1978).

The cells of endothelium exhibit presence of prominent nuclei, radial cell expansion and sometimes endopolyploidy. These features are in common to the cells of tapetum present in an anther. Hence, the endothelium is also called the **integumentary tapetum**. The cytological features shared between the endothelium and the tapetum suggest similarity of function of both tissues (Maheshwari 1950). Species in which the nucellus does not degenerate, the embryo sac may receive nutrients directly from the nucellus. However, in ovules where the nucellus dissolves or degenerates around the embryo sac, the embryo sac becomes contiguous with the inside of the inner integument. Thus, the endothelium differentiating from the inner layer of the integument is believed to be involved in supply of nutrition to the growing embryo sac. For instance, in *Penstemon gentianoides* (Scrophulariaceae) the embryo sac remains surrounded by the endothelium except in the micropylar region. The endothelium along with micropylar integumentary cells was observed to play a role in transport of metabolites into the embryo sac (Dane et al. 2007). The endothelium might also play the role as a barrier and a protective tissue between the integument and the embryo sac as seen in *Helianthus annus* (Newcomb 1973).

In several hybrid plants of Solanaceae, endothelium proliferates until seeds are mature (Chaban et al. 2020). In normally developing ovules, the endothelium separates form the integument after fertilization and develops as an independent tissue. During seed development it displays secretory activity causing lysis of dying cells of the integument to provide space for the growth of the embryo. However, if fertilization fails to occur, the endothelium may separate and proliferate leading to the death of the embryo sac and formation of a pseudo-embryo occurs which substitutes for the embryo. Pseudo-embryos are often detected in the ovaries of parthenocarpic tomato fruits.

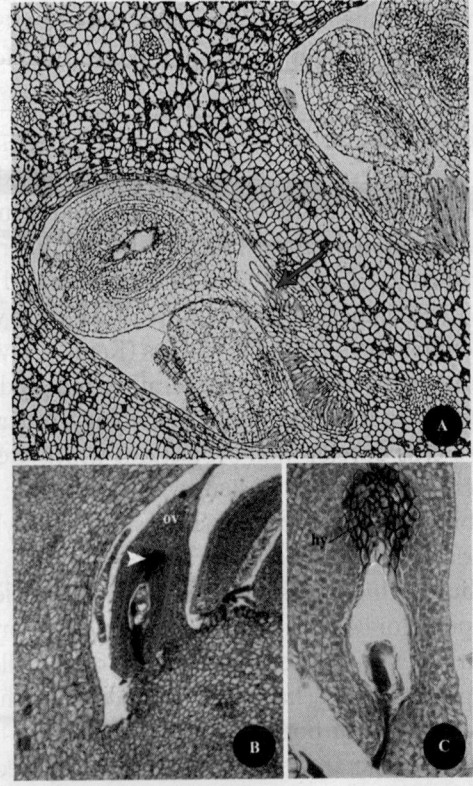

Figure 5.6 A. Obturator (arrows) originating from funiculus in *Aegle marmelos*. **B.** Hypostase (arrowhead) seen at the base of embryo sac (red box) in *ovule of Oroxylum indicum* (**ov**). **C.** Enlarged view of embryo sac in *Oroxylum indicum* where compactly arranged cells of hypostase (**hy**) may be seen.

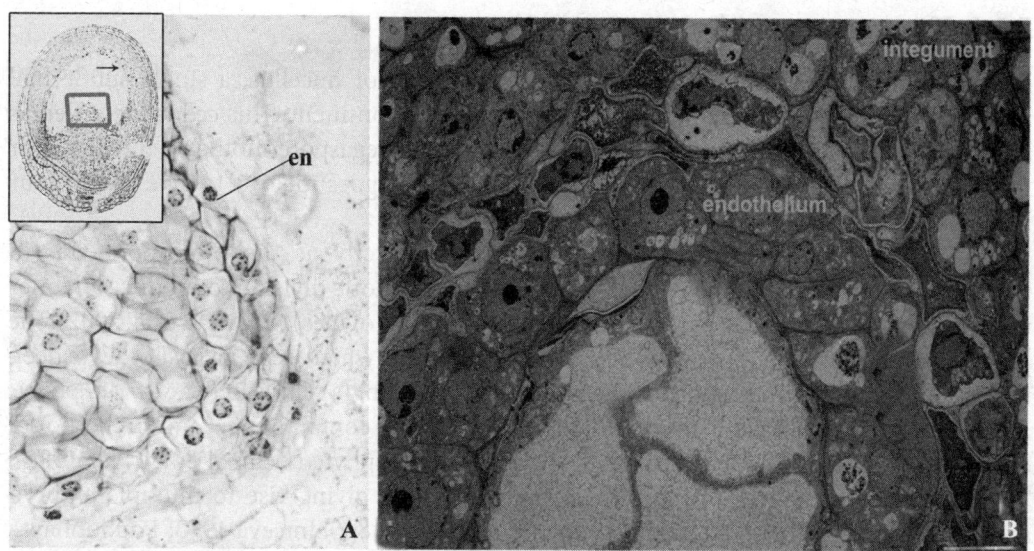

Figure 5.7 A. Endothelium (**en**) around the embryo sac (**es**) in *Brassica* sp. Inset: Enlarged view of endothelium to show cells with prominent nuclei (arrow). **B.** Transmission electron micrograph of endothelium layer of *S. lycopersicum*. Note the prominent nuclei in the endothelial cells.
(*Source*: B. Chaban et al. 2020, published under CC BY License.)

Box 5.3

Differences between Endothelium and Endothecium

Endothecium	Endothelium
• Endothecium is found in the anther.	• Endothelium is found in the ovule.
• It is one of the wall layers of the anther, inner to the epidermis.	• It is a specialized layer of cells derived from the innermost layer of inner integument.
• Cells show radial thickenings, which contain high proportion of α-cellulose.	• Cell walls show cuticlar thickenings.
• Endothecium helps in dehiscence of anthers.	• Endothelium is involved in the nutrition of the female gametophyte.

5.6 Development of Female Gametophyte

The process of development of the female gametophyte (or megagametophyte or embryo sac) takes place inside an ovule. The whole development process can be divided into two chief stages: megasporogenesis and megagametogenesis. The megasporogenesis is the process of formation of a reduced megaspore from either megasporocyte or the megaspore mother cell present in the nucellus, by meiosis. This is followed by mitotic divisions, nuclear migration, and cytokinesis to produce mature embryo sac with female gamete (egg cell); and the process is called megagametogenesis.

5.6.1 Megasporogenesis

In the developing ovule; shortly after the demarcation of nucellus, a single subdermal nucellar cell increases in size and its nucleus becomes prominent. This cell is identified as an 'archesporial cell' which ultimately gives rise to the megaspore mother cell (MMC) or megasporocyte. The resultant megasporocyte then undergoes meiosis to give rise to four haploid megaspores.

The archesporial cell may sometimes function directly as the megasporocyte (in tenuinucellate ovules) or it may undergo one anticlinal mitotic division to produce an inner megasporocyte and an outer parietal cell (in crassinucellate ovules) (Bouman et al. 2002). The parietal cell may undergoes a series of transverse divisions which push megasporocyte deep into the nucellus. The archesporium may not always be a single cell, as in *Brassica campestris* and soybean, which have a multicellular archesporium containing several archesporial cells, out of which only one gives rise to the megagametophyte (Kennell & Horner 1985). There are a few instances of more than one megaspore giving rise to multiple embryo sacs in an ovule, e.g., *Phellodendron amurense* (Starshova & Solntseva 1973; Poddubnaya-Arnoldi 1976), early-divergent angiosperms such as Schisandraceae (Friedman et al. 2003; Williams & Friedman 2004) and Hydatellaceae (Rudall et al. 2008). Eventually, only one of the embryo sac survives. Interestingly, *mem* mutants in *Arabidopsis* are also known to have similar double gametophytes wherein mutations affect the archespore selection and megaspore mother cell specification, leading to the initiation of two gametophytes in the same ovule (Schmidt et al. 2011). While the development of more than one megasporocyte from a multicellular archesporium is a common phenomenon in angiosperms (Maheshwari 1950), development of multiple gametophytes from a single megasporocyte has not been reported (Grossniklaus & Schneitz 1998).

5.6.2 Selection of Functional Megaspore and Types of Embryo Sac Development

During megasporogenesis, megasporocyte (or MMC) undergoes meiosis to produce a reduced megaspore. In most taxa, meiosis is followed by cytokinesis, resulting in four reduced megaspore cells. However, in some angiosperms cytokinesis and cell wall formation show variation due to which the number of megapores that eventually contribute to the formation of the mature female gametophyte varies. On the basis of number of megaspores contributing to the development of embryo sac, three types of embryo sac development are known:

- *Monosporic type*: This is the most common type of embryo sac found in angiosperms (Maheshwari 1950; Yadegari & Drews 2004) wherein both meiosis I & II are followed by cytokinesis and cell wall formation, resulting in four megaspores. The spindles of the first meiotic division in a MMC are oriented parallel to the micropylar-chalazal axis of the nucellus. The subsequent wall formation takes place perpendicular to this axis, resulting in a dyad of megaspores. Both the cells of the dyad undergo a second meiotic division and formation of transverse walls, resulting in a linear arrangement of four megaspores. Eventually, only one megaspore out of the four survives (*either at chalazal or micropylar end*) and the surviving megaspore is called the functional megaspore (FM). The other three non-functional megaspores degenerate. The type of embryo sac which develops from a single, uninucleate megaspore is called *monosporic type* of embryo sac implying development of the embryo sac from a uninucleate haploid megaspore (Fig. 5.8). The FM grows in size and the non-functional megaspores are eventually crushed by the expanding FM. Commonly, the megaspore which is closest to the chalaza enlarges and is functional, e.g., *Polygonum*. However, in rare instances as in *Oenothera*, the megaspore closest to the micropyle becomes functional. The megaspores most commonly show linear arrangement after completion of meiosis; however, other arrangements are also seen such as tetrahedral in *Arabidopsis* (Webb & Gunning 1990), and T-shaped tetrads in maize and *Hippophae rhamnoides* (Russell 1978; Mangla et al. 2015). In *Aristolochia*, five different types of megaspore arrangements are seen; these being linear, tetrad, T- shaped, decussate and isobilateral.

- **Bisporic type**: In species with the bisporic type of embryo-sac development, cytokinesis after the first meiotic division in a MMC results in the formation of a dyad, i.e., two cells (Fig. 5.8). However, one of the cells of the dyad degenerates and further development of embryo sac occurs from the other cell of the dyad alone. The second meiotic division in the functional cell is not followed by cytokinesis. Both the haploid megaspore nuclei which are vertically arranged participate in subsequent female gametophyte development. Thus, such development is designated as *bisporic type* of embryo sac development as the functional megaspore is binucleate. In most cases, the upper or the micropylar cell of the dyad degenerates and embryo sac develops from the lower dyad cell, e.g., *Allium*, *Trillium*. A variation of this occurs in *Endymion hispanicus* where the embryo sac develops from the micropylar cell of dyad.

The monosporic or bisporic megasporogenesis of angiosperms is marked by a well-defined pattern of callose deposition (comprised of β-1,3-glucan units). A boundary of callose is first seen around megaspore mother cell at meiotic prophase I and later, callose also accumulates in the transverse walls and around dyads and tetrads. This callose subsequently disappears at the stage of functional megaspore or at the beginning of the differentiation of the embryo sac (Rodkiewicz 1970).

- **Tetrasporic type**: Another pattern of megasporogenesis is where the megasporocyte undergoes meiosis without cytokinesis and all the four megaspore nuclei lie in a common cytoplasm. The type of embryo sacs which develop from a tetranucleate megaspore (also known as **coenomegaspore**) is designated as *tetrasporic type* (Fig. 5.8).

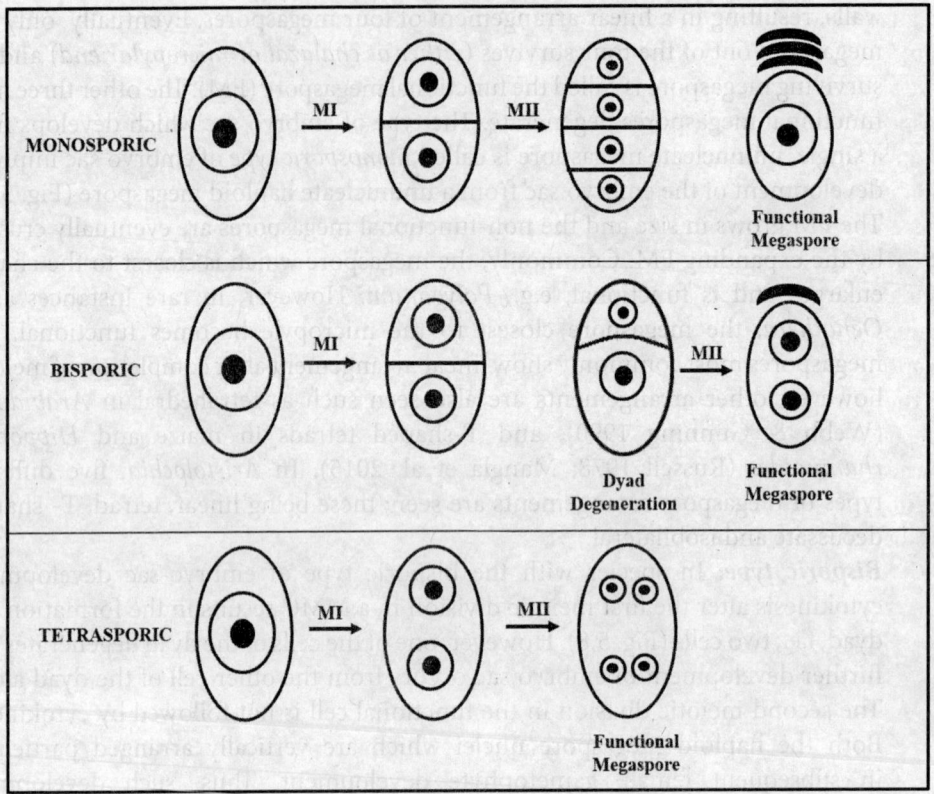

Figure 5.8 Different patterns of megasporogenesis as seen in angiosperms. Micropylar end is towards top of page.

The different cells of a mature embryo sac have different levels of genetic heterogeneity, depending on whether one, two, or all four megaspore nuclei contribute to their formation. In the monosporic type of development all the cells of a mature embryo sac are genetically identical since they all are the products of one megaspore nucleus. On the other hand, cells in mature embryo sac following bisporic and tetrasporic type of development are genetically dissimilar as either two or four megasporial nuclei contribute to the embryo sac development respectively. As all the four megaspore nuclei participate in the formation of an embryo sac, a tetrasporic embryo sac is genetically more heterogeneous than a bisporic embryo sac. To emphasize on the nuclear nature of embryo sac formation, Battaglia (1983) revised the nomenclature of embryo sac from monosporic to monokaryosporic, bisporic to dikaryosporic and tetrasporic to tetrakaryosporic which has been adopted in some embryological texts.

The selection of functional megaspore is under the control of hormone signaling which occurs in the surrounding diploid sporophytic tissue. Evidence suggests that cytokinin and auxin signaling preferentially produced in the chalaza is needed for the formation of a functional megaspore (Bencivenga et al. 2012; Cheng et al. 2013). Other than cytokinin and auxin signaling, Glycosylphosphatidylinositol (GPI)-anchored arabinogalactan protein 18 (AGP18) is also required for selection and survival of the functional megaspore (Demesa-Arevalo & Vielle-Calzada 2013). Reduced activity of AGP18 in the functional megaspore results in female gametophyte development; while overexpression causes the development of extra functional megaspores. Once the functional megaspore is selected, cellular organization starts to demarcate it. Most of the organelles preferentially accumulate in the chalazal most cell after the first meiotic division (Schulz & Jensen 1971; Russell 1978). Also, plasmodesmata are evident only between the functional megaspore and the nucellus (Willemse & Bednara 1979; Wilms 1980; Cass et al. 1985). Initially the callose appears in the cell walls of megasporocyte and following meiosis it is seen in the walls of the megaspores. However, after meiosis, callose wall starts thinning and eventually disappear from the walls of functional megaspore (Rodkiewicz 1970). In contrast, callose is retained in the walls of non-functional megaspores, which probably blocks the nutrients supply from the nucellus (hence degenerate). Thus, only the functional megaspore is able to receive nutrients. Consequently, the functional megaspore not only inherits a richer cytoplasm but is also better equipped to receive nutrients from the maternal tissues.

Box 5.4

Differences between Monosporic, Bisporic and Tetrasporic Embryo Sac Development

Monosporic	Bisporic	Tetrasporic
• Both meiotic divisions are accompanied by cell wall (plate) formation.	• Only one meiotic division is accompanied by cell wall (plate) formation.	• None of the meiotic division is accompanied by cell wall (plate) formation.
• Meiosis of the diploid megaspore mother cell produces four haploid megaspores.	• Meiosis of the diploid megaspore mother cell produces two binucleate megaspores.	• Meiosis of the diploid megaspore mother cell produces one tetranucleate megaspore.
• Embryo sac develops from a uninucleate megaspore.	• Embryo sac develops from a binucleate megaspore.	• Embryo sac develops from a tetranucleate coenomegaspore.
• Cells of a monosporic embryo sac are genetically identical as they arise from one meiotic product.	• Cells of a bisporic embryo sac are not genetically identical as they arise from two different meiotic products.	• Cells of the tetrasporic embryo sacs are most genetically diverse as they arise from four different nuclei.
• Example: *Polygonum, Oenothera*	• Example: *Allium, Endymion*	• Example: *Plumbago, Plumbagella*

Box 5.5

Embryo Sac Development in Podostemaceae

Podostemaceae is a family of aquatic angiosperms with many unique embryological characters. For long, embryo sac development in the family was considered exclusively of bisporic type. After the first meiotic division in the megaspore mother cell; the upper cell of the dyad (towards micropylar end) degenerates, and the female gametophyte develops from the lower cell (towards chalazal end). Since, embryo sac develops from the lower cell of the dyad it was considered to be bisporic in origin (Chiarugi 1933; Maheshwari 1947). However, after a thorough reinvestigation of embryo sac development it was found that the embryo sac development in Podostemaceae can be either monosporic or bisporic. In all the species, the upper cell of the dyad degenerates and the female gametophyte develops from the lower cell. After the second meiotic division in the lower cell of the dyad two vertically arranged megasporial nuclei are formed. However, in species like *Vanroyenella plumosa, Hydrobryum griffithii, Zeylanidium olivaceum, Indotristicha ramosissima* embryo sacs develop from the upper nucleus alone and the lower nucleus degenerates (Chaudhary et al. 2014). Since the embryo sac arises from only one megasporial nucleus of the lower cell of the dyad development is considered monosporic and not bisporic. Whereas in other species like *Polypleurum munnarense, P. dichotomum, Hydrobryopsis sessilis, Willisia selaginoides* and *Zeylanidium johnsonii* both the nuclei divide and give rise to embryo sac. Thus, the embryo sac is of bisporic type in these species. This underlines the concept that the type of embryo sac development is designated on the basis of the number of megaspore nucleus/nuclei that participate in the formation of an embryo sac.

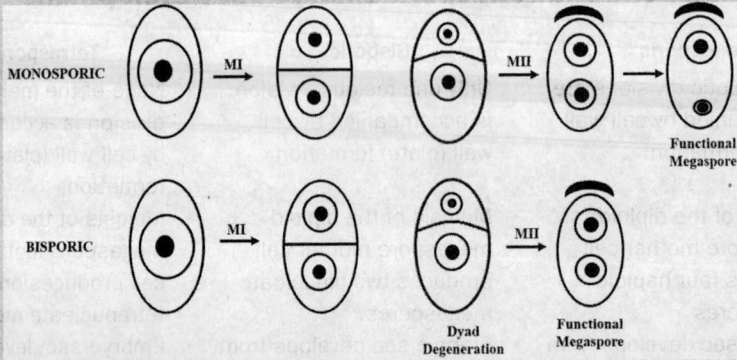

Figure 5.9 Embryo sac development in Podostemaceae can be either monosporic or bisporic depending on number of nucleus/nuclei in the functional megaspore. **MI**: Meiosis I, **MII**: Meiosis II. Micropylar end is towards top of the page.

5.6.3 Megagametogenesis

Megagametogenesis is the process of development of embryo sac from the functional megaspore. It includes mitotic divisions followed by the positioning of the resultant nuclei, cellularization and eventually organization of mature embryo sac. Just as variations are seen in the formation of functional megaspore (megasporogenesis); embryo sac development (megagametogenesis) in angiosperms also varies. The characteristics that vary include number of mitotic divisions, placement of resulting nuclei, and patterns of cellularization and organization of mature embryo sac.

A typical embryo sac of angiosperms is an eight nucleate, seven celled structure; formed after three rounds of mitotic divisions in the megaspore nucleus (Maheshwari 1950; Endress 2011). The resulting organized cells are: an egg cell associated with two synergids (*forming the egg apparatus*) at the micropylar end, a large central cell with two nuclei, and three antipodals cells present at the chalazal end (Fig. 5.10). One of the two sperm cells carried by the pollen tube fertilizes the egg cell resulting in the formation of a zygote, and the other fuses with the nuclei of the central cell (double fertilization), giving rise to the endosperm.

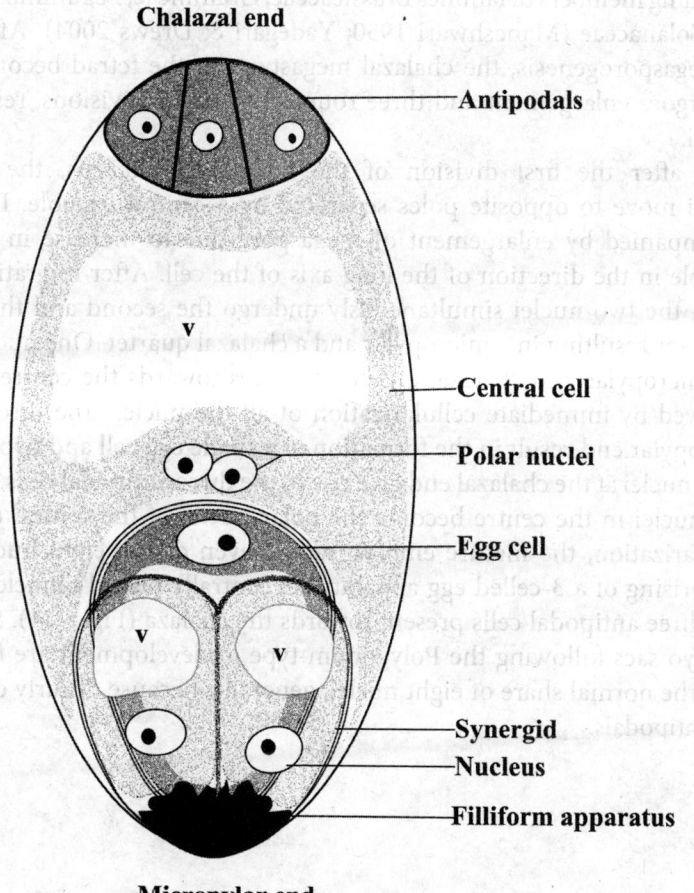

Chalazal end

Antipodals

Central cell

Polar nuclei

Egg cell

Synergid

Nucleus

Filliform apparatus

Micropylar end

Figure 5.10 Diagrammatic sketch of 7-celled, 8-nucleate Polygonum-type of mature embryo sac. Such type of embryo sac is most common among angiosperms. Note the egg cell and synergids share common walls. v: vacuole. (*Synergids are drawn as semi-transparent cell to comprehend their association with egg cell.*) **See Color Plates (page 479).**

5.6.4 Different Types of Embryo Sac

As discussed earlier, the type and organization of gametophyte development is dependent on the number of megaspore nuclei involved and is broadly classified into three main types: monosporic, bisporic and tetrasporic. These types are further categorized into various sub-types and named after the genus in which they were first described (Fig. 5.12).

5.6.4.1 *Monosporic Embryo Sacs*

There are two types of monosporic embryo sacs, viz. Polygonum and Oenothera type.

- *Polygonum type:* (Monosporic 8-nucleate): The most widely studied developmental pattern of female gametophye development is the Polygonum-type (first observed in *Polygonum divaricatum*). It is exhibited in more than 70% of flowering plants including members of families Brassicaceae, Gramineae, Leguminoseae, Malvaceae and Solanaceae (Maheshwari 1950; Yadegari & Drews 2004). After completion of megasporogenesis, the chalazal megaspore of the tetrad becomes functional, undergoes enlargement and three rounds of mitotic divisions, resulting in eight nuclei.

  Soon after the first division of the megaspore nucleus, the two daughter nuclei move to opposite poles separated by a central vacuole. Division is also accompanied by enlargement of megaspore due to increase in size of central vacuole in the direction of the long axis of the cell. After migration to opposite poles, the two nuclei simultaneously undergo the second and the third mitotic divisions resulting in a micropylar and a chalazal quartet. One nucleus each from the micropylar and chalazal quartet migrates towards the centre of cell. This is followed by immediate cellularization of all the nuclei. The three nuclei at the micropylar end result in the formation of a single egg cell and two synergids. The three nuclei at the chalazal end give rise to the three antipodal cells. The remaining two nuclei in the centre become the polar nuclei of the central cell. Thus, post cellularization, the mature embryo sac is seven celled, eight-nucleate structure comprising of a 3-celled egg apparatus, a centrally placed binucleate central cell and three antipodal cells present towards the chalaza (Fig. 5.11). Sometimes, the embryo sacs following the Polygonum-type of development are found with less than the normal share of eight nuclei, generally, because of early degeneration of the antipodals.

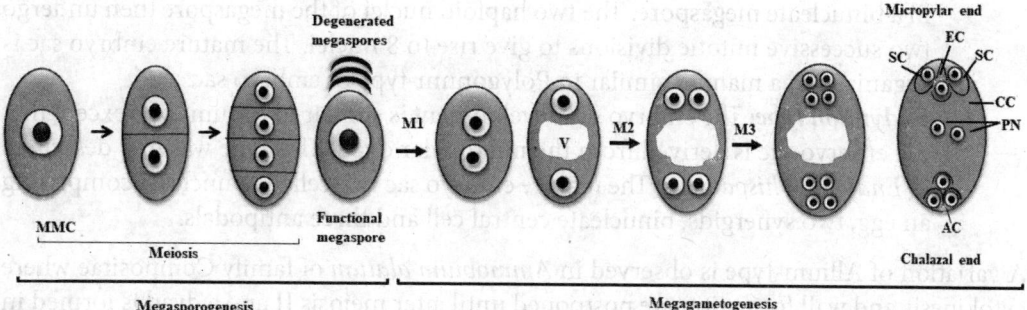

Figure 5.11 Schematic diagram of Polygonum-type of embryo sac development. **MMC:** Megaspore mother cell, **M:** Mitosis, **V:** Vacuole, **EC:** Egg cell, **SC:** Synergid cell, **CC:** Central cell, **PN:** Polar nuclei, **AC:** Antipodal cell.

The final organization of a mature embryo sac in *Amboralla trichopoda*, a basal angiosperm, needs a special mention here. Initially ontogeny of its embryo sac corresponds to that of Polygonum-type and after two round of mitosis, 8 nuclei get arranged in the egg apparatus, a central cell and three antipodals. Interestingly post cellularization, one of the synergid divides laterally and forms two daughter cells. This leads to presence of three synergids and one egg cell at the micropylar end (4-celled egg apparatus).Thus, the mature embryo sac is 8-celled, 9-nucleate structure (Friedman & Ryerson 2009).

- *Oenothera type* (Monosporic 4-nucleate): It is a 4-nucleate embryo sac instead of the 8-nucleate embryo sac seen in Polygonum-type. This type was first described in *Oenothera lamarckiana* where four nuclei are organized into an egg apparatus and a uninucleate central cell. This type of embryo sac is usually formed by the micropylar megaspore of the tetrad (in contrast to chalazal megaspore as in *Polygonum*). After the first mitotic division, the two resulting nuclei migrate to the micropylar pole and undergo a second division to give rise to four nuclei. Thus, all the nuclei remain situated in the micropylar part of the embryo sac. The resulting embryo sac lacks a second polar nucleus and antipodal cells. The Oenothera-type of embryo sac is a characteristic feature of family Onagraceae. This type of embryo sac development can also be seen in Nymphaeaceae and Schisandraceae where it is designated as either Nuphar- or Schisandra-type (Friedman & Ryerson 2009).

5.6.4.2 *Bisporic Embryo Sacs*

In bisporic embryo sacs two haploid megaspore nuclei take part in embryo sac formation. Bisporic embryo sacs are of the following two sub-types:

- *Allium type*: The megaspore mother cell undergoes meiosis I to form a dyad of cells. The upper dyad cell soon degenerates and the embryo sac is derived from the chalazal dyad cell alone. The lower cell of the dyad completes meiosis II resulting in a binucleate megaspore. The two haploid nuclei of the megaspore then undergo two successive mitotic divisions to give rise to 8 nuclei. The mature embryo sac is organized in a manner similar to Polygonum-type of embryo sac.

- *Endymion type*: The embryo sac development is similar to Allium-type except that the embryo sac is derived from the micropylar dyad. This type was first described in *Endymion hispanicus*. The mature embryo sac is 7-celled/8-nucleate comprising an egg, two synergids, binucleate central cell and three antipodals.

A variation of Allium-type is observed in *Ammobium alatum* of family Compositae where cytokinesis and wall formation are postponed until after meiosis II and a dyad is formed in which each cell is binucleate (Davis 1966). A special situation occurs in *Schisandra chinensis* where the female gametophyte is bisporic in origin but the final organization of the embryo sac is similar to the Oenothera-type.

5.6.4.3 *Tetrasporic Embryo Sacs*

Tetrasporic embryo sacs are derived from a tertranucleate megaspore also known as coenomegaspore. Before the start of mitosis, the four nuclei in a coenomegaspore organize themselves into any one of the three types of arrangements:

a. Two nuclei at the micropylar end and two at the chalazal end (2+2)
b. One nucleus at the micropylar end, one at the chalazal end, and one laterally on each side (1+1+1+1)
c. One nucleus at the micropylar end and three at the chalazal end (1+3)

After this repositioning, the nuclei may or may not undergo fusion before mitosis. Based on the type of nuclear configuration before mitosis; the presence or absence of nuclear fusion and the number of mitotic divisions in the functional megaspore, seven principal types of tetrasporic embryo sacs are recognized namely *Adoxa, Peperomia, Penaea, Drusa, Fritillaria, Plumbagella* and *Plumbago* (Maheshwari, 1950). These seven types of tetrasporic embryo sacs are described below:

NO FUSION OF NUCLEI

- *Adoxa type*: It was first described in *Adoxa moschatellina*, where post-meiosis, the nuclei rearrange in 2+2 organization. This is followed by a single mitotic division in the coenomegaspore to form an 8-nucleate embryo sac. Following cellularization the embryo sac comprises seven cells including an egg apparatus, three antipodals and a central cell with two polar nuclei. The other example of Adoxa-type embryo sac is *Sambucus*.
- *Peperomia type*: This type of embryo sac is 16 nucleate and it was first described in *Peperomia pellucida*. In this type, the four nuclei of the coenomegaspore rearrange in 1+1+1+1 configuration and undergo two successive mitotic divisions resulting in 16 nuclei. These nuclei initially remain dispersed in the sac and soon undergo

reorganization and cellularization. At maturity, an egg and a single synergid at the micropylar end, and 8 nuclei in the central cell (which later fuse to form an octoploid secondary nucleus) can be seen. The 6 peripheral cells finally degenerate.

- *Penaea type*: In Penaea-type, the nuclei in coenomegaspore take up a tetrapolar arrangement (1+1+1+1) and undergo two mitotic divisions to form a 16 nucleate embryo sac. Initially, the 16 nuclei are arranged in groups of four; one at the micropylar end, one at the chalazal end and one along each side wall. Later, during cellularization, a cell wall is formed only around three nuclei of each quartet while the fourth nucleus remains free and moves to the centre. Three cells in each quartet form the egg apparatus consisting of two synergids and an egg cell. Thus, four polar groups each with three cells and a centrally placed tetranucleate central cell are formed. This type of embryo sac development is characteristic of the family Penaeaceae, and is also seen in some members of the Malphigiaceae and Euphorbiaceae. A modification of this type is seen in some species of *Acalypha* where occasionally 5 or 6 polar nuclei are formed. These additional nuclei are contributed by one or both the lateral quartets which consequently consist of an egg and a synergid only (Mukherjee 1957).

- *Plumbago type*: This type of embryo sac development was first seen in *Plumbago capensis* where the 4 nuclei of the coenomegaspore are also arranged cross-wise. However, the nuclei in this type undergo only a single round of mitotic division resulting in 8 nuclei arranged in 4 pairs. One of the nuclei of the micropylar pair develops into a lenticular egg cell whereas the other nucleus increases in size and migrates to the centre to function as a polar nucleus. One nucleus from each of the remaining three pairs also undergoes a slight increase in size and moves towards the centre to function as polar nuclei. The remaining 3 nuclei either degenerate or occasionally either one, two or all three of them cellularize to form cells which persist and assume egg cell like appearance. Thus, a Plumbago-type of embryo possesses a single egg cell and a tetranucleate central cell. There are no synergids and antipodals.

- *Drusa Type*: Seen first in *Drusa oppositifolia*, this type is characterized by a 1+3 arrangement of four nuclei of the coenomegaspore (1 at micropylar end and 3 at chalazal end). Two mitotic divisions lead to 2+6 and 4+12 nuclear stages of female gametophyte. Out of the 4 micropylar nuclei, two synergids, an egg and upper polar nucleus are formed. Rest of the 12 chalazal nuclei give rise to a lower polar nucleus and 11 antipodals. However, eleven antipodal cells are rarely seen due to degeneration of nuclei at the chalazal pole. For example, *Ulmus* with Drusa-type embryo sacs usually possesses fewer than eleven antipodal cells (Walker 1950).

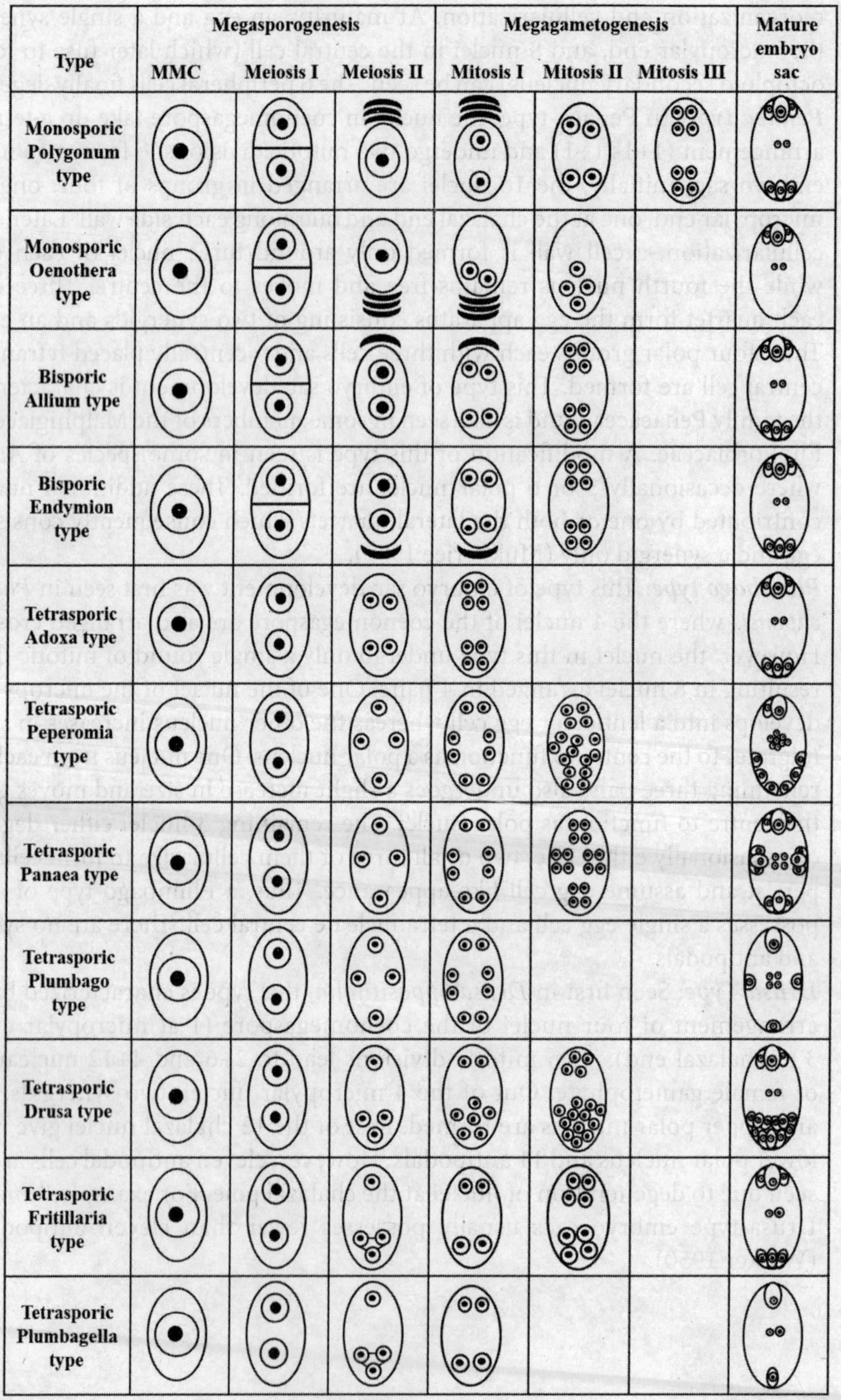

Figure 5.12 Major types of embryo sac development seen in angiosperms. Micropylar end is towards the top of page.

FUSION OF NUCLEI

- *Fritillaria type*: In some tetrasporic taxa, the three nuclei of the coenomegaspore fuse to form a triploid nucleus. In such taxa, the nuclei get organized in a 1+3 arrangement in the coenomegaspore and three nuclei at chalazal end fuse to form a triploid nucleus. The fusion of three chalazal nuclei of the coenomegaspore was first described by Bambacioni in 1928 in *Fritillaria persica* and is known as **Bambacioni's effect**. The fusion of the 3 chalazal megaspore nuclei may take place when they are either in the prophase stage or in early metaphase of mitosis (Agarwal 1950). Thus, after a first round of mitosis, 2 haploid nuclei are formed at the micropylar end and 2 triploid nuclei at the chalazal end. Second round of mitotic division results in the formation of 8 nuclei, of which the 4 micropylar nuclei are haploid whereas the 4 nuclei at the chalazal end are triploid. Thus, mature embryo sac consists of a haploid egg cell and synergids forming the egg apparatus, three triploid antipodals and a tetraploid secondary nucleus formed by the fusion of one haploid and one triploid polar nucleus. Fritillaria-type of embryo sacs are characteristic of many members of the Liliaceae and several species of *Piper*.

- *Plumbagella type*: In *Plumbagella micrantha*, the 4 megaspore nuclei take up a 1+3 arrangement and the three megaspore nuclei at the chalazal end fuse to form a triploid nucleus. However, unlike in the Fritillaria-type, there is only a single mitotic division which results in 2 haploid micropylar nuclei and the 2 triploid chalazal nuclei. One of the haploid nuclei at the micropylar end organizes into an egg cell and one of the triploid nuclei at the chalazal end organizes into a single antipodal cell. The remaining haploid and triploid nuclei fuse to form a tetraploid secondary nucleus. Thus, the mature embryo sac is three-celled consisting of an egg cell, a central cell with a haploid and triploid polar nucleus and a triploid antipodal cell. Notably, synergids are absent in this type of embryo sac.

5.6.4.4 *Embryo Sac Development in Chrysanthemum cinerariaefolium*

Embryo sac development in *Chrysanthemum cinerariaefolium* was described by Martinoli in 1939. Embryo sac development in *C. cinerariaefolium* is also tetrasporic in origin. However, it is more or less isolated from the other types of tetrasporic embryo sacs that have been described so far. Hence it has been named the Chrysanthemum cinerariaefolium-type of embryo sac development. After meiosis, the megaspore nuclei take up 1+2+1 arrangement such that there is 1 nucleus at each pole and 2 nuclei in the centre. The two central nuclei are separated from the polar nuclei by vacuoles and may either fuse to form a single diploid nucleus or retain their position without undergoing fusion. Based on the absence or presence of fusion between the two central nuclei; embryo sac configuration can be of either of the following types:

Type I: The two central nuclei fuse to form a diploid nucleus such that there are three nuclei in the female gametophyte; one haploid nucleus at each pole and one central diploid nucleus. Two mitotic divisions give rise to twelve nuclei, a haploid quartet at each end and one diploid quartet in the centre. The micropylar quartet gives rise to an egg apparatus and the upper polar nucleus and the chalazal quartet give rise to 4 antipodals, all of which are haploid. The four central diploid nuclei separate, one functions as the other polar nucleus and the remaining three organize themselves into antipodal cells. Occasionally, embryo sacs with less than 12 nuclei are seen because either the central diploid nucleus undergoes only one mitotic division or there is failure of divisions at the chalazal end (Fig. 5.13).

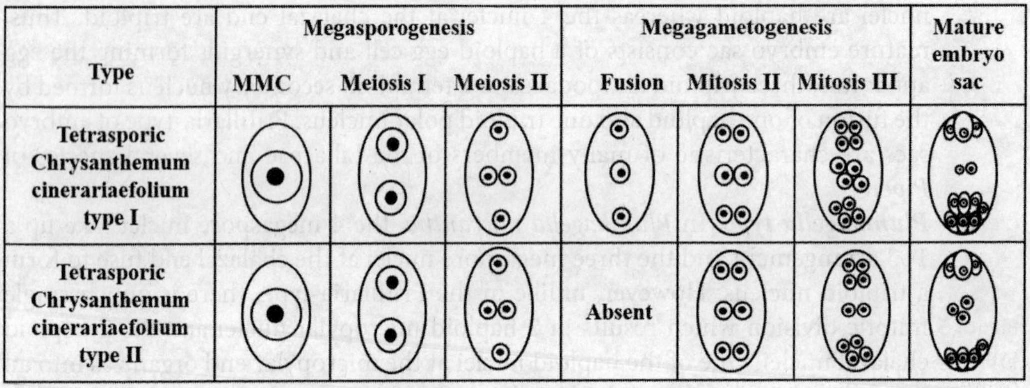

Type	Megasporogenesis			Megagametogenesis			Mature embryo sac
	MMC	Meiosis I	Meiosis II	Fusion	Mitosis II	Mitosis III	
Tetrasporic Chrysanthemum cinerariaefolium type I							
Tetrasporic Chrysanthemum cinerariaefolium type II				Absent			

Figure 5.13 Embryo sac development in *Chrysanthemum cinerariaefolium.* Micropylar end is towards the top of the page.

Type II: The two central megaspore nuclei do not fuse but only lie in contact with each other and do not undergo any further divisions. On the other hand, the micropylar and the chalazal megaspore nuclei divide twice forming 4 nuclei at each pole. This results in an embryo sac that contains ten nuclei. Out of the four nuclei of the micropylar quartet, three organize into egg apparatus and the fourth nucleus along with the two central nuclei functions as the polar nucleus. All the 4 nuclei from the chalazal quartet organize as antipodals. Sometimes the chalazal nucleus divides either once or fails to divide at all resulting in 7 or 8 nucleate female gametophyte, respectively (Fig. 5.13).

5.7 Cellular Anatomy and Ultrastructure of Cells of Embryo Sac

Embryo sacs comprise four groups of cells that function in fertilization, embryogenesis, and nutrition of embryo. Of these, the minimum complement of cells that are required to effect double fertilization is termed as the **Female Germ Unit** (FGU) (Dumas et al. 1984). The FGU of a typical angiosperm is composed of an egg cell, two synergids, and a central cell (Fig. 5.14). The components of the FGU are involved in facilitating double fertilization. They attract and receive the pollen tube, cause the sperm cells to be discharged, and transport and promote the fusion of one sperm cell with the egg cell and the other sperm cell with the central cell (Dumas & Mogensen 1993).The ultrastructure of each cell type of an embryo sac is described in detail in following sections.

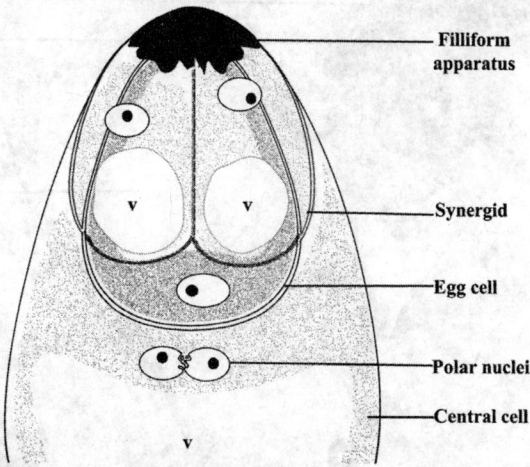

Figure 5.14 Female germ unit: assemblage of an egg cell, the two synergids and a central cell. The two synergids remains connected with each other by means of plasmodesmata present in a common wall between them. Also note that the egg cell and the synergids also share a common cell wall. v: vacuole (*Synergids are drawn as semi-transparent cells to understand their association with egg cell.*)

5.7.1 Egg Cell

It is located at the micropylar end of the embryo sac along with synergids, and together they comprise the egg apparatus. The mature, receptive egg cell is usually similar in shape to the synergids but is typically placed at lower level than synergids. The egg cell is separated from synergids, by either complete cell walls, partial cell walls or the plasmalemma alone (Fig. 5.14 and 5.15). In *Capsella*, the wall covers the entire length of the egg cell. In cotton, the cell wall extends only half way up the cell, leaving the remaining half of the cell without a cell wall. Here, the cell is surrounded by plasma membrane only. The micropylar cell wall of the egg contains cellulose microfibrils and acidic polysaccharides. The chalazal part of the cell wall is typically interrupted by a few areas of contacts between the plasma membrane of egg cell and synergids; and as well as, between egg cell and central cell (Russell 1993). Chalazal end of the egg cell wall lacks fibrillar components but has characteristic electron-dense bodies between the FGU cells (Sumner & Caeseele 1989; Russell 1993). The electron-dense bodies are present at intervals and are seen to be associated with microtubules. These bodies appear to maintain a constant distance between the cells, apparently stabilizing the egg-central cell boundary before fertilization (Cass & Karas 1974).

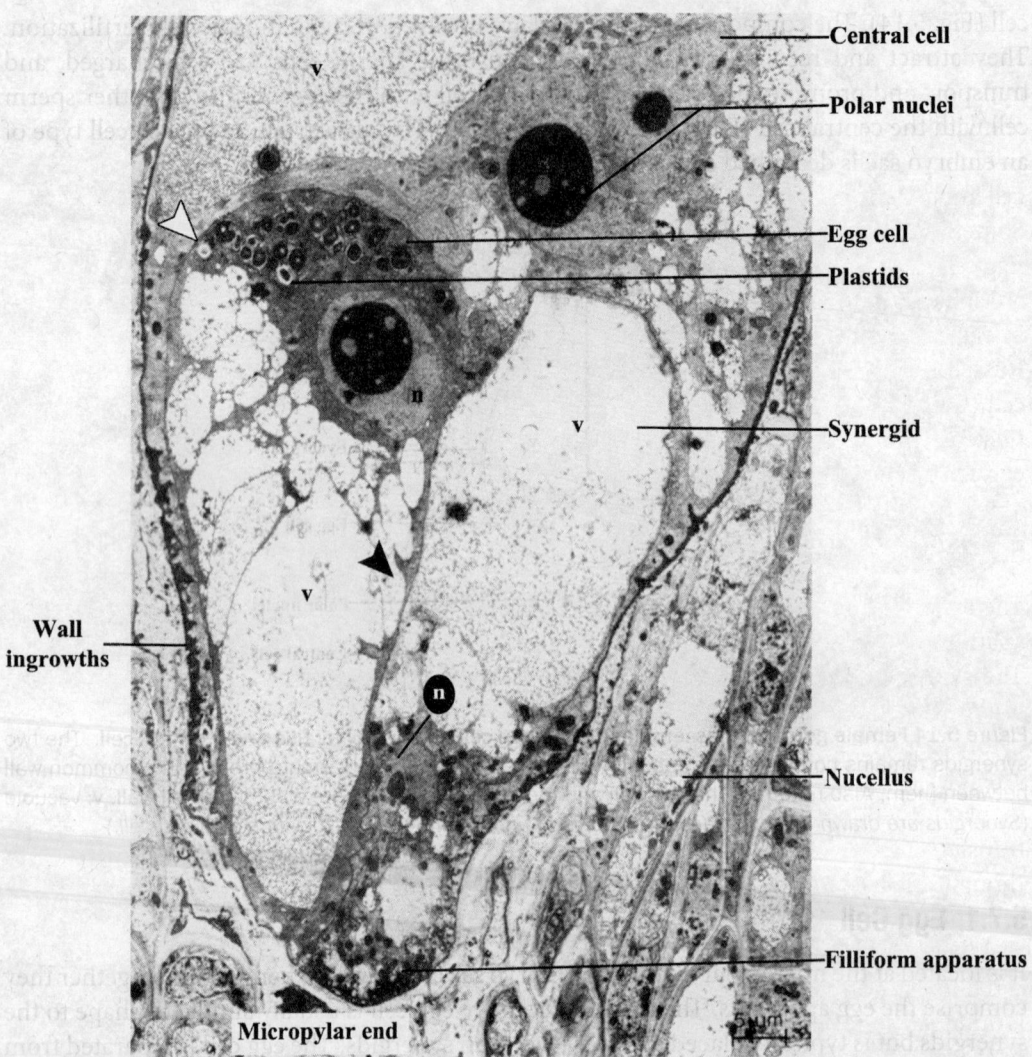

Figure 5.15 Transmission electron micrograph of egg apparatus and central cell. A synergid is visible in the section. Note the reverse polarity of egg cell and synergid. v: vacuole, n: nucleus. (*Source:* Wilms 1981, *reproduced with permission.*)

The distribution of cytoplasm within the egg cell is highly polarized as the nucleus and most of the cytoplasm is limited to the chalazal end whereas a large vacuole is present at the micropylar end (Fig.5.14 and 5.15; Jensen 1965; Schulz & Jensen 1968). Alternatively, the nucleus may be located centrally surrounded by a large number of smaller vacuoles. The egg cell in cotton, which is up to 100μ long, consists of a large vacuole in the centre and very little cytoplasm towards the periphery. On the other hand, in *Capsella*, the micropylar vacuole occupies half of the egg cell (Jensen 1973). Unlike its variable position, the egg nucleus appears regular in form and structure, with a large size and a large nucleolus indicating extensive amplification of rDNA and an exceptionally high potential for the production of ribosomes (Alberts et al. 1989).

The ultrastructure of the egg cytoplasm varies considerably among the angiosperms. Mostly, the cytoplasm of egg cell contains plastids, mitochondria, endoplasmic reticulum, and lipid bodies, but is poor in Golgi bodies. Variability exists in the distribution and abundance of these organelles. The number of plastids in the egg varies from as much as 730 in *Plumbago* (Russell 1987) to 8 -12 in *Daucus* species (Boblenz et al. 1990). Plastids in the egg cell are usually distributed around the nucleus but the size and shape varies among species. Sometimes plastids are filled with starch grains which may serve as a dynamic energy reserve. Egg cell cytoplasm is rich in mitochondria. These are typically perinuclear in distribution and are spherical to roundly ellipsoidal with poorly developed cristae. Number of mitochondria in egg cell range from 1000 to 2500 in *Impatiens* to 40,000 or more in *Plumbago* (Huang & Russell 1992). *Plumbago* is the only flowering plant known to contain microbodies in the egg cell (Cass & Karas 1974). The microtubular cytoskeleton of the egg cell is densest near the nucleus with random orientation of microtubules. Abundance of various organelles in the egg cell cytoplasm is indicative of it being a physiologically active cell.

5.7.2 Synergids

The synergids are also cytologically distinct and strongly polarized cells. They are placed a little above the egg cells to increase the access of and penetration by pollen tubes. The synergids are unusual in possessing a pronounced thick cell wall structure at their micropylar end. These are numerous finger-like projections of cell wall into the synergid cytoplasm known as filiform apparatus (Figs. 5.14, 5.15 and 5.16; Huang & Russell 1992). The ultrastructure of the filiform apparatus suggests it to be a type of transfer wall common in plants, moving substances from one cell to another (Gunning & Pate 1969). These transfer walls greatly increase the surface area of the plasma membrane of the cell. It is thought that the filiform apparatus mediates the transport of molecules into and out of the synergid cells (Huang & Russell 1992). Chemically, the filiform apparatus is rich in pectins and hemicellulose.

Unlike most angiosperms, the filiform apparatus in the Cabombaceae and Nymphaeaceae is under-developed or absent. In *Nymphaea gardneriana*, the micropylar end of the synergids develops a rudimentary filiform apparatus with slight inward projections and in *Cabomba caroliniana*, the synergids are devoid of a filiform apparatus (Zini et al. 2016). Also, there are plants like *Plumbago* and *Plumbagella* where embryo sacs lack synergids. In such plants, the egg cell shows the appearance of a filiform apparatus (Cass & Karas 1974). The filiform apparatus is organized similar to that in the synergids of other flowering plants.

Distribution of organelles in the cytoplasm of synergids is also polarized, typically containing a large vacuole or vacuoles of various sizes at the chalazal region of the cell, while the nucleus lies either in the middle of the cell or towards the filiform apparatus. Thus, with respect to positioning of nucleus, vacuole and other organelles, synergids exhibit polarity opposite to that of egg cells (Figs. 5.14, 5.15 and 5.16). Unlike egg cell, synergid nuclei are normally smaller, denser, and more irregular in form and may lack nucleoli (Russell 1993). The ultrastructural appearance of synergid reflects an active metabolism with numerous, well-formed mitochondria (Jensen 1974). The vacuoles of synergids contain massive quantities of calcium (Jensen 1965). Nearly 50% of the dry weight of the synergid vacuole is calcium (Chaubal & Reger 1992).

Box 5.6

Differences between Synergid and Egg Cell

Synergid	Egg cell
• Synergid is the cell of the embryo sac which receives pollen tube.	• Egg cell is the female gamete.
• Synergids are characterized by the presence of finger like projections called filiform apparatus.	• Egg cell lacks filiform apparatus except in some species like *Plumbago* and *Plumbagella*.
• Nucleus and most of the cytoplasm of synergid cell lies either towards middle or near the filiform apparatus.	• Nucleus and most of the cytoplasm is limited to the chalazal end.
• Synergid nuclei are normally smaller, denser, and more irregular in form and may lack or contain small nucleoli.	• Egg nucleus is bigger and with a large nucleolus.
• A large vacuole or vacuoles of various sizes are present at the chalazal region of the cell.	• A large vacuole is present at the micropylar end of the egg cell.

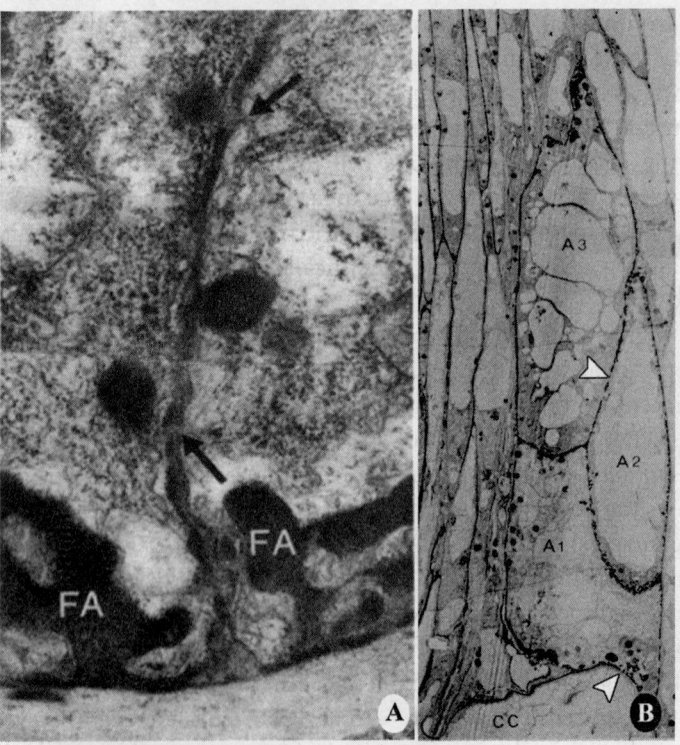

Figure 5.16 Transmission electron micrographs. A. Enlarged view of a part of synergids with filiform apparatus (FA) and common synergid cell wall with plasmodesmata (arrows). **B.** Three antipodals (**A1, A2, A3**). The antipodals and central cell (**cc**) remained connected with plasmodesmata (arrowheads) present in their common cell walls. (*Source:* Wilms 1981, *reproduced with permission.*)

The synergid cells are necessary for pollen tube guidance in the ovule, as shown by Higashiyama et al. (2001) using a laser ablation technique in *Torenia*. The synergids are also essential for the cessation of pollen tube growth and release of the sperm cells (Jensen et al. 1985). Details of pollen tube attraction by synergids are provided in chapter 7. The life span of synergids is typically short. The first synergid degenerates as soon as it comes in physical contact with the pollen tube. The disintegration of the second synergid is initiated after the fertilization of egg cell.

Box 5.7

Polarity in Embryo Sac

Throughout embryo development, the female gametophyte exhibits a polarity along its micropylar-chalazal axis. The polarity is observed right from the start of megasporogenesis and may have important consequences for the development of the embryo after fertilization. The micropylar-chalazal polarity of the female gametophyte always reflects the earlier-established distal-proximal axis of the diploid ovular tissue, indicating that extrinsic maternal factors provide the positional information necessary for reproducing the polarity in the female gametophyte (Bencivenga et al. 2011). The arrangement of microtubules, as detected by immunofluorescence, indicates a role of the cytoskeleton in distribution and positioning of cytoplasmic contents during megasporogenesis and megagametogenesis (Huang et al. 1990; Webb & Gunning 1990). The final organization of the embryo sac may reflect the establishment of heterogeneities in the cytoplasm during megasporogenesis that are enhanced during megagametogenesis (Russell 1993). It is speculated that the positional Information for embryo sac, is either inherited from the megasporocyte or the ovule may provide cues via intercellular contacts. The differences in the nutritional status at the two ends of the embryo sac are believed to influence the development and positioning of the egg and synergids. Willemse (1981) considered the presence of intercellular contacts between the developing megagametophyte and the ovule to be important for establishing a polar nutrient flow. Recent results suggest the involvement of the plant hormone auxin as a potential source for a maternally derived morphogenetic gradient. Work by Pagnusset et al. (2009) shows that embryo sac development is accompanied by accumulation of auxin in the micropylar region and both the generation and the transduction of a micropylar-chalazal auxin gradient are critical for the patterning of the female gametophyte. Mechanical interactions within the ovule have also been considered to affect embryo sac development. Lintilhac (1974) proposed a model for ovule and embryo sac development in which mechanical interactions between the nucellus and integuments influence development of the megaspore and embryo sac. In this model, the architecture of the integuments and nucellus produces a region within the nucellus that is relatively free of stresses. The functional megaspore occupies this region and may thus be in a better position than the other megaspores to develop and grow without physical constraints.

5.7.3 Central Cell

The central cell participates in double fertilization to initiate endosperm development, a feature which distinguishes the angiosperms from all other plant groups. It is the largest cell in the embryo sac and usually has two large conspicuous nuclei called the polar nuclei (Fig. 5.14 and 5.15). The polar nuclei or central cell nuclei originate at both the micropylar and chalazal ends of the coenocytic megagametophyte. During organization of the embryo sac one nucleus from each pole migrates to the centre. After cellularization, these two nuclei lie adjacent to each other, thus forming an axis almost perpendicular to the chalazal-micropylar axis that is very close to the egg nucleus (Mansfield et al. 1991). Depending on the species, the two polar nuclei either completely fuse like in cotton, *Capsella, Arabidopsis* or only partially fuse before fertilization like in maize, wheat and rice (Jensen 1965; Christensen et al. 1997).

Ultrastructural analyses show that the central cell is surrounded by a cell wall and a plasma membrane with plasmodesmata occurring in the common walls between the central cell and other cells of the megagametophyte (Han et al. 2000). In *Capsella*, and cotton, the central cell wall is continuous except at the chalazal end of the egg cell and synergids. At this end the cell wall remains irregular and, in some areas entirely absent (Jensen 1965). The central cell cytoplasm has an extensive endoplasmic reticulum (ER), is rich in mitochondria, dictyosomes, microbodies, polysomes, chloroplasts with well-developed grana and starch and lipid reserves (Fig. 5.17).The polar nuclei are huge and contain enormous nucleoli indicating high degree of metabolic activity (Schulz & Jensen 1973) which is thought to be correlated with the function of the absorption, metabolism, and transport of nutrients from the adjacent maternal tissue to the egg. Wall ingrowths like in filiform apparatus may also develop in the central cell of some species like soybean (Folsom & Cass 1992).

5.7.4 Antipodal Cells

At the chalazal end of the ovule are located a set of ephemeral cells called the antipodal cells (Fig. 5.10). Although usually haploid, there are reports of variable number of nuclei in antipodals in *Artemisia* and *Phyllis* (Fagerlind 1936), and also varying degree of polyloidy as seen in *Caltha palustris* (Graft 1941). Multinucleate antipodals can also be observed in *Zea mays, Stackhousia, Tagetes, Chrysanthemum* and *Papaver*. In certain Gentianaceae and Poaceae, the number of antipodals is variable and the number may go up to as many as 300 antipodals as in *Sasa paniculata* (Yamaura 1933). The antipodal cells are smallest in size among the cells of a female gametophyte and organelles like the vacuole are almost undetectable. The ultrastructure of the cytoplasm of antipodals reveals the presence of many mitochondria, ribosomes and RER (Newcomb 1973) and wall protrusions. The wall between antipodals and central cell thickens with maturity (Fig. 5.16 B; Newcomb 1973) and plasmodesmata occur between the antipodals and the central cell (Schulz & Jensen 197). Of all the cells of the embryo sac, only the antipodal cells have no established functions. It has been proposed that in many plants, the antipodal cells undergo a programmed cell death during embryo sac maturation or even prior to fertilization (Heydlauff & Gross-Hardt 2014). However, in many cereals, including maize and rice, the antipodal cells continue to proliferate following fertilization (Diboll & Larson 1966). Ultrastructural studies have reported invaginated cell walls of antipodal cells. These invaginations are present towards the surrounding nucellar cells leading to the hypothesis that in maize and other cereals; the antipodals may function as transfer cells to transport the nutrient from the surrounding sporophytic cells into the embryo sac (Maeda & Miyake 1997). Study by Song et al. (2014) has also shown the persistence of antipodals in *Arabidopsis* using female gametophyte-specific fluorescent reporters. It was reported that rather than undergoing programmed cell death and degeneration at the mature stage of female gametophyte, the antipodal cells persist beyond fertilization in *Arabidopsis*.

5.8 Unique Ovules and Embryo Sacs

- **Embryo sac haustoria:** Haustorium is a projection or an outgrowth that can penetrate other tissues to absorb nutrients. Several haustorial structures from cells of the embryo sac are known. These structures may develop from the megaspore, synergids, antipodals cells or endosperm (Schnarf 1929). Embryo sac haustoria are known to perform nutritive functions for the embryo sac. Occurrence of embryo sac haustorium is an embryological characteristic of many families like Rubiaceae and Santalaceae. The initiating cell or tissue, shape, size, course of development, and time of development of haustoria varies among families. It may start developing at two, four or eight nucleate stages of embryo sac or may even develop from the one of the cells of the mature embryo sac like the synergids (e.g., Compositae), central cell (e.g., Rubiaceae) or antipodals (e.g., Poaceae). In some families more than one type of haustoria are present, e.g., in Saxifragaceae and Lentibulariaceae, three types and in Crassulaceae and Santalaceae, four types are reported. A unique condition is observed in *Quinchamalium chilense* (Santalaceae) where extensive synergid and antipodal haustoria develop in the same embryo sac (Agarwal 1962). In *Quinchamalium*, the tips of both the synergids extend up to the base of the ovary at the 4-nucleate stage of the embryo sac. The haustoria can further elongate along the vascular supply of the style and, finally, reach up to one-third the length of the style. At the chalazal end, the antipodal nuclei do not organize into independent cells leading to the formation of an antipodal chamber with three nuclei. The tip of the antipodal chamber elongates and passes through the funiculus, eventually reaching the placenta branching therein.
- **Podostemaceae:** Podostemaceae is a unique embryological family of angiosperms. The defining embryological features of the family are: (a) 4-celled/ 4-nucleate condition of embryo sac, (b) occurrence of single fertilization (only syngamy) and no endosperm formation, (c) presence of nucellar plasmodium, also called 'pseudo embryo sac,' (d) pollen grains in dyads (except in subfamily Tristichoideae), (e) presence of suspensor haustoria, (h) lack of a plumule and a radicle in a mature embryo (Mohan Ram & Sehgal 2007) *(details given in chapters 7, 9 and 10)*.

Typically, an organized embryo sac of Podostemaceae is 4-celled, consisting of two small synergids, a large egg cell and a central cell harboring a polar nucleus. However, in several genera of the family, early degeneration of central cell has been reported leading to the formation of a novel 3-celled/3-nucleate mature female gametophyte. This comprises of only two synergids and an egg cell; not known to occur in any other angiosperm family (Fig. 5.17 A; Sehgal et al. 2011; Chaudhary et al. 2014). Female gametophytes with inverted polarity (egg apparatus towards chalazal end) are also noted in many species of Podostemaceae (Fig. 5.17 B, Nagendran et al.1977; Chaudhary et al. 2014).

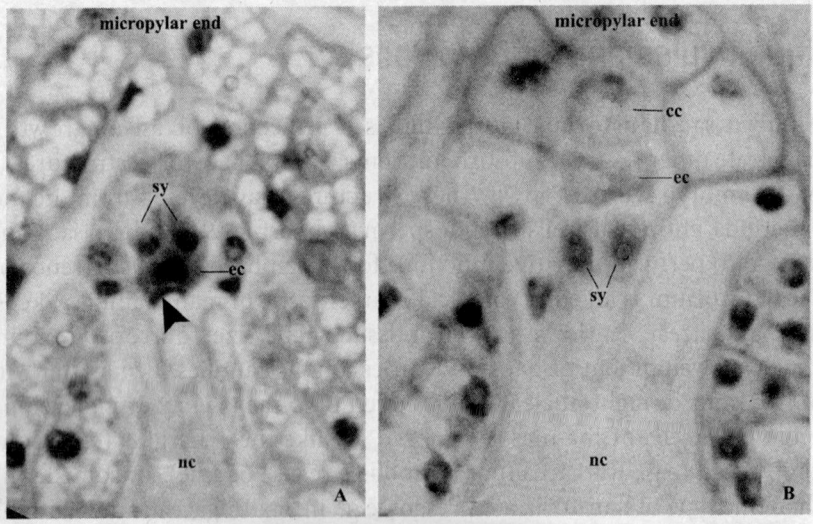

Figure 5.17 **Mature embryo sac in a Podostemaceae. A.** 3-celled mature embryo sac in *Podostemum subulatum*, after central cell degeneration. Degenerated central cell can be seen as crescent shaped body (arrowhead). **B.** Mature embryo sac (4-celled) with reverse polarity (in *Polypleurum munnarense*). **sy:** synergid; **ec:** egg cell; **cc:** central cell.

- **Naked embryo sacs:** Whereas in majority of the angiosperms, the embryo sac is located inside the ovule, there are a few species in which embryo sacs protrude out from the micropyle of the ovule and are termed as naked embryo sacs, e.g., some species of *Philadelphus*, *Thesium*, *Galium*, *Utricularia*, *Vandellia*, and *Torenia* (Maheshwari 1950). Out of these, naked embryo sacs of *Torenia fournieri* have been extensively studied. The mature embryo sac of *Torenia* protrudes through the micropyle placing the synergids, egg cell and part of the central cell outside the ovule but within the ovary locule adjacent to the placenta (Tiwari 1983). In the ovary of *T. fournieri*, the surface of the placenta is covered with mucilaginous material with the micropylar portion of the naked embryo sac being immersed within it. Pollen tubes grow through the mucilaginous material on the surface of the placenta and enter the naked embryo sac through the region of the filiform apparatus of the synergids (Higashiyama et al. 1997). Due to this feature, *Torenia* acts as a model system for making observations of double fertilization in the living tissue. Pollen tube penetration and subsequent fertilization events occur within the portion of two synergids, egg and central cell, which remains outside the body of the ovule and can be clearly observed. Higashiyama et al. (1997) used direct observation of the naked embryo sac to establish a course of events for fertilization in *T. fournieri* by documenting changes occurring within the egg and central cell.

- **Loranthaceae:** Some very unusual features of ovary structure and embryology are seen in the family Loranthaceae as normal ovules are absent. In many species the ovarian cavity is occupied by a central mound or a column called the mamelon, which may be homologous with an axile placenta. In some cases, the base of the mamelon is lobed, and these lobes may be homologous with ovules. In other cases, the mamelon is simple without any lobes while in several others, the mamelon is completely absent such that the ovarian cavity is hardly more than a small dilation of the base of the stylar canal. The sporogenous tissue is, located either in the mamelon lobes or at the base of the ovarian cavity. The resulting embryo sacs, elongate and move out of ovarian cavity and rise up in the stylar canal to various levels, so that fertilization occurs in the style, sometimes even close to the base of the stigma. The length of the embryo sacs in Loranthaceae varies from 16 mm as in *Helicanthes elastic* (Johri et al. 1957) to 48 mm in *Moquiniella rubra* (Johri & Raj 1969). *M. rubra* shows the longest Polygonum-type of 8-nucleate embryo sac in Loranthaceae, where four to nine embryo sacs develop concurrently. At the 4-nucleate stage, the tip of the embryo sac (with the egg apparatus) extends into the style and stigma and then curves backwards and becomes hooked.

Glossary

Amphitropous ovule: The type of ovule in which the curvature of the ovule extends into the nucellus, thereby making it horse-shoe shaped.

Anatropous ovule: Ovules which are completely inverted because of the extensive growth and curvature of the funiculus, such that the micropyle lies very close to the funiculus.

Archesporial cell: It is a cell from which a sporocyte differentiates in a sporangium.

Archesporium: Term used for a group of cells from which archesporial cell develops. In most cases, archesporium is unicellular and directly behaves as the archesporial cell.

Bambacioni effect: It is the fusion of 3 chalazal nuclei of a coenomegaspore before the first division (either in the prophase stage or in early metaphase) during the formation of embryo sac in *Fritillaria*.

Campylotropous ovule: The ovule is characterized by a curved morphological axis, such that the alignment between the chalaza and micropyle is lost and body of the ovule appears curved or bent.

Chalaza: It is the median part of the ovule extending from the base of the integuments to the point of attachment of funiculus.

Circinotropous ovule: The type of ovule in which funiculus grows extensively and curves 360° to assume the shape of a circle and as a consequence micropyle lies opposite the placenta.

Crassinucellate ovule: The ovules with sub-hypodermal sporogenous cell. Such ovules possess a more than one cell layer thick nucellus around the meiocyte.

Endostome: The part of micropyle formed by the inner integument.

Endothelium: An additional layer of cell differentiating from the inner epidermis of the inner integument of ovule which provides nourishment to the embryo-sac.

Epistase: A secondary group of cells found between the integuments at the apex of the nucellus in maturing seeds of some species.

Exostome: The portion of the micropyle formed by the outer integument.

FGU: A collective term use to define structural and functional association of egg cell, two synergids, and central cell. It is the minimum number of cells required for double fertilization.

Funiculus: A stalk-like structure which attaches ovule to the placenta in the ovary.

Hemitropous ovule: The ovule in which the morphological axis is placed transversely at right angle to the funicle and parallel to the placenta.

Hypostase: A well-defined but irregularly outlined group of thick-walled cells situated between the bases of the two integuments and the vascular bundle entering from the raphe.

Mamelon: A central mound or column found in the ovarian cavity of Loranthaceae which may be homologous with an axile placenta.

Megagametogenesis: The process of development of female gametophyte or embryo sac from the functional megaspore is known as megagametogenesis.

Megasporogenesis: It is the process of development of four haploid megaspores from the megaspore mother cell/megasporocyte by meiosis.

Micropyle: It is a small opening in the ovule formed between integuments at the micropylar end through which pollen tubes enter the ovule.

Nucellus: It is the central part of the ovule or the megasporangium that encloses the female gametophyte.

Obturator: An outgrowth near the micropyle of ovule originating from the placenta or funicle or integument or style, presumed to guide the pollen tube to the micropyle.

Orthotropous ovule: Ovules which lack any curvature such that the micropyle lies opposite to the placenta at an angle of 180° from it.

Ovule: It is a specialized structure consisting of a megasporangium enveloped by one or two integuments within which female gametophyte or megagametophyte develops.

Tenuinucellate ovule: The ovules in which the sporogenous cell/megaspore mother cell is hypodermal and a single-layered nucellar tissue covers the meiocyte.

Key Questions

Q5.1   Differentiate between the following:
   a. Endothecium and Endothelium
   b. Synergid and Egg cell
   c. Bisporic and Tetrasporic type of embryo sac development
   d. Male gametophyte and Female gametophyte
   e. Male sporophyte and Female sporophyte

Q5.2   Describe the development of Plumbagella-type of embryo sac using diagrams.

Q5.3   Discuss the significance of callose in plant reproduction.

Q5.4   Write notes on:
a. Types of ovules
b. Female Germ Unit
c. Polygonum-type of female gametophyte development
d. Megasporogenesis
e. Megagametogenesis

Q5.5   Do synergids and egg cells exhibit reverse polarity? Justify your answer.

Q5.6   Give one example of a species or family where the following may be observed:
a. Embryo sac without synergids
b. Mamelon
c. Naked embryo sac
d. Embryo sac with reverse polarity/arrangement
e. Numerous antipodals
f. Egg cell with filiform apparatus
g. Naked ovules
h. Embryo sac haustoria
i. Circinotropous ovule
j. Hypostase
k. Pseudo-embryo sac
l. Four celled egg apparatus

Q5.7   Fill in the blanks:
a. The contents of pollen tube are discharged in ....cell of the embryo-sac.
b. True ovules are absent in family...........................
c. ....is a group of cells present right below the embryo sac and above funiculus.
d. Filiform apparatus is present in the egg cell of ....
e. ....is an example where mature embryo sac is 8-celled and 9-nucleate structure.

Q5.8   Draw well labelled diagrams of the following:
a. L.S. of anatropous, bitegmic, crassinucellate ovule exhibiting a Polygonum-type of embryo sac
b. Ultrastructure of egg apparatus
c. Anatropous, bitegmic, tenuinucellate ovule exhibiting an Oenothera-type of embryo-sac.

Practicals

Exercise 5.1: To study different types of ovules

Ovules are specialized sporophytic structures that house the female gametophyte 'the embryo sac'. The ovule is attached to the placenta by means of a stalk called the funiculus. Typically, an ovule is comprised of a megasporangium called the nucellus enclosed by one or two protective layers called integuments. A small opening on the surface of the ovule formed by the integuments is called a micropyle. Opposite to the micropyle of an ovule is the chalaza, representing the basal part of the ovule. In most of the angiosperms , the ovules become curved during development due to growth of the funiculus. However, the degree of ovule curvature is variable among different taxa and is an important embryological characteristic for each species. So, the degree of curvature becomes an important basis for classification of ovules. On the basis of angle of curvature and placement of the micropyle of the ovule with respect to the placenta, ovules are classified into various types. To determine the type of ovule in a species, longitudinal sections of the ovary are cut or if the ovary is large enough it is dissected for visual examination. Characteristic features of different types of ovules are given below.

Longitudinal section (L.S.) of an orthotropous ovule *(Fig. 5.18 A)*

- Ovules are not curved and the micropyle lies opposite to the chalaza.
- Micropyle is located at an angle of 180° from the placenta.
- Presence of vasculature which traverses from the base of the funiculus to the chalazal-nucellar region in a straight line.
- Nucellus is straight or unbent.
- Such ovules are found in Amborellaceae, Nymphaeaceae, Piperaceae, Polygonaceae.

L.S. of an anatropous ovule *(Fig. 5.18 B)*

- The micropyle is positioned near the placenta because of the extensive growth of the funiculus and curvature of the long axis by 180°.
- The integuments, nucellus and chalaza are located parallel to the axis of the funiculus.
- Vascular strand traverses from the base of the funiculus and curves 180° towards the chalazal-nucellar region.
- Nucellus is straight or unbent.
- This is the most common type of ovule in angiosperms, e.g., Rutaceae, Asteracaeae.

L.S. of a hemitropous ovule

- Hemitropous ovule is intermediate in curvature between the anatropous and orthotropous ovule.
- The micropyle–chalazal axis is straight and is at 90° to the funiculus.
- The micropyle–chalazal axis is parallel to the placenta.
- Vascular strand traverses from the base of the funiculus and curves 90° towards the chalazal-nucellar region.
- Nucellus is straight or unbent.
- Example: *Ranunculus, Primula*

L.S. of a campylotropous ovule *(Fig. 5.18 C)*

- A part of morphological axis is curved such that the alignment between the chalaza and micropyle is lost, i.e., not in a straight line.
- Curvature in the upper part of the body of ovule places the micropyle close to the funiculus.
- Nucellus is bent along the lower side.
- Vascular strand is straight.
- Example: Chenopodiaceae, Brassicaceae, Capparaceae

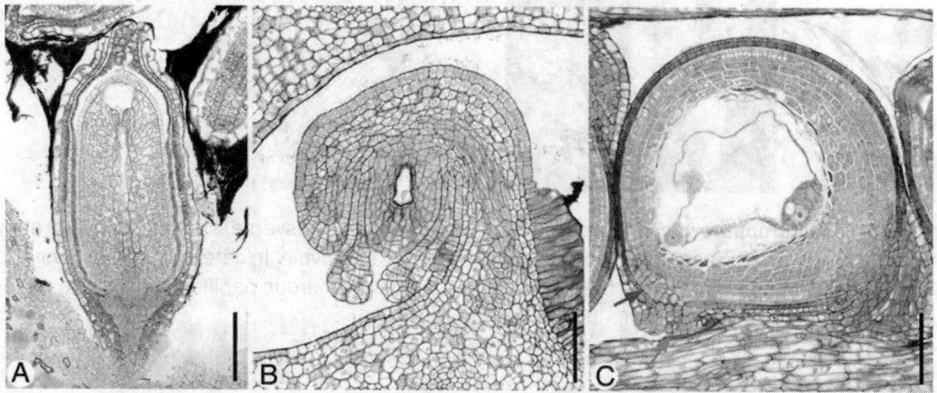

Figure 5.18 Types of ovules based on curvature. Showing median longitudinal sections of ovules; using microtomy. **A.** Orthotropous ovule (*Barclaya rotundifolia*). **B.** Anatropous ovule (*Asimina triloba*). **C.** Campylotropous ovule (*Hypecoum pendulum*). Zig-zag micropyle partly formed by the outer integument and partly formed by the inner integument (arrow). (*Source:* Endress 2011, *reproduced with permission.*)

L.S. of an amphitropous ovule

- Curvature of the morphological axis of ovule, also extending to the nucellus such that the alignment between the chalaza and micropyle is lost, i.e., not in straight line.
- Nucellus is bent both along the lower and upper side making it horse-shoe shaped.
- Vascular strand is straight.
- Example: Crossosomataceae, Papaveraceae, Alismataceae.

L.S. of a circinotropous ovule *(Fig. 5.19)*

- Due to the continuous growth of the funiculus, the ovule curves initially, then becomes anatropous (straightens) and then curves again to face upwards as in orthotropous ovules.
- Funiculus curves 360° and assumes shape of a circle.
- Such ovules are characteristic of Cactaceae, also found in Plumbaginaceae.

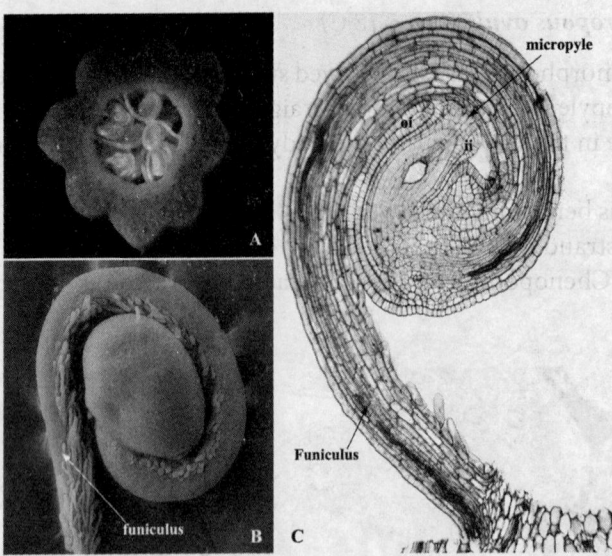

Figure 5.19 Circinotropus ovule in *Epiphyllum phyllanthus*. A. Transverse section of ovary showing circinotropus ovules. **B.** Scanning electron micrograph of the ovule in lateral view. **C.** longitudinal section of circinotropous ovule. Note the elongated funiculus. Numerous papillae can also be observed on funiculus. oi: outer integument; ii: inner Integument.

Exercise 5.2 To study special ovular structures

Other than the typical parts of an ovule, there may sometimes be present certain tissues or special auxiliary structures in it. These tissues or structures are specialized to perform certain functions which help in proper development of ovules and also fertilization. To study these structures longitudinal sections of ovule are cut and identification is done on the basis of characteristic features as discussed below (refer to Figs. 5.6 and 5.7).

Endothelium

- It is the innermost cell layer of the inner integument and is different from the other cell layers.
- Cells are radially elongated with thickened walls and show very prominent nuclei.
- It is separated from embryo sac by two layers of cuticle.
- The thickness and continuity of cuticle may vary.
- In ovules with endothelium, the nucellus is much reduced and degenerates before the embryo sac reaches maturity.
- The endothelium is usually contiguous with the developing embryo sac.
- The endothelium is involved in the supply of nutrition to the growing embryo sac. It acts as a protective layer in absence of nucellus. It also functions as a barrier between integuments and embryo sac.
- Commonly seen in members of Solanaceae, Brassicaceae.

Obturator

- An outgrowth near the micropyle originating either from the carpel, placenta, funicle, integument or style.
- It is composed either of thin-walled compact tissue or is made up of loose hairy outgrowths.
- Cells are elongated with dense cytoplasm and are surrounded by surface exudate which is carbohydrate and protein rich.
- It provides mechanical and chemical guidance to the growing pollen tube to reach the micropyle.
- Example: *Aegle marmelos, Rhododendron, Prunus.*

Epistase

- A secondary group of cells often found between the integuments at the apex of the nucellus.
- Usually seen as a layer of vertically elongated cells opposite to the micropyle.
- The inner tangential walls of cells are lignified and thickened.
- Cells are rich in starch grains and have thick polysaccharide rich walls.
- In some species, it may form a cap-like structure of cutinized cells during advanced stages of embryo development.
- The functions of epistase are not clearly known but it is speculated that they play an adaptive role in the secretion of chemotropic substances to direct the pollen tube growth towards the female gametophyte.
- Example: Onagraceae, Nymphaeaceae.

Hypostase

- A well-defined but irregularly outlined group of thick-walled cells at the chalazal end of the ovule.
- It is situated between the bases of the two integuments and is directly on top of the end of the vascular bundle.
- Sometimes it extends into the nucellus up to the base of the megagametophyte.
- It is connected more or less directly with the inner integument.
- The cells are thick-walled due to lignification. They are heavily cutinized and poor in cytoplasmic content.
- Hypostase prevents the downward growth of the embryo sac and supplies water and nutrients to embryo sac from the vascular strand.
- It is seen in families like Onagraceae, Bignoniaceae, Elaeagnaceae and Liliaceae.

Exercise 5.3: To study monosporic, bisporic and tetrasporic types of embryo sac development

The process of development of haploid megaspores as a result of meiosis in the megasporocyte is known as megasporogenesis. One of the resulting haploid megaspores then undergoes megagametogenesis and gives rise to an embryo sac. The overall process of megasporogenesis is same among angiosperms. But the mode in which cytokinesis and cell wall formation take place shows considerable variation, leading to a varied number of meiotic products contributing to the formation of female gametophyte. Based on the number of haploid nuclei taking part in embryo sac formation, three types of embryo sac development are seen: monosporic, bisporic and tetrasporic. In monosporic types only one haploid nucleus is involved in the formation the embryo sac while in the bisporic and tetrasporic types, two and four haploid nuclei are involved, respectively. The type of embryo sac development is characteristic for a species and is an important embryological trait. Female gametophytes are deeply embedded in the female reproductive organs which make them quite inaccessible to study. Therefore, to understand the development of an embryo sac in a plant, serial sections of ovules are cut and observed through a microscope. Given below is a sequential account of different types of embryo sac development with accompanying figures for reference.

Monosporic embryo sac development *(Fig. 5.20 A)*

A. One of the nucellar cells enlarges and becomes prominent due to large nucleus and cytoplasm of low density as compared to surrounding nucellar cells. This cell is known as megaspore mother cell (MMC). Depending on the type of ovule the MMC may be present just below the nucellar epidermis (tenuinucellate ovule) or may be separated from nucellar epidermis by few layers of intervening cells (crassinucellate ovule). The given figure shows a crassinucellate type of ovule.

B. MMC undergoes the first meiotic division followed by wall formation resulting in two cells with one haploid nucleus each. The cell towards the micropyle is called the micropylar dyad cell while the one towards chalaza is known as the chalazal dyad cell.

C. Both the cells then undergo the second meiotic division to form a tetrad of megaspores. The cytoplasm in the megaspores is transparent as compared to the surrounding nucellar cells.

D. Out of the four resultant megaspores, three degenerate and only one survives. It is known as the functional megaspore (FM). In the given figure, the three micropylar megaspores degenerate (arrow) and only the chalazal-most megaspore remains functional. The FM participates in gametogenesis to form embryo sac.

E. This type of embryo sac development in which a uninucleate megaspore gives rise to the embryo sac is known as monosporic type.

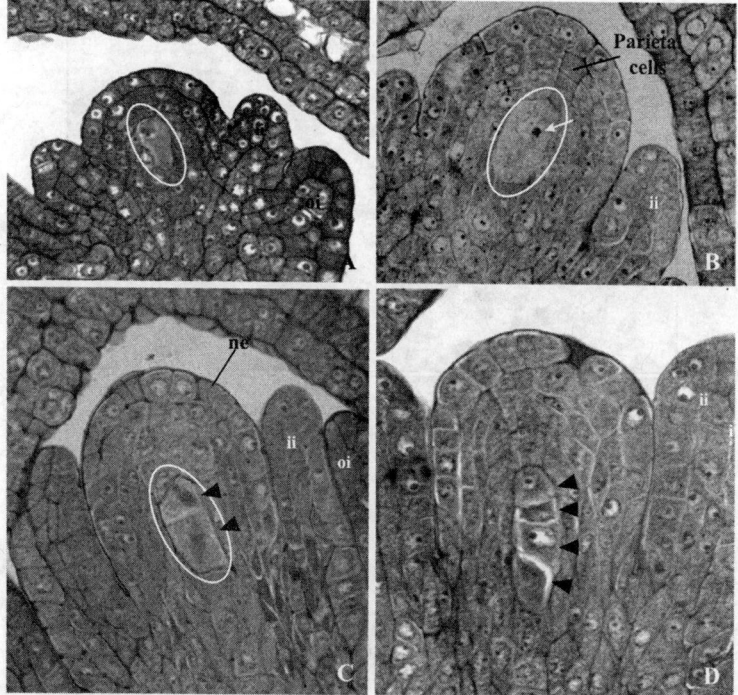

Figure 5.20 Monosporic type of embryo sac development in *Hippophae rhamnoides*. A. Archesporial cell. **B.** Enlarged MMC differentiated within nucellus in a bitegmic, crassinucellate ovule. **C.** Dyad of cells (encircled and arrowhead) after meiosis I. **D.** Linear tetrad of megaspore (arrowheads). Chalazal megaspore survives and becomes functional megaspore. **ii**: inner inregument, **oi**: outer integument.

Bisporic embryo sac development (Fig. 5.21)

A. One of the nucellar cells enlarges and differentiates as a megaspore mother cell (MMC).

B. MMC undergoes the first meiotic division followed by wall formation resulting in two cells with one haploid nucleus each. The cell towards the micropyle is called micropylar dyad cell while the one towards chalaza is known as chalazal dyad cell.

C. One of the cells of the dyad degenerates and this is usually the micropylar cell. In the given figure the micropylar cell of dyad is showing degeneration and the remaining chalazal cell remains functional.

D. The chalazal cell then undergoes the second meiotic division but it is not followed by cell wall formation.

E. This results in a megaspore with two haploid nuclei in a common cytoplasm. The two nuclei are separated by a vacuole.

F. The megaspore with two haploid nuclei is the functional megaspore and both of its nuclei contribute towards the formation of an embryo sac; hence, the embryo sac development here is of the bisporic type.

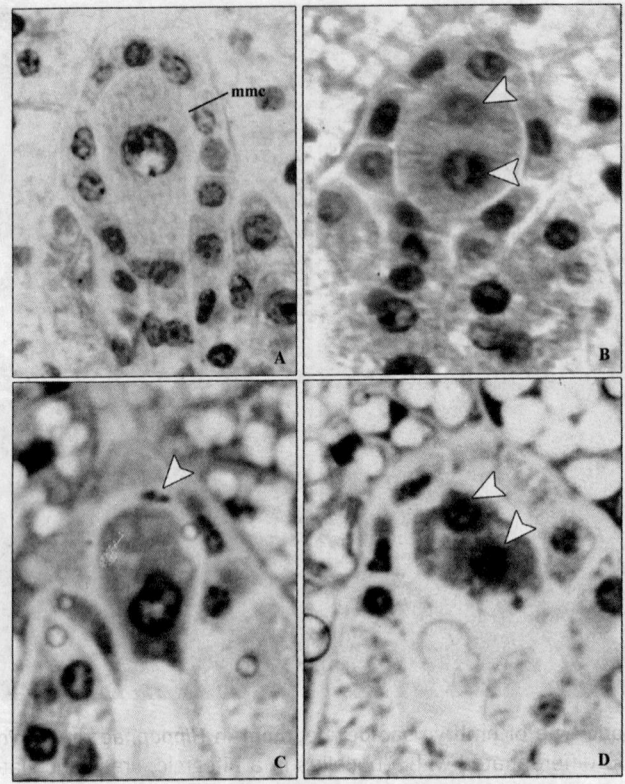

Figure 5.21 Bisporic type of embryo sac development in *Podostemum* sp. **A.** A prominent MMC in a tenuinucellate ovule. **B.** Dyad of cells (arrowheads) after meiosis I. **C.** Micropylar cell degenerates and is seen as a crescent shaped structure (arrowhead). **D.** Functional megaspore showing two megasporial nuclei after meiosis II.

Tetrasporic embryo sac development *(Fig. 5.22)*

A. One of the nucellar cells enlarges and differentiates as megaspore mother cell (MMC).

B. The MMC undergoes meiosis without any wall formation such that a single cell with four haploid nuclei is formed.

C. This resultant megaspore with four nuclei in a common cytoplasm is the functional megaspore and is called a coenomegaspore.

D. All the four nuclei of the coenomegaspore contribute towards the embryo sac formation; hence, the embryo sac development is of tetrasporic type.

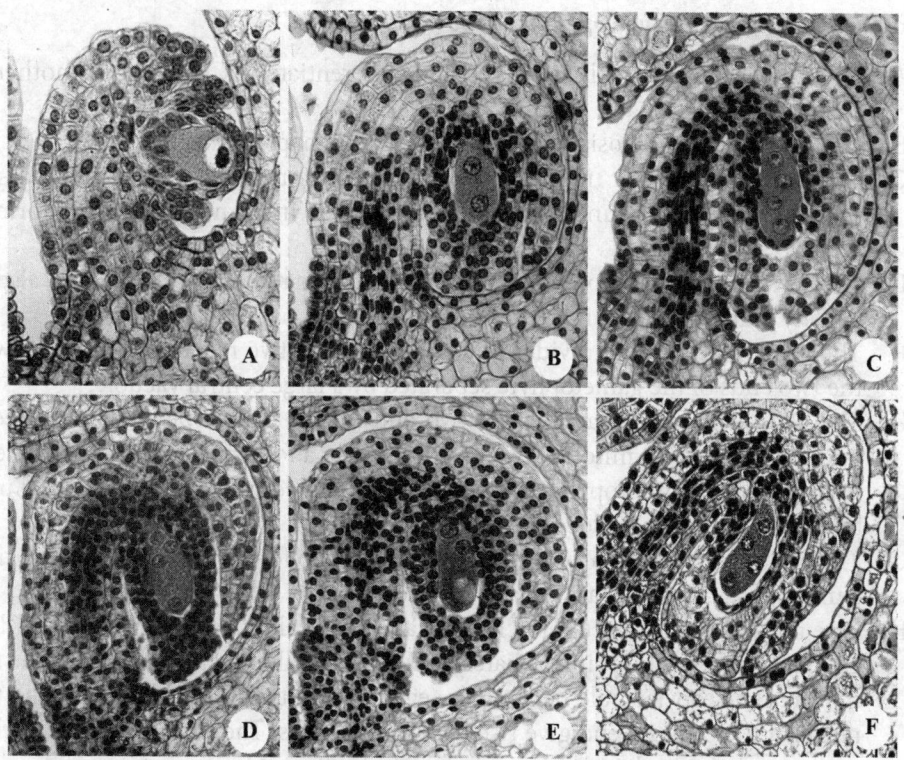

Figure 5.22 Tetrasporic type of embryo sac development in Lily A. Enlarged MMC differentiated in a bitegmic tenuinucellate ovule. **B.** Two haploid nuclei in a common cytoplasm after meiosis I. **C.** A coenomegaspore with four haploid nuclei in common cytoplasm. **D.** *Drusa, Fritillaria* type nuclei (3+1) arrangement. **E.** Adoxa-type nuclei (2+2) arrangement. **F.** *Penaea, Plumbago* type of nuclei arrangement (1+1+1+1). **See Color Plates (page 480).** (*Source*: A–E: University of Wisconsin, Stevens Biology Lab; F: Wikipedia commons.)

Exercise 5.4: To study Polygonum-type of embryo sac development

The process of embryo sac development is divided into two stages: megasporogenesis and megagametogenesis. Megasporogenesis is the process of formation of four haploid megaspores from a megasporocyte after meiosis. Subsequent mitotic divisions, nuclear migration, and cytokinesis occur in the functional megaspore to produce the mature embryo sac, a process known as megagametogenesis. Based on the mode of megasporogenesis and megagametogenesis, the embryo sac development can vary in different angiosperms. The most widely studied developmental pattern of female gametophye development is the Polygonum-type, which is named after the genus *Polygonum divaricatum*. It is exhibited in more than 70% of flowering plants including members of families Brassicaceae, Gramineae, Leguminoseae, Malvaceae and Solanaceae. The Polygonum-type of embryo sac follows a monosporic mode of development. Mature embryo sacs contain eight nuclei, organized into seven cells. These resulting cells are the egg cell, two synergids, a large central cell with two polar nuclei or one secondary nucleus, and three antipodals. An egg cell together with two synergids form the egg apparatus at micropylar end. The steps involved in formation of Polygonum-type of embryo sac along with a figure for reference (Fig. 5.23) are given below.

Monosporic (Polygonum) type of embryo sac development

A. One of the nucellar cells enlarges and differentiates as megaspore mother cell (MMC).

B. MMC undergoes meiosis resulting in a tetrad of haploid megaspores.

C. The three micropylar megaspores degenerate (arrow) and the chalazal-most megaspore remains functional and participates in gametogenesis to form the embryo sac.

D. The functional megaspore enlarges with the formation of vacuoles and also, the density of its cytoplasm increases.

E. The megaspore nucleus divides mitotically and the two nuclei are pushed towards the two ends of the gametophyte separated by a large central vacuole.

F. The two nuclei after reaching the two ends of the gametophyte, simultaneously undergo the second mitotic division resulting in a four-nucleate gametophyte; the two nuclei at the micropylar end remained separated from the two chalazal nuclei by a large central vacuole.

G. The four nuclei undergo a third mitotic division to form eight nuclei giving rise to a micropylar and a chalazal quartet of nuclei.

H. One nucleus each from the micropylar and chalazal quartet migrates towards the centre of the cell. This is immediately followed by cellularization such that the three nuclei at micropylar pole become the egg cell and two synergids, while the three nuclei at the chalazal end give rise to antipodal cells. The remaining two nuclei in the centre become the polar nuclei of central cell.

Exercise 5.5: To study the ultrastructure of egg cell, synergids and central cell

An embryo sac comprises four types of cells that function during fertilization, embryogenesis, and nutrition of the embryo. The characterization of cells of female gametophyte has been carried out in many species using electron microscopy. Ultrastructural studies have thrown light on the cellular components of individual cells of an embryo sac and also shown how each cell is different from the other. Ultrastructural data also offers an opportunity to understand the interdependence of these cells during their maturation. Observations of cells of an embryo sac highlight the fact that they are polarized, with an organelle-rich cytoplasm at one pole and a vacuole at the other (Jensen 1972) (refer to Figs. 5.14 and 5.15).

Ultrastructure of an egg cell

• The egg cell is located at the micropylar end of the embryo sac along with synergids and together they comprise the egg apparatus.

• It is separated from synergids, by either complete cell walls, partial cell walls or the plasmalemma alone.

• The distribution of cytoplasm within the egg cell is highly polarized.

• Nucleus along with most of the cytoplasm is limited to the chalazal end whereas a large vacuole is present at the micropylar end.

• Nucleus is large in size with a large nucleolus indicating extensive amplification of rDNA.

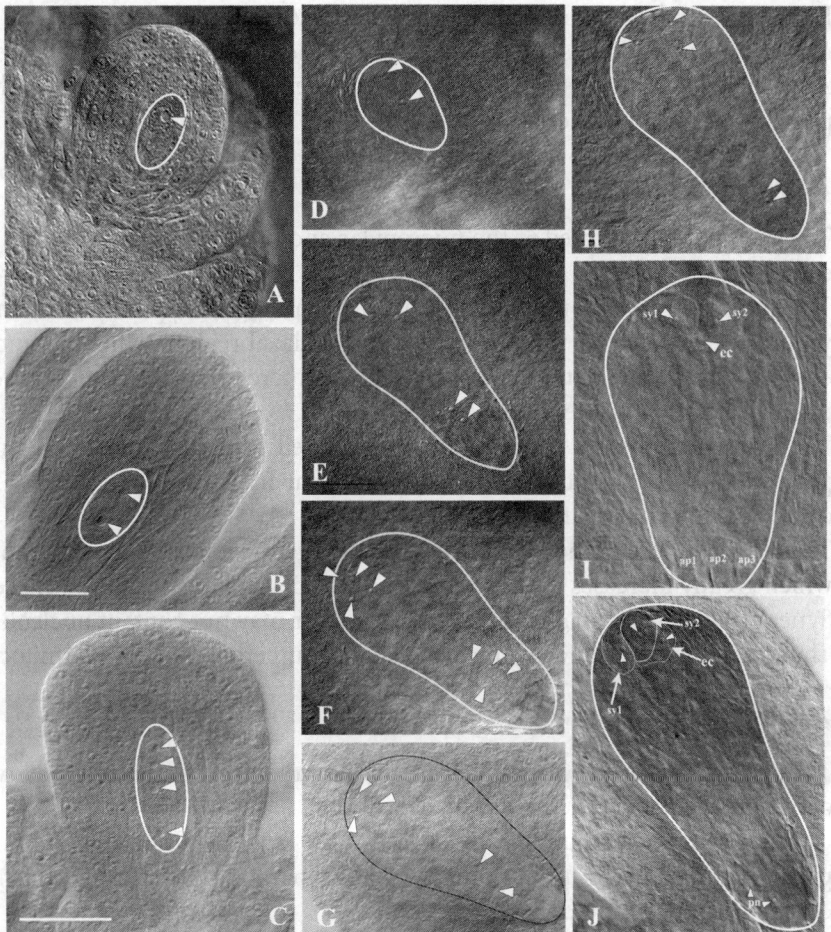

Figure 5.23 Polygonum-type of embryo sac development in *Hippophae rhamnoides* studied by method of 'ovule clearing'*. *Nuclei are marked with arrowheads.* **A–C: Megasporogenesis:** **A.** Ovule with an enlarged megaspore mother cell (encircle). **B.** Ovule at dyad stage (encircled) of megasporogenesis after Meiosis I. The two nuclei are noticeable at the opposite ends of the dyad. **C.** A linear tetrad (encircled) of megaspores after Meiosis II. The division is accompanied with wall formation between the megaspores. Note that the megaspore at the chalazal end is the largest and is functional. *The stages can be seen (and thus compared) with Fig. 5.20.* **D–H. Megagametogenesis:** **D.** First mitotic division (2-nuclear stage of female gametophyte). **E.** Immature embryo sac after the second mitotic division, showing two nuclei at each of the opposite ends (4-nuclear stage of female gametophyte after 2nd round of mitosis). **F.** Embryo sac with eight nuclei (after 3rd round of mitosis): four nuclei (arrow heads) are located towards the micropylar end and four at chalazal end. **G.** Part of embryo sac showing migration of a nucleus from the micropylar end to the chalazal end of ovule. **H.** Three nuclei arranged at the micropylar end and two in the center (as polar nuclei) before the complete organization of embryo sac. **I.** Early organization of egg apparatus (egg cell:**ec** and two synergids:**sy**) can be seen at micropylar end while three antipodals (**ap1**, **ap2** and **ap3**) are visible at chalazal end, the two polar nuclei are not visible in this plane. **J.** Mature embryo sac with an egg apparatus (two synergids, **sy1** and **sy2**; and one egg cell) at the micropylar end and 2 polar nuclei (**pn**). Antipodals have degenerated.

(*: *The method of ovule clearing has been described in Chapter 11 on Polyembryony and Apomixis.*)

- Egg cell cytoplasm is rich in organelles. It contains large number of plastids, mitochondria, endoplasmic reticulum, and lipid bodies.
- Abundance of various organelles in egg cell cytoplasm indicates that it is physiologically a very active cell.

Ultrastructure of synergids

- The synergids are located at the micropylar end of the embryo sac along with the egg cell.
- The synergids are strongly polarized cells. These cells typically contain a large vacuole or vacuoles of various sizes at the chalazal region, while the nucleus lies either in the middle of the cell or towards the micropyle. Thus, synergids exhibit a polarity that is opposite to that of the egg cell.
- The synergids possess a pronounced thick cell wall and numerous finger-like projections of cell wall into their cytoplasm at their micropylar end. These are called the filiform apparatus.
- The ultrastructure of the filiform apparatus suggests it to be a type of transfer wall which greatly increases the surface area of the plasma membrane of the cell.
- Synergid nuclei are smaller, denser, and may or may not contain nucleoli.
- The cytoplasm reflects an active metabolism with numerous, well-formed mitochondria.

Ultrastructure of a central cell

- The central cell is the largest cell in the embryo sac and usually has two large conspicuous nuclei called polar nuclei.
- Central cell is surrounded by a cell wall and a plasma membrane, and plasmodesmata occur in the common walls separating the central cell from other cells of the megagametophyte.
- The central cell cytoplasm is rich in mitochondria, dictyosomes, microbodies, polysomes, endoplasmic reticulum, starch & lipid reserves and chloroplasts with well-developed grana.
- The polar nuclei are huge and contain enormous nucleoli indicating high degrees of metabolic activity.

Exercise 5.6: To estimate number of ovules in a flower

Number of ovules per flower varies among different species. In some families, the number of ovules per flower may be a characteristic feature (example: four per flower in Lamiaceae) but mostly it varies within a family. Ovule numbers can range from one (e.g., Ulmaceae, Elaeagnaceae) to several hundred (e.g., Orchidaceae, Solanaceae, Bignoniaceae). Number of ovules per flower can sometimes be indicative of the pollination mode of a species. For example: Plants that occur in low densities or those that rely on specialist pollinators may have higher ovule numbers like Orchidaceae. Ovule number has to be estimated for calculating Pollen-Ovule ratios (P/O) which is an indicator of the breeding system of a species. Estimation of number of ovules in species with few ovules can be done easily by dissection of ovary and counting the number of ovules. However, in species with higher number of ovules given protocol may be followed:

Suggested species: Oroxylum indicum, Delonix regia, Cassia sp., *Vinca rosea, Solanum nigrum,* and *Phlox sp.*

Material Required

Glass vial, any liquid detergent, micropipette, graph paper, glass slide, blade, stereo zoom microscope.

Procedure

1. Take a wide mouth glass vial and pour some water in it.
2. Put a drop of liquid detergent in the glass vial and shake well to mix detergent.
3. Dissect an ovary from a fresh flower.
4. Longitudinally cut open the ovary into few segments depending upon its size without damaging the ovules inside.
5. Take a segment of ovary and carefully dislodge the ovules in the glass vial. Repeat with the other segment/s of ovary such that all the ovules are in the glass vial.
6. Take a glass slide and place a graph paper below it. Gently shake the glass vial and put a drop of solution from it using micropipette over the glass slide.
7. Count and note down carefully the number of ovules in each square of the graph paper using stereomicroscope. Repeat till solution in the glass vial is finished.

The above protocol can be used provided that the number of ovules in a flower are limited and countable. However, in plants where ovule number is too large, then the protocol should be modified as follows:

1. Take a wide mouth glass vial and pour 5 ml water in it.
2. Put a drop of liquid detergent in the glass vial and shake well to mix detergent.
3. Dissect ovary from a fresh flower.
4. Longitudinally cut open the ovary into few segments depending upon its size without damaging the ovules inside.
5. Take a segment of ovary and carefully dislodge the ovules in the glass vial. Repeat with the other segment/s of ovary such that all the ovules are in the glass vial.
6. Take a glass slide and place a graph paper below it. Gently shake the glass vial and put 0.5 ml of solution from it using a Pasteur pipette (glass dropper) over the glass slide.
7. Count and note down carefully the number of ovules present in 0.5 ml of solution using stereomicroscope.
8. Calculate the number of ovules/flower use the following formula:
    (Number of ovules in 0.5 ml/0.5) × 5.

Precautions

1. Cut the ovary walls gently without damaging the ovules.
2. Scrape ovules from ovary wall gently into the solution.
3. Shake the solution every time before taking it.

Exercise 5.7: To assess the receptivity of ovules using Toluidine Blue 'O' test

In many multi-ovulate species, it is commonly found that not all the ovules develop into seeds. One of the reasons cited for this is non-receptivity of the ovules at the time of arrival of pollen tubes (Sengupta & Tandon 2010). Thus, the state of ovules when pollen tubes approach them is important for successful fertilization and seed set. Pollen tubes receive many cues to make their way towards the ovules. One of the attractants is a protein and polysaccharide rich exudate secreted in the micropylar region by receptive ovules. The presence of such exudate can be detected by staining with toluidine blue 'O' which is used as a test for ovule receptivity.

Material Required

Fresh flowers of different species (five each), toluidine blue 'O' stain, phosphate buffer, glass slide, blade, stereo zoom microscope.

Procedure

1. Take a fresh flower and dissect the ovary.
2. Take a microslide and put a drop of phosphate buffer.
3. Carefully dissect all the ovules in the drop of phosphate buffer.
4. Dry out the buffer using filter paper and immediately replace it with a drop of toluidine blue 'O'.
5. Gently lower a cover glass and observe all the ovules for a bluish magenta spot near the micropyle.
6. Count the number of receptive and non-receptive ovules and record your observations.
7. Repeat the procedure with other flowers of the same species to calculate an average number or percentage of receptive ovules in a flower.

Observations

Ovules with bluish magenta spots near the micropyle are receptive. While taking the observation care should be taken that the cut end of the funiculus which also develops color due to injury is not mistaken for a micropylar spot (Fig. 5.24).

Precautions

1. Ovules should be dissected carefully without causing any damage.
2. Ovules should not dry while replacing phosphate buffer with toluidine blue 'O'.
3. Scoring should be done carefully avoiding false positives (color developed at the cut end of the funiculus).

Preparation of Toludine Blue 'O' (100 ml)

Prepare 0.1M phosphate buffer (pH- 6.5). To 90ml of phosphate buffer add 0.025g (0.025 %) TBO dye salt. Mix well. Adjust the pH of solution (after adding TBO) to 4.4 using 5N HCl and make the final volume 100ml with 0.1M phosphate buffer.

Note: The dye solution stains the electrode of pH meter, thus throughly wash the electrode after use.

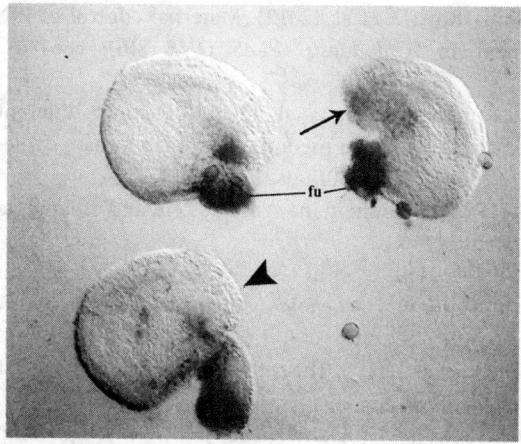

Figure 5.24 Arrow indicates the receptive ovule with bluish staining at micropyle while ovule marked with arrowhead is not receptive. Note funiculus (**fu**) is deeply stained blue. **See Color Plates (page 480).**

Bibliography

Abe K. (1971). Contributions to the embryology of the family Orchidaceae. V. Development of the embryo sac in *Oreorchis patens*. *The Science Reports of the Tohoku University IV (Biology)* 35: 219–24.

Agarwal, J. S. (1950). The embryo sac of *Fritillaria liliacea*. *Proceedings of Indian National Science Academy* 16: 87–92.

Agarwal, S. (1962). Embryology of *Quinchamalim chilense* Lato. *Plant Embryology: A Symposium* (New Delhi: CSIR), pp. 162–9.

Alberts, B. Bray, D. Lewis, J. et al. (1989). *Molecular Biology of the Cell,* 2nd ed. New York: Garland.

Almeida, O. J. G., Sartori-Paoli, A. A. and de Souza, A. (2010). Flower morpho-anatomy in *Epiphyllum phyllanthus* (Cactaceae). *Mexican Journal of Biodiversity* 81: 65–80.

Arbeloa, A. and Herrero, M. (1987). The significance of the obturator in the control of pollen tube entry into the ovary in peach (*Prunus persica*). *Annals of Botany* 60: 681–5.

Bambacioni, V. (1928). Ricerchesullaecologia e sullaembriologia di *Fritillaria persica* L. *Annals of Botany* 18: 7–37

Battaglia, E. (1983). Embryological questions: discussion of the concepts of spore, sporogenesis and apospory in relation to the female gametophyte of angiosperms. *Annals of Botany* (Roma) 41: 1–25.

Bencivenga, S., Simonini, S., Benkova, E. et al. (2012). The transcription factors BEL1 and SPL are required for cytokinin and auxin signaling during ovule development in*Arabidopsis*. *The Plant Cell* 24: 2886–97.

Boblenz, K., Nothnagel, T. and Metzlatt M. (1990). Paternal inheritance of plastids in the genus *Daucus*. *Molecular Genetics and Genomics* 220: 489–91.

Bouman, F., Cobb, L., Devente, N., et al. (2002). *The Seeds of Gentianaceae, Gentianaceae: Systematics and Natural History.* Cambridge University Press, pp. 498–572.

Cass, D. D., and Karas, I. (1974). Ultrastructural organization of the egg of *Plumbago zeylanica*. *Protoplasma* 81: 49–62.

Cass, D. D., Peteya, D. J. and Robertson, B. L. (1985). Megagametophyte development in *Hordeum vulgare*. 1. Early megagametogenesis and the nature of cell wall formation. *Canadian Journal of Botany* 63: 2164–71.

Ceccato, L., Masiero, S., Sinha Roy, D. et al. (2013). Maternal control of PIN1 is required for female gametophyte development in *Arabidopsis*. *PLoS ONE* 8(6): e66148. doi.org/10.1371/journal. pone.0066148.

Chaban,. I. Baranova, E., Kononenko, N. et al. (2020). Distinct differentiation characteristics of endothelium determine its ability to form pseudo-embryos in tomato ovules. *International Journal of Molecular Science* 21: 12. //doi.org/10.3390/ijms21010012.

Chaubal, R. and Reger, B. J. (1992). Calcium in the synergid cells and other regions of pearl millet ovaries. *Sexual Plant Reproduction* 5: 34–46.

Chaudhary, A., Khanduri, P., Tandon, R. et al (2014). Central cell degeneration leads to three-celled female gametophyte in *Zeylanidium lichenoides* Engl. (Podostemaceae). *South African Journal of Botany* 91: 99–106.

Chen, Y. H., Li, H. J., Shi, D. Q., et al. (2007). The central cell plays a critical role in pollen tube guidance in *Arabidopsis*. *Plant Cell* 19: 3563–77.

Cheng, C. Y., Mathews, D. E., Schaller, G. E. and Kieber J. J. (2013). Cytokinin-dependent specification of the functional megaspore in the *Arabidopsis* female gametophyte. *The Plant Journal* 75: 642–55.

Christensen, C. A., King, E. J., Jordan, J. R. et al (1997). Megagametogenesis in *Arabidopsis* wild type and the Gf mutant. *Sexual Plant Reproduction* 10: 49–64.

Dane, F., Olgun, G. and Ekici, N. (2007). Ultrastructure of endothelium in ovules of *Penstemon gentianoides* Poir. (Scrophulariaceae) at mature embryo sac phase. *Acta Biologia Hungary* 58: 225–33.

Davis, G. L. (1966). *Systematic Embryology of the Angiosperms*. New York: Wiley.

Demesa–Arévalo, E. and Vielle–Calzada, J. (2013). The classical arabinogalactan protein AGP18 mediates megaspore selection in *Arabidopsis*. *The Plant Cell* 25: 1274–87.

Diboll, A. G. and Larson, D. A. (1966). An electron microscopic study of the mature megagametophyte in *Zea mays*. *American Journal of Botany* 53: 391–402.

Dumas, C., Knox, R. B., McConchie, C. A. and Russell, S. D. (1984). Emerging physiological concepts in fertilization: what's new? *Plant Physiology* 15: 17–20.

Dumas, C. and Mogensen, C. H. (1993). Gametes and fertilization: maize as a model system for experimental embryogenesis in flowering plants. *The Plant Cell* 5(10): 1337–48.

Endress, P. K. (2001). Origins of flower morphology. *Journal of Experimental Zoology* 29: 105–15.

Endress, P. K. (2006). Angiosperm floral evolution: morphological developmental framework. *Advances in Botanical Research* 44: 1–61.

Endress, P. K. (2011). Angiosperm ovules: diversity, development, evolution. *Annals of Botany* 107: 1465–89.

Esau, K. (1977). *Anatomy of Seed Plants*. New York: John Wiley and Sons.

Folsom, M. W. and Cass, D. D. (1990). Embryo sac development in soybean: cellularization and egg apparatus expansion. *Canadian Journal of Botany* 68: 2135–47.

Friedman, W. E., Gallup, W. N. and Williams J. H. (2003). Female gametophyte development in *Kadsura*: implications for Schisandraceae, Austrobaileyales, and the early evolution of flowering plants. *International Journal of Plant Sciences* 164: S293–S305.

Friedman, W. E. and Ryerson, K. C. (2009). Reconstructing the ancestral female gametophyte of angiosperms: insights from *Amborella* and other ancient lineages of flowering plants. *American Journal of Botany* 96: 129–43.

Gasser, C. S., Broadhvest, J. and Hauser B. A. (1998). Genetic analysis of ovule development. *Annual Review of Plant Physiology and Plant Molecular Biology* 49: 1–24.

González, M. V., Coque, M. and Herrero, M. (1996). Pollen–pistil interaction in kiwifruit (*Actinidia deliciosa*; Actinidiaceae). *American Journal of Botany* 83: 148–54.

Grossniklaus, U. and Schneitz, K. (1998). The molecular and genetic basis of ovule and megagametophyte development. *Seminars in Cell and Developmental Biology* 9: 227–38.

Guilford, V. B. and Fisk, E. L. (1952). Megasporogenesis and seed development in *Mimulus trigrinua* and *Torenia fournieri*. *Bulletin of Torrey Botanical Club* 79: 6–24.

Gunning, B. E. S. and Pate J. S. (1969). 'Transfer cell' – plant cells with wall ingrowths specialized in relation to short distance transport of solutes – their occurrence, structure and development. *Protoplasma* 68: 107–33.

Han, Y. Z., Huang, B. Q., Zee, S. Y. and Yuan, M. (2000). Symplastic communication between the central cell and the egg apparatus cells in the embryo sac of *Torenia fournieri* Lind. before and during fertilization. *Planta* 211: 158–62.

Heydlauff, J. and Groß–Hardt, R. (2014). Love is a battlefield: programmed cell death during fertilization. *Journal of Experimental Botany* 65: 1323–30.

Higashiyama, T. and Yang, W. C. (2017). Gametophytic pollen tube guidance: attractant peptides, gametic controls, and receptors. *Plant Physiology* 173: 112–21.

Huang, B. Q. and Russell, S. D. (1992). Female germ unit: organization, reconstruction and isolation. *International Review of Cytology* 140: 233–93.

Igersheim. A. and Endress P. K. (1998). Gynoecium diversity and systematics of the paleoherbs. *Botanical Journal of the Linnean Society* 127: 289–370.

Jensen, W. A., Ashton, M. E. and Beasley C. A. (1985). Pollen tube–embryo sac interaction in cotton. In *Pollen: Biology and Implications for Plant Breeding*. New York: Elsevier Biomedical, pp. 67–72.

Jensen, W. A. (1965). The ultrastructure and composition of the egg and central cell of cotton. *American Journal of Botany* 52: 781–97.

Jensen, W. A. (1973). Fertilization in flowering plants. *Bioscience* 23: 21–7.

Jensen, W. A. (1974). Reproduction in flowering plants. *Dynamic Aspects of Plant Ultrastructure*. New York: McGrawHill, pp. 481–503.

Johansen, D. A. (1928). The hypostase: its presence and function in the ovule of the Onagraceae. *Proceedings of National Academy of Sciences of the USA* 14: 710–13.

Johri, B. M., Agrawal, J. S. and Garg, S. (1957). Morphological and embryological studies in the family Loranthaceae I. *Helicamheselastica* (Desr.) *Dans. Phytomorphology* 7: 336–54.

Johri, B. M. and Raj, B. (1969). Morphological and embryological studies in the family Loranthaceae–XII. *Moquiniella rubra* (Spreng. f.) Balle. Österreichischebotanische *Zeitschrift* 116: 475–85.

Kapil, R. N. and Tiwari, S. C. (1978). The integumentary tapetum. *Botanical Review* 44: 457–90.

Kasahara, R. D., Portereiko, M. F., Sandaklie–Nikolova, L. et al.(2005). MYB98 is required for pollen tube guidance and synergid cell differentiation in Arabidopsis. *Plant Cell* 17: 2981–92

Kennell, J. C. and Horner, H. T. (1985). Megasporogenesis and megagametogenesis in soybean, *glycine max*. *American Journal of Botany* 72: 1553–64.

Lin, S., Splittstoesser, W. E. and George, W. L. (1983). A comparison of normal seeds and pseudo-embryos produced in partenocarpic fruits of 'Severianin' tomato. *Horticultural Science* 19: 45–53.

Lintilhac, P. (1974). Differentiation, organogenesis and the tectonics of cell wall orientation. II. Separation of stresses in a two-dimensional model. *American Journal of Botany* 61: 135–40.

Losada, J. L. and Herrero, M. (2017). Pollen tube access to the ovule is mediated by glycoprotein secretion on the obturator of apple (*Malus* × *domestica*, Borkh). *Annals of Botany* 119: 989–1000.

Maeda, E. and Miyake, H. (1997). Ultrastructure of antipodal cells of rice (*Oryza sativa*) before anthesis with special reference to concentric configuration of endoplasmic reticula. *Japanese Journal of Crop Science* 66: 488–96.

Maheshwari, P. (1950). *An Introduction to the Embryology of Angiosperms*. New York: McGraw–Hill.

Mangla, Y., Chaudhary, M., Gupta, H. et al. (2015). Facultative apomixis and development of fruit in a deciduous shrub with medicinal and nutritional uses. *AoB PLANTS*: plv098. doi.org/10.1093/aobpla/plv098.

Mansfield, S. G., Briarty, L. G. and Erni, S. (1991). Early embryogenesis in *Arabidopsis thaliana* I. The mature embryo sac. *Canadian Journal of Botany* 69: 447–60.

Mohan Ram, H. Y. and Sehgal, A. (2007). Podostemaceae – an evolutionary Enigma. In *Proceedings of National Seminar on Evolutionary Biology and Biotechnology*, Kolkata. Zoological Survey of India, pp. 37–46.

Mukherjee, P. K. (1958). The female gametophyte of *Acalypha malabarica* Muell. with a brief discussion on the Penæa type of embryo-sac. *Journal of Indian Botanical Society* 37: 504–8.

Nagendran, C. R., Arekal, G. D. and Subramanyam, K. (1977). Embryo sac studies in three Indian species of *Polypleurum* (Podostemaceae). *Plant Systematics and Evolution* 128: 215–26.

Newcomb, W. (1973). The development of the embryo sac of sunflower *Helianthus annuus* after fertilization. *Canadian Journal of Botany* 51: 879–90.

Pagnussat, G. C., Alandete-Saez, M., Bowman, J. L. and Sundaresan, V. (2009). Auxin-dependent patterning and gamete specification in the *Arabidopsis* female gametophyte. *Science* 324: 1684–9.

Palser, B. F., Rouse, J. L. and Williams E. G. (1989). Coordinated timetables for megagametophyte development and pollen tube growth in *Rhododendron nuttallii* from anthesis to early post fertilization. *American Journal of Botany* 1: 1167–202.

Poddubnaya–Arnoldi, V. A. (1976). *Cytoembryology of Angiosperms*. Moscow: Nauka

Raghavan, V. (2003). Some reflections on double fertilization, from its discovery to the present. *New Phytologist* 159: 565–83.

Reiser, L. and Fischer, R. L. (1993). The ovule and the embryo sac. *The Plant Cell* 5: 1291–301.

Rodkiewicz, B. (1970). Callose in cell walls during megasporogenesis in angiosperms. *Planta* 93: 39–47.

Rudall, P. J., Remizowa, M. V., Beer, A. S. et al. (2008). Comparative ovule and megagametophyte development in Hydatellaceae and water lilies reveal a mosaic of features among the earliest angiosperms. *Annals of Botany* 101: 941–56.

Russell, S. D. (1978). Fine structure of megagametophyte development in *Zea mays*. *Canadian Journal of Botany* 57: 1093–110.

Russell, S. D. (1987). Quantitative cytology of the egg and central cell of *Plumbago zeylanica* and its impact on cytoplasmic inheritance patterns. *Theoretical and Applied Genetics* 74: 693–9.

Russell, S. D. (1993). The egg cell: Development and role in fertilization and early embryogenesis. *The Plant Cell* 5: 1349–59.

Schmidt, A., Wuest, S. E. Vijverberg, K. et al. (2011). Mother cell uncovers the importance of RNA Helicases for plant germline development. *Plos Biology* 9(9): e1001155.

Schneitz, K., Hülskamp, M., Kopczak, S. D. and Pruitt R. E. (1997). Dissection of sexual organ ontogenesis: a genetic analysis of ovule development in *Arabidopsis thaliana*. *Development* 124: 1367–76.

Schulz, P. and Jensen W. A. (1971). *Capsella* embryogenesis: the chalazal proliferating tissue. *Journal of Cell Science* 8: 201–27.

Schulz, R. and Jensen, W. A. (1968). *Capsella* embryogenesis: The synergids before and after fertilization. *American Journal of Botany* 55: 541–52.

Sehgal, A., Khurana, J. P., Sethi, M. and Hussain, A. (2011). Occurrence of unique three celled megagametophyte and single fertilization in an aquatic angiosperm – *Dalzellia zeylanica* (Podostemaceae–Tristichoideae). *Sexual Plant Reproduction* 24: 199–210.

Sengupta, S. and Tandon, R. (2010). Assessment of ovule receptivity as a function of expected brood size in flowering plants. *The International Journal of Plant Reproductive Biology* 2: 51–63.

Shamrov, I. I. (2018). Diversity and typification of ovules in flowering plants. *Wulfenia* 25: 81–107.

Simpson, M. G. (2010). *Plant Systematics*. Elsevier Academic Press.

Song, X., Yuan, L. and Sundaresan, V. (2014). Antipodal cells persist through fertilization in the female gametophyte of *Arabidopsis*. *Plant Reproduction* 27: 197–203.

Starshova, N. P. and Solntseva, M. P. (1973). Characterization of the embryo sac in *Phellodendron amurense. Rupr. Botanicheskii Zhurnal* 58: 11.

Sumner, M. J. and Van Caeseele, L. (1989). The ultrastructure and cytochemistry of the egg apparatus of *Brassica campestris. Canadian Journal of Botany* 67: 177–90.

Takhtadzhian, A. (1991). *Evolutionary Trends in Flowering Plants.* New York: Columbia University Press.

Tiwari, S. C. (1983). The hypostase in *Torenia fournieri* Lind: a histochemical study of the cell walls. *Annals of Botany* 51: 17–26.

Walker, R. I. (1950). Megasporogenesis and development of megagametophyte in *Ulmus. American Journal of Botany* 37: 47–52.

Webb, M. C. and Gunning B. E. S. (1990). Embryo sac development in *Arabidopsis thaliana*: Megasporogenesis, including the microtubular cytoskeleton. *Sexual Plant Reproduction* 3: 244–56.

Willemse, M. T. M. and Bednara, J. (1979). Polarity during megasporogenesis in *Gasteria verrucosa. Phytomorphology* 29: 156–65.

Williams, J. H. and Friedman, W. E. (2004). The four-celled female gametophyte of *Illicium* (Illiciaceae; Austrobaileyales): implications for understanding the origin and early evolution of monocots, eumagnoliids, and eudicots. *American Journal of Botany* 91: 332–51.

Wilms, H. J. (1981). Ultrastructure of the developing embryo sac of spinach. *Acta Botanica Neerlandica* 30: 75–99.

Wilms, H. J. (1980). Development and composition of the spinach ovule. *Acta Botanica Neerlandica* 29: 243–60.

Yadegari, R. and Drews, G. N. (2004). Female Gametophyte Development. *The Plant Cell Online* 16 (suppl_1): S133–S141.

Yamaura, A. (1933). Karyologische and embryologische Studien uber einige Bambusa–Arten (Vor-laufigeMitteilung). *Botanical Magazine (Tokyo)* 47: 551–5.

Zhou, Q. Y., Jin, X. B. and Fu D. Z. (2004). Developmental morphology of obturator and micropyle and pathway of pollen tube growth in ovary in *Phellodendron amurense* (Rutaceae). *Acta Botanica Sinica* 46: 1434–42.

Zini, M. L., Galatti, B. G., Ferrucci, M. S. and Zarlavsky, M. (2016). Ultrastructural study of the female gametophyte and the epistase in Cabombaceae and Nymphaeaceae. *Flora–Morphology Distribution Functional Ecology of Plants* 220. 10.1016/j.flora.2016.02.006.

6

Pollination

Flowers of Bombax ceiba are pollinated by multiple pollinators including insects, birds, bats and squirrels.

6.1 Introduction

Once a flower is fully developed, most of the plants display their sex organs to carry out reproduction. This is accomplished by opening of the flower bud which is known as floral anthesis. Opening of flower is followed by pollination, a very important step in plant reproduction. Pollination is the transfer of pollen from an anther of a flower to a stigma of the same or a different flower. The transfer of pollen grains to the conspecific (belonging to the same species) stigma is the primary step in reproduction or seed formation. Pollination also forms the basis for genetic heterogeneity in plants. Study of methods of pollination and subsequent fertilization in plants is referred to as **pollination biology**.

Considering the fact that plants are immobile, most angiosperm species have to rely on external agents for the transfer of pollen from anther to stigma. Pollination services are rendered by various biotic and abiotic agencies and their presence is essential for optimal reproductive performance of a flowering plant. In case of biotic pollination, the plant attracts a particular type of pollinator and when the same pollinator visits the next flower there are chances that its pollen is carried to another flower of the same species. In exchange for this service, the pollinator gets access to the food (pollen and nectar) offered by the flowers. Thus, by visiting a particular type of flower, the pollinator gets important food resources and the plant gets pollinated in return. Several groups of insects, birds, bats and other animals are dependent on plants for their nutritional needs, especially during their breeding seasons. Such mutually beneficial relationship between the angiosperms and the pollinators has led to co-evolution and adaptations among these groups over millions of years. Plants and pollinators have co-evolved and undergone changes in their physical forms to increase the chances of successful interaction.

An array of floral features like color, form, nectar and scent exhibited by angiosperms is associated with different modes of pollination. Such correlation between floral features and pollinating agency is referred to as **pollination syndrome**. The pollination services provided by biotic and abiotic agents are not only essential for fertilization but are also required to maximize dispersal distances and increase genetic variability in plants. The present chapter gives an all-encompassing understanding of different modes of pollination seen in flowering plants and also outlines how flowering plants and their pollinators have co-evolved to benefit each other.

6.2 Types of Pollination: Self and Cross

Pollination can be broadly classified into two types: Self-pollination and Cross-pollination. Transfer of pollen grains from an anther to a stigma of the same flower or a different flower on the same plant is called self-pollination. However, when the pollen grains are transferred from the anther to the sigma of a different flower on a different plant of the same species, it is referred to as cross-pollination. Most flowers expose their stigmas and stamens at maturity and are called **chasmogamous flowers**, e.g., *Brassica*. In contrast, there are plants which produce flowers which do not open or undergo anthesis. Such flowers are known as **cleistogamous flowers** and this phenomenon of producing closed flowers is called cleistogamy, e.g., *Commelina* sp., *Viola* sp. Occurrence of cleistogamous or chasmogamous flowers in some species is dependent on environmental parameters like temperature and light, e.g., *Portulaca oleracea*. Some members of Acanthaceae (e.g., *Ruellia* sp.) are known to produce chasmogamous flowers in summers and cleistogamous flowers in winters. Self-pollination is obvious in cleistogamous flowers but chasmogamous flowers can be either self- or cross-pollinated. Some species produce both closed and open flowers which are morphologically similar, a phenomenon known as pseudocleistogamy, e.g., *Impatiens capensis* produces both chasmogamous and cleistogamous flowers on aerial shoots.

The mode of pollination determines the genetic relatedness of the gametes (whether self-related or not) which will undergo fertilization. Thus, pairing of individuals during pollination controls the transfer of genes from one generation to the next. The genetic relatedness between individuals undergoing gametic fusion also indicates the mating system of a species. Thus, based on the source of pollen, the following categories of pollination are recognized:

i. **Autogamy:** Pollen transfer from anther to the stigma of the same flower (*intrafloral selfing*) (Fig. 6.1 A).

ii. **Allogamy**: Pollen transfer from anther to the stigma of another flower of the same or a different plant. Further, allogamy can be of two types:
- **Geitonogamy:** Pollen transfer from anther to the stigma of another flower present on the same plant (*interfloral selfing*) or to a flower present on another plant of the same clone (ramet) (Fig. 6.1 B).
- **Xenogamy:** Pollen transfer from an anther to the stigma of a different plant (genet) (Fig. 6.1 C).

As both autogamy and geitonogamy occur on the same plant, they qualify for being considered as self-pollination. Only xenogamy is considered as true cross-pollination. Autogamy may or may not require pollen transferring agents (pollinators). However, for geitonogamy and xenogamy the presence of pollinating agents is essential. Some authors consider geitonogamy as a type of cross-pollination, since there is a requirement for an external agent for transfer of pollen to the stigma of another flower, though on the same plant.

6.2.1 Self-Pollination

Self-pollination is mostly seen in hermaphrodite/bisexual flowers and monoecious plants. In self-pollinated plants, anther dehiscence and attainment of stigma receptivity occur at the same time. For an efficient transfer of pollen grains, anthers and stigmas of within-flower pollinated plants are usually located in close proximity of each other. Geitonogamy or within-plant pollination requires a pollinating agent for transfer of pollen among flowers of the same plant, e.g., *Ipomopsis aggregata, Malva moscata*. About 70–80% flowering plants are self-pollinated. Self-pollination is often seen as an adaption to ensure reproduction under difficult ecological conditions. A major evolutionary advantage of self-pollination is that reproduction is ensured even when relatively few (or even one) individuals are present in a population. However, self-pollination leads to inbreeding which reduces the genetic diversity in a population. Inbreeding can result in the accumulation of deleterious alleles which reduces the biological fitness of a population, a phenomenon known as inbreeding depression.

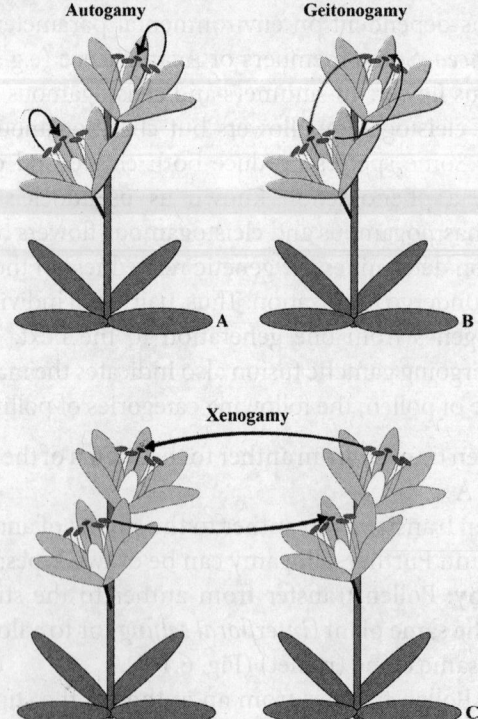

Figure 6.1 Types of pollination. A. Autogamy: Pollen grains are transferred from anther to the stigma of the same flower **B.** Geitonogamy: Pollen grains are transferred from anther to the stigma of the different flowers on the same plant **C.** Xenogamy: Transfer of pollen grains from anther to the stigma of the flowers borne on different plants of the same species.

Flowering plants display various floral contrivances in order to facilitate self-pollination. For example, *Valeriana wallichii* (Valerianaceae), conducts special stylar movements for self-pollination. Here, the style gets deflexed towards one or the other anther of the same or even adjacent flowers for collecting pollen. It becomes bow-shaped towards an anther to trap self-pollen and can develop an even sharper bend to capture pollen on a corolla lobe (Verma & Sharma 2010). The anthers in bisexual flowers of *Holcoglossum amesianum*, (Orchidaceae), bend (upwards) in order to insert the pollinia into the stigmatic cavity. This bending of stamens, occurs against gravity and can be considered an adaptation to ensure self-pollination in windless, drought-stricken and insect-limited habitats of some orchids. Similarly, anthers in some members of Podostemaceae clasp the stigma and dislodge the self-pollen grains to achieve self-pollination (Khanduri et al. 2014; Fig. 6.2 A).

Obligate self-pollination before the anthesis of flower has been reported from aquatic plants as a method to evade contact with water, e.g., *Utricularia inflexa* var. *stellaris* (Khosla et al. 1998), *Hydrobryopsis sessilis* (Sehgal et al. 2009) and *Podostemum ceratophyllum* (Philbrick 1984). This type of self-pollination which occurs during the bud stage and the flower eventually opens later is referred to as **pre-anthesis cleistogamy**. In *Utricularia inflexa*, the anthers are in close proximity to the stigma in bud stage. The stamens lie just beneath the stigmatic lip which later recurves and arches over the anthers in the form of a hood. The filaments elongate to bring the anthers in contact with the stigma. The anthers dehisce and the pollen mass, in which many pollen show *in situ* germination, gets deposited on the stigma (Fig. 6.2 B) followed by pollen tube penetration into the stigma (Khosla et al. 1998). Self-pollination is also frequently seen in plants where the floral architecture occludes the access of pollinators. For example, in *Commelina,* the flowers are generally small and have dull corollas, with the androecium and gynoecium remaining enclosed by the petals. Hence, the pollination agents are not able to access the stamens and the pistils. A unique type of pollination, called **internal geitonogamy** is observed in the monotypic aquatic family Callitrichaceae. In submerged flowers of *Callitriche heterophylla,* the pollen grains germinate within the intact undehisced anthers and the pollen tubes grow through the filament, wading through the vegetative tissue, ultimately entering the ovary of pistillate flowers through the base (Philbrick & Bernardello 1992).

6.2.2 Cross-Pollination

Compared to self-pollination, cross- pollination has certain evolutionary advantages. Cross-pollination promotes genetic heterogeneity within a population which in turn enhances the phenotypic variability of a species. The greater extent of genotypic and phenotypic variability is of evolutionary significance as it provides better adaptability to plants through a range of environmental conditions and also increases the chances of surviving any evolutionary change. However, outcrossing requires transfer of gametes between individuals through either an abiotic or a biotic agency. In the absence or scarcity of pollinators, sexual reproduction may fail to occur.

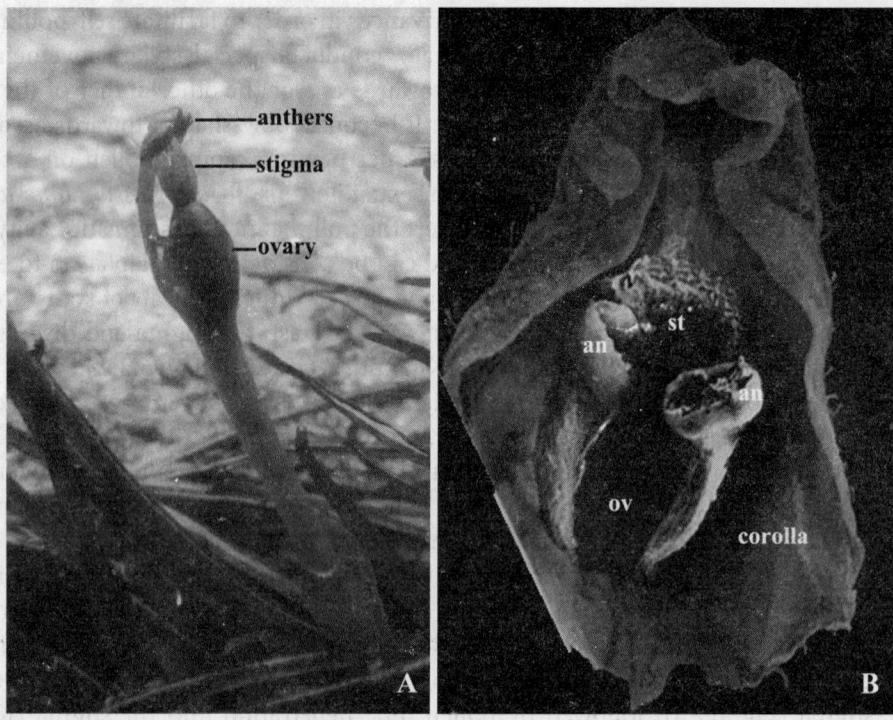

Figure 6.2 Self-Pollination. A. A Podostemaceae member showing clasping of stigma by anthers before dehiscence. **B.** Scanning electron micrograph of *Utricularia inflexa* flower showing pre-anthesis cleistogamy. Note: anthers have reached the level of stigma and are undergoing dehiscence (part of the corolla has been removed to expose the sex organs).

Most of the flowering plants (~90%) are hermaphroditic, yet the proportion of ovules fertilized by self and outcross pollen varies widely among species, ranging from predominant self-fertilization to exclusive outcrossing (Barrett & Eckert 1990). However, some plant species can undergo selfing and outcrossing in the same individual. For example, *Viola* sp. (Violets) and *Clarkia* sp. bear two types of flowers i.e. chasmogamous flowers; which undergo cross-pollination and cleistogamous flowers in which self-pollination occurs. Also, there are some plants in which both self- and cross-pollination may occur within the same flower like in *Myosurus* sp. In this plant, numerous small pistils are borne on a receptacle which elongates with maturity. When the receptacle is small, some of the pistils get pollinated by pollen from surrounding anthers (selfing). Gradually, as the receptacle elongates, pistils become positioned much above the level of the anthers. The remaining unpollinated pistils are then pollinated with cross-pollen brought by insects. Another example of a plant with a mixed pollination system is *Wrightia tomentosa*. The extent of selfing and outcrossing may vary between species. Such mixed pollination strategy is regarded as an adaptive advantage because it ensures reproduction regardless of availability of pollinators via self-pollination and also promotes some outcrossing to increase overall genetic variation.

Box 6.1

Differences between Self and Cross Pollination

Self-Pollination	Cross-Pollination
• It is the transfer of pollen grains from anther to the stigma of the same flower or another flower of the same plant.	• It is the transfer of pollen grains from anther to the stigma of a different flower of another plant of the same species.
• Generally, no external agent is involved in pollination.	• Either a biotic (like insects, birds) or an abiotic (wind, water) agent is involved in pollination.
• Anthers and stigmas mature at the same time.	• Anthers and stigmas may or may not mature at the same time.
• Lesser number of pollen grains are produced as pollination is targeted.	• Substantial quantities of pollen grains are produced as there is no guarantee of pollination.
• Self-pollination can occur in both cleistogamous as well as chasmogamous flowers.	• Cross-pollination can occur only in chasmogamous flowers.
• Self-pollination often leads to inbreeding depression and formation of homozygous off-springs.	• Cross-pollination leads to the formation of heterozygous off-springs which exhibit hybrid vigor.

6.3 Mechanisms to Promote Cross-Pollination

Whenever the disadvantages of inbreeding depression overtake the advantages associated with selfing; outcrossing is thought to be favored in plant populations. Plants resort to a variety of mechanisms to enhance the likelihood of outbreeding or cross-pollination which include the following (Dafni 1992; Barrett 2002):

*   ***Dicliny or Dioecy:*** It is an extreme adaptation that favors cross-pollination. The plant produces flowers of only one type, i.e., flowers are unisexual with either male or female sex organs. Thus, individual plants have either male (staminate) or female (pistillate) flowers imposing outcrossing as the only means of pollination. However, there are many intermediate states exhibited by many dioecious plants such as gynodioecy (where the population comprises some individual plants bearing only pistillate flowers and others with hermaphrodite or bisexual flowers), androdioecy (where the population comprises some individual plants with only staminate flowers and others with bisexual flowers) and trioecy (population with all three types of individual plants: those with only staminate flowers, those with only pistillate flowers and those with bisexual flowers). These alternative mechanisms allow a mix of selfing along with outcrossing in order to provide reproductive assurance.
*   ***Dichogamy:*** In certain hermaphrodite species, the anthers and stigma do not mature at the same time. This temporal separation of maturation of male and

female organs in a bisexual flower which prevents self-pollination is known as dichogamy. In some species, anthers mature and dehisce before the stigma attains maturity or becomes receptive. Such a condition is known as **protandry**. Thus, the pollen from such a flower will not be able to pollinate the pistil of the same flower but can pollinate another flower which may be present on same or on another plant where the stigma is receptive. In *Cactus* and some members of Asteraceae, the anthers dehisce much before the stigma attains maturity or becomes receptive to receive the pollen grains. In some species like *Aristolochia* and *Magnolia,* the stigma becomes receptive much before anther dehiscence and such a condition is called **protogyny**. Under such circumstances, autogamy cannot occur and cross-pollination is the only alternative (Fig. 6.3 A).

- *Herkogamy*: In some plant species, cross-pollination is favored by the spatial separation of male and female organs in a flower. In such species, the position of anther and stigma becomes a barrier to self-pollination and this phenomenon of spatial separation of anther and stigma is known as herkogamy. In members of Caryophyllaceae and Capparaceae, the stigma projects much above the anthers, thereby preventing the access of pollen from same flower. Other example is *Gloriosa superba* where the anthers dehisce at a distance from the stigma, such that pollen does not land on the stigma (Fig. 6.3 B). In some Liliaceae members, the style curves and moves away from dehiscing anthers to get away from possible self-pollination. In some species, anthers are elevated due to elongation of filaments but the style elongates much later. By the time, stigma and style mature and reach the level of anthers, there is no viable pollen left in them, hence pollination is affected by pollen from younger anthers of other plants, e.g., *Penstemon, Ribes.*

Figure 6.3 A. A protogynous flower of *Magnolia* sp. Note the stigmas are expanded (arrow), while stamens are still not dehisced. **B.** A flower of *Gloriosa superba*. Note the stigma is spatially separated from the anthers. **See Color Plates (page 481).**

- **Heterostyly:** Heterostyly is a type of herkogamy in which plants exhibit polymorphic styles. In a heterostylous species, two or more forms of flower are seen which differ in heights at which the stamens and style are positioned. Based on the number of types of flowers produced in a species, heterostyly is of two types: distyly or tristyly. In distylous species, two types of flowers occur; these being pin and thrum types. While the inward flowers (also called pin morphs) possess a long style and short stamens, the thrum flowers (also called thrum morphs) possess a short style and long stamens. This reciprocal arrangement of sex organs increases the chances of pollination between the morphs rather than within the same flower (or same morph). For instance, in *Primula vulgaris*, an insect visiting a thrum morph is likely to get pollen deposited on its body in such a position that facilitates pollination when it visits a pin morph. A more complex condition is tristyly, where flowers with three lengths of styles and stamens (long style, mid-way style and short style) are present. It is estimated that heterostyly is prevalent in at least 28 angiosperm families and is widely associated with animal pollination.

- **Self-incompatibility:** In many species, self-pollination do not result in fertilization, either because of the inability of pollen to germinate on stigma or because of inhibited growth of pollen tubes in the style. Such species are called self-incompatible because the gametes derived from same parent (or same genotype) are unable to fertilize the same genotype. In such cases pollen from genetically distant conspecific plant are required to accomplish pollination and fertilization. More details of self-incompatibility are provided in chapter 8.

6.4 Modes of Pollination

Pollination can be biotic or abiotic depending on the agency or mode involved in transfer of pollen grains from anther to stigma. Wind and water are the abiotic pollination agencies whereas animals, insects, birds are the biotic pollinators. Almost 85% of our plants are dependent on biotic pollinators for pollination and rest are pollinated by abiotic means (Ollerton et al. 2017). Renner & Ricklefs (1995) have estimated that out of the 13,500 genera of angiosperms, nearly 12,000 genera are insect pollinated, 874 are pollinated by wind or water, about 500 genera have species pollinated by birds, and 250 are pollinated by bats. Flowers have variously adapted for shape, size, structure, color, and odor to complement their particular pollinating agent(s). The correlation between the floral traits of plants and corresponding pollinating agent is often referred to as pollination syndrome. The suites of floral traits play a crucial role in dispersal of pollen of plant species that reproduce strictly by means of cross-pollination, and also in some cases of self-pollination. The apparent strength and generality of such correlations indicate plant-pollinator co-evolution.

Box 6.2

Other Mechanisms to Promote Cross-Pollination

- **Enantiostyly:** It is the presence of mirror-image (*enantio:optical*) flowers in a species, where the style shows bending either to the left side or the right side of the floral axis. It is also a type of herkogamy where the curvature of the style is opposite the curvature of at least one stamen. Like in heterostyly, enantiostyly results in the preferential deposition of pollen on one side of the pollinator's body. For example, any insect visiting a left-handed flower would tend to pick up pollen on the right side of its body. When the same insect visits a right-handed flower, it is more likely to pollinate it. Enantiostyly is widely distributed among 13 families of flowering plants including monocotyledons and dicotyledons. Some examples of genera and families where enantiostyly has been reported are *Heteranthera multiflora* (Pontederiaceae), and *Wachendorfia* sp. (Haemodoraceae).

- **Heterodichogamy or Flexistyly:** This type of floral strategy combines both herkogamy and dichogamy. One example is *Alpinia* sp. (tropical gingers, Zingiberaceae) where the populations are composed of two types of plants based on two style morphs. Out of the two, one shows anther dehiscence in the morning and the other in the afternoon. Morph with anther dehiscence in the morning has style curved upwards, so that stigmas are spatially separated from anthers and do not contact the pollinator. After the completion of the male functional phase, styles bend downwards into a position where stigmas can contact pollinators. Contrastingly, the other morph has downward bending style and stigma in the morning but the anther dehiscence takes place in the afternoon when the style curves upwards; hence reducing the interference with pollen donation. These stigmatic movements in the two floral phenotypes are synchronous and pollination occurs only between floral forms. The flexistyly is reported from several genera belonging to about 11 families of angiosperms. *Kingdonia uniflora* (Circaeasteraceae) and *Platycarya* (Juglandaceae) are examples of flexistyly.

- **Inversostyly:** Inversostyly is a lesser known polymorphism in which the floral morphs display reciprocal vertical positioning of anthers and stigma without alteration in organ height. It has been reported in *Hemimeris racemosa* (Scrophulariaceae) which is pollinated by oil-collecting bees (Pauw, 2005). In *H. racemosa* there are two style morphs with styles deflected either upwards or downwards and two stamens located in the opposite position within flowers. When bees visit the two morphs, the pollen gets deposited at two different locations on the underside of their bodies, which facilitates cross-pollination. The polymorphism resembles enantiostyly as both conditions involve flowers that differ in stylar orientation. However, inversostyly is distinct from enantiostyly because of the vertical rather than the horizontal positioning of sex organs.

Source: Based on Barrett 2002; 2010.

6.5 Abiotic Pollination

6.5.1 Pollination by Wind: Anemophily

Anemophily is a form of pollination where pollen is transported by air currents from one plant to another. Approximately 18% of the families and 10% of the angiosperm species are dependent on wind for pollination (Ackerman 2000), including grasses and cereal crops like wheat, rice, corn, rye, barley and oats, the infamous allergenic ragweeds (members of Asteraceae) and many deciduous trees such as mulberry, oak and maples. The ecological setup that favors wind pollination is a dry environment with low precipitation and low humidity conditions facilitating anther dehiscence and pollen dispersal. Therefore, wind pollinated species are usually found at higher latitudes and elevations, especially in temperate areas. They are rare in the tropics, especially in lowland rainforests (Regal 1982). Wind-pollinated species typically inhabit relatively open terrains with low species diversity so that there is little obstruction to pollen flow by intervening vegetation. Anemophilous plants usually occur in dense populations as it increases the probability of reaching receptive stigmatic surfaces. Evolutionarily, anemophily is often associated with dioecy or dichogamy, which encourage outcrossing (Renner & Ricklefs 1995). Wind pollinated plants often display a set of reproductive traits (Table 6.1) that enhance male (pollen dispersal) and female (pollen capture) function. These traits are referred to as wind pollination syndrome (Ackerman 2000).

Wind pollinated flowers are typically simple, green or dull colored with inconspicuous flowers and a reduced perianth. Characteristic exerted, versatile stamens, sticking out of the flower are common in anemophilous flowers, enabling pollen grains to disperse even in mildly blowing wind. The stigmas of anemophilous species are usually feathery which enhances the probability of catching pollen grains present in the surrounding air (Fig. 6.4). Wind pollinated plants produce flowers in huge numbers thereby increasing the overall surface area of stigmas to catch the pollen. Pollen grains of wind pollinated species are light, dry and generally small in size in the range of 17-58 µm (except corn where they measure 90-100 µm) reflecting the aerodynamic requirements. Small size of the pollen grains facilitates their removal from anthers and allows them to travel to farther distances (Niklas 1985).

Once the pollen is released to the wind, the chance of a particular pollen grain landing on a compatible stigma is small. Therefore, pollen grains in wind pollinated plants are produced in huge quantities. Anemophilous pollen are generally unornamented and do not contain pollenkitt substances, removing the possibility of clumping. Pollen grains being very low in protein content are of very low nutritional benefit to pollinators. Also, wind pollinated plants do not waste any resources in the production of nectar or any other rewards and attractants such as color or scent for pollinators. Reduction in the number of ovules per ovary is another characteristic feature of wind-pollinated species, sometimes as low as just one per ovary in many families such as Cyperaceae, Juglandaceae and Poaceae. This reduction is probably to save the plant resources, as chances of anemophilous pollen reaching the right stigma are bleak. Also, more flowers with fewer ovules rather than one flower with many ovules can increase the probability of capturing more pollen grains.

The position of the flowers is also instrumental in capturing the airborne pollen. Inflorescences in most wind pollinated species are exposed and are not covered by the foliage. Generally, flowering in anemophilous plants occurs in a leafless canopy as seen in *Salix*, mulberry and *Hippophae* sp. Highly condensed inflorescence is another feature of wind pollinated species which accounts for effective dispersing and trapping of pollen. The flowers remain arranged in long, pendent catkins or on long inflorescences dangling from branches so that pollen are easily shaken lose by the wind. Plant height and relative positioning of male and female reproductive organs also play an important role in pollen dispersal and mating opportunities in wind pollinated species. While the taller plants disperse pollen in a more

Figure 6.4 Wind pollination syndrome in different plants. A. Catkin-like inflorescence with male flowers of a *Salix* tree. Such inflorescence enhances the pollen dispersal. **B.** Exposed feathery stigma and versatile anthers (arrow) in a grass (*Chloris* sp.) species. **C.** Inflorescence of monoecious *Plantago* sp. showing exposed anthers and stigmas. Note that all these species have reduced perianth.

efficient manner, the shorter plants with small heights are better in capturing pollen. Location of male reproductive organs above the females promotes efficient pollen transfer (as seen in *Carex pedunculata*). In *Pennisetum*, the anthers are placed on long filaments, thus promoting pollen dispersal even in the absence of a strong air current. In maize, the male inflorescences (tassels) are borne terminally on the plant while female inflorescences (cobs) are present laterally at lower levels. Pollen here is large and whenever the wind blows, the tassels are shaken and pollen are released, eventually falling down vertically because of their weight. The stigmas which we see as the silk of the cobs, project beyond the leafy envelopes, and are long, receptive and ready to catch the falling pollen.

Many anemophilous species have evolved special strategies to disperse pollen through wind, raising their chances of successful pollination and reproduction. *Cornus canadensis* and *Morus alba*, utilize a catapult mechanism of anther dehiscence. *Cornus canadensis*, commonly known as bunchberry dogwood, opens its flowers explosively and ejects pollen in the air at an amazing speed of 0.5 m/s to a remarkable height of 2.5 cm (more than ten times the size of the flower). Such explosive floral anthesis relies on an abrupt release of stored elastic energy generated due to turgor pressure created within the anthers.

Box 6.3

Reproductive Traits and Ecological Conditions Associated with Wind Pollination

Reproductive traits	
Stigma	Feathery
Pollen: ovule ratio	High
Pollen diameter	10-50 µm
Pollen ornamentation	Smooth with reduced or absent pollen kit
Stamen filaments	Long
Nectaries	Absent or reduced
Fragrance	Absent or reduced
Perianth	Absent or reduced
Flower Type	Usually unisexual
Inflorescence type	Often condensed, pendulous, catkin-like
Inflorescence position	Away from the vegetation
Ecological conditions	
Wind speed	Low to moderate
Humidity	Low
Precipitation	Infrequent
Plant density	Moderate to high
Neighboring Vegetation	Open

6.5.2 Pollination by Water: Hydrophily

Another kind of abiotic pollination is hydrophily which involves water as a vector in the transportation of pollen. About 3% of all angiosperm families are hydrophilous (Ackerman 2000). However, it is noteworthy that aquatic habitat does not guarantee water mediated pollen transfer. Aquatic families, where pollination takes place above the surface of water are not considered hydrophilous as the species could be self-pollinated (Podostemaceae), pollinated by wind (Cyperaceae, Hydatellaceae, Juncaceae, and Poaceae), or even by insects as in *Nelumbo* and *Nymphaea*. Thus, the term hydrophily strictly applies to a condition where water carries pollen from anther to stigma of a flower. Hydrophily is distinguished on the basis of whether pollen is transported on the surface of water or beneath the water surface. Pollination occurring on the surface of water is called **ephydrophily** whereas underwater pollination is called **hyphydrophily**. The only type of pollination occurring in water is hyphydrophily which is also referred to as true hydrophily.

During ephydrophily, the surface and wind currents cause either pollen or detached staminate flowers or anthers to 'raft' towards emergent or slightly submerged carpellate flowers and inflorescences (Cook 1988). After reaching the female flowers, the pollen or anthers either touch exposed stigmas directly; or the pollen or anthers dip into the meniscus around partly submerged carpellate flowers/inflorescences, subsequently depositing pollen on stigmas (Sculthorpe 1967; Cook 1988). For instance, in *Vallisneria*, the female plants produce flowers under water which later float to the surface of water due to elongation of the pedicel. The male plants also, produce hundreds of flowers under water which detach from the mother plant and float to the surface of water. With the help of water currents, male flowers reach the female flowers and transfer sticky pollen from their anther to the stigma of female flowers. Since water is the ultimate carrier of pollen to the stigma, *Vallisneria spiralis* is the most commonly studied example of hydrophily (Fig. 6.5 A). In *Elodea*, the pollen grains are directly liberated on to the surface of water to be carried by water currents to the stigma. In *Hydrilla*, a unique mechanism is observed where female flowers have three sepals and three petals which enclose a long style bearing three stigmas and an ovary in the base of a hypanthium. Due to an elongation of the hypanthium, the female flower ascends to the surface of the water. During this movement, the perianth segments remain closed over the stigmas and open only at the water surface to form a wide funnel which floats with its rim touching the surface of water. The walls of this funnel hold back the water and prevent the wetting of the stigmas. Thus, the wine glass like perianth of the female flower houses the stigmas which lie just below the level of water but are open to air above. In the meantime, the male flowers detach from the plant and subsequently rise to the surface of the water and float where the perianth segments uncurl. The anthers dehisce explosively and spread pollen up to 20 cm around the open female flowers. For effective pollination to occur, pollen grains get into the air stream and drop vertically to reach the stigma.

Hyphydrophily, or underwater pollination, is relatively uncommon in angiosperms, and largely restricted to the monocotyledons (Les et al. 1997). Sea grasses like *Halophila* and *Thalassia* exhibit hyphydrophily. These grasses have spherical, wettable pollen which are shed as a coherent mass along with thecal slime. They float freely in water and on coming across a receptive stigma, the slime sticks the pollen to the stigmatic papillae and eventually the pollination is accomplished under water. In *Najas minor*, a monoecious species, the flowers are borne singly in the axils of the leaf. Male flowers are positioned above the carpellate ones and bear spherical to ovoid pollen which are heavier than water. Such that they sink and fall on the stigma present below bringing about pollination. A very interesting mechanism of hyphydrophily is reported in free-floating, submerged aquatic *Ceratophyllum*. The highly reduced monoecious flowers are generally borne singly at a node. The male flowers are aggregations of sessile stamens (20–25 in number) whereas the female flowers do not have a stigma but only long styles which are receptive for their entire length. As the stamens mature, they abscise and rise to the surface of the water. More than 10% of the pollen germinate inside the stamen and put out long pollen tubes. The abscised stamens with elongating pollen tubes float on the surface of water for a day or two and then anthers dehisce to liberate the pollen tubes. Due to the weight of the pollen tubes, the stamens start sinking in the water. While descending they may come across the styles of female flowers and pollinate them.

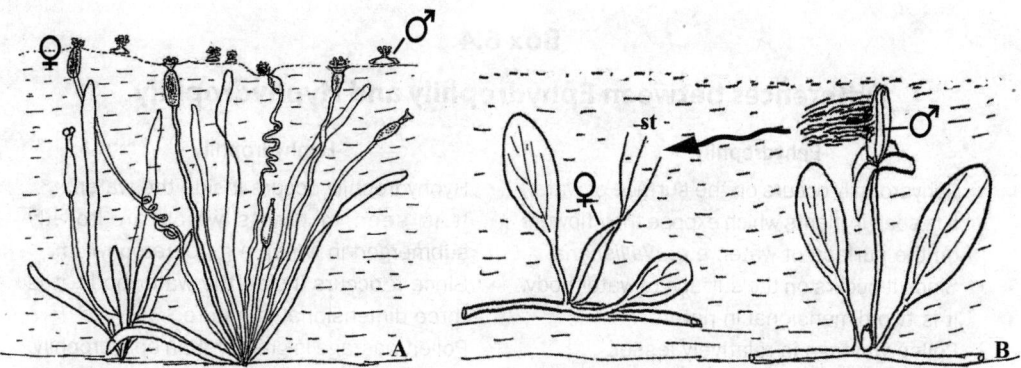

Figure 6.5 A. Ephydrophily in *Vallisneria* sp. Note the floating male flowers (♂) and female flowers (♀) on the water surface. **B.** Hyphydrophily in *Halophila* sp. Male flowers (♂) release pollen grains in slime mass in the water which with help of water currents (arrow) comes across the stigmas (st; ♀: female flowers) and pollination is accomplished.

Water in the form of rain, has also been described as a pollination agent in some terrestrial plants. When it rains, the flowers get filled with water up to a certain level. The pollen grains float in this water to reach the exposed stigmas which are also at the level of water. This type of pollination is called **Ombrophily** and is seen in the orchid, *Acampe rigida*.

It is essential for all forms of hydrophily, that pollen is not destroyed due to wetting before reaching the stigma. To circumvent precocious hydration there are a number of potential waterproofing mechanisms in the pollen wall of hydrophilous plants like thin, reduced and elastic exine with a greater number of apertures (sometimes to the order of omni-aperturate conditions). Floatation is also facilitated by presence of starch grains in the pollen, which provide buoyancy. Pollen grains in hydrophilous plants are of various sizes and shapes, with mostly spherical, ellipsoidal, reniform (boomerang-shaped), and filiform (filamentous or needle shaped) pollen found in various groups (Ackerman 2000). The largest pollen is found in the seagrass *Amphibolis*, which has filamentous pollen, about 6 mm in length and ~20 μm in diameter (Ducker & Knox 1976). Like anemophily, hydrophily is an inherently wasteful process due to non-directionality of pollen. The 3-dimensionality of hyphydrophily leads to high pollen wastage, which may be lesser in ephydrophily because of the confinement of pollen grains to a two-dimensional water surface.

Hydrophilous species exhibit many similarities with anemophilous species in floral syndromes like frequent dicliny, reduced perianth, high pollen production, reduced pollen exine and enlarged receptive stigmatic surface. However, there are significant differences between them with respect to morphology and ultrastructure of pollen and stigma. Unlike anemophily where pollen dispersal distances can be up to several hundred kilometres, the pollen dispersal distances in hydrophiles are entirely dependent on population size and the water body they inhabit and usually do not exceed a few metres.

Box 6.4

Differences between Ephydrophily and Hyphydrophily

Ephydrophily	Hyphydrophily
• Ephydrophily occurs on the surface of water.	• Hyphydrophily occurs inside the water.
• It is seen in plants which expose their flowers on the surface of water, e.g., *Vallisneria*.	• It is seen in plants which are totally submerged in water, e.g., *Ceratophyllum*.
• Since it occurs on the surface of water body, it is two dimensional in nature.	• Since it occurs inside the water body, it is three dimensional in nature.
• Pollen wastage is relatively lesser.	• Pollen wastage is greater than ephydrophily.
• Ephydrophillous plants usually have water repellent pollen.	• Hyphydrophilous pollen grains can endure wetting.

Box 6.5

Differences between Biotic and Abiotic Pollination

Biotic Pollination	Abiotic Pollination
• Pollination carried out by various biotic agents like insects, birds, bats.	• Pollination carried out by abiotic agencies i.e. wind and water.
• Flowers are large, showy and bright in color.	• Flowers are small, dull and inconspicuous.
• They produce copious amounts of nectar and fragrance.	• As no pollinators are required, flowers do not invest much in nectar and fragrance
• Relatively less number of pollen grains are produced.	• Huge amounts of pollen are produced because of non-directionality of pollen movement.
• Pollen grains are large and sticky.	• Pollen grains are generally small and light.
• Pollen exine is sculptured and mostly contains pollenkitt.	• Pollen exine is highly reduced.
• Low pollen-ovule ratio.	• High pollen-ovule ratio.
• Receptive stigmatic surface is comparatively smaller.	• Receptive stigmatic surface is usually enlarged.

6.6 Biotic Pollination

6.6.1 Pollination by Insects: Entomophily

The insect-plant association is one of the oldest and most thoroughly studied association. According to Endress (1994), insects were the original pollinators of early flowering plants, and all the non-insect pollinators came much later in the evolution. Preserved gymnosperm pollen in insect guts provide direct evidence that insects were consuming pollen as far back as the Permian, much before the appearance of angiosperms. The early associations between angiosperms and insects were more generalized as insects were still in the process of developing adaptations for flower feeding. Wasps, moths, thrips, beetles and flies were the main pollinators in earlier times. The parallel radiation of angiosperms and the more plant-dependent insects such as bees and butterflies followed in the Mid-Cretaceous and into the Tertiary (Grimaldi 1999; Grimaldi & Engel 2005) periods, during which time mutations occurred in both plants and insects. In many cases, these mutations were selectively advantageous. Mutations that favored the plants were those that increased the recognisability of a flower for an insect; such as - flower color, size, shape, fragrances and those that caused insects to revisit primarily for; rewards such as sugary nectar or protein rich pollen and stamens. Similar adaptive mutations occurred in insects that increased their capacity to recognize the most nutritious flowers from a distance. As a result, congruent rise of flowering plants and insects took place and the flowers possessing specific combinations of floral traits appeared to be visited more frequently by corresponding pollinator taxa. This co-evolution not only occurred between flowers and insects, but also occurred between flowers and birds, and flowers and bats. A whole range of insects from bees, wasps, butterflies, beetles, moths, flies to ants and even cockroaches have been reported to facilitate pollination in plants.

6.6.1.1 *Bee and Wasp Pollination*

Of all flowering plants pollinated by insects, nearly 80% are pollinated by bees. Pollination by bees is called **Melittophily**. Bees are instrumental in pollinating most of our crop species, ornamental plants and are considered as generalist pollinators (those who visit several species). Bees pollinate many members of Asteraceae, Brassicaceae, Burseraceae, Clusiaceae, Euphorbiaceae, Fabaceae, Orchidaceae, Rosaceae, Rutaceae and Sapotaceae. Bee pollinated species are characterised by the presence of showy, brightly colored flowers, usually blue or yellow (Fig. 6.8 A-C). These flowers often have markings called "nectar guides" that are visible in the ultraviolet wavelengths which help these insects to easily locate nectaries such as in *Taraxacum* (common dandelions), *Digitalis purpurea*, *Jacaranda mimosifolia*, and many orchids. Also, bees avoid red flowers because they possess a trichromatic visual system which is sensitive to ultraviolet, blue and green wavelengths (Menzel & Backhaus 1991) and cannot distinguish red color. Melittophilous flowers have a sweet, delicate fragrance and produce copious amounts of nectar which is hidden but not very deep in the flower. Such flowers vary in size and shape, open during the day, have concealed sex organs, and suitable landing platform for the pollinators.

Pollinator bees (honeybees, bumblebees and orchid bees) differ in their size, tongue length and pollination behavior (solitary or colonial). Bees can range in size from 2-32 mm, with the most common honey bee being about 12mm long. Some plants can only be pollinated by bumblebees as their pollen is more or less firmly held by anthers, which require a buzz (sonication) to be shaken out of the anther. Bumblebees grab on to the flowers and vibrate their flight muscles at a frequency that causes dislodging of pollen grains from anthers as seen in *Solanum melongena* (Wanigasekara & Karunaratne 2012). Bees usually carry and transport pollen in corbiculae or pollen baskets. A corbicula is a slightly curved area surrounded by hair located on the hind legs of bees. During collection, the pollen are moistened with nectar, and packed into the corbiculae, which act as baskets for holding the pollen (Fig. 6.8 B-C).

Salvia exhibits a specialized floral mechanism for bee pollination known as the **turn-pipe mechanism** or the **staminal lever mechanism**; (Regine 2004; Reithet al. 2007). The bi-lipped corollas of salvias have two stamens with two lobes each. The stamens are attached to the corolla tube. The upper lobe of the stamen is fertile and is separated by an elongated connective tissue from the lower sterile lobe. The fertile lobes are under the hood of the upper lip of the corolla and the sterile parts of both the anthers join to form a sterile plate, placed just above the lower lip of the corolla. When an insect lands on the lower lip, the fertile lobe automatically comes down to touch the back of the bee and thereby deposits pollen grains. The flowers of *Salvia* are protandrous, and on maturity their stigmas bend down and touch the backs of bees which have already visited other flowers and thus receive pollen grains (Fig. 6.6).

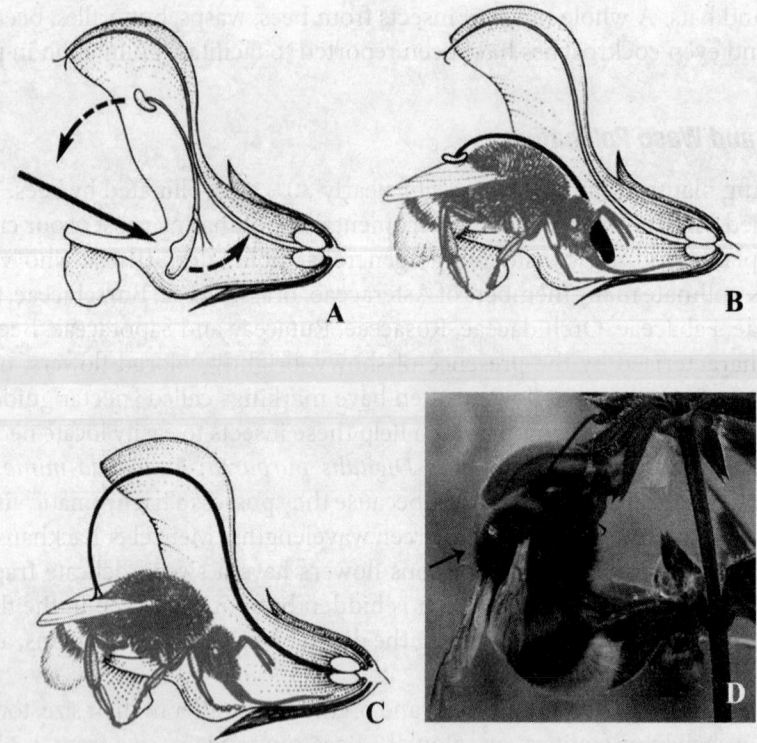

Figure 6.6 Staminal Lever mechanism of pollination in *Salvia pratensis*. A–C. Longitudinal section of flowers. **A.** Lever like stamens (arrow: sterile anther or plate, broken arrows depict the direction of movement of anthers) and distantly placed stigma. **B.** A bee pushing the plate or sterile anther towards the interior side such that the fertile anther bends down to thorax and dislodge the pollen grains. **C.** A flower with mature stigma. On entry of bee; stigma touches the pollen laden dorsal thorax of bee. This transfers the pollen from bee to stigma. **D.** A bee visiting *S. pratensis* flower where the dorsal thorax is touching the stigma (arrow). Pollen mass can be seen on bee's dorsal thorax as a yellow spot. **See Color Plates (page 481)**. (*Source*: A–C Reith et al. 2007, *reproduced with permission*; D. Koch et al. 2007, published under CC BY 4 license.)

Several species of Anacardiaceae, Burseraceae and Simaroubaceae are pollinated by wasps. Striking examples of wasp pollination by means of sexual deceit are seen in family Orchidaceae. Several orchid species resort to sexual deception by producing scents in the flowers that mimic sex pheromones to attract pollinators. A beautiful example of this form of deception is seen in the orchid *Chiloglottis trapeziformis*, where flowers release a compound mimicking the odor of the sex pheromone normally produced by female thynnine wasps (*Neozeleboria cryptoides*). This scent of sex pheromone lures male thynnine wasps to mate with the flower. In this attempt of pseudo-copulation, male wasps pick up pollen from the flower and transfer it to a next conspecific flower when once again lured by the pheromone-mimicking odor (Bower 1996). Similarly, *Telipogon peruvianus* flowers release volatile compounds that attract four species of male tachinid flies (Martel et al. 2016).

6.6.1.2 *Fly Pollination*

Pollination by flies is called **Myophily**. Such flowers are often red-brown in color, have a fleshy texture and emit a characteristic odor of rotten meat (e.g., Sterculiaceae and Aristolochiaceae). *Aristolochia*, is a typical example of myophily. Here, the pollination by flies is also referred to as the **Fly-trap mechanism** of pollination (Fig. 6.7). The large zygomorphic flowers of *Aristolochia* have perianth consisting of three united, tubular lobes.

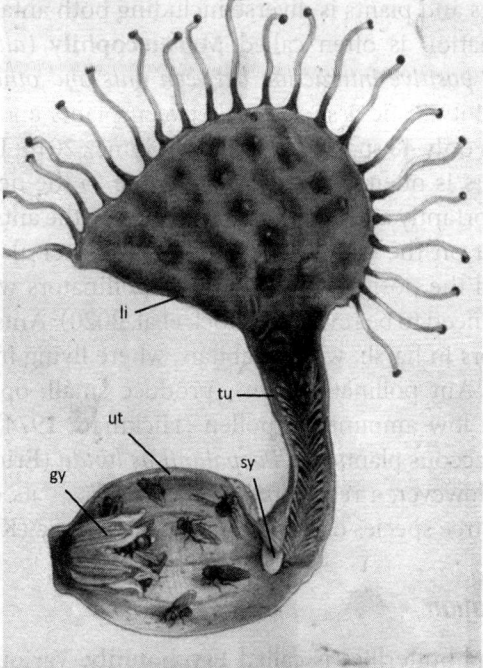

Figure 6.7 Fly-trap mechanism in *Aristolochia* sp. Gynostemium has stamen on the outside and stigma on the inside. Note the number of flies trapped inside utricle and the deflexed perianth tube hairs preventing the exit of trapped flies. **Li:** limb, **tu:** tube, **sy:** syrinx, **ut:** utricle, **gy:** gynostemium. **See Color Plates (page 482).** (*Source:* Bliss et al. 2013, *reproduced with permission.*)

The perianth tube is lined inside with strigose downward-pointed hairs, which facilitate the entry of flies into the chamber of the flower, but restrict their exit. The basal part of the perianth tube is swollen and is called utricle. It surrounds the fused styles, stigmas, and anthers, collectively known as the gynostemium. The utricle terminates into an expanded limb which is often colorful and thought to visually attract pollinators. At the time of anthesis, flies are attracted by the strong odor emitted by the flowers. Light penetrating the thin tissues at the base of the corolla tube attracts the flies to the stigmas and the stamens. Once the flies have reached the base of the flower or the utricle, their exit is prevented by a lining of hinged hair. Flowers show protogyny, with stigmas being receptive at the time of anthesis. The anthers dehisce later in the day or the next day. Following anther dehiscence, the thorax of the flies becomes loaded with sticky pollen grains. Eventually, the hairs on the inner wall of the perianth tube wither, allowing the trapped fly to exit. When the same fly gets attracted by another flower, chances are that it will transfer the pollen grains on to the stigma of that flower. Most of the species of *Aristolochia* are reported to be pollinated by saprophagous flies of different families, including Anthomyiidae, Chloropidae, Milichiidae, Phoridae, Sarcophagidae and Syrphidae (Murugan et al. 2006). Furthermore, the pollinators oviposit in the flowers, and their larvae grow on the fallen, decaying flowers on the ground, reflecting a symbiotic relationship between the plant and the pollinator.

6.6.1.3 *Ant Pollination*

Association between ants and plants is diverse including both antagonistic and mutualistic interactions. Ant pollination is often called **Myrmecophily** (*although myrmecophily is a broader term used for positive interaction between ants and other organisms like plants, arthopods and fungi*). Mutualistic association between plants and ants is rather rare and has been documented in only 48 species (de Vega & Gómez 2014; Delnevo et al. 2020). The activity of ants in flowers is often deemed negative due to the nectar robbing, pollinator repulsion and most importantly antimicrobial secretions by the ants, which has been shown to have a negative effect on the viability of pollen (Beattie et al. 1985). However, recent reports have highlighted the positive role of ants as pollinators which indicate that these negative generalizations need to be revisited (Stock et al. 2020). Ants usually serve as reliable pollen carriers/pollinators in harsh, windy habitats, where flying insects are less active and ant population is high. Ant pollinated plants produce small, open flowers with readily accessible nectaries and low amounts of pollen (Hickman, 1974). Pollination by ants is mostly reported in herbaceous plants like *Paepalanthus lundii* (Eriocaulaceae), *Chamorchis alpina* (Orchidaceae). However, a recent report demonstrates its occurrence in *Syzygium occidentale*, an endemic tree species of the Western Ghats, India (Kuriakose et al. 2018).

6.6.1.4 *Butterfly Pollination*

Pollination by moths and butterflies is called **Psychophily.** Various plants of families like Boraginaceae, Rubiaceae, Asteraceae and Verbenaceae are pollinated by butterflies. Butterfly pollinated flowers are usually large, showy, pink or purple, mildly scented with a wide landing area. They also possess nectar guides with nectaries hidden in narrow tubes or spurs, to be reached by long tongue or proboscis of butterflies (Fig. 6.8 E). Butterfly pollinated flowers

---

Box 6.6

Differences between Turn-pipe Mechanism and Fly-trap Mechanism

Turn-pipe mechanism	Fly-trap mechanism
• This is observed in bee pollinated species.	• This is observed in fly pollinated species.
• Flowers are bright in color with a sweet delicate fragrance and high amounts of nectar as in *Salvia*.	• Flowers are dull in color and foul-smelling as in *Aristolochia*.
• Flowers open during the day.	• Flowers open at night.
• Presence of a bi-lipped corolla with two partially sterile anthers forming a plate near the lower lip of corolla for bees to land. Fertile halves are separated from plate by elongation of connective. The two halves together form a lever or a turn-pipe.	• Corolla is tube-shaped with deflexed hair lining the corolla tube. The deflexed hair facilitate entry of flies in the corolla tube but prevent their exit.
• The bee lands on lower lip of corolla to extract nectar and pushes against the sterile plate. This automatically brings down the fertile anthers to touch the back of the bee, thereby depositing pollen. When these pollen laden bees visit other flowers, their backs rub against the stigmas and effect pollination. Hence this type of pollination is called **Turn-pipe or Lever Mechanism**.	• Since the fly gets trapped within the corolla tube, this mechanism is called the **Fly trap Mechanism**. In its futile attempts to move out of the corolla tube, the fly rubs against the stigma and effects pollination if it already has pollen on it.

---

offer nectar as reward than pollen, as generally butterflies are unable to digest pollen. Most studies have indicated that butterflies are generalist pollinators and visit a variety of plants. However, some of the exclusively butterfly pollinated plants are *Caesalpinia pulcherrima*, *Asclepias syriaca*, *Aesculus californica*, *Platanthera ciliaris*, *Phlox* sp., *Cadaba fruiticosa*, *Gloriosa superba* (Reddy & Bai 1984; Daniels et al. 2020).

6.6.1.5 *Moth Pollination*

Moths are predominantly nocturnal visitors and are usually restricted to nectar containing flowers which are dull yellow, or white. Such flowers stand out in fading evening light or even at night. Flower visiting moths are divided in two groups: large hawkmoths (Sphingidae) and small settling moths (mainly Noctuidae). Large hawkmoths have a high metabolic rate, often forage by hovering in front rather than alighting on the flowers, and usually have proportionally longer proboscis (Faegri & Pijl 1979). Pollination by hawkmoths is called **Sphingophily** and flowers pollinated by them are usually large, tubular or brush-like. Settling moths fly slowly between flowers, and seek nectar by landing and walking on the inflorescences. Pollination by settling moths is called **Phalaenophily**. The phalaenophilous flowers are usually small, and are often present in aggregated inflorescences (Faegri & Pijl 1979). *Yucca* is a classic example of moth-pollinated species with obligate pollination

Figure 6.8 Insect pollination. A–C. Bee pollination. Bees foraging on **A.** *Escholozia* sp., **B.** *Papaver* sp. **C.** *Calendula* sp. Insets (A and B) show bees carrying pollen deposits or pollen basket (arrow). **D.** Moth pollination in *Wrightia tomentosa*. Note the moth has inserted its proboscis inside the flower (arrow, anther) to take away the nectar. **E.** A hedge blue butterfly foraging on a capitulum for nectar. Note proboscis is inside the tubular disc floret. **See Color Plates (page 483).**

mutualism existing between yuccas (Agavaceae) and yucca moths (Lepidoptera). The adult moth pollinates yucca flowers and her larvae feed on the developing seeds. Yucca is pollinated by a moth known as *Tegeticulla yuccasella*. Each spring, adult moths emerge from underground cocoons and male and female moths mate on yucca plants. When a female is ready to lay eggs, she visits a yucca flower to collect pollen. While collecting the sticky pollen with her tentacles, the female moth keeps packing the pollen into a ball and sticking it under her head. When same moth flies off to another freshly opened yucca flower to lay her eggs inside the ovary, she scrapes a small amount of pollen from the sticky ball and packs it into tiny depressions within the style, thus effecting pollination. Before leaving the flower, she marks it with a pheromone (a chemical other moths can sense) pointing to other moths that the flower has already been visited. When the eggs hatch, the larvae feed on yucca seeds within the fruit. Typically, there are more seeds than the larvae in a particular flower can feed on. When the larvae finish eating, they burrow out of the fruit – usually during rains.

Wrightia tomentosa, a member of Apocynaceae is pollinated both by hawkmoths and settling moths. In order to consume nectar, a moth inserts its proboscis through the widest part of a slit present between the connivent anthers. While withdrawing its proboscis, it slides along the tapered region of the anther and hence fresh pollen gets deposited on it (Fig. 6.8 D). When the moth visits the next flower, the stigmatic papillae with sticky secretions are able to scrape off the pollen from the proboscis and this leads to pollination (Barman et al. 2018).

6.6.1.6 *Beetle Pollination*

Pollination by beetles is often regarded as being the least-specialized among animal pollination. Beetle-pollinated flowers (e.g., in Annonaceae, Araceae, Lauraceae, Magnoliaceae, Myristicaceae, Palmae) are usually large and bowl shaped with exposed sex organs, greenish or off-white, with or without strong fruity or a foul smell, without nectar guides, producing huge quantities of pollen grains, and have easily accessible rewards. Pollination by beetles is called **Cantharophily.**

6.6.2 Pollination by Birds: Ornithophily

According to a recent estimate, over 900 species of birds pollinate about 500 genera of plants (Anderson 2016). Bird pollination or **Ornithophily** is reported in 25 families of flowering plants like Bignoniaceae, Bombacaceae, Cannaceae, Costaceae, Ericaceae. Some common bird pollinated species are *Bauhinia variegata*, *Erythrina variegata*, *Butea monosperma* (Fabaceae), *Bombax ceiba*, *Chorisia speciosa* (Bombacaceae), *Spathodea latifolia* (Bignoniaceae) and the Rhododendrons. Ornithophilous flowers are characterized by their large size to accommodate birds. They are often red or scarlet so that these brightly colored flowers can be readily noticed by birds from a distance. Nonetheless, a range of other colors like orange, yellow, white and violet have also been reported in few bird pollinated plants (Cronk & Ojeda 2008). The petals of such flowers often fuse to form urn-shaped or tubular structures so that the birds can put in their beaks to lap up the nectar. Also, the tubular corolla excludes insects or other animals that may destroy flowers. Bird pollinated flowers are devoid of any smell or odor and produce copious amounts of nectar with the help of nectaries present at the base of the ovaries, as a reward for birds. Flowers with generalist bird pollinators tend to have a diluted nectar whereas those having specialist bird pollinators (those who visit one or a very few plant species like sunbirds or hummingbirds) have relatively concentrated nectar. The most nectar-feeding birds are found in the New World.

Birds involved in pollination have some attributes exclusive to them like long distance flights, great visual ability and strong spatial memory. Hummingbirds (Trochilidae), sunbirds (Nectariniidae), honey-eaters (Meliphagidae) and lorikeets (Psittaculidae) have evolved as major groups of specialist pollinators. Other than these, short-billed birds such as weavers, bulbuls and starlings are only occasional pollinators. Morphologically, pollinator birds possess long and curved beaks, brushy tongues and are capable of hovering flight or perching on the flower. The hovering behavior is found mainly in hummingbirds, which collect nectar without landing on the plant. Therefore, plants pollinated by humming birds usually have hanging or pendant flowers. To accommodate hummingbirds, flowers are often positioned in such a way that the birds can comfortably feed while hovering without brushing their wings on adjacent stems or leaves. The exact angle of the blooms decides which hummingbird species will pollinate it as different species have different bill shapes. Unlike hummingbirds, the perching birds land on stems, leaf stalks, adjacent branches, and flower buds. Flowers and inflorescences of plants pollinated by perching birds are typically sturdier through the formation of sclerenchyma or collenchyma tissue in various floral parts. Sunbirds are common perch birds which land on an appropriate spot and access nectar by dipping their sharp bills into the top of flowers and inserting their brush tipped tongues deep into the bottom of the flower. Sunbird-pollinated flowers are typically long, tubular, and red-to-orange in color, showing convergent evolution with many hummingbird-pollinated flowers but are sturdier. Small herbaceous plants may be pollinated by birds that perch on the ground, and usually their flowers are oriented vertically erect as seen in *Gastrolobium praemorsum* (Keighery 1982). Perching is a more widespread foraging mechanism than hovering (Cronk & Ojeda 2008).

Pollination mechanisms seen in ornithophilous species are an outcome of both spatial and temporal association of the reproductive organs with respect to the position of pollinating birds. For instance, in *Ipomopsis,* a ring is formed by five stamens such that pollen are deposited all around the base of the birds' beaks. Bird-pollinated *Salvia* species transfer pollen either by means of a staminal lever mechanism or in species with immovable stamens, the distance between nectar and pollen is large resulting in pollen getting readily deposited on the bird's feathered head while foraging for nectar.

There are several examples where a relationship exists between flowering season of plants and breeding season of birds when the requirement of food is maximum. Trees in Fabaceae and Bombacaceae shed their leaves to maximize the display of flowers during the breeding season of pollinator birds. Nectar serves as an inducement for birds to visit the flowers and affect pollination. Some bird pollinators share a generalized relationship with the plants, as many bird-pollinated flowers are also visited by insects and bats. *Butea monosperma* receives seven species of birds, but only the purple sunbird and the three-striped squirrel are effective pollinators (Tandon et al. 2003). It is noteworthy that a single plant species may be visited by many types of birds but not every bird is a pollinator. To qualify as pollinator for a species, the bird must access the sex organs of the flowers and carry and transfer the pollen load. Birds which just visit the flowers or access the flower for pollen and nectar but do not transfer the pollen load on stigma are referred to as *floral visitors*. Many birds, insects and other flower visitors are known to remove nectar from flowers through a hole pierced or bitten in the corolla without providing any pollination service. They are referred to as *nectar robbers*. For instance, Singh et al. (2014) reported that in *Tecomella undulata*, a member of Bignoniaceae, all floral visitors are not pollinators. The bright showy flowers of *T. undulata* attract two species of bulbul (*Pycoronotus* sp.) and one species of sunbird (*Nectarinia asiatica*). The purple sunbirds rob the flowers of nectar by piercing through the corolla tube but are not instrumental in the pollination. On the other hand, the two species of bulbul carry out pollination by foraging deep inside the corolla tube. While inside the corolla tube, their head comes in contact with both the dehisced anthers as well as the stigma, ultimately affecting pollination (Fig. 6.9).

Box 6.7

Summary of Features of Biotic and Abiotic Pollination

Biotic pollination:

- It is carried out by living organisms like bees, birds, moths, butterflies, bats, ants.
- 80% of all pollination is carried out by biotic agents.
- Biotic Pollination is dependent on the availability of the pollinators.
- Flowers are generally bright in color.
- Copious amounts of nectar are available in such flowers as reward for pollinators.
- Generally, the flowers are sturdy and large so as to allow the pollinator to land on them. They can also be leathery in texture.
- Common examples are *Ceiba pentandra, Butea monosperma, Alstonia*.

Abiotic pollination:

- Pollination is carried out by non-living agencies like wind and water.
- 20% pollinations are carried out by abiotic agencies.
- Pollination is not constrained by the availability of the pollinators.
- Flowers are generally dull in color and not very ornate.
- There is absence of nectar in such species.
- There is reduction in perianth parts, stigmas and anthers as species do not need to attract pollinators.
- It is an inherently wasteful process due to non-directionality of pollen and therefore, pollen are produced in huge quantities.
- Common examples are *Helianthus, Vallisneria, Zostera*.

Figure 6.9 Ornithophily in the bright orange-red and tubular flowers of *Tecomella undulata*. A. Sunbird robbing flowers for nectar. Inset: a hole (arrow) made in corolla tube for stealing nectar. **B.** *Pycnonotus leucotis* (Bulbul) is a legitimate pollinator of *T. undulata*. While ingesting the nectar present in corolla tube, the reproductive organs touch the head of the bird and pollen transfer occurs (arrow). **See Color Plates (page 483)**. (*Source*: Singh et al. 2014, published under CC BY 4 License.)

6.6.3 Pollination by Bats: Chiropterophily

Pollination by bats is known as **Chiropterophily**. About 528 species (28 species in India) of angiosperms in 250 genera of 67 families are reportedly pollinated by nectar-feeding bats, which are important pollinators in lowland, moist or wet, tropical forests (Fleming et al. 2009). Only two families (out of a total 18 families) of bats, i.e. Pteropodidae and Phyllostomidae have been reported as specialized nectar-feeders or pollinators (Fleming & Muchhala 2008). Specialized nectar bats navigate and search for flowers in the dark by sending out high-pitched calls and listening to the echoes that bounce back. They can distinguish nectar-containing flowers from background clutter through their sense of echolocation. Morphologically, they have an elongated rostrum, dentition that is reduced in size and number of teeth, and a long tongue tipped with hair-like papillae which they use to collect nectar rapidly during brief flower visits (Freeman 1995).

Bat-pollinated plants, are almost exclusively trees, bearing flowers/inflorescence away from foliage, such as projected above the canopy, suspended on long stalks (flagelliflory), or emerging from branches or trunk (cauliflory), or borne on deciduous trees after they have dropped their leaves (Faegri & van der Pijl 1979). The flowers are large and tubular and may be radially symmetrical, often of the shaving-brush type and dull in color (white or green). They have a wide opening and often emit a musty/foetid odor while opening in the evening, night or early morning (nocturnal anthesis). The chiropterophilous flowers produce large volumes of hexose-rich nectar as high as 15 ml per flower per night, to compensate for highly metabolic flights of bats. Bats use smell, and echolocation to determine flower characteristics and depend on their excellent spatial memory to visit the same trees repeatedly. Their long narrow tongues enable them to reach nectar concealed deep in the flowers and during the search they get loaded with pollen. A nectar feeding bat makes around 80–100 flower visits during one night and needs to consume about 1 mg sugar or 5 µl nectar with 20% sugar concentration at each flower visit (Fleming et al. 2009). Other than nectar, some bats are known to consume and digest the pollen grains and some even feed on insects present in the flowers. Some examples of bat pollinated trees are *Adansonia digitata*, *Bauhinia variegata*, *Bombax Ceiba* (Raju et al. 2005, Fig. 6.10), *Kigellia pinnata* and *Oroxylum indicum* (Vikas et al. 2009).

6.6.4 Pollination by Snail: Malacophily

Pollination by snails is called **Malachophily**, a rare phenomenon reported in *Volvulopsis nummularium* (family Convolvulaceae, commonly known as the morning glory family), a prostrate rainy-season weed, which is also visited by honey bees. Flowers open in the morning and last only for half a day. Majority of the pollination is carried out by bees. However, on rainy days, when bees are not active, the snails are the exclusive pollinators (Sarma et al. 2007) (Fig. 6.11).

Figure 6.10 Bat Pollination. A frugivore bat feasting on nectar in a *Bombax ceiba* flower. During foraging the pollen grains gets deposited on their fur and snout. **See Color Plates (page 484)**.

Figure 6.11 Snail pollination in *Volvulopsis nummularium*. **See Color Plates (page 484)**.

6.6.5 Some Unusual Pollinators

Over and above the pollinators mentioned in the previous sections, some other groups of animals that have also been documented as pollinators are non-flying mammals like marsupials, primates, rodents and some reptiles like lizards. Primate pollination is best documented on the island of Madagascar where black and white ruffed lemurs are the main pollinators of traveler's palm, *Ravenala madagascariensis* (Kress et al. 1994). Flowers of traveler's palm are large and are protected by tough bracts that have to be forcibly opened by a pollinator, which is remarkably done by lemurs. While lapping nectar using their long snouts and tongues, they collect pollen on their muzzle and fur, and then transport it to the next flower. Some other well-known examples of non-flying mammal pollinators are mice, pollinating Pagoda lily (Wester et al. 2009); and, honey possum, a tiny Australian marsupial pollinating *Banksia* (Hackett & Goldingay 2001).

Box 6.8

Summary of Floral Syndrome Associated with Various Pollinator Groups

Pollinator group	Pollination type	Floral characters and examples
Bees	Melittophily	Flowers are highly variable in morphology, brightly colored and have concealed sex organs. They are mild in odor and are often zygomorphic with landing platforms and nectar guides. Nectar although hidden, does not lie deep in the flower and so, is easy to access. Examples: Members of Fabaceae, Orchidaceae, Rosaceae, Rutaceae
Fly	Myophily	Flowers are small-large, reddish-brown with a fleshy texture. They emit a characteristic rotten odor and are without nectar guides. Fly-trap mechanism is exhibited by some myophilous species. Examples: Members of Anacardiaceae, Sterculiaceae
Moths	Phalaenophily (Small settling moths) and Sphingophily (Large hawkmoths)	Flowers are pale and heavily scented, exhibit nocturnal blooming and may be horizontal or pendent. Nectar is deeply hidden in long tubular corolla or spurs. Examples: Members of Agavaceae, Apocyanaceae
Butterfly	Psychophily	Flowers are large, showy, pink or purple, mildly scented with a wide landing area. Possess nectar guides with nectaries hidden in narrow tubes or spurs. Offer more nectar than pollen. Examples: Members of Boraginaceae, Verbenaceae
Ants	Myrmecophily	Small, open flowers with readily accessible nectaries and low amounts of pollen. Examples: Members of Eriocaulaceae, Orchidaceae
Beetles	Cantharophily	Flowers are unspecialized, usually large, either cylindrical or bowl shaped, with exposed anthers and stigma (sex organs), may produce strong odor, without nectar guides, produce huge numbers of pollen grains and rewards are easily accessible. Examples: Members of Annonaceae, Arecaeae
Birds	Ornithophily	Flowers are robust possessing long tubular corolla, large amounts of nectar (of low sugar concentration), variously colored (often red) and without odor. Examples: Members of Bombacaceae, Bignoniaceae
Bats	Chiropterophily	Flowers are large and tubular, borne away from foliage, dull in color with wide flower opening. They often emit a musty/foetid odor and show nocturnal blooming. Examples: Members of Bombacaceae, Bignoniaceae and Myrtaceae

Although rare, lizards are also known to pollinate plants. Pollination by lizards is more common on islands than on mainland because of their high density on islands due to lower predation risk. However, Crozier et al. (2019) documented pollination of *Guthriea capensis* (commonly known as 'Hidden Flower') by the Drakensberg Crag Lizard in South Africa which is a rare report of a continental plant pollinated by reptiles. The lizard is attracted by the unusual scent of visually cryptic flowers. It feeds copiously on nectar and carries pollen on its snout.

It is well known that insect mediated pollination is mostly carried out by pollinators belonging to four insect orders: Coleoptera, Diptera, Hymenoptera and Lepidoptera. There are instances of other group of insects effecting pollination in some plants. These groups include cockroaches, true bugs, net-winged insects, wasps and crickets. With new evidence emerging and scientists still discovering new associations between plants and the pollinators, it will not be surprising if these rare pollinators turn out to be commonly involved in specialized plant–pollinator systems.

Box 6.9

Habitat Fragmentation, Climate Change and Pollination

Non availability or decline of the prospective pollinators at the time of flowering is a major cause of concern for the successful breeding of any species. Due to habitat loss and fragmentation many plants have become confined to small, highly isolated populations. One of the major consequences of fragmentation is the change in the foraging patterns of pollinators and pollinator behavior. If the density of plants in a particular area becomes low due to fragmentation, the availability of pollinator and its visitation frequency are also negatively affected. Sometimes, this results in complete breakdown of pollination mutualisms that may even trigger the local extinction of plants. For instance, *Elaeocarpus ganitrus* (Elaeocarpaceae) is a threatened species from northeast India and present only in fragmented habitats. Such conditions have led to sparse pollination and high rates of abortion of ovules in the species (Khan et al. 2005). In many flowering plants, all the flowers do not result in fruits or seeds. One of the major causes for this is pollen limitation that occurs when plants produce fewer fruits or seeds than they would with adequate pollen receipt. Pollen limitation has also been widely reported and often interpreted as an evidence for insufficient pollinator visitation in fragmented habitats. Declining populations of pollinators in a habitat may lead to a global pollination crisis affecting not only the plant populations but also the organisms that either directly or indirectly rely on them.

Both industrialization and urbanization activities are also causing imbalance in plant pollinator associations. Exposure to city and street lights and the noise of road traffic have been shown to hamper navigational abilities of pollinators, eventually reducing their populations. Increasing environmental pollution is also posing a serious concern for biodiversity, as it affects growth and reproduction of both plants and animals. Although most severe effects of pollution are on seed germination and seedling survival, pollination too is affected in polluted locations. Pollen viability, crucial for reproductive success of plants, is considered to be highly sensitive to pollutants. Floral scents released by plants to attract pollinators are highly volatile and easily get chemically altered by pollution. As the majority of our crop plants are pollinated by insects, the use of uncontrolled insecticides has proved to be hazardous even for the beneficial pollinating insects especially bees.

Climate change is emerging as another serious threat to the plant-pollinator relationship. Temperature influences the timing and duration of biological processes such as growth, development, flowering and senescence, alone or in combination with other factors. Unusual temperature fluxes, precipitation and seasonal variation can have serious concerns for the pollination success of a plant. Visible symptoms of climate change have been seen in the form of alteration in phenology of many plant species. For instance, precocious flowering in response to rise in temperature has been reported in *Rhododendron arboruem*. These phenological shifts can decouple the relationship between the plant and the pollinators such that when the flowering is either delayed or takes place earlier due to shift in phenology, pollinators are unavailable. As a consequence, the relationship between plant and pollinator is disrupted.

6.7 Ambophily

There are several species which show mixed pollination by insects as well as the wind. Such a condition is known as **Ambophily** which is seen in species of *Plantago* sp. (Plantaginaceae), *Salix* (Salicaceae), *Mallotus* sp. (Euphorbiaceae) and *Discaria chacaye* (Rhamnaceae). Evolutionarily, ambophily is considered a transitional stage during evolution of wind or insect pollination (Culley et al. 2002). Some studies have shown that ambophily can be advantageous in environments where conditions favoring either wind or biotic pollination are uncertain. Thus, insect pollination may compensate for uncertainty of wind in anemophily or on the other hand, wind pollination may compensate for decline in pollinator visits.

6.8 Pollen Storage

As the pollen grains are used widely in reproductive biological studies and in other applied fields of biology, it is very important that the pollen be available all through the year, irrespective of the flowering season of the plants. Thus, it becomes very important to store the pollen grains for extended periods of time under suitable conditions to maintain their viability, so that they can be used as and when required. Pollen storage is useful for breeding programmes, genetic conservation, pollination limitation studies and breeding system experiments.

Some of the experimental studies where pollen grains are required are:

1. Studies on various aspects of pollen biology, biotechnology and pollen allergy.
2. Preserving germplasm for future use or international exchange.
3. Overcoming cross-ability barriers imposed due to temporal and spatial isolation of parent species.
4. For studying the mechanism of self-incompatibility.
5. Eliminating the need to grow pollen parents continuously in breeding programmes.
6. Implementing supplementary pollinations in order to sustain yield.

6.8.1 Methods of Pollen Storage

The four important methods which are employed for storage of pollen grains are:

- **Storage under low temperature:** This method employs low temperature and low relative humidity (RH) conditions. Pollen grains are collected in small vials and placed in an air tight desiccator with a drying agent like dry silica gel to maintain a low relative humidity. If the pollen grains are to be stored for short term like for a few weeks or months, then the sealed desiccators are kept in a refrigerator or a deep freezer. If the storage has to be for more than a year, then the pollen are stored at sub-freezing temperature of -20°C or even less. Pollen grains of cereal crops are not able to withstand desiccation and hence, require high RH under refrigerated conditions. To overcome this problem, a method called 'Pollen dryer' developed by Barnabas & Kovacs (1997) is employed, where air with 20-40% humidity at a temperature of 20°C is blown through pollen.

- **Storage of vacuum dried/ freeze dried pollen:** Freeze drying involves rapid freezing of pollen (-60–80°C) and gradual removal of water under sublimation. Such pollen grains are exposed to cooling and vacuum drying at the same time and later stored at sub-zero temperature. For effective use of this method, the duration of drying and subsequent rehydration has to be optimized. This method is used for a number of species (except the cereals) for long term storage. In recent years, this method has become unpopular because of the availability of the cryopreservation technique, which is simpler and more efficient.

- **Cryopreservation:** The storage at ultralow temperature (-196°C) in liquid nitrogen is known as cryopreservation. Here, the pollen grains are dried so that their water content goes below a threshold level and then are stored in liquid nitrogen at -196°C. Under ultra-low temperatures, the pollen viability is retained for a very long period of time as the pollen undergo negligible metabolic changes at such low temperatures. Like storage under low temperature, cryopreservation of cereal pollen grains uses a 'pollen dryer' technique. Using this technique, the cryopreserved pollen of *Secale cereale, Triticum aestivum* and *Zea mays* maintain viability for up to 10 years. Pollen drying can also be achieved using microwaves or sunlight, especially for bee collected pollen.

- **Storage in organic solvents:** Pollen grains dried over silica can be stored in various organic solvents in air tight vials. Commonly used solvents are acetone, benzene, ethanol, ether, chloroform and phenol. Although these solvents are considered toxic to various living organisms, they enhance the viability of pollen grains which may further affect fertilization. Pollen grains are recovered by filtration or evaporation of the solvent and eventually used for pollination or other similar experiments. Pollen grains of lily have been shown to be capable of not only germinating but also effecting fertilization, even after 10 years of being in ethanol. Extensive studies by Jain & Shivanna (1988; 1989; 1990) have shown that non-polar solvents such as hexane, cyclohexane and diethyl ether are more effective for pollen storage than the polar solvents.

6.8.2 Utility of Pollen Storage: Some Examples

- *Phalaenopsis* is the genus of Orchidaceae that is immensely popular as potted plant and cut flower worldwide due to long shelf life. About 60 species of the *Phalaenopsis* are native to tropical Asia and the larger islands of the Pacific Ocean. Economically valuable, plants of *Phalaenopsis* are extensively used for the development of further novel varieties. For a breeding program, only superior parents are selected for cross-hybridization which may not be available all the time. Hence, properly stored pollen grains become an alternative for both cross and manual pollination in plant improvement programs. Recently Yuan et al. (2018) evaluated storage behavior of pollen from a *Phalaenopsis* hybrid at different temperatures, (room temperature, 4°C, –20°C, and –80°C) for periods of up to 96 weeks. The viability of stored pollen grains was assessed by TTC staining, *in vitro* germination and manual pollination. The study highlighted the fact that stored pollen (at all these temperatures) showed viability for a minimum of 4 weeks and were also capable of successful pollination. Pollen grains lost their viability after the duration of 4 weeks when stored at room temperature. Storing at lower temperature helps in maintaining prolonged viability for up to 40 weeks when stored at 4°C, and up to 96 weeks when stored at –20°C and –80°C.

- The double-coconut palm (*Lodoicea maldivica*, family: Arecaceae) of Seychelles is one of the most interesting plant species in the world. The seed of this palm resembles two coconuts fused together and is the largest seed in the world. *Lodoicea maldivica* is a dioecious species with separate male and female plants. A single female plant of double coconut was raised in the Acharya Jagadish Chandra Bose Indian Botanic Garden (AJCBIBG), Botanical Survey of India (BSI), Howrah, using seeds obtained from Seychelles in 1894. It is the only living plant of this type in India. The plant bloomed for the first time in 1988 but in the absence of male flowers, no fruits could be obtained. To bring about fruit set, pollen grains were brought in an ice box from the Peradeniya Botanic Garden, Sri Lanka, where both male and female trees exist, in October, 2006. The pollen grains were maintained in a deep freezer (–10°C or below) until the female flowers became receptive. However, even after repeated artificial pollinations from 2006-2012, no success was achieved. Then another set of pollen grains was received from Nong Nooch Tropical Garden, Thailand in August 2013 which was stored in a deep freezer (–10°C or below) till the female flowers became receptive. On 17 August 2013, artificial pollination was conducted with stored pollen in three out of seven female flowers in a newly emerged inflorescence which showed receptivity. The artificial pollination proved successful as the flowers yielded fruits (Hameed 2016). Thus, international pollen exchange using pollen storage methods can prove highly beneficial for *ex situ* conservation of globally threatened species.

Glossary

Ambophily: Occurrence of both insect and wind pollination in a species.

Autogamy: Transfer of pollen from anther to the stigma of the same flower.

Chasmogamous flower: Flowers that open at maturity and expose their sex organs in order to facilitate pollination.

Cleistogamous flower: Flowers that do not open at maturity to expose their sex organs, hence remaining as floral buds.

Floral foragers: Insects, animals or birds which visit the flowers or access the flower for pollen and nectar but are not instrumental in pollination i.e. do not transfer pollen.

Floral rewards: These are attractions made available to pollinators and foragers. The pollen and nectar together constitute the chief floral rewards.

Geitonogamy: When the transfer of pollen occurs from anther to the stigma of another flower borne on the same plant

Genet: A single organism or a group of organisms that are derived from a single zygote and are genetically distinct from others.

Nectar robbers: Floral visitors that steal or rob nectar from flowers through a hole pierced or bitten in the corolla without providing any pollination service.

Pollen basket: It is curved area surrounded with hairs on the hind legs of some bee species and is also called corbicula. With the help of corbicula, the bees collect and carry pollen grains to their hives.

Pollination syndrome: A correlation between floral traits (such as the color and shape of corolla) and the corresponding pollinating agent.

Pre-anthesis cleistogamy: A form of self-pollination prior to opening of flower. Anther dehiscence and pollen deposition on stigma occurs well before the floral anthesis and ensures obligate selfing.

Ramet: A physiologically distinct organism that is a part of genetically similar group of individuals.

Xenogamy: Transfer of pollen from anther to the stigma of another flower present on a different plant.

Key Questions

Q6.1 **Differentiate Between:**
   a. Biotic and Abiotic pollination
   b. Turn-pipe mechanism and Fly-trap mechanism of pollination
   c. Autogamy and Xenogamy
   d. Chasmogamy and Cleistogamy
   e. Hydrophily and Anemophily
   f. Ephydrophily and Hyphydrophily
   g. Ornithophily and Entomophily
   h. Self-pollination and Cross-pollination

Q6.2   Fill in the blanks:
a. Pollen are light in weight in .... species.
b. Dichogamy promotes ....pollination.
c. .... pollination leads to hybrid vigor.
d. Transfer of pollen from anther to the stigma of another flower of the same plant or of another plant of the same clone is called ....
e. *Viola* and *Clarkia* sp. bear both .... and .... flowers.
f. Water repellent pollen grains are found in ....
g. .... bees shake the pollen out of anthers by their buzzing action.
h. .... is an example of pollination by moths.
i. .... birds hover over the flowers to collect nectar.
j. Shaving-brush type of flowers, are found in species pollinated by ....
k. Cryopreservation is a common method of storing ....
l. Buzz pollination is a characteristic feature of family........................

Q6.3   Discuss wind pollination syndrome highlighting the similarities between wind and water pollination.

Q6.4   Discuss various modes of pollen storage. Provide a suitable example.

Q6.5   Give an account of various means/mechanisms promoting cross-pollination.

Q6.6   Self-pollination is known to provide reproductive assurance, then why have many species evolved mechanisms of cross-pollination?

Q6.7   Provide an account of various mechanisms favoring self-pollination with suitable examples.

Practicals

Exercise 6.1: To study the time of anther dehiscence

The information on the time of anther dehiscence is essential for collection of fresh pollen grains in order to perform wide variety of tests including manual pollination. There are some generalizations with respect to time of anther dehiscence and the mode of pollination, such as the anthers in wind pollinated flowers usually dehisce during the early hours of the day. It has also been observed that the bee and butterfly pollinated flowers open at ~8 am, as this group of insects become less active in bright sunlight. On the contrary, the moth and bat pollinated flowers exhibit anther dehiscence at dusk.

To observe time of anther dehiscence, the following steps are required:

1. Tag the flowers at different developmental stages on the basis of their size. For easy identification one may place the floral buds and flowers on a graph paper.
2. If feasible, also dissect the anthers from all the marked stages and observe their size.
3. Observe the tagged flowers at intervals of one day at least. These observations are species dependent and will vary. Sometimes staggered observations all through the day might be required. Record the time of anther dehiscence and mark the corresponding anther developmental stage.

To make this exercise more effective one may check the various stages of pollen (in anthers of marked stages) development by staining with acetocarmine. The mature pollen grains at a particular stage provide clues for anther maturity and subsequent anther dehiscence. Here, it must be noted that anther dehiscence might take place before the floral anthesis, especially in selfed plant species.

Exercise 6.2: To study the histochemistry of floral rewards

Pollen grains and nectar are the main rewards offered by a flower to the foragers and pollinators. These rewards are rich in various metabolites and help to fulfil the energy requirement of the pollinators and foragers. The pollen grains of biotically pollinated species are sticky due to presence of lipids and proteins on their exine as well as in their cytoplasm. Plants produce a range of nectar in terms of its quality and quantity. There are reports which have suggested nectar also contains phenolics and alkaloids. Flowers which possess phenolics are pollinated by specific group of insects and these compounds may serve as a repellent for other insects. Bird pollinated flowers offer copious quantities of nectar rich in sugar (but diluted). On the other hand, species with abiotic mode of pollination do not produce nectar and their pollen grains are of a dry-type characterized by absence of pollen kit substances and presence of starch in their cytoplasm. Thus, the study of chemical properties of floral rewards can provide clues to the mode of pollination and also the type of biotic pollinators involved. There are simple histochemical tests that can be employed to study the chemical nature of pollen grains and nectar. Some of the procedures are described here.

I. Pollen grains reserve materials and pollen histochemistry

Material Required
Flowers of different species, Potassium Iodide solution, Coomassie Brilliant Blue G solution (CBBR G), Sudan Black B solution, Double Distilled Water (DDW), glass slides, coverslips, glass droppers, needles, light microscope

A. Test for starch (Dafni 2005)

Procedure

1. Add hydrated pollen grains onto a glass microslide.
2. Add a drop of Iodine solution to the pollen grains and mix well.
3. Place a coverslip and observe slide under microscope.

Result
Brown color in pollen grains shows presence of starch (Fig. 6.12 A).

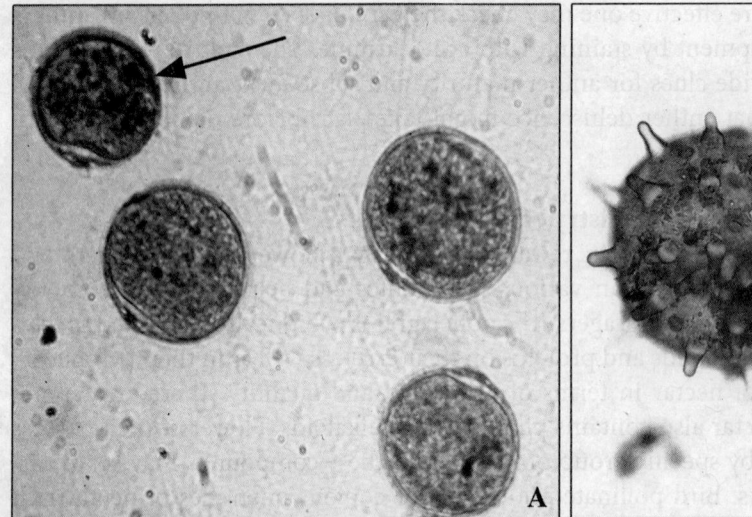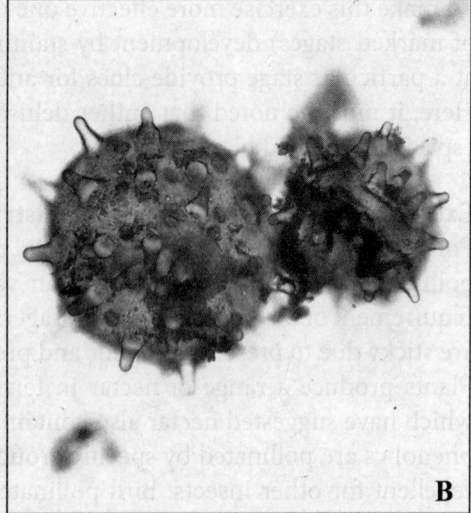

Figure 6.12 Pollen grains of A. *Hippophae rhamnoides* showing presence of starch. **B.** *Hibiscus rosa-sinensis* stained blue with Coomassie Brilliant Blue G solution indicating the presence of proteins. **See Color Plates (page 485).**

B. Test for proteins (Dafni 2005)

Procedure

1.   Add hydrated pollen grains onto a glass microslide.
2.   Add a drop of Coomassie Brilliant Blue G solution to the slide and mix well.
3.   Place a coverslip and observe slide under microscope.

Result

Blue color in pollen grains shows the presence of proteins (Fig. 6.12 B).

C. Test for lipids (Dafni 2005)

Procedure

1.   Add hydrated pollen grains onto a glass microslide.
2.   Add a drop of Sudan Black B solution to the slide and mix well.
3.   Place a coverslip and observe slide under microscope.

Result

Brown-Black color in pollen grains shows presence of lipids

Preparation of Test solutions

   a.   Preparation of Iodine Solution: To 100 ml of DDW add 0.2 g of KI and 1 g of I_2 and stir the solution well.

b. Preparation of CBBR G Solution: To 100 ml mixture of 45% methanol and 10% acetic acid, add 0.25 g of the CBBR G dye. Before use, dilute one part of this solution by adding 3 parts of 20% sucrose solution.

c. Preparation of Sudan Black B Solution: Prepare saturated solution of Sudan Black B in 95% ethanol and filter the solution before use.

Precautions

1. Always use hydrated pollen grains for the tests.
2. Use clean needle for mixing the pollen grains with stain solution.
3. Use only glass droppers to pour the stain.
4. Always use freshly prepared solution of CBBR G and Sudan Black B.

II. Nectar Analysis

For nectar analysis, one has to collect fresh nectar. It is very easy to collect nectar from large sized flowers (e.g., *Bombax ceiba*) and from flowers which produce large amounts of nectar. Collection can be easily done by using glass dropper or a pipette. However, in bee pollinated or moth pollinated flowers nectar is present inside the corolla tube. The collection of nectar from such flowers can be done using simple glass capillaries, micropipettes and calibrated capillaries. Such calibrated capillaries aid in quantification of the nectar amount as well. If only qualitative analysis is the objective, then Whatman's paper wick can also be used. As the nectar shows presence of carbohydrates (sucrose), lipids and proteins, its quality may get degraded on long term storage in liquid form. Thus, nectar samples can be preserved for long duration by absorption of collected samples on the Whatman's paper disc followed by its storage in a desiccator. These paper discs with nectar sample can be utilized for nectar analysis. The methods for nectar analysis provided here are basic and one may perform these tests easily in the field.

A. Sugar quantification in nectar using hand held refractometer (Dafni 2005)

Principle: A refractometer is a simple instrument used for measuring concentrations of aqueous solutions. When light enters a liquid, it changes its direction. The phenomenon is called refraction. Refractometers measure the degree to which the light bends (the angle of refraction). The refraction angle is then correlated with established refractive index (nD) values by the refractometer. Using these values, one can determine the concentrations of solutions.

Procedure

1. Collect nectar drop using glass or microcapillary.
2. Transfer a drop of nectar to the refractometer screen without touching it.
3. Close the screen.
4. Read the scale of the refractometer through eyepiece.

B. Colorimetric test for quantitative estimation of amino acid content in nectar (Dafni 2005)

Preparation of Histidine Scale

1. Prepare Histidine solution of 3.9 mg/ml concentration in 20% sucrose solution. This solution would correspond to the **concentration 10** on standard scale.
2. Prepare a series of 50% dilutions down to scale 1 (~49um).
3. Mark 11 spots on a Whatman No.1 filter paper.
4. Put 1µl of each Histidine concentration (1–10) on each numbered spot and sucrose solution on spot 11. This strip of paper serves as the standard scale.
5. Spray 0.2% Ninhydrin solution (prepared in acetone) over the strip. Leave the paper for 2 hrs at room temperature or at 65°C for 30–40 minutes.
6. The marked spots will develop lighter to darker blue-purple coloration (as per concentration of spots).

Procedure

1. Make a nectar spot on filter paper disc. Dry it and spray it with 0.2% Ninhydrin solution (in acetone).
2. Leave the paper for 24 hrs at room temperature or at 65°C for 20–30 minutes.
3. Compare the color intensity with the standard scale and note the concentration.

C. Colorimetric/qualitative test for phenolics in nectar (Dafni 2005)

Procedure

1. Place a nectar spot on a filter paper disc and dry it.
2. Cover the nectar spot with Folin-Ciocalteu's phenol reagent (0.5N) and allow it to dry.
3. Add a drop of 20% Na_2CO_3.

Result

Blue/Brown color indicates presence of phenols.

D. Colorimetric/qualitative test for alkaloids in nectar (Dafni 2005)

Procedure

1. Place a nectar spot on a filter paper disc and dry it.
2. Add a drop of Dragendorff's reagent to the dried nectar spot.

Result

Orange/Red color in yellow background indicates the presence of alkaloids.

Preparation of Dragendorff's reagent

1. Solution A: 0.85 g of basic Bismuth subnitrate in 50 ml of 20% Acetic acid.
2. Solution B: 20 g KI in 50ml water.
3. Mix equal volumes of Solution A and Solution B. Store in refrigerator in dark bottle. Before use dilute the mix 20 times with cold 20% acetic acid.

E. Colorimetric/qualitative test for proteins in nectar (Dafni 2005)

Procedure

1. Place a nectar spot on a filter paper disc and dry it.
2. Immerse filter paper disc with dried nectar spot in 0.1% solution of Bromophenol blue (prepared in methanol) for 30minutes.
3. Rinse the disc in 5% Acetic acid (3 times)

Result

Green/Blue color indicates the presence of proteins

Precautions

1. Refractometer must be calibrated before use.
2. Do not touch the screen of the refractometer with the glass capillary.
3. Hold the Whatman disc (with nectar spot) using clean forceps.
4. Ninhydrin is carcinogenic. Hence should be sprayed only after donning protective gloves and mask.
5. Folin-Ciocalteu's phenol reagent is light sensitive, so must be stored in amber colored bottles in refrigerator (4°C).
6. Dragendorff's reagent must be freshly prepared.

Exercise 6.3: To study the floral features of various species and discuss them in context of pollination syndrome

Having read about the numerous features which are associated with the specific types of pollinators and different modes of pollination, it is time to move on to studying these floral features in the field and correlate them to a putative mode of pollination based on the pollination syndrome exhibited by the plant. You may select plants of your choice or select from the list given below:

Suggested plants: *Zea mays* (maize), Wheat, Sunflower, *Bombax ceiba*, *Brassica* sp., *Tridax procumbens*, *Datura stramonium*, Mulberry

Observe the flowers and note down the following characteristics:

Inflorescence type:
Flower shape:
Flower size:
Flower symmetry:
Flower color:
Flower odor:
Position of stamens:
Pollen type (dry / sticky):
Presence of nectar:
Quantity of nectar:

On the basis of the characteristics of flowers you may think about, speculate on, and discuss the possible modes of pollination.

Bibliography

Ackerman, J. D. (2000). Abiotic pollen and pollination: ecological, functional, and evolutionary perspectives. *Plant Systematics and Evolution* 222: 167–85.

Anderson, J. T. (2016). Plant fitness in a rapidly changing world. *New Phytologist* 210: 81–7.

Barman, C., Singh, V. K., Das, S. and Tandon, R. (2018). Floral contrivances and specialized pollination mechanism confer strong influence to elicit mixed-mating in *Wrightia tomentosa* (Apocynaceae). *Plant Biology* 20: 546–54.

Barnabas, B. and Kovacs, M. (1997). Storage of pollen. In K. R. Shivanna and V. K. Sawhney, eds, *Pollen Biotechnology for Crop Production and Improvement*. New York: Cambridge University Press, pp. 293–314.

Barrett, S. C. H. and Eckert, C. G. (1990). Current issues in plant reproductive ecology. *Israel Journal of Botany* 39: 5–12.

Barrett, S. C. H. (2002). The evolution of plant sexual diversity. *Genetics* 3: 274–84.

Barrett, S. C. H. (2010). Understanding plant reproductive diversity. *Philosophical Transactions of the Royal Society: B-Biological Sciences* 365: 99–109.

Baumann, R. M., Claßen-Bockhoff, R. G. and Speck, T. (2007). New insights into the functional morphology of the lever mechanism of *Salvia pratensis* (Lamiaceae). *Annals of Botany* 100: 393–400.

Beattie, A. J., Turnbull, C. T., Hough, T. et al. (1985). The vulnerability of pollen and fungal spores to ant secretions: evidence and some evolutionary implications. *American Journal of Botany* 72: 606–14.

Bliss, B. J., Wanke, S., Barakat, A. et al. (2013). Characterization of the basal angiosperm *Aristolochia fimbriata*: a potential experimental system for genetic studies. *BMC Plant Biology* 13: 13.

Bower, C. C. (1996). Demonstration of pollinator-mediated reproductive isolation in sexually deceptive species of *Chiloglottis* (Orchidaceae: Caladeniinae). *Australian Journal of Botany* 44: 15–33.

Cook, C. D. K. (1988). Wind pollination in aquatic angiosperms. *Annals of the Missouri Botanical Garden* 75: 768–77.

Cronk, Q. and Ojeda, I. (2008). Bird-pollinated flowers in an evolutionary and molecular context. *Journal of Experimental Botan* 59: 715–27.

Crozier, L. G., McClure, M. M., Beechie, T. et al. (2019). Climate vulnerability assessment for Pacific salmon and steelhead in the California Current Large Marine Ecosystem. *PLoS ONE* 14(7): e0217711.

Culley, T. M., Weller, S. G. and Sakai, A. K. (2002). The evolution of wind pollination in angiosperms. *Trends in Ecology and Evolution* 17: 361–9.

Dafni, A., Kevan, P. G. and Husband, B. C. (2005). *Practical Pollination Biology*. Cambridge (Canada): Enviroquest Limited.

Dafni, A. (1992). *Pollination Ecology*. Oxford: Oxford University Press.

Daniels, R. J., Johnson, S. D. and Peter, C. I. (2020). Flower orientation in *Gloriosa superba* (Colchicaceae) promotes cross-pollination via butterfly wings. *Annals of Botany* 125: 1137–49.

de Vega, C. and Gómez, J. M. (2014). Polinización por hormigas: conceptos, evidencias y futuras direcciones. *Ecosistemas* 23: 48–57.

Delnevo, N., van Etten, E. J., Clemente, N. et al. (2020). Pollen adaptation to ant pollination: a case study from the Proteaceae. *Annals of Botany* 126: 377–86.

Ducker, S. C. and Knox, R. B. (1976). Submarine pollination in seagrasses. *Nature* 263: 705–6.

Endress, P. K. (1994). *Diversity and Evolutionary Biology of Tropical Flowers*. New York: Cambridge University Press.

Faegri, K. and van der Pijl L. (1979). *The Principles of Pollination Ecology*. Oxford: Pergamon Press.

Fleming, T. H., Geiselman, C. and Kress, J. W. (2009). The evolution of bat pollination: a phylogenetic perspective. *Annals of Botany* 104: 1017–43.

Fleming, T. H. and Muchhala, N. (2008). Nectar-feeding bird and bat niches in two worlds: pantropical comparisons of vertebrate pollination systems. *Journal of Biogeography* 35: 764–80.

Freeman, P. W. (1995). Nectarivorous feeding mechanisms in bats. *Biological Journal of the Linnean Society* 56: 439–63.

Grimaldi, D. and Engel, M. S. (2005). *Evolution of the Insects.* Cambridge: Cambridge University Press.

Grimaldi, D. (1999). The co-radiations of pollinating insects and angiosperms in the Cretaceous. *Annals of Missouri Botanical Garden* 86: 373–406.

Hackett, D. J. and Goldingay, R. L. (2001). Pollination of *Banksia* spp. by non-flying mammals in north-eastern New South Wales. *Australian Journal of Botany* 49: 637–44.

Hameed, S. S. (2016). Artificial pollination and fruit set in double coconut growing in India. *Current Science* 110: 976–8.

Hanna, C., Foote, D. and Kremen, C. (2014). Competitive impacts of an invasive nectar thief on plant–pollinator mutualisms. *Ecology* 95: 1622–32.

Hanna, C., Naughton, I., Boser, C. et al. (2015). Floral visitation by the Argentine ant reduces bee visitation and plant seed set. *Ecology* 96: 222–30.

Hannah. L., Steele, M., Fung, E. et al. (2017). Climate change influences on pollinator, forest, and farm interactions across a climate gradient. *Climatic Change* 141: 63–75.

Hickman, J. C. (1974). Pollination by ants: a low-energy system. *Science* 184: 1290–2.

Ibarra-Isassi, J. and Oliveira, P. S. (2018). Indirect effects of mutualism: ant–treehopper associations deter pollinators and reduce reproduction in a tropical shrub. *Oecologia* 186: 691–701.

Jain, A. and Shivanna, K. R. (1988). Storage of pollen grains in organic solvents: effect of solvents on pollen viability and membrane integrity. *Journal of Plant Physiology* 132: 499–501.

Jain, A. and Shivanna K. R. (1989). Loss of viability during storage is associated with changes in membrane phospholipid. *Phytochemistry* 28: 999–1002.

Jain, A. and Shivanna, K. R. (1990). Membrane state and pollen viability during storage in organic solvents. In S. K. Sinha, P. V. Sane, S. C. Bhargava and P. K. Agrawal, eds, *Proceedings of International Congress of Plant Physiology.* New Delhi: Society of Plant Physiology and Biochemistry, pp. 1341–9.

Jain, A. and Shivanna, K. R. (1990). Storage of pollen grains of *Crotalaria retusa* in oils. *Sexual Plant Reproduction* 3: 225–7.

Keighery, G. J. (1982). Bird-pollinated plants in Western Australia. In J. M. Powell and A. J. Richards, eds, *Pollination and Evolution.* Sydney: Royal Botanic Gardens, pp 77–90.

Khajuria, A., Verma, S. and Sharma, P. (2010). Stylar movement in *Valeriana wallichii* DC. – a contrivance for reproductive assurance and species survival. *Current Science* 100: 1143–4.

Khan, M. L., Bhuyan, P. and Tripathi, R. S. (2005). Effects of forest disturbance on fruit set, seed dispersal and predation of Rudraksh (*Elaeocarpus ganitrus* Roxb.) in northeast India. *Current Science* 88: 133–42.

Khanduri, P., Chaudhary, A., Uniyal, P. L. and Tandon, R. (2014). Reproductive biology of *Willisia arekaliana* (Podostemaceae), a freshwater endemic species of India. *Aquatic Botany* 119: 57–65.

Khosla, C., Shivanna, K. R. and Mohan Ram, H. Y. (1998). Pollination in the aquatic insectivore *Utricularia inflexa* var. *stellaris. Phytomorphology* 48: 417–25.

Koch, L., Lunau, K. and Wester, P. (2017). To be on the safe site–ungroomed spots on the bee's body and their importance for pollination. *PLoS ONE* 12(9): e0182522. https://doi.org/10.1371/journal.pone.0182522.

Kress, J. W., Schatz, G. E., Andriani, F. A. and Morland, H. S. (1994). Pollination of *Ravenalamadagascariensis* (Strelitziaceae) by lemurs in Madagascar: evidence for an Archaic Coevolutionary System? *American Journal of Botany* 81: 542–51.

Kuriakose, G., Sinu, P. A. and Shivanna, K. R. (2018). Floral traits predict pollination syndrome in *Syzygium* species: a study on four endemic species of the Western Ghats, India. *Australian Journal of Botany* 66: 575–82.

Les, D. H., Cleland, M. A. and Waycott, M. (1997). Phylogenetic studies in Alismatidae, II: evolution of marine angiosperms (seagrasses) and hydrophily. *Systematic Botany* 22: 443–3.

Mangla, Y. and Gupta, C. K. (2014). Love in the air-wind pollination: ecological and evolutionary considerations. In R. Kapoor, I. Kaur and M. Koul, eds, *Plant Reproductive Biology and Conservation*. Delhi: I. K. International, pp. 234–45. ISBN: 978-93-82332-90-9.

Mangla, Y. and Tandon, R. (2014). Pollination ecology of Himalayan sea buckthorn, *Hippophae rhamnoides* L. (Elaeagnaceae). *Current Science* 106: 1731–5.

Martel, C., Cairampoma, L., Stauffer, F. W. and Ayasse, M. (2016). *Telipogon peruvianus* (Orchidaceae) flowers elicit pre-mating behavior in *Eudejeania* (Tachinidae) males for pollination. *PLoS ONE* 11: e0165896. https://doi.org/10.1371/journal.pone.0165896.

Menzel, R. and Backhaus, W. (1991). Color vision in insects. In Gouras, P., ed., *Vision and Visual Disfunction*. London: Macmillan, 1991, pp. 262–88.

Muchhala, N. (2006). The pollination biology of *Burmeistera* (Campanulaceae): specialization and syndromes. *American Journal of Botany* 93: 1081–9.

Niklas, K. J. (1985). The aerodynamics of wind pollination. *The Botanical Review* 51: 329–83.

Ollerton, J. (2017). Pollinator diversity: distribution, ecological function, and conservation. *Annual Review of Ecology, Evolution and Systematics* 48: 353–76.

Pauw, A. (2005). Inversostyly: a new stylar polymorphism in an oil-secreting plant, *Hemimeris racemosa* (scrophulariaceae). *American Journal of Botany* 92: 1878–6.

Philbrick, C. T. (1984). Aspects of floral biology, breeding system, and seed and seedling biology of *Podostemum ceratophyltum* (Podostemaceae). *Systematic Botany* 9: 166–74.

Raju, S. A. J., Rao, P. S. and Rangaiah, K. (2005). Pollination by bats and birds in the obligate outcrosser *Bombax ceiba* L. (Bombacaceae), a tropical dry season flowering tree species in the Eastern Ghats forests of India. *Ornithological Science* 4: 81–7

Reddi, C. S. and Bai, G. M. (1984). Butterflies and pollination biology. *Proceedings of Indian Academy of Science (Animal Science)* 93: 391–6.

Regal, P. (1982). Pollination by wind and animals: ecology of geographic patterns. *Annual Review of Ecology and Systematics* 13: 497–524.

Regine, C. B., Speckb, T., Twerasera, E. et al. (2004). The staminal lever mechanism in *Salvia* L. (Lamiaceae): a key innovation for adaptive radiation? *Organisms, Diversity and Evolution* 4: 189–205.

Reith M., Baumann, G., Regine, C. B. and Speck T. (2007). New insights into the functional morphology of the lever mechanism of *Salvia pratensis* (Lamiaceae). *Annals of Botany* 100: 393–400.

Renner, S. S. and Ricklefs, R. E. (1995). Dioecy and its correlates in the flowering plants. *American Journal of Botany* 82: 596–606.

Sarma, K., Tandon, R., Shivanna, K. R. and Mohan Ram, H. Y. (2007). Snail-pollination in *Volvulopsis nummularium* . *Current Science* 93: 826–31.

Sculthorpe, C. D. (1967). *The Biology of Aquatic Vascular Plants*. Edward Arnold, London.

Sehgal, A., Sethi, M. and Mohan Ram H. Y. (2009). Development of the floral shoot and preanthesis cleistogamy in *Hydrobryopsissessilis* (Podostemaceae). *Botanical Journal of the Linnean Society* 159: 222–36.

Shivanna, K. R. (2003). *Pollen Biology and Biotechnology*. New Delhi: Oxford and IBH Publishing Co. Pvt. Ltd. ISBN 81-204-1579-5..

Singh, V. K., Barman, C. and Tandon, R. (2014). Nectar robbing positively influences the reproductive success of *Tecomella undulata* (Bignoniaceae). *PLoS ONE* 9 (7): e102607. https://doi.org/10.1371/journal.pone.0102607.

Stock, W. D., Byrne, M., DelNevo, N. et al. (2020). Pollen adaptation to ant pollination: a case study from the Proteaceae. *Annals of Botany*. https://doi: 10.1093/aob/mcaa058.

Tandon, R., Shivanna, K. R. and MohanRam H. Y. (2003) Reproductive biology of *Butea monosperma* (Fabaceae). *Annals of Botany* 92: 715–23.

Thakur, P. and Bhatnagar, A. K. (2013). Pollination constraints in flowering plants–human actions undoing over hundred million years of co-evolution and posing an unprecedented threat to biodiversity. *The International Journal of Plant Reproductive Biology* 5: 1–45.

Vikas, G. M., Tandon, R. and Mohan Ram H. Y. (2009). Pollination ecology and breeding system of *Oroxylum indicum* (Bignoniaceae) in the foothills of the Western Himalaya. *Journal of Tropical Ecology* 25: 93–6.

Wanigasekara, R. W. M. U. M. and Karunaratne, W. A. I. P. (2012). Efficiency of buzzing bees in fruit set and seed set of *Solanum violaceum* in Sri Lanka. *Psyche 2012*: 1–7. https://doi:10.1155/2012/231638.

Wester, P., Stanway, A. and Pauw A. (2009). Mice pollinate the Pagoda Lily, *Whiteheadia bifolia* (Hyacinthaceae)–first field observations with photographic documentation of rodent pollination in South Africa. *South African Journal of Botany* 75: 713–19.

Wylie,. R. B. (1917). The pollination of *Vallisneria Spiralis. Botanical Gazette* 63: 135–45.

Yuan, S. C., Chin, S. W., Lee, C. Y. and Chen, F. C. (2018). *Phalaenopsis* pollinia storage at sub-zero temperature and its pollen viability assessment. *Botanical Studies* 59 (1). doi:10.1186/s40529-017-0218-2.

7

Pollen–Pistil Interactions and Fertilization

In flowering plants, there is an elaborate screening process for pollen grains and pollen tubes, present at different levels in the pistil.

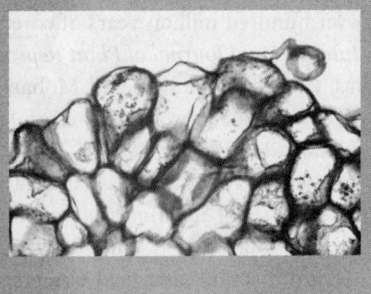

7.1 Introduction

In the previous chapter, we learnt that pollination results in transferring of pollen grains from the anther to the stigma of a flower. Such a transfer or landing of pollen grains initiates a series of events that involve a continuous exchange of signals between the haploid pollen and the diploid maternal tissue of the pistil. These events and interactions which include pollen selection or rejection, pollen hydration, pollen tube growth, its nourishment and entry into ovules and embryo sac are recognized as the **pollen–pistil interactions**, (Herrero & Hormaza 1996; Shivanna 2003; Lora et al. 2016). The pollen–pistil interactions result in screening and selection of conspecific/homospecific pollen (of same species) from the heterospecific pollen (pollen of other species) ensuring fertilization only between the conspecific male and female gametes.

In a pistil, while stigma is the landing platform and recipient of the pollen, the style is the conduit for the transfer of non-motile male gametes to the embryo sac seated in the ovules with the help of pollen tubes. The process of delivery of non-motile sperm to the egg via a pollen tube is known as *Siphonogamy*. It is regarded as a key innovation in the course of evolution of angiosperms that has allowed flowering plants to carry out sexual reproduction on land without the need for water. Flowering plants have an elaborate screening process for selecting the right pollen and the pollen tubes. This system works at different levels in the pistil. The selection of pollen starts at the stigma itself which allows only the compatible pollen grains to germinate while the incompatible ones are rejected. The selected pollen grains then germinate and put forth pollen tubes which grow in the style, where again a competition takes place to select the best mate. The pollen tubes travel at a rate specific to each species to ultimately reach the ovule. A study by Williams (2008) covering about 130 seed plant families and 717 taxa suggested that the time interval between pollination and fertilization ranges between 15 minutes to >12 months in angiosperms. This time interval is referred to as the *fertilization interval.*

The first pollen tube to reach the embryo sac bursts inside one of the synergids and discharges the two male gametes (or sperms) along with some pollen cytoplasm. One of the two sperm cells fuses with the egg cell nucleus and the event of fusion of a male sperm (gamete) with the egg cell nucleus is known as syngamy, also called *fertilization*. In angiosperms, the second male gamete also undergoes fusion with the polar nuclei present in central cell of female gametophyte resulting in a second event of fertilization. These two events of fertilization together are known as **double fertilization**, a unique and key feature of angiosperms. It was first described by Nawaschin (1898) in *Lilium martagon* and *Fritillaria tenella*. The phase or duration between the landing of pollen grain on the stigma, till the pollen tube arrival at embryo sac is known as the *progamic phase* (Williams 2008; Mangla et al. 2013).

The whole process of screening of the right male partner for fertilization in flowering plants has been illustrated through a beautiful analogy by Shivanna (2016). For a simplified understanding, he compared the structural entities of a pistil to the residence of the female partner with a series of locked doors; the first door being at the stigma and the others along the corridor in the style. Pollen–pistil interactions for selection of an appropriate male partner are summarized by him as "*For successful accomplishment of fertilization, the pollen must be able to open each of these locks with the matching keys. All species which are closely related may have several matching keys to open these locks on the pistil. In the evolutionary progression, the species which were once closely related become more and more distant from each other and hence lose the matching keys to the locks on the pistil. As the divergence increases, the ability of the pollen to open the locks on the pistil decreases, leading to rejection of hetero-specific pollen. Sometimes, the pollen from very closely related species is able to open the pistil locks and even germinate and eventually reach the ovule but, in all likelihood, the fertilization may not occur. In case, the fertilization is accomplished, the embryos are aborted and the endosperm development fails preventing the formation of inter-specific hybrids.*"

Pollen–pistil interactions are of utmost importance in sexual reproduction and seed formation. A better understanding of the biology of pollen–pistil interactions has direct relevance for plant breeding as well. This chapter gives an account of pollen–pistil interactions and fertilization in the angiosperms with accompanying details of structure of stigma and style, pollen tube guidance, pollen tube entry and fertilization.

7.2 Stigma and Style

For a better understanding of the pollen–pistil interactions, knowing the structural details of stigma and style is imperative. Morphologically, stigmas exhibit much variation, even greater than the pollen grains (Fig. 7.1). On the basis of presence or absence of exudate at the time of pollination (and maturity), the stigmas in flowering plants have been broadly categorized into two types: **wet**, and **dry** (Heslop-Harrison & Shivanna 1977). In a wet stigma, a viscous exudate or secretion called *Extra Cellular Matrix* (ECM) is present on the stigmatic surface. ECM is an assortment of extracellular molecules secreted by stigmatic cells. Irrespective of the morphology of the stigma, ECM is invariably present on the receptive surface of 'wet' stigma (Heslop-Harrison & Shivanna 1977). It is a highly heterogenous layer consisting of proteins, carbohydrates, amino acids, and sometimes even phenols and its composition varies among families. In Solanaceae, the stigma secretions are primarily lipidic in nature containing complex mixtures of long-chain saturated and unsaturated triacylglycerides whereas, the secretions are aqueous and carbohydrate-rich in Liliaceae. ECM plays a vital role in pollen–pistil interactions and is crucial for pollen capture and adhesion. On making contact with a wet stigma, the pollen (any pollen that comes in contact with the stigma, irrespective of species) is quickly trapped by surface tension and gets immersed within the ECM. Epidermal cells of wet stigmas are devoid of a continuous cuticle, making entry of pollen tubes into the stigma relatively easier and unhindered. Wet stigmas can be seen in families like Liliaceae (*Lilium*), Rutaceae (*Citrus, Aegle marmelos*), Leguminosae (*Pisum*), Solanaceae (*Petunia*), Orchidaceae, Burseraceae (*Commiphora* sp.).

The dry stigmas, on the other hand, are devoid of any exudate and instead show presence of a hydrated layer called the pellicle, over a continuous cuticle as seen in Brassicaceae (*Arabidodpsis, Brassica*), Poaceae (*Oryza sativa*), Papaveraceae (*Papaver rhoeas*), and Elaeagnaceae (*Hippophae rhamnoides*) (Fig. 7.2). The pollen tube penetration in species with dry-stigmas requires a little more effort on the part of the pollen tubes as they have to secrete hydrolytic enzymes such as cutinase to breach a continuous cuticle, in order to penetrate the stigma (Shivanna 2003). The surface of both wet and dry types of stigmas may include papillae (*Olea* sp., *Utricularia, Ulmus* sp.) or may be smooth (as in *Hippophae* sp.). The former type of stigma is referred to as *papillate stigma* and the latter as *non-papillate stigma*. For example, in grasses, the stigma is dry and papillate but in *Citrus* species, it is non-papillate and wet. Papillate stigmas are further divided into unicellular or multicellular and uniseriate or multiseriate type. Family Asteraceae exhibits the presence of an intermediate type of stigma, called the semi-dry stigma, as an exceptional case (Hiscock et al. 2002).

Important correlations can be drawn between the stigma type and the type of pollen produced by a plant indicating co-evolution of pollen and stigma structures. For example, 3-celled pollen generally occur in species with dry stigmas, and 2-celled pollen mostly occur in species with wet stigmas (Heslop-Harrison & Shivanna 1977). The mature stigma also contains several enzymes like peroxidases, phosphatases and nonspecific esterases (Heslop-Harrison & Shivanna 1985; Shivanna 2003). The presence of exudates (in case of wet stigma) along with high activity of these enzymes marks the maturity of stigma, which is also referred to as *stigmatic receptivity*. After pollination, stigma loses its receptivity as indicated by low or nonexistent activity of stigma surface enzymes.

The style is the connection between the stigma and the ovary. Most of the pollen tube growth occurs in a specialized tissue called the transmitting tissue that connects the stigma, the style, and the ovary. The transmitting tissue in basal angiosperms is called *compitum*, a common pollen tube transmitting tract of all carpels. Two types of styles have been characterized in angiosperms – **solid** and **hollow styles** (Fig. 7.3). Solid style, also known as closed style, is characterized by a core of transmitting tissue composed of narrow elongated cells. Genera such as *Primula, Nicotiana, Arabidopsis, Brassica, Petunia,* and *Wrightia* (Fig. 7.3 A-C) exhibit solid styles. The cells of transmitting tissue remain fixed end to end through cell walls which are traversed by plasmodesmata. Laterally, the layers or files of transmitting tissue cells are separated from each other by intercellular spaces, which are filled with ECM secreted by cells of transmitting tissue. These stylar exudates serve as a source of nourishment and guidance to the growing pollen tubes. Stylar ECM is also a mixture of heterogenous compounds such as pectins, carbohydrates, proteins, glycoproteins and often lipids. Like stigmatic ECM, the content of the stylar ECM varies among species. In *Brassica napus, Helianthus annuus,* and *Gossypium hirsutum,* abundant Ca^{2+} stores have also been detected in the ECM of transmitting tissues (Zheng et al. 2019).

Figure 7.1 Types of Stigmas. Dry stigma in: **A.** *Bombax ceiba.* **B.** *Papaver* sp. **C.** Semi-dry bifid stigmas in *Launea* sp. (Asteraceae). Wet stigma in: **D.** *Citrus* sp. **E.** *Lilium* sp. **F.** *Crateva adansonii* (*Source:* Mangla & Tandon 2011, *reproduced with permission*). **G.** Scanning electron micrograph showing a part of dry, non-papillate stigma (*Hippophae rhamnoides*). **H.** Papillate stigma of *Crotolaria* sp. Note the structural variability in stigma among different species. **See Color Plates (page 485).**

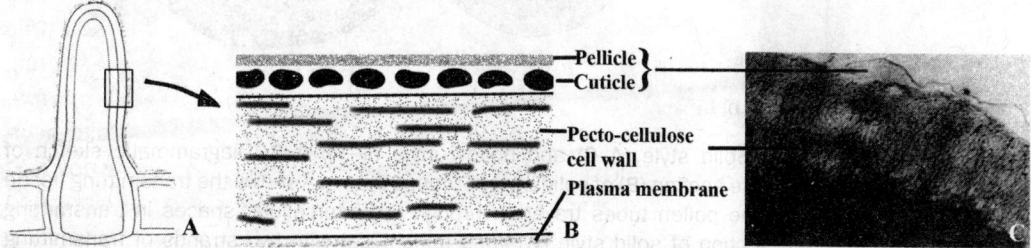

Figure 7.2 A. Diagrammatic sketch of a papilla in dry stigma. **B.** Details of boxed layers in figure A. Note the discontinuous cuticle layer (*Source:* Adapted and modified after Shivanna 1977). **C.** Transmission electron micrograph showing cuticle-pellicle layer over a stigmatic cell (*Hippophae rhamnoides*).

The hollow or open style comprises a canal called **stylar canal** which is lined by a few layers of glandular cells called the canal cells. These cells exude secretions which may fill the canal as observed in *Lilium* sp., *Aegle marmelos, Tecomella undulata* (Bhardwaj & Tandon 2013; Shivanna 2020) (Fig. 7.3 D-F) or remain as a layer on their inner tangential walls as seen in *Ornithogalum caudatum* (Tilton and Horner 1980).

In solid styles, pollen tubes either grow through the intercellular space of transmitting tissue or thickened cell walls of the transmitting tissue or grow between the plasma membrane and the cell wall of the transmitting tissue cells. In hollow styles, pollen tubes either grow through the secretion in the canal or secretion accumulated between the cuticle and the wall of the epithelial cells. In pistils with solid-style like *Primula*, pollen tubes enter the cuticle of the stigmatic papillae and grow down between the cuticle and pecto-cellulosic wall, ultimately entering the ECM of the style. In pistils with hollow-styles like *Lilium*, pollen tubes first grow

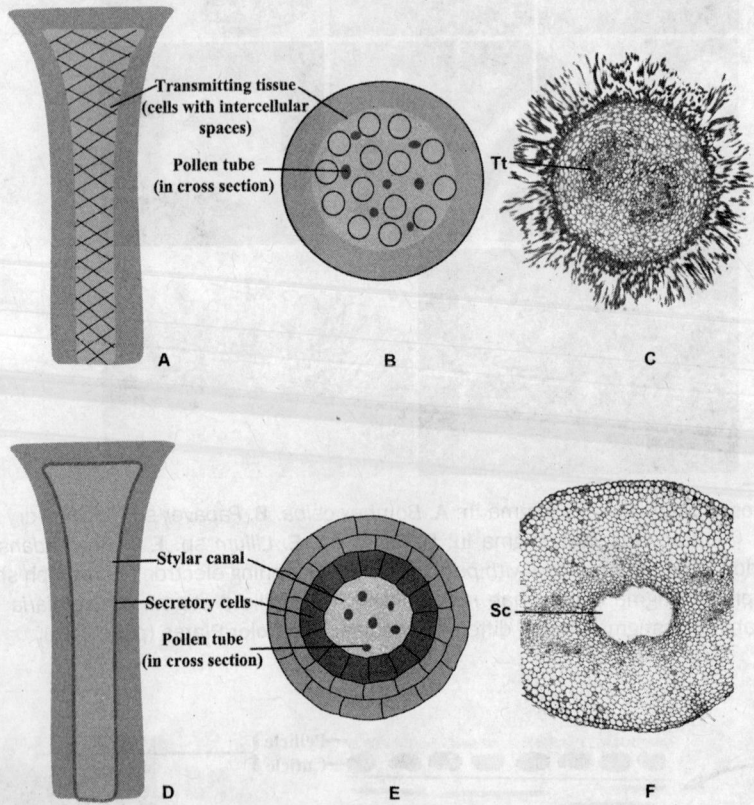

Figure 7.3 Types of Style. Solid style (**A–C**) and hollow style (**D–F**). **A–B.** Diagrammatic sketch of longitudinal (**A**) and transverse section (**B**) of solid style. Stripped area in A shows the transmitting tissue with intercellular spaces. The pollen tubes traverse through the intercellular spaces in transmitting tissue (**B**). **C.** Transverse section of solid style (*Wrightia tomentosa*) with two strands of transmitting tissue tracts (**Tt**). **D–E.** Diagrammatic sketch of longitudinal (**D**) and transverse section (**E**) of hollow style. The pollen tubes travel through stylar canal (**E**). **F.** Transverse section of style of (*Tecomella undulata*) showing a stylar canal (Sc). **See Color Plates (page 486).** (*Source:* A–B, D–E: Adapted and modified after Shivanna 2016.)

in the stigmatic tissue and then glide along the surface of stylar canal which is filled with the secretion. In some species with hollow styles like *Gladiolus, Crocus,* the mucilaginous and polysaccharidic secretion gets accumulated between the cell wall and the cuticle of canal cells. The pollen tube penetrates the cuticle and grows through this secretion (Heslop-Harrison & Heslop-Harrison 1985; Gotelli et al. 2017). In many species (e.g., *Brassica, Lilium*), the growing pollen tubes initiate the degeneration of adjoining cells of stigma and the transmitting tissue. The content after degeneration serves as a source of nourishment and guidance to growing tubes. Degeneration also removes the physical constraint of space for numerous pollen tubes (Heslop-Harrison & Heslop-Harrison 1985; 1987).

There are a few species with an intermediate condition of style, known as **semi-closed style**, where the stylar canal is surrounded by one or a few layers of transmitting tissue, e.g., *Hovenia dulcis, Vigna adenantha* (Castro & Agulló 1998; Gotelli et al. 2017). Pollen tubes in semi-closed styles generally grow through the intercellular matrix of transmitting tissue, or even may grow sometimes through the secretion in the canal as observed in *Ziziphus mucronata* (de Graaf et al. 2001; Gotelli et al. 2017).

Box 7.1

Differences between Stigma and Style

Stigma	Style
• Stigma provides a platform for landing, hydration, recognition and germination of pollen grains.	• Style primarily serves as a conduit between the stigma and the ovary, for the growth and movement of the pollen tubes.
• Stigma is highly variable in its morphology and can be wet or dry in nature.	• Not much variability is seen in the morphology of style and structurally it is of either solid or hollow type.
• In evolutionary sequence of pistil, stigma evolved first.	• During evolution of pistil, style evolved after stigma.

7.3 Pollen–Pistil Interactions

Pollen–pistil interactions cover all the sequential events starting from pollination till the entry of pollen tubes into the ovules (Shivanna 2020b). The pollen grains landing on the stigma can be either from the same species only (conspecific or homospecific) or some may be from different species (heterospecific) as well. The pistil is equipped to distinguish a homospecific pollen from a heterospecific pollen and has the prerogative to allow or disallow a pollen to germinate, allowing only the homospecific pollen to germinate and effect fertilization.

Once pollen land on the stigma, they get hydrated and release pollen wall components on the stigma itself. The surface components of the stigma communicate with the surface components released by the pollen and a recognition reaction occurs between them. Recognition activates the machinery within pistil which facilitates pollen germination and pollen tube growth. Under optimal pollination conditions, the number of pollen grains landing on the stigma is much more than the number required to fertilize all the ovules of an ovary. This leads to a competition between pollen grains for germination. Only the most vigorous pollen grains are able to germinate and deliver their pollen tubes to the stigma and subsequently into the style and the ovary. Competition amongst the pollen tubes also occurs in the style which selects and eventually allows the most vigorous pollen tubes to enter the ovules to accomplish fertilization. Therefore, merely being the right type of pollen grain does not ensure entry into the ovule. The less vigorous pollen grains in which germination is delayed, may reach the ovary when all the ovules are already fertilized. This type of competition is considered healthy for the fitness of the progeny in angiosperms (Erbar 2003; Hiscock & Allen 2002).

All these events starting from the landing of pollen on the stigma till their entry into the embryo sac are vital for successful formation of the zygote. At each stage there are interactions between the haploid pollen and the diploid pistil involving extensive chemical communication. These events may be divided into the following six stages (Shivanna 2003; Hiscock & Allen 2008; Johnson et al. 2019; Fig. 7.4):

i.   Pollen capture and adhesion
ii. Pollen hydration
iii. Germination of the pollen followed by production of the pollen tube
iv. Penetration of the stigma by the pollen tube
v.   Growth of the pollen tube through the stigma and style
vi. Entry of the pollen tube into the ovule and discharge of the sperm cells

Each stage is discussed separately in the following sections.

7.3.1 Pollen Capture and Adhesion

The first stage of pollen–pistil interactions is the capture and adhesion of pollen by the stigmatic surface. In wet stigma, the pollen capture and adhesion are non-specific as the pollen get easily trapped in stigmatic exudates. However, pollen capture and adhesion on dry stigma is strictly regulated involving a series of complex interactions between the pollen coat substances and the cuticle-pellicle layer. Pollen adhesion in the dry type of stigma can be divided into two stages; pollen capture, and cross-linking. Pollen capture involves the exine wall and the pollen coat proteins. Studies using pollen grains which lack a pollen coat, pollen grains with disordered exine and purified exine fragments which bind to the surface of the stigma confirm this (Swanson et al. 2004). Exine also helps in recognition of conspecific pollen grains, e.g., the pollen grains of *A. thaliana* adhere to their own stigma with high affinity mediated by exine; while they bind poorly to stigmas from other families, and even to stigmas from related *Brassica* species, indicating a species-specific selection.

Such specificity lowers the probability of access of stigma by illegitimate pollen. After the pollen capture, the lipid and proteinaceous components of the pollen coat are mobilized on stigma and form an interface between the pollen wall and the stigmatic surface. This leads to cross linking between components of the pollen coat and the stigmatic surface and the formation of a pollen foot at the point of contact with the stigma. The *pollen foot* is the site where the proteins, lipids and carbohydrates of the pollen coat and stigma mix for the first time (Zheng et al. 2018). The proteins on the stigma surface (cuticle-pellicle layer) act as receptors for pollen coat proteins and this interaction plays an important role in the recognition and strong adhesion of compatible pollen. Observations on cross-linking during adhesion in *Brassica oleracea* show two cross-linking pairs of stigma and pollen coat proteins namely, S locus glycoprotein (SLG) and pollen coat protein, class A1(PCPA1) and S locus glycoprotein (SLG)-like receptor1 (SLR1) and pollen coat protein, class A2(PCP-A2) (Edlund et al. 2004). Apart from proteins, the aromatic compounds, lipids, and pigments of pollen coat are also instrumental in adhesion (Chapman & Goring 2010). This is exemplified by the mutant and incompatible pollination studies in *Brassica* where mutant pollen which lack lipids in their pollen coat adhere poorly to the stigma and neither form a pollen foot nor do they hydrate.

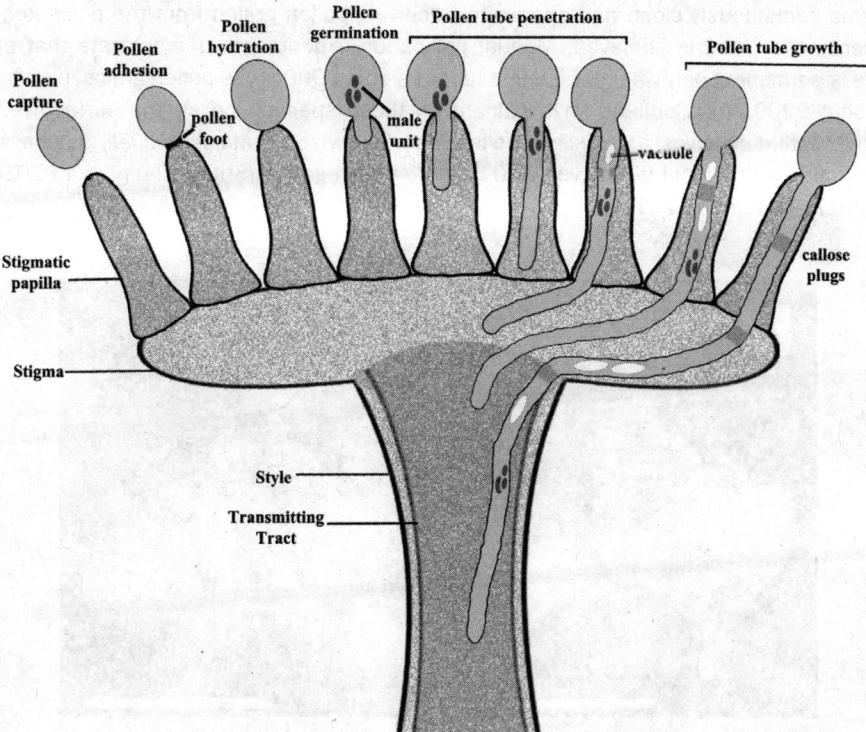

Figure 7.4 Diagrammatic representation of steps involved in pollen–pistil interactions. (*Source*: Modified after Hiscock & Allen 2008.)

Box 7.2

Sensitive Stigma in Flowering Plants

The stigmas in some species show sensitivity towards pressure applied on them in the form of touch. All such stigmas are referred to as **touch sensitive stigmas**. The existence of sensitive stigmas was described as early as in 1841 by Henderson and in 1861 by Kabsch (Newcombe 1922). Such a pressure is what is exerted by a pollinator in each of its visits. The behavior of touch senstive stigmas consist of three successive events: temporary closure, reopening and permanent closure. After being touched by the pollinator, the stigma closes temporarily and reopens swiftly or gradually, if the pollen load received is not enough. This closing and opening of stigma will continue till it receives a sufficient quantity of pollen, after which stigma closes permanently. (Newcombe 1922; 1924; Jin et al. 2017). The examples of sensitive stigmas can be seen in families Bignoniaceae (*Tecoma radicans, Catalpa bignonioides, Oroxylum indicum, Kigelia pinnata*), Scrophulariaceae (*Mimulus aurantiacus*), Martyniaceae (*Martynia* sp., *Diplacus* sp.), Lentibulariaceae (*Utricularia vulgaris*), Linderniaceae (*Torenia fournieri*), and Phrymaceae (*Majus miquelii*).

A very interesting mechanism involving sensitive stigma is seen in members of the family Bignoniaceae. In *Oroxylum indicum* and *Kigelia pinnata,* the stigma is bi-lipped and the ovary bears numerous ovules (~400-600). In these species, a single visit of pollinator (bats) is not sufficient to deliver the pollen load to ensure fertilization of numerous ovules. Thus, the lips of stigma continuously close and open after receiving pollen grains from the pollinator, till a sufficient pollen load is achieved. Manual pollination experiments demonstrate that stigma closure is permanent only when the stigma receives about 900 cross pollen grains in *O. indicum* and about 9200 cross pollens in *K. pinnata*. In these species the stigma surface remains receptive until it receives a sufficient pollen load. Likewise, in *Majus miquelii* stigma shows opening and closing, until it receives 600 or more pollen grains (Sritongchuay et al. 2010; Jin et al. 2017; Raina et al. 2017).

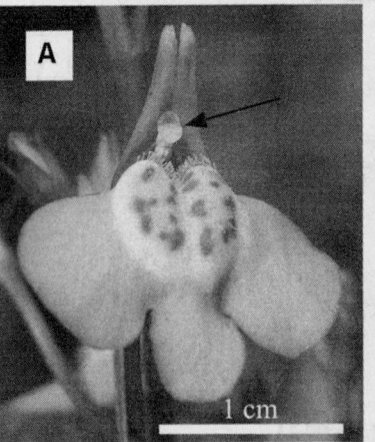

Figure 7.5 Flower of *Mazus miquelli* (A) with an open bi-lipped stigma (arrow) and (B) with a closed stigma. **See Color Plates (page 486).** (*Source*: Jin et al. 2017, published under CC BY 4.0 License.)

7.3.2 Pollen Hydration

Mature pollen grains are metabolically inactive and released as desiccated units from anthers with water content ranging from 15 to 35%. They may undergo further dehydration after release from the anther or during transfer by the vectors. For pollen to be able to germinate and deliver pollen tubes, hydration of the desiccated pollen grains on the stigma surface is an essential step. On the wet stigma; water, nutrients and other small molecules from the stigmatic exudate are readily available which get rapidly transported into the pollen to hydrate it, negating any chances of species specificity. However, this step of pollen–pistil interactions is highly regulated in species with dry stigma where species specific hydration takes place. For hydration in dry stigma, the waxy cuticle present on the stigmatic surface releases fluids due to a localized change in the cuticle structure triggered by compatible pollen grains. The lipids in the pollen coat reorganize to create a capillary system between the stigma and the pollen, through which water transport takes place from the stigma surface to the pollen grain (Murphy 2006). This reorganization of lipids in the pollen coat during pollen foot formation is called *coat conversion* (e.g., in *B. oleracea* and *Arabidopsis*). However, it is not known how the hydrophobic lipids of the pollen coat become channels for water transport.

Other than the lipids, proteins from both, stigmatic surface and pollen coat also aid in pollen hydration. On the stigmatic side, there is an evidence for a plasma membrane-localized aquaporin-like protein controlling the flow of water from the stigma to the pollen grain as seen in Brassicaceae (Dixit et al. 2001). In the pollen coat, the most abundant proteins are glycine-rich oleosin proteins (GRPs) which are known to help in hydration. GRPs contain a lipid binding domain and help in reorganization of lipids for diffusion of water during pollen foot formation. The sequences of the GRP proteins are divergent across the members of Brassicaceae, supporting these proteins as the candidates for species specific hydration or mate discrimination (Mayfield et al. 2001; Fiebig et al. 2004).

Enzymes like lipases in the pollen coat of dry stigma of *Arabidopsis* also help in hydration. (Mayfield et al. 2001; Edlund et al. 2004). Such enzymes modify stigmatic surface to allow penetration of pollen tubes into the pistil. Besides the external components, the internal signaling pathways of pollen also contribute to its own hydration by regulating the reactive oxygen species (ROS) levels (Zheng et al. 2019; refer to Box 7.4 for details).The duration of rehydration of the pollen on the stigma is variable among species. In *A. thaliana*, this duration is 15 minutes whereas it is more than 60 minutes in *Brassica oleracea* (Chapman & Goring 2010).

7.3.3 Pollen Germination

Next stage is pollen germination which is marked by the formation of a pollen tube that enters the stigmatic tissue. Pollen germination can be divided into two stages: first is polarization of the pollen cytoplasm to support extension of the pollen tube, and second, is the rupturing of the exine wall by the pollen tube while emerging from the pollen grain. Polarization of pollen cytoplasm occurs within minutes of hydration of the pollen, transforming non-polar pollen grain into a highly polarized entity. Pollen organizes its cytoplasm in such a way that it supports a tubular structure which includes: (i) formation of filamentous cytoskeletal structures covering the nuclei; (ii) reorientation of the large vegetative cell nucleus facilitating its entry in the extending tube before the generative cells; (iii) assemblage of mitochondria and polysaccharide particles at tube tip; (iv) simultaneous translocation of secretory vesicles and deposition of callose at the site on the plasma membrane of pollen where the pollen tube will emerge; (v) Splitting of intine to allow the exit of pollen tubes. Mutant-based evidences suggest that the pollen coat lipids are also important in pollen germination. In the mutants lacking a lipid-rich pollen coat, absence of pollen polarity and formation of disoriented pollen tubes is seen (Micheli 2001; Zheng et al. 2018; 2019). It has been observed that tricellular pollen grains show quicker onset of pollen germination and subsequent tube growth than bicellular pollen. This may be because the tricellular pollen are usually dispersed in a more hydrated and metabolically active state as compared to bicellular pollen and therefore, do not undergo most of the re-hydration process (Williams & Brown 2018; Pacini & Dolferus 2019).

In most of the species, the pollen tube emerges through the apertures, and in some it comes out directly by breaking the exine wall. The rupturing of the exine wall for the pollen tube emergence requires weakening of exine facilitated by enzymatic digestion and tearing by local turgor pressure (Edlund et al. 2004). In both the conditions, Ca^{2+} concentration increases in the pollen cytoplasm and a tip-focused gradient of Ca^{2+} forms at the potential germination site after hydration. In tobacco, the Ca^{2+} accumulation in pollen cytoplasm is initially localized within small vacuoles. However, as germination proceeds, the small vacuoles fuse to form large vacuoles creating turgor pressure to push the cytoplasm to form pollen tube. In some species (e.g., tobacco, lettuce), pollen also absorb the Ca^{2+} present in the stigma in the form of precipitates (Ge et al. 2009). Furthermore, downstream signaling components of Ca^{2+} such as calmodulin and calmodulin-like proteins are also found to be involved in pollen germination (Zheng et al. 2018).

7.3.4 Penetration of Stigma by Pollen Tube

After passing through the exine wall, the pollen tubes penetrate the physical barrier of the stigma and enter the stylar tissues. Pollen tube invasion of the stigmatic surface requires secretion of a number of enzymes by the pollen. Activity of acid phosphatases, ribonucleases, esterases, amylases, and proteases has been localized in pollen intine and tube (Knox & Heslop-Harrison 1970). Esterases, particularly those known as cutinases, are important for penetrating the stigma cuticle. For hydrolysis of pectin in the stigma cell wall, pollen expresses genes encoding pectin esterase and pectate lyase (Kim et al.1996; Wu et al. 1996). A pollen–specific polygalacturonase is also detected at the pollen tube tip in *Brassica* (Dearnaley & Daggard 2001). The genetic regulation of pollen tube penetration remains poorly understood. However, recently a novel gene, *O-FUCOSYLTRANSFERASE1* (*AtOFT1*) has been identified in *Arabidopsis* which plays a key role in pollen tube penetration through the stigma-style interface. This gene encodes a Golgi-localized protein indicating its potential role in cellular glycosylation events in pollen–pistil interactions (Smith et al. 2018).

7.3.5 Pollen Tube Growth

Pollen tubes show directed growth through the stigmatic tissue and style until they reach ovules to deliver male gametes to the embryo sac (Taylor & Hepler 1997; Lord & Russell 2002). As mentioned earlier, pollen tube grows through transmitting tissue in solid styles or glides through canal in the hollow styles. Transmitting tract cells and canal cells secrete free sugars, amino acids, glycolipids, glycoproteins, polysaccharides, lipids, and proteins into the extracellular matrix, providing nutrients to growing pollen tubes. Pollen tube growth is a high ATP-consuming process which involves a precise coordination between the cytoskeleton, the cytoplasmic organelles and membrane vesicles that deliver membrane and cell wall material for growth (Chebli et al. 2013). The growth of pollen tube takes place at a very high rate of elongation; indeed, pollen tube is the fastest growing plant cell known. This rapid elongation of pollen tubes is essential for male reproductive success. Pollen tube growth rates in early flowering taxa (e.g., *Amborella, Nuphar, Austrobaileya.*) range from ~0.02 μm^{s-1} to 0.1600 μm^{s-1} while the rate of pollen tube growth in higher angiosperms can be upto 2.8 μm^{s-1} as in maize or 0.2-0.3 μm^{s-1} as in lily (Williams 2008). The key characteristics of pollen tube growth are: i) the growth of the pollen tube is polar, ii) the polar growth is limited to its tip, and iii) pollen tubes undergo unidirectional growth.

7.3.5.1 *Pollen Tube Cytoplasm*

With the emergence of the pollen tube almost entire content of the pollen grain moves into the tube which includes vegetative cell and sperm cells and various organelles. This content gets separated from the remainder of the pollen by a callose plug. As the pollen tube grows, the callose plugs form at regular intervals and positions, starting at the base of the tube, and the regions behind the callose plugs become vacuolated. The vacuoles assist in the polar growth of the pollen tube by pushing the cytoplasm to the apex of the tube; such that the cytoplasm remains concentrated in the apical portion of the pollen tube regardless of its length (Franklin-Tong 1999; Fig. 7.6). The formation of vacuoles in the tube cytoplasm (behind the tip) is also significant in maintaining the tube turgor pressure during growth which is considered critical (Zheng et al. 2018).

In the growing pollen tube several zones can be marked (Fig. 7.6 and 7.7 A). The tip of the pollen tube is peculiar in having organelles with low refractivity and this zone looks 'clear' under a microscope. It is commonly referred to as the **clear zone.** The starch-containing amyloplasts are absent in this zone. It is further divided into an apical and a sub-apical zone. The apical region is typically of inverted cone-shaped zone (also known as apical dome) and contains endoplasmic reticulum elements and numerous vesicles. The subapical region of the clear zone comprises Golgi apparatus and mitochondria. The region behind the clear zone is called the **granular zone.** It is characterized by the presence of larger organelles such as amyloplasts and vacuoles which show higher refractivity under the microscope (Lancelle & Hepler 1992; Hepler & Winship 2015; Hafidh & Honys 2016).

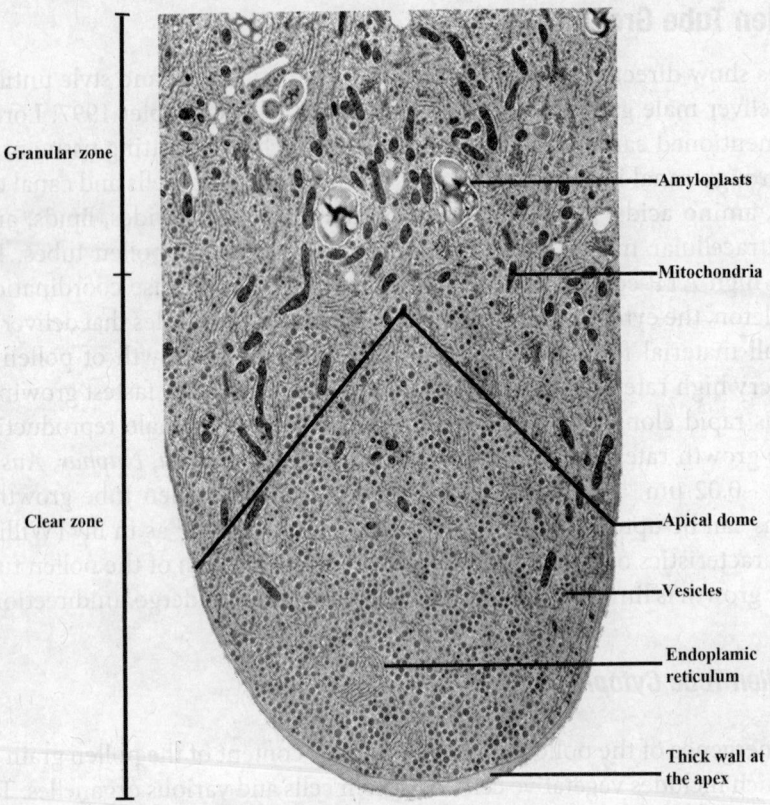

Granular zone

Amyloplasts

Mitochondria

Clear zone

Apical dome

Vesicles

Endoplamic
reticulum

Thick wall at
the apex

Figure 7.6 Ultrastructure (transmission electron micrograph) of a pollen tube in lily. In the apical dome (or apex) abundant small vesicles can be seen which contribute wall material. Note the thickest cell wall present at the extreme apex which gradually becomes thin towards the flanks of the dome. (*Source*: After Lancelle & Hepler 1992, *reproduced with permission*.)

An indispensable need of the growing pollen tube is the presence of active cytoplasmic streaming to deliver material needed for growth occurring at the tip. This is accomplished by a functional actinmyosin-based cytoskeleton. The actin microfilaments are organized in parallel bundles differentiated into cortical and central bundles in the subapical zone of pollen tube. The cortical filaments transport vesicles towards the tip whereas the central filaments are dedicated for the basipetal vesicular transport. Hence, these bundles of actin filaments and their myosin motors work in opposite directions and the resultant vesicular flow looks like a reversed fountain (Fig. 7.7 B). In the apical zone, filaments are randomly oriented and form a mesh like structure (Chebli 2013; Hafidh & Honys 2016; Bascom et al. 2018).

The other biochemical mechanism that drives the pollen tube elongation is a steep Ca^{2+} gradient within the pollen tube tip. Pollen tubes exhibit calcium concentration in the range of micromolars in a few micrometres of the apical region while the rest of the tube contains normal calcium levels of ~ 100 nM (varying between species). Ca^{2+} gradient is not detected in pollen tubes that are not actively growing, hinting that Ca^{2+} participates in determining the orientation of the tube growth. The high amount of Ca^{2+} in the immediate vicinity of the tip apex occurs through the influx of extracellular calcium with the help of calcium-permeable channels (Holdaway-Clarke & Hepler 2003). A recent study has demonstrated a tight correlation between calcium, and actin; in the tip-oriented growth of pollen tube. It was shown that a calcium gradient is essential for actin accumulation at the tip. If there is high cytosolic calcium, it promotes disassembly of a tip focused actin while a low calcium concentration promotes assembly (Bascom et al. 2018). The excess accumulation of calcium at the tip also induces apical swelling due to irregular vesicle fusion at the pollen tube tip.

Studies of *in vitro* pollen tube growth have also highlighted the essential roles of ions and pH dynamics in tip growth, indicating the importance of active and spatially localized ion transporters like H^+ion channels. The cytoskeletal changes and other cytoplasmic events involved in the tip growth also require extracellular cues which are transduced to the cytoplasm. These cues are thought to be perceived by Receptor-like kinases (RLKs) which are localized at the plasma membrane of pollen tubes. Two such pollen–specific RLKs: LePRK1 and LePRK2 have been identified in tomato which are thought to play a role in pollen tube growth (Chebli 2013; Zheng et al. 2018).

7.3.5.2 *Pollen Tube Wall*

The pollen tube wall is primarily made of two types of pectins, the secreted esterified pectins and the de-esterified pectins (polygalacturonic acid pectin). The apex or tip of the pollen tube is made up of only esterified pectins (single layer). Apart from maintaining cellular integrity the esterified pectins also provide elasticity to the tip which helps in pollen tube growth under internal turgor pressure (Zheng et al. 2019). The lateral walls and the shank (i.e., the central part of the pollen tube) contain de-esterified pectins along with callose and cellulose. The de-esterified pectins become crosslinked by Ca^{2+}, resulting in a stiff insoluble pectate gel that provides support to the growing tube. In the lateral walls of pollen tubes, three sub-layers are distinct. These being, the outermost layer of de-esterified pectins (Ca^{2+} crosslinked), the middle pecto-cellulosic fibrillar layer and the inner most callose sheath. This three-layered lateral wall enhances the mechanical stability of the pollen tube, prevents cell wall expansion in response to turgor pressure and maintains the cylindrical shape of pollen tubes (Chebli et al. 2012).

During pollen tube growth, new cell wall is continually formed at the growing tip. Cell wall synthesis at the tip is the result of an interplay between vesicular trafficking, exocytosis, and endocytosis of pollen wall material. Golgi-derived secretory vesicles supply cell wall material comprising of polymerized and esterified pectins (synthesized in the Golgi) along with material for plasma membrane extension to the site of growth. These vesicles (also called p-particles) are delivered to the tip of pollen tube via cytoplasmic streaming, which depends on cortical actin filaments that are oriented parallel to the long-axis of the pollen tube. The vesicles carrying cell wall components fuse with the plasma membrane at apical dome (exocytosis). The rate of fusion of vesicles with plasma membrane is always higher than that required for wall synthesis. Thus, the quantity of material delivered by exocytosis exceeds requirements for pollen tube growth and the excess is retrieved and recycled through lateral endocytosis (at the flank) of the pollen tube. This constant anterograde and retrograde vesicle trafficking mediates growth at tip and recycling of plasma membrane components through endocytic internalization (Fig. 7.7 B and C) (Chebli et al. 2013; Onelli & Moscatelli 2013; Grebnev et al. 2017). This exo/endocytosis is a Ca^{2+}-dependent process that requires a GTPase which is involved in the pulling and detachment of the endocytic vesicle from the plasma membrane. Some studies on pollen tubes in *N. tabacum* have also found the presence of a special compartment of the trans-golgi network, located at the interface between the apical dome and the adjacent regular cytoplasm of pollen tube. This trans-golgi network compartment appears ideally positioned to mediate the recycling of membrane material by endocytosis. Additionally, this trans-golgi network also helps in uptake of signalling molecules from the pistil confining polarity indicators to the tip region (Stephan et al. 2014; Grebnev et al. 2017).

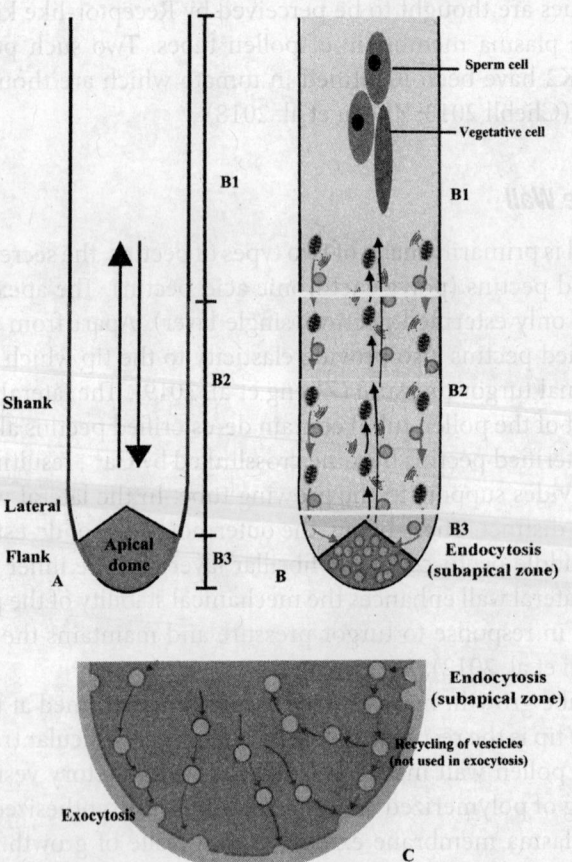

Figure 7.7 Diagrammatic representation of pollen tube and organelle trafficking in it. A. Different zones in pollen tube: B1, B2 and B3. B1: Part of pollen tube carrying sperm cells and vegetative nucleus. B2: Shank and lateral region enabling the transport of various organelles and vesicles to the tip. B3: Pollen tube tip. The arrows oriented in opposite direction indicate movement of organelles and vesicles. **B.** Movement of vesicles, brown arrows indicate the reverse fountain of vesicles mediated by cytoskeleton. **C.** Enlarged view of apical dome and flank of pollen tube showing the processes of exocytosis (of vesicles carrying esterified pectins) and endocytosis of vesicles. **See Color Plates (page 487).** (*Source*: Modified after Grebnev et al. 2017; Chebli et al. 2013.)

7.4 Pollen Tube Guidance: Cues from Sporophytic Tissue, Stylar Exudates and Ovule

The female gametophyte or embryo sac is deeply seated in the ovule which itself is located inside the ovary. Pollen tubes are commissioned with the task of carrying non-motile male gametes all the way from stigma to embryo sac in the ovule. Irrespective of the distance between the stigma and the embryo sac and whether the pistil is upright or drooping, the pollen tubes always follow a predetermined path in the pistil. To reach the ovules, pollen tubes follow an array of cues from the pistil. Pollen tube guidance by the pistil can be divided into two phases; pre-ovular and ovular. As the name suggests, pre-ovular guidance is the navigation provided to the pollen tubes from the stigma to the ovary by the sporophytic tissue of the pistil (except ovule) whereas ovular guidance refers to the precise navigation provided to a particular pollen tube by an ovule.

• ***Pre-Ovular Guidance:*** The pre-ovular guidance includes mechanical, chemotropic, and physiological cues derived from the sporophytic tissue of the pistil (except the ovule) (Mizuta & Higashiyama 2018). A combination of these factors work in different species for pollen tube guidance and successful delivery of the pollen tube to the ovary. There are some molecules which have been identified as chemo-attractants of pollen tube in stigma and style. A chemocyanin called plantacyanin localized in the stigma of Lily, has been thought to promote directional pollen tube growth. The stigmatic cysteine-rich adhesins (SCAs) of Lily bind to the pollen tube to allow plantacyanin to come in contact with plasma membrane and effect directional growth of the pollen tube (Kim et al. 2003). SCA like protein and plantacyanin also facilitate pollen tube guidance in *Arabidopsis*. In tomato, the cysteine-rich Stigma Specific Protein 1 (STIG1); which is secreted in the exudate at the stigma surface and style has also been shown to stimulate pollen tube growth (Huang et al. 2014). The exudates present in the intercellular spaces of solid style as well as those present in the canal of the hollow style are rich in several types of proteins, which guide the pollen tubes in their journey from the stigma to the ovary. Out of these, arabinogalactan glycoproteins (AGPs) play major role in guiding pollen tubes through the style (Cheung et al. 1995; 2008). Recent studies have shown AGPs are abundantly present in stigmatic exudates, transmitting tissues, pollen grains and pollen tubes in *Arabidopsis, Brassica*, and *Nicotiana* (Nguema-Ona et al. 2012). During growth of pollen tube, the pollen tube AGPs are sent to the apex, where they play a role in signalling processes related to pollen tube guidance (Zang et al. 2011). In *Nicotiana tabacum*, AGPs called TTS (Transmitting Tissue-Specific) proteins serve as attractant and promote pollen tube growth. TTS proteins display a gradient of increasing glycosylation from top to the bottom of *Nicotiana* styles which pollen tubes perceive and then grow in that direction (Cheung et al. 1995; 2008).

The Ca^{2+} gradient and oscillation in the pollen tube are also essential for pollen tube growth and guidance; oscillation being a kind of mechanical cue. Therefore, Ca^{2+}

channels on the plasma membrane of pollen tubes appear to play an important role in pollen tube guidance. These calcium channels get modulated by the brassino-steroids and γ-aminobutyric acid (GABA) present in pistil. Recent findings also indicate the importance of the endoplasmic reticulum (ER), ion homeostasis, and protein processing in pollen tube guidance (Guan et al. 2014).

- ***Ovular Pollen Tube Guidance:*** The movement of pollen tubes is also precisely guided by ovules. Ovular guidance not just attracts a pollen tube but also prevents the entry of multiple pollen tubes (pre- and post-fertilization) in the ovule. This is crucial in multi-ovular pistils like *Arabidopsis thaliana* where the ovary encloses ~60 ovules and >100 pollen tubes may grow in the transmitting tract (Mizuta & Higashiyama 2018). This indicates that pollen tube guidance is a very specific navigation system and facilitates precise one-to-one coupling between ovules and pollen tubes (Losada & Herrero 2017; Mizuta & Higashiyama 2018). It has also been proposed that the entry of second pollen tube in fertilized ovule is inhibited by the programmed cell death (PCD) of the persistent (remaining) synergid. (Mizuta & Higashiyama 2018).

Ovular guidance system can be further divided into two stages: *funicular guidance* and *micropylar guidance*. After traversing through the style, the pollen tubes emerge at or near the placenta (a phenomenon also known as pollen tube emergence). The funicular guidance begins from the surface of placenta going up to the funiculus while the micropylar guidance starts from the entrance of the micropyle and continues till entry in the embryo sac. The funicular guidance in *Arabidopsis* has been shown to be mediated through peptide signalling involving Phytosulfokine (PSK), a plant peptide growth factor and two mitogen-activated protein kinases, MPK3 and MPK6 (Guan 2014). Ovular appendages like the obturator also play a role in pollen tube guidance at the funiculus. Various studies have suggested that the cells of obturator produce surface exudates and provide mechanical as well as chemical guidance to the growing pollen tube.

Micropylar pollen tube guidance has been intensively studied and better understood than funicular guidance. Russell (199) suggested that the nucellar cells along the micropyle or the synergid cells are the source of these chemotropic secretions. The first evidence that the synergid cells in the embryo sac secrete a pollen tube attractant came from *in vitro* fertilization assays in *Torenia fournieri*, where ovules with ablated synergids did not attract pollen tubes (Higashiyama et al. 2001). It has been demonstrated that pollen tube attractants (LUREs, cysteine-rich polypeptides) and their receptors (PRK6 and MDIS–MIK1) play a key role in micropylar guidance in *Arabidopsis* and *Torenia*. In maize, a small peptide, the EGG APPARATUS1 (ZmEA1), has also been reported to participate in micropylar pollen tube guidance (Zheng et al. 2018). High levels of calcium in the degenerating synergids of wheat and pearl millet have also been suggested to play some role in chemotropic pollen tube attraction (Shivanna 2003). Recent studies have also established that the interaction between central cell and egg cell is also important for the function of ovular guidance, at least in *A. thaliana* (Higashiyama & Yang 2017). The development of the female gametophyte also has an effect on the ovular pollen tube guidance. Delayed development of female gametophyte in *Arabidopsis*, leads to defects in micropylar guidance of pollen tube. In a mutant showing impaired female gametophyte development in *Arabidopsis*, pollen tubes lose their way and fail to enter the micropyle, and sometimes, even two pollen tubes are simultaneously attracted by the mutant female gametophytes (Losada & Herrero 2017; Mizuta & Higashiyama 2018).

7.5 Pollen Tube Entry into Ovule and Embryo Sac

After growing through the style, pollen tubes enter the ovary and continue to grow along the placenta. Once the pollen tubes reach the ovule, they may enter the ovule, through the micropyle or chalaza or even the funiculus. The most common type of entry of pollen tubes into the ovule is through the micropyle and is called as *porogamy* (Fig. 7.8 A and B). It was first observed in Lily. The entry of a pollen tube into the ovule from the chalazal region is known as *chalazogamy* and was first observed by Treub (1981) in *Casuarina* and is less commonly seen. Other examples include *Juglans, Quercus* and *Betula* (Sogo et al. 2004) (Fig. 7.8 C). When the pollen tubes enter into the ovule through the middle portion of the ovule, either through the funiculus or through any of the integuments, the entry is referred to as *mesogamy* (Fig. 7.8 D). It is seen in *Populus* and *Cucurbita,* where the pollen tube enters through the integuments, while in *Pistacia*, it enters through the funiculus.

7.6 Double Fertilization

In the majority of angiosperms, the pollen tube is known to enter the embryo sac through the micropyle. After entering through the micropyle, the pollen tube invades one of the synergids through its filiform apparatus. While in the synergid, pollen tube further grows for a while before finally discharging all its content into it. The pollen tube is not seen to penetrate any other cell of the embryo sac besides the synergid, (neither the egg nor the central cell). The mode of pollen tube discharge varies from species to species. In cotton, the discharge occurs through a pore which develops at the back of the pollen tip facing the egg. In *Capsella*, the discharge occurs by rupturing the pollen tube at its tip. In *Epidendrum*, the pollen tube forms a cap over the micropylar end of the two synergids and then sends a fairly narrow tube through the filiform apparatus of one of the synergids and releases its content from its tip (Dresselhaus et al. 2017; Fig. 7.9 A and B). Once discharged, the sperms are carried to the extreme chalazal end of the synergid. Out of the two sperm cells or nuclei, one sperm cell makes contact with the plasma membrane of the egg cell and the other one enters into the central cell. The plasma membranes of the sperm cell and the egg fuse to form a bridge and the sperm nucleus passes over this bridge. The final stage of fertilization in flowering plants is the fusion of one sperm nucleus with the egg nucleus and that of the second sperm nucleus with the two polar nuclei. Thus, of the two male gametes released

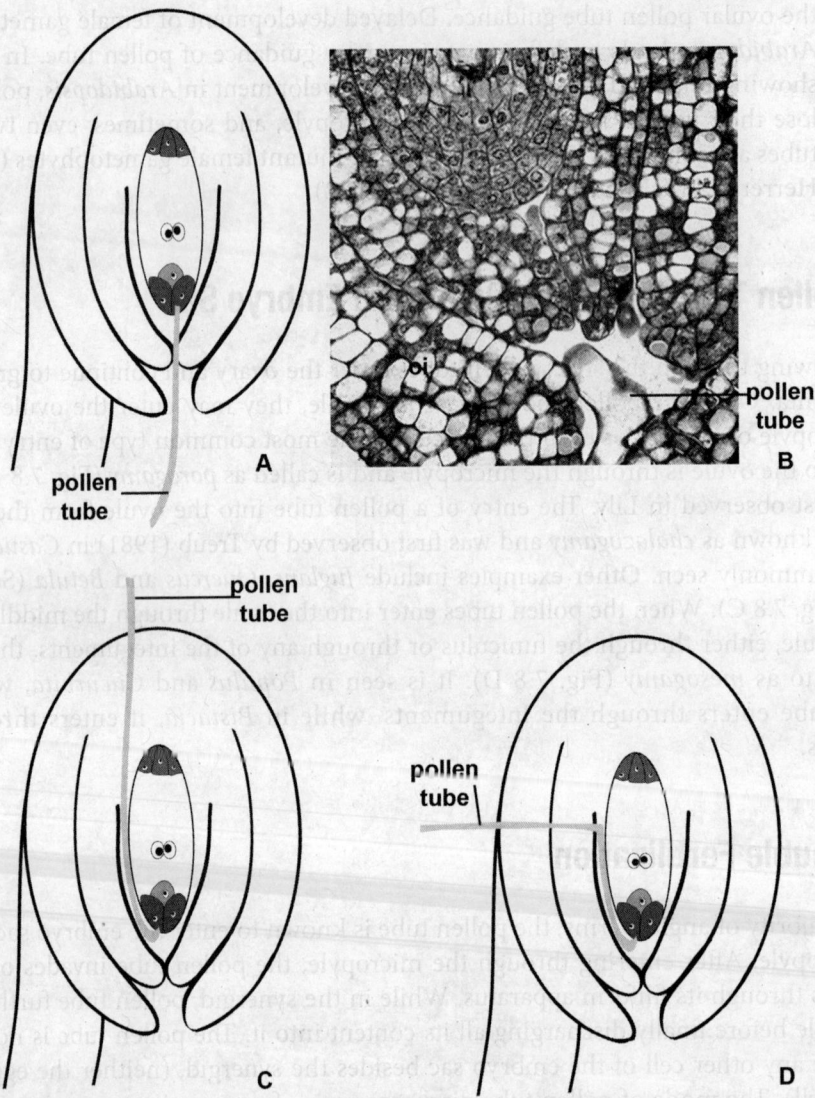

Figure 7.8 Entry points of pollen tube in the ovule. A–B. Porogamy. **B.** A part of longitudinal section of ovule showing micropylar entry of pollen tube. **C.** Chalazogamy. **D.** Mesogamy. **See Color Plates (page 488).**

by the pollen tube into the synergid, one fuses with the egg nucleus to form a zygote and the other fuses with the two polar nuclei present in the central cell to form the primary endosperm nucleus (PEN). These two fusion events are unique to the flowering plants and are referred to as **Double-fertilization**. The fusion between the egg cell nucleus and the sperm cell nucleus to form zygote is called syngamy while the fusion of two polar nuclei with one sperm nucleus is called triple fusion (fusion of three nuclei: 2 polar nuclei plus one sperm nucleus). The zygote develops into an embryo and the primary endosperm nucleus develops into an endosperm which eventually nourishes the embryo.

The ultrastructural studies in cotton suggest that the sperm nuclei are passively carried by cytoplasmic streaming to the egg cell and to the polar nuclei in the central cell. Once the sperm nucleus is close to the egg nucleus, fusion begins with the joining of the outer membranes of the two nuclear envelopes (Jensen 1973) (Fig. 7.9 C). Initially, the fusion occurs in limited areas of contact but later the entire membranes of the two nuclei fuse forming many small bridges between themselves. Eventually the two nuclei fuse completely to form a zygote. The nucleus of the zygote is thus surrounded by a membrane composed of both the sperm and the egg nuclear envelope. Similarly, the second sperm fuses with the two polar nuclei. Initially, these two polar nuclei fuse with each other by the joining of their endoplasmic reticulum attached to outer membranes of their nuclear envelopes. In the next stage, many bridges are formed between the two nuclei (i.e., secondary nucleus), but the process is arrested at this stage till the arrival of the second sperm to form the primary endosperm nucleus (Fig. 7.9 A–C).

Recent studies have identified several kinases which are responsible for the bursting of pollen tube and fertilization, e.g., ANXUR receptor-like kinases (ANX1, ANX2), BUDDHA'S PAPER SEAL (BUPS1, BUPS2) and ERULUS (ERU) (Schoenaers et al. 2017). Mutation studies established that the lack of such kinases may lead to aberrant fertilization and low seed set. In *Arabidopsis*, two new mediators of female-specific kinases required for fertilization, identified are HERCULES RECEPTOR KINASE 1 (HERK1), and the AT5G59700 (ANJEA/ANJ). HERK1 and ANJEA are together responsible for pollen tube reception and bursting. These kinases act redundantly at the filiform apparatus of the synergids to control pollen tube growth, arrest and burst (Trigo et al. 2019).

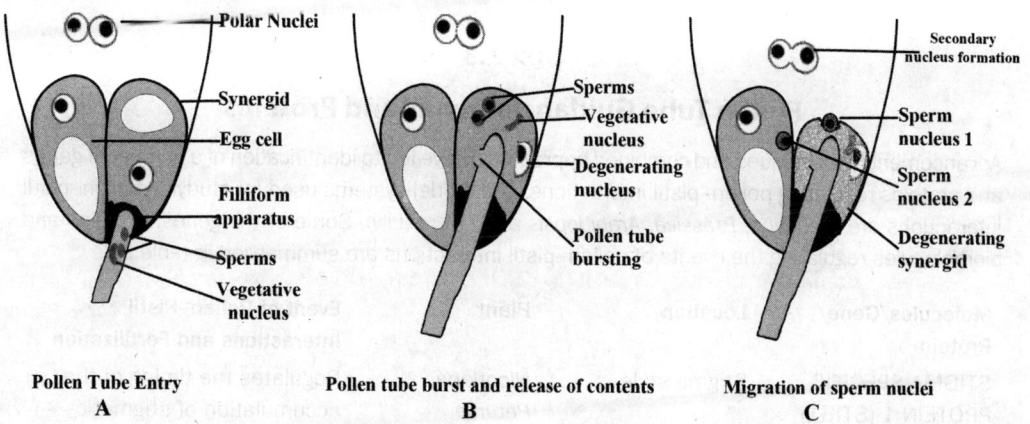

Figure 7.9 A. Pollen tube reception in the synergid. B. Pollen tube discharge by subterminal pore. C. Sperm nuclei migration for fertilization with egg cell and central cell. See Color Plates (page 488).

7.7 Unique Cases

- **Polyspermy:** Fertilization of the egg cell or the central cell by more than one sperms or male gametes is called polyspermy. Occasionally, more than one pollen tube enters the ovule, a phenomenon termed as **polytubey** which can lead to polyspermy of either egg cell or central cell or both. Polyspermy is of rare occurrence in egg cells and may result in the development of viable, triploid progeny that contains three sets of chromosomes (Chaudhary 2020). Polyspermy happens more frequently in central cells resulting in defects in the endosperm ploidy due to more than one copy of paternal DNA. This disrupts the development of endosperm and is called triploid block which ultimately leads to abortion of the seed (Grossniklaus 2017; Köhler et al. 2010; Scott et al. 2008).

- **Podostemaceae:** Out of the several unique embryological features found in Podostemaceae, lack of double fertilization and consequently endosperm deserves special mention. Although two functional male gametes are produced in Podostemaceae, only one fuses with the egg cell and forms the embryo. The other male gamete degenerates and hence the second event of fertilization and consequent endosperm are lacking in the family. Looking into the reasons for the absence of double fertilization; Sehgal et al.(2011) and Chaudhary et al. (2014) reported that the central cell nucleus degenerates before the arrival of pollen tubes and therefore, due to absence of a central cell there is no second event of fertilization. Since, endosperm does not develop in the seed, another tissue known as a pseudo embryo sac or nucellar plasmodium (formed by dissolution of cells of nucellus) nourishes the developing embryo (*see chapter 9 for details*).

Box 7.3

Pollen Tube Guidance: Genes and Proteins

Advancement in techniques and combined approaches have led to identification of a number of genes and proteins regulating pollen–pistil interactions. The model systems used for studying pollen–pistil interactions are *Nicotiana, Brassica, Arabidopsis* and *Lilium* (Lily). Some of the genes, proteins and biomolecules regulating the events of pollen–pistil interactions are summarized in table below:

Molecules/Gene/ Protein	Location	Plant	Event of Pollen–Pistil Interactions and Fertilization
STIGMA-SPECIFIC PROTEIN 1 (STIG1)	Stigma-style	*Nicotiana, Petunia*	Regulates the timing of the accumulation of stigmatic and stylar exudates
TURGOR REGULATION DEFECT 1 (TOD1) gene	Pollen tube	*Arabidopsis*	Maintenance of turgor pressure in pollen tubes
O-FUCOSYL TRANSFERASE1 (AtOFT1)	Pollen tube	A novel gene in *Arabidopsis*	Participates in pollen tube penetration through the stigma-style

Chemocyanins and Plantocyanins (blue copper proteins, homologs of chemocyanins	Stylar transmitting tissue	*Arabidopsis*	Involved in Pollen adhesion and pollen tube guidance
Calcium (Ca^{2+})	Multiple Stigmatic papillae, Pollen tube tip.	*Arabidopsis, Nicotiana, Lilium*	Pollen recognition and directs tube growth in style
Arabinogalactans	Stigma-style transmitting tissue	*Arabidopsis, Nicotiana, Lilium*	Pollen recognition and directs tube growth
Cysteine- rich adhesins (SCA)/ LTP5 (SCA-like lipid transfer protein 5)	Stigma-style transmitting tissue	*Lilium* sp.	Pollen tip growth and guidance
CHX21 and CHX23	Localized in the endoplasmic reticulum (ER) of the pollen tube	*Arabidopsis*	These proteins regulate pollen tube emergence and regulate local pH. Alter actin polymerization for the reorientation of the pollen tube growth direction.
AtLURE1 (defensin-like DEFL; peptides,)	Synergids	*Arabidopsis, Zea mays*	Peptides guide tube growth and regulate the pollen tube bursting in synergid.
4-O-methyl-glucuronosyl residue of arabinogalactan polysaccharide (AMOR)	Style	*Torenia fournieri*	Induces competency in the pollen tube to react to LURE-dependent guidance cues.
MITOGEN ACTIVATED PROTEIN KINASES (MPK3) and (MPK6)	Funiculus	*Arabidopsis*	Pollen tube guidance at the funiculus
TURAN and *EVA / ZmEA1* (or *EGG APPARATUS1*)	Egg apparatus	*Arabidopsis, Zea mays*	Pollen tube guidance at micropyle
ROS (Reactive Oxygen species)	Transmitting tissue and embryo sac	*Arabidopsis*	Pollen tube growth, rupture of pollen tube and fertilization dependent on Ca^{2+}
FERONIA/SIRÉNE (FER/SRN); ANXURE 1 and 2 (ANX1 and ANX2) (All are receptor-like kinases)	Synergids (most prominently in the filiform apparatus)	*Arabidopsis*	Pollen tube reception; thought to trigger pollen tube rupture in synergids

MYB family (MYB97, MYB101 and MYB120)	Localized in the pollen tube nucleus	*Arabidopsis*	Activates other genes responsible for pollen tube-synergid communication and induces degeneration of synergid
GAMETE EXPRESSED 2 (GEX2)	Localized in the sperm cell surface (a protein)	*Arabidopsis*	First surface proteins identified in flowering plants to be involved in gamete attachment and recognition
EC1 (EGG CELL 1) (Small cysteine-rich protein)	Egg cell	*Arabidopsis*	Putative role in gamete activation before fusion. Induces accumulation of storage vesicles in the egg cell before fertilization
GENERATIVE CELL SPECIFIC1/ HAPLESS2 (GCS1/ HAP2)	Generative Cell	*Lilium longiflorum, Arabidopsis*	Best characterized gamete interaction protein in plants, required for regulating gamete fusion and fertility
LORELEI (LRE)	Anchored to the synergid membrane	*Arabidopsis*	Prevents polytubey
EIN3-EIN2/EIL2-dependent	Activated after fertilization of the egg cell	*Arabidopsis*	Ethylene-signalling cascade; Causes synergid degeneration and blocks multiple pollen tube attraction
MYB98 (a transcription factor)	Specific expression in the synergid cells	*Torenia fournieri, Arabidopsis*	Mutant shows abnormal filiform apparatus, and this inhibits the secretion of attractants. Hence, loss of micropylar guidance.
CCG, CBP1 and GEX3	Predominantly expressed in the central and/or egg cell	*T. fournieri, Arabidopsis*	Proteins play a role in micropylar guidance of pollen tube
FER (Feronia receptor kinase)	Filiform apparatus of the synergid cells	*Arabidopsis*	Enhances pollen tube reception
Nodulin-like proteins (ENODLs, or ENs),	Synergid cells	*Arabidopsis*	Enhances reception and bursting of pollen tube

Box 7.4

Reactive Oxygen Species and Reactive Nitrogen Species: Role in Pollen–Pistil Interactions

The reactive oxygen species (ROS) and reactive nitrogen species (RNS), play a role in diverse signalling pathways in plants (e.g., stomatal closure, adventitious root development, root hair growth, and host-pathogen interaction). Such varied functions of ROS and RNS molecules, has invited considerable attention from plant biologists to investigate their role in plant reproduction. In the last two decades, a number of studies have determined the role of ROS and RNS in reproductive processes like pollen tube growth and pollen–stigma interactions. In growing pollen tubes, the ROS, and RNS are found to be spatially localized at the pollen tip (e.g., *Nicotina* sp.), and in subapical zones (e.g., *Lilium* sp.) respectively. ROS (produced by pollen NADPH oxidase) are thought to help in polarized tube growth while RNS, particularly NO is involved in the directional growth. Studies have also indicated that NO signalling mediates the attraction of pollen tubes by ovules (Hiscock & Allen, 2008).

Recent investigations have demonstrated the presence of ROS and NO in the stigma at different developmental stages of species, e.g., *Senecio squalidus, Helianthus, Glycine max, Corylus avellana, Elaeocarpus hainanensis, Michelia alba, Brassica, Arabidopsis, Magnolia, Pisum,* and *Citrus.* These investigations have established that ROS and NO are players in pollen–stigma interaction and also provide protection to developing stigma from pathogens. An analysis of 10 different species from monocot and about 22 species from eudicot families (e.g., *Zea mays, Lilium, Ulmus, Calendula, Artemisia*) have indicated that NO scavenges ROS at the site of pollen adhesion. Hence, it appears to be playing a role in pollen–stigma interaction. An investigation in sunflower has also shown more ROS accumulation in stigmatic papillae prior to pollination and higher NO accumulation in post pollination stages. Hence, ROS and NO exhibit reverse trends during pollen–stigma interaction (Bright et al. 2009; Sharma & Bhatla 2013; Zafra et al. 2016). ROS scavenging is also correlated well with the increased activity of enzymes like superoxide dismutases and peroxidases on stigma (Dafni & Maués 1998; Zafra et al. 2016).

Recent investigation (Liu et al. 2021) in *Arabidopsis*, elucidated the distinct role of ANJEA-FER (AT5G59700-FERONIA) receptor kinase complex in pollen hydration. This complex may be considered as a stigmatic gatekeeper *(earlier such kinases were known to play a role in pollen tube bursting and fertilization).* ANJ-FER receptor kinase complex, perceives the RAPID ALKALINIZATION FACTOR peptides RALF23 and RALF33 and induces reactive oxygen species (ROS) production in the stigmatic papillae. After pollination, POLLEN COAT PROTEIN B-class peptides released from the pollen coat bind with ANJ-FER complex and reduce ROS production on stigma. This facilitates hydration of pollen grains. Another recent study (Podolyan et al. 2021) in lily, has indicated that oxygen radicals affect ionic zoning, membrane potential, and pH gradients in lily pollen tubes. Hydroxyl radicals cause depolarization all along the tube while oxygen radicals incite hyper-polarization and cytoplasm-alkalinization. This has given rise to the fresh concept where ROS may be considered as a pollen tube growth stimulator.

7.8 *In vitro* Pollen Tube Germination

Our knowledge of the physiology and biochemistry of pollen germination and pollen tube growth largely comes from *in vitro* studies. It is known that the pollen grains have a store of sugar, starch, lipids and phytic acid in them. These storage products get mobilized and play an important role in germination and also in the initial stages of pollen tube growth.

In vitro pollen germination addresses many basic questions in sexual reproduction and is particularly useful in hybridization programmes. For example, *in vitro* pollen germination is a reliable method to test the viability of pollen which is an important factor before carrying out artificial pollinations. Many pollen germination media ranging from simple sugars to complex mixtures of vitamins, growth regulators and various minerals have been standardized. Pollen germination media differ from species to species and sometimes even for different crop varieties of the same species. In general, successful pollen germination requires many ions and components in different concentrations depending on species. Among the various pollen germination media, the one developed by Brewbaker and Kwack (1963) is the most widely used. Key components of a pollen germination medium are:

- **Sucrose/Carbohydrates:** Pollen development and tube growth are high energy-consuming processes where carbohydrates act as the main energy source. Thus, exogenous sugars are important for *in vitro* pollen germination, especially sucrose. In addition they provide and maintain proper osmoticum, not only for germination of pollen but also for the sustained growth of the pollen tube (Lagera et al. 2017).

- **Calcium:** Calcium has also been shown to be necessary for pollen grain germination, pollen tube growth, sugar synthesis and accumulation (Muengkaew et al. 2016). When a small number of pollen grains are used for *in vitro* germination, the rate of germination and elongation is usually lower in comparison when large numbers of pollen grains are used. This is often due to the diffusion of calcium from pollen grains into the aqueous medium, leaving a very low endogenous level of calcium, which is insufficient for germination. However, when a large number of pollen grains are used, diffused calcium gets trapped between the pollen grains thereby furnishing the required level of calcium needed for germination. This is called the *crowding effect* or *population effect* and rates of germination can be raised either by adding more pollen, or by adding Ca^{2+} to the medium.

This requirement of Ca^{2+} for pollen germination was confirmed by the work of Brewbaker & Kwack (1963) conducted on 86 species belonging to 39 plant families. External Ca^{2+} is also known to affect pollen tube growth as lower Ca^{2+} concentrations lead to excess accumulation of vesicles at the tip, inducing apical swelling. Increased Ca^{2+} concentrations accelerate vesicle fusion at the tip of the pollen tube and hence enhance the growth of tube (Steinhorst & Kudla 2013; Zheng et al. 2019). However, pollen grains of some plants can germinate on media without Ca^{2+} too. It is presumed that in such plants, some Ca^{2+} stored in the pollen wall is released into the medium during pollen hydration.

- **Boron:** Boron in the form of boric acid or borate is also known to influence pollen germination and pollen tube growth. Pollen grains in most flowering plants are deficient in boron. This deficiency is overcome by moderate quantities of boron present on the stigma and also in the style. During *in vitro* studies, supplying exogenous boron, mimics the stigma and stylar supply and enhances pollen germination and pollen tube growth. Boron is also known to prevent the bursting of pollen tubes and plays a significant role in biosynthesis and translocation of carbohydrates. Recent studies have highlighted the role of boron in pectin esterification and cross-linking. Hence, boron deficiency directly hampers pollen tube elongation. Experiments on lily pollen grains have indicated that presence of boron stimulates the plasma membrane H^+-ATPase, which in turn supports germination and growth of pollen tubes.
- **PEG:** Polyethylene glycol (PEG) of various molecular weights is also used to improve pollen germination and tube growth *in vitro* in a number of species. Tandon et al. (1999) reported the effects of boron, calcium, potassium and magnesium in association with PEG to improve *in vitro* germination and tube growth in oil palm pollen. This formulation led to more than 10 times longer pollen tubes than those obtained in the standard sucrose and boric acid medium. A novel *in vitro* pollen germination medium was standardized by Vikas et al.(2012) for *Azadirachta indica* by optimizing the concentration of Brewbaker & Kwack's medium components in combination with different concentrations of sucrose and polyethylene glycol (PEG). This method enhanced the germination of pollen grains many fold. In general, addition of 5%–10% PEG (MW 4000) in pollen germination medium provides good results.
- **Flavanols:** Small aromatic molecules such as carotenoids, flavonoids, jasmonates, phenolic acids, brassino-steroids, and most of the phytohormones are detected in pollen grains. Out of these, flavanols, a specific class of flavonoids, have been found to play an important role in pollen germination. Mostly all pollen grains contain flavonols, often in very high quantities. They are synthesized in the tapetum, released into the locule, and taken up by the developing gametophyte. Flavonols are believed to stimulate pollen tube growth because in flavonol deficient plants like *Petunia*, pollen grains fail to germinate, or produce a functional tube. Stigmatic exudate also provides kaempferol, a flavonol aglycone during pollen germination. Thus, adding flavonols to pollen germination medium may enhance *in vitro* pollen germination.
- **Physical factors:** Increasing temperature is also known to accelerate the rate of pollen germination as well as pollen tube growth. Optimum temperature for pollen grains to germinate is between 20–30°C. Humidity is another physical factor which can affect pollen germination and viability. Studies on various species of *Acacia* have shown that high relative humidity in combination with low temperatures for long periods is beneficial to polyad viability and pollen tube growth. In *Quercus* sp. high air temperatures and low relative humidity in air during floral development negatively affects the pollen germination and tube growth under *in vitro* conditions. In contrast, the strawberry cultivars show pronounced reduction in their percentage germination at a temperature of 30°C and relative humidity of 75–85%. Thus, the optimum requirement of temperature and humidity varies from species to species.

Glossary

Aquaporins: Aquaporins are water-transporting molecular channels, which are present either in the plasma membrane or the vacuolar membrane or both.

Arabinogalactans: Arabinogalactans (AGP) are non-enzymatic, cell surface hydroxyproline-rich glycoproteins.

Crowding effect/Population effect: Higher percentage of germination of pollen grains under *in vitro* conditions when present in clumps/cluster.

Double fertilization: In flowering plants, one sperm cell fertilizes the egg cell (syngamy) and another sperm cell fertilizes the central cell (triple fusion). These two events of fertilization are together known as double fertilization.

Egg cell: Egg cell is the female gamete that fuses with a sperm cell to form an embryo.

Extracellular matrix: A collection of extracellular molecules rich in carbohydrates, lipids and proteins, secreted by plant cells such as those of stigma and style. It serves as a source of pollen tube nourishment and guidance in pistil.

Fertilization interval: The time interval between pollination and fertilization.

Funicular guidance: A mechanism of pollen tube guidance from the surface of the placenta to the funiculus.

Micropylar guidance: It is the mechanism of pollen tube guidance into the micropyle.

Pollen foot: An interface between the stigma and the pollen formed by the mobilization of the lipids and the proteinaceous component of pollen coat after pollen capture on stigma.

Pollen load: The total number of pollen grains transferred on to the stigma by the pollinator during its visit.

Pollen tube emergence: The emergence of the pollen tube from the transmitting tract before entering the placenta.

Pollen tube rupture: Bursting of pollen tube wall at its tip or side wall to discharge the two sperm cells and pollen cytoplasm upon arrival at synergid.

Pollen tube: A protuberant cylindrical cell emerging from the germinating pollen grain that contains the sperm cells (generative cell) and vegetative cell.

Polyspermy: Fertilization of an egg cell by multiple sperms/male gametes.

Polytubey: The entry of multiple pollen tubes into an embryo sac.

Reactive nitrogen species: Reactive molecules derived from nitrogen metabolism, mainly the nitric oxide (NO).

Reactive oxygen species: Molecules derived from the metabolism of oxygen such as hydrogen peroxide or superoxide radical.

Stigma Cysteine-rich Adhesins (SCAs): The specific type of proteins associated with pectins produced in the sporophytic tissue of pistil. SCAs enter into the pollen tube tip via endocytosis and serve as a signal for directional growth of pollen tube tip and also act as 'extracellular glue' for pollen tubes.

Stigma: It is a part of the pistil which serves as a site for landing/capture of pollen and provides salubrious conditions for pollen germination.

Style: Median portion of the pistil connecting the stigma and the ovary.

Syngamy: The event of fusion of male and female gamete is known as syngamy.

Trans-golgi network: A series of interconnected tubules and vesicles at the trans-face of the Golgi stack.

Transmitting tissue tract: A tissue present in the style constituted of cylindrical cells with intercellular spaces that forms a passage (tract) for the movement of the pollen tube in the pistil.

Triple fusion: The fusion of a sperm cell with two polar nuclei of the central cell.

Key Questions

Q7.1 Differentiate between:
  a. Stigma and Style
  b. Solid and Hollow Style
  c. Wet and Dry Stigma
  d. Polyspermy and Polytubey
  e. Syngamy and Double fertilization

Q7.2 Write a short note on:
  a. Factors required for *in vitro* pollen germination
  b. Pollen capture and adhesion
  c. Pollen germination
  d. Pollen tube wall
  e. Role of exine held proteins and lipids in pollen–pistil interactions
  f. Crowding or Population effect
  g. Penetration of stigma by pollen tube

Q7.3 Fill in the blanks:
  a. The contents of the pollen tube are discharged in ....cell of the embryo sac.
  b. Recognition-rejection reaction takes place at...................
  c. .... plays a vital role in pollen–pistil interactions and is crucial for pollen capture and adhesion.
  d. Granular zone of the pollen tube contains ....
  e. ....is the entry of the pollen tube in the ovule through the integuments.
  f. The...................guidance begins from the surface of the placenta and goes up to the funiculus in the ovule.
  g. ....are shed in a partially dehydrated state from the flowers.
  h. The................of stigma is a highly heterogeneous layer consisting of proteins, carbohydrates, amino acids and sometimes even phenols.
  i. There is absence of transmitting tissue in ....style.
  j. Species with .... pollen have a tendency to disperse pollen in a more hydrated and metabolically active state.

Q7.4 What are the different stages of pollen–pistil interactions in a compatible pollination? Elaborate on pollen capture, adhesion and hydration.

Q7.5 "Dry and wet stigma behave differently during pollen–pistil interactions." Do you agree with the statement? Justify your answer.

Q7.6 Elaborate on the structure of a pollen tube with suitable illustrations.

Q7.7 Enumerate the functions of Ca^{2+} in pollen–pistil interactions.

Q7.8 Pollen tube shows polar growth. Explain the statement with the help of suitable diagrams.

Practicals

Exercise 7.1: To study the stigma and the styles of different flowers

Stigma is a part of the pistil which serves as a site for landing or capture of pollen and pollen germination whereas the style is the median portion connecting the stigma and the ovary. A male gametophyte has to communicate with these two structures before reaching the female gametophyte for fertilization. There are different types of stigma and style seen in flowering plants. Structural differences between these types have a pronounced effect on the possible interactions between the male gametophyte and the sporophytic tissues of the pistil. Thus, studying the structural details of stigma and style is important in order to understand pollen–pistil interactions and fertilization. Nonetheless, stigma type is also associated with pollination mechanism as studied in chapter 6.

Materials required

Flowers of randomly selected plants, forceps, needles, safranin, slides and microscope.

Suggested flowers: *Crotolaria, Petunia, Nicotiana, Impatiens, Bombax ceiba, Brassica, Chorisia, Lilium, Hibiscus, Citrus.*

Procedure

1. Collect freshly opened flowers and carefully excise the pistil.
2. Observe the stigma under a stereomicroscope and describe its morphology.
3. Record whether the stigma is of the wet or dry type on the basis of presence or absence of exudate. Also record whether the stigma is papillate or non-papillate.
4. Observe style for its morphology.
5. Cut transverse section of the style in the middle portion, stain the section and mount.
6. Observe and record the nature of the style (solid or hollow).

Observation: Write your observations in the sample table provide below.

Flower	Stigma			Style	
	Shape	Papillate/Non-papillate	Wet/Dry	Long/short or any other feature	Hollow/ Solid
Brasssica					
Petunia					
Lilium					
Crotolaria					

Exercise 7.2: To study the stigma receptivity by cytochemical localization of non-specific esterases and peroxidase activity

Stigma receptivity is the ability of the stigma to support pollen germination and pollen tube growth following pollination. Receptivity of the stigma is crucial for the success of pollination. A non-receptive stigma at the time of pollination cannot provide the required conditions for pollen grains to germinate and may result in reproductive failure. Therefore, studying stigma receptivity is important for reproductive characterization of a species. The mature stigma contains several enzymes such as peroxidases, phosphatases and nonspecific esterases. By performing simple histochemical localization of such enzymes on stigma and comparing their activity levels at different development stages of flower, one can determine the stage of stigmatic receptivity. In general, stigma gradually attains receptivity during its development and receptivity declines after pollination. The characterization of the developmental stage with maximum stigmatic receptivity is a prerequisite in breeding experiments, manual pollination, and artificial pollination. Such tests are also imperative to assess the male and female functions of the flowers at a particular time. For an instance, by performing these simple tests one may determine dichogamy.

Principle

For the cytochemical localization of non-specific esterases on stigma α-naphthyl acetate serves as substrate for non-specific esterases. The latter hydrolyses α-naphthyl acetate and produces a colorless compound 'α-naphthol' which forms a red-brown insoluble complex with Fast Blue B. In this reaction Fast Blue B acts as a coupling agent (Shivanna & Rangaswamy 1992). The intensity of the red-brown color in a stigma after treatment indicates the level of receptivity. In the cytochemical detection of peroxidase activity on stigma surface, hydrogen peroxide is a substrate for peroxidases. The oxygen liberates in the form of bubbles from the stigma surface through the action of peroxidases. The faster the rate of bubble formation, the greater is the stigma receptivity (Fig. 7.10).

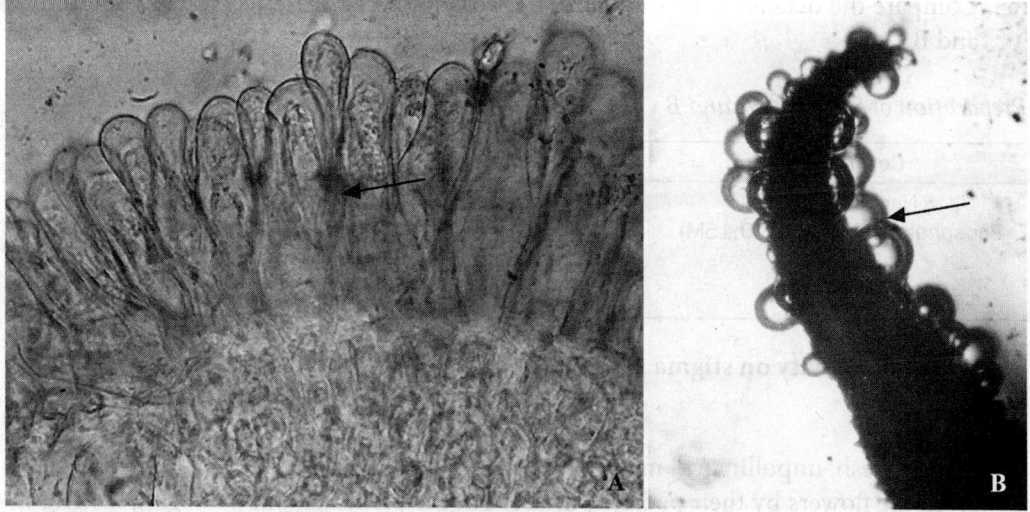

Figure 7.10 Stigma receptivity tests. A. Receptive stigma showing accumulation of red complex in papillae (arrow); tests positive for non-specific esterases. **B.** Receptive stigma (of *Bombax ceiba*) showing activity of peroxidase. Arrow indicates the liberation of oxygen bubbles. **See Color Plates (page 489).**

Materials Required

Chemicals: ạ-naphthyl acetate, phosphate buffer (pH-7, 0.15M), sucrose, Fast Blue B, distilled water, benzidine, hydrogen peroxide, ethanol, glycerol

Glassware: measuring cylinder, petri plates, glass dropper, conical flasks, glass slides

Equipment: Weighing balance, stereo zoom microscope, compound microscope, forceps, needle

Suggested flowers: *Crotolaria, Petunia, Nicotiana, Impatiens, Bombax ceiba, Brassica, Chorisia, Lilium, Hibiscus, Citrus.*

A. Cytochemical localization of non-specific esterases on stigma surface

Procedure

1. Collect fresh, unpollinated, mature flowers of different species.
2. Hold the flowers by their pedicels; dissect out the pistil without damaging the stigma and style.
3. Take a few drops of Solution A in a cavity slide and Solution B in another cavity slide.
4. Dip only one stigma into each solution, without damaging the other parts of pistil. Incubate the preparations for 20–30 minutes. Note the pistils must be at same developmental stage for comparison.
5. Remove the pistils from each solution.
6. Rinse twice with phosphate buffer in watch glass or Petri dish. Make whole mount preparations or cut longitudinal sections of the stigma and mount in glycerol.
7. Observe the preparations under compound microscope.
8. Compare the details of the stigmatic surfaces (color change) incubated in Solution A and B

Preparation of Solutions A and B

Component	Solution A	Solution B (control)
ạ-Naphthyl acetate	5 mg	0mg
Phosphate Buffer (pH-7, 0.15M)	10 ml	10 ml
Sucrose	10–15%	10–15%
Fast blue B	25 mg	25 mg

B. Peroxidase activity on stigma surface

Procedure

1. Collect fresh, unpollinated, mature flowers of different species.
2. Hold the flowers by their pedicels; dissect out the pistil without damaging the stigma and style.
3. Take a few drops of Solution A in one cavity slide and Solution B in another.
4. Dip only the stigma into each solution, without damaging the other parts of pistil.

5. Incubate the preparation for 2–3 minutes.
6. Observe the preparation under stereo zoom or dissection microscope.
7. Count the number of bubbles emerging out from the stigma surface.

Preparation of Test solutions*

With substrate (A): 1% Benzidine in 60% Alcohol: H_2O_2: Water (4:11:22)

Control (B): 1% Benzidine in 60% Alcohol: Water (4:33)

*Note: First prepare 60% ethanol solution. Then prepare 1% solution of Benzidine in it. Benzidine is not completely soluble and remains suspended.

Precautions

1. The test solutions must be prepared fresh.
2. Do not damage the pistil and touch the stigma.
3. Always use a control while performing these tests.

Exercise 7.3: To test germinability of pollen grains under *in vitro* condition through various methods

Theory

In vitro pollen germination is an important exercise to study the requirements of pollen germination of a species. Pollen germination under *in vitro* conditions is also a test to estimate pollen viability. There are various chemical and physical factors which play an important role in the germination of pollen. Pollen germination medium differs from species to species and requires standardization.

Materials Required

Cavity block with glass lid, Improvised Humidity Chamber (IHC), sucrose, boric acid, calcium, potassium nitrate, magnesium sulphate, Tris salt, PEG 400, distilled water, conical flasks, glass droppers, measuring cylinder, spatula, forceps, needle, weighing balance, compound microscope.

Suggested flowers: *Vinca rosea, Impatiens, Crotolaria, Tradescantia*

I. Cavity slide technique/Hanging drop method (Shivanna & Rangaswamy 1992):

This method is used for pollen germination in liquid media.

Procedure

1. Dissect out the undehisced mature anthers from a flower.
2. Keep in IHC for 15–20 minutes.
3. Squeeze pollen grains from hydrated anthers on a glass slide.
4. Place a drop of pollen germination medium* over the glass plate and add hydrated pollen in it. Gently disperse the pollen grains into the medium using needle.
5. Apply Vaseline/petroleum jelly on the margins of the glass plate.

6. Place cavity block on the glass plate, taking care that the drop of medium must be in the centre.
7. Invert the setup gently in a way that the drop sticking to glass slide will hang in the cavity.
8. Incubate the culture setup for about 20–30 minutes.
9. Observe under compound microscope at low magnification (4X).

As an alternate, one may use coverslip and cavity slide for the hanging drop set-up.

II. Sitting drop culture

This method is also used for pollen germination in liquid media.

Procedure

1. Dissect out the undehisced mature anthers from a flower.
2. Keep in IHC for 15–20 minutes.
3. Squeeze pollen grains from hydrated anthers on a clean glass slide.
4. Add a drop of pollen germination medium* over the pollen grains.
5. Mix pollen grains in germination medium with the help of a clean needle. Keep the culture in IHC for 30–40 minutes.
6. Observe under compound microscope.

Result
The viable pollen grains show the emergence of pollen tube (Fig. 7.11).

Preparation of pollen germination medium
The simplest medium for pollen germination comprises of only sucrose. The concentration of sucrose may vary 5-10%. The other media may include sugar, calcium nitrate and boric

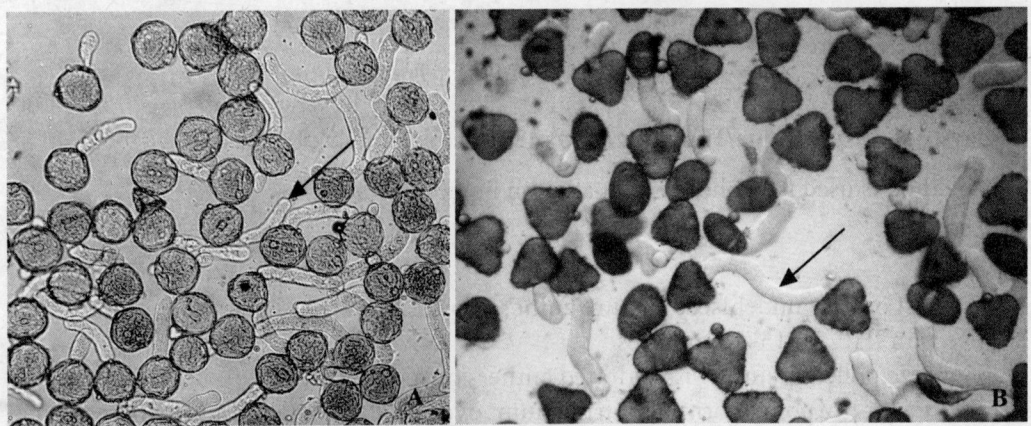

Figure7.11 Pollen germination in Brew-Baker and Kwack's medium under *in vitro* conditions. **A.** *Vinca rosea*; **B.** *Bombax ceiba*. Arrow indicates a pollen tube. **See Color Plates (page 489).**

acid to which poly ethylene glycol, vitamins, amino acids, and growth regulators may be added. The composition of some of the widely used media is given below:

	Brewbaker & Kwack medium	Roberts medium	PEG medium
Sucrose	10%	20%	-
Boric acid	100 mg/l	10 mg/l	100 mg/l
Calcium	300 mg/l ($CaNO_3$)	362 mg/l ($CaCl_2$)	-
Potassium nitrate	100 mg/l	100 mg/l	-
Magnesium Sulphate	200 mg/l	-	-
Tris	-	60–130 mgl/1	-
PEG 400	-	-	0.1 to 1.1 M

Precautions

1. Avoid contamination of germination media with vaseline or petroleum jelly.
2. Pollen grains must be hydrated before use.
3. Germination medium must be prepared afresh.
4. Do not add excess medium over glass lid or slide.

Exercise 7.4: To study the pollen tube path in the pistil by decolorized aniline blue staining
The growing pollen tubes form callose plugs and localization of such callose plugs can be used for studying pollen tube path in the pistil. Callose can be localized by staining pollinated stigmas with a dye called 'Aniline blue' which in decolorized state (at pH-9.2) specifically stains callose. When such stained pistils are observed under a fluorescence microscope, callose shows fluorescence.

Material Required

Flowers of two different species/genera (e.g., *Brassica, Catharanthus, Petunia, Crotolaria, Nicotiana*

Glassware and such: Glass vials/ screw cap bottle (30–50 ml), glass slides, coverslips, measuring cylinder, a pair of forceps, needles, blade

Chemicals: Agar, sucrose, acetic alcohol (1:3, v/v), 1N NaOH, or 1N KOH, double distilled water,

Equipment: weighing balance, hot plate, hot air oven (with thermostat), fluorescence microscope

Procedure
Both open pollinated and manually pollinated pistils can be used for the study. Procedure using open pollinated pistils is discussed here. For details of pollen tube growth in manually pollinated pistils refer to chapter 7.

Step 1 Collection and fixation

1. Collect the pollinated pistils and fix them in acetic alcohol for 24 hours.

Step 2 Tissue clearing and staining

1. Wash the fixed/stored pistils two-three times with double distilled water.
2. Transfer the washed pistils to another screw cap bottle containing 1N NaOH and fasten the cap.
3. Incubate the bottle/vial in hot air oven at 50–60°C for about 1–3 hrs. Check at regular intervals the extent of softening of pistil tissue. The tissue must not over-soften. (*Note: Here one may increase the concentration of NaOH to 5N but in that case duration of incubation should be decreased.*)
4. After appropriate duration, wash (twice) the pistils carefully with double distilled water.
5. Transfer the washed pistils to screw cap bottle containing decolorized aniline blue solution. Cover the bottle with aluminium foil and incubate the pistils with dye at 4°C for 24 hrs.
6. Mount the pistil over a slide in aniline blue solution and place the coverslip. Observe under UV-fluorescence microscope (FITC filter or Ex filter 450–490{Blue range}) (Fig. 7.12).

(*Note: For better observations of pollen tubes/growth, slightly press the coverslip to spread the pistil tissue.*)

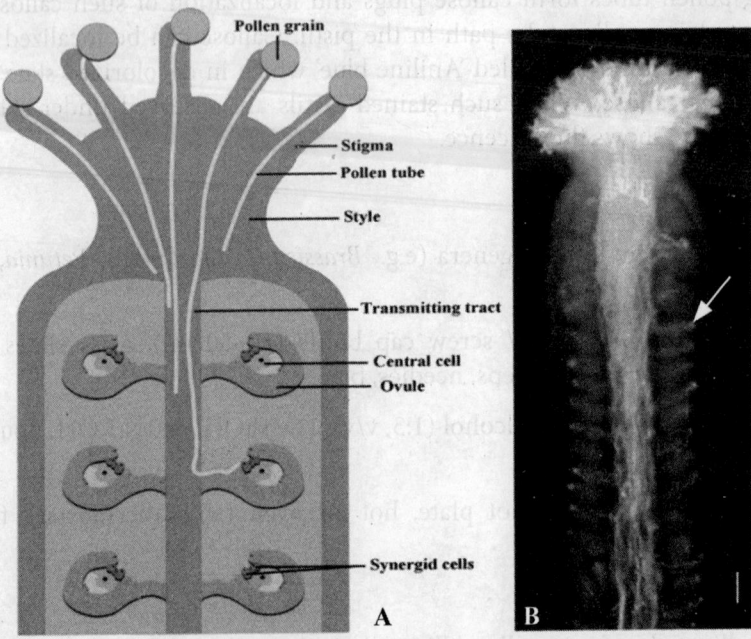

Figure 7.12 A. Diagrammatic sketch of pollen tube path in *Arabidopsis thaliana*, a species with a solid style. **B.** Aniline blue stained pistil showing fluorescing pollen tubes. Pollen tube entry in to the ovules (arrow) can also be seen. **See Color Plates (page 489).** (*Source:* Zheng et al. 2018, published under CC BY 4.0 License.)

Preparation of decolorized Aniline Blue solution
Mix 0.005% Aniline Blue (water soluble) in 0.05M Na_2HPO_4. Adjust the pH to 11.0 with 1N NaOH. Filter the solution and store in dark bottle in refrigerator.

Precautions

1. Avoid over clearing or softening of tissue in NaOH.
2. NaOH must be completely washed off before adding aniline blue.
3. Pistils must be completely immersed in the aniline blue solution.

Bibliography

Arbeloa, A. and Herrero, M. (1987). The significance of the obturator in the control of pollen tube entry into the ovary in peach (*Prunus persica*). *Annals of Botany* 60: 681–5.

Bascom, C. S., Hepler, P. K. and Bezanilla, M. (2018). Interplay between ions, the cytoskeleton, and cell wall properties during tip growth. *Plant Physiology* 176: 28–40.

Beck-Pay, S. L. (2012). The effect of temperature and relative humidity on *Acacia mearnsii* polyad viability and pollen tube development. *South African Journal of Botany* 83: 165–71.

Bhardwaj, V. and Tandon, R. (2013). Self-incompatibility and post-fertilization maternal regulation cause low fecundity in *Aegle marmelos* (Rutaceae). *Botanical Journal of Linnean Society* 172: 572–85.

Bosch, M. and Hepler, P. K. (2005). Pectin methylesterases and pectin dynamics in pollen tubes. *Plant Cell* 17: 3219–26.

Brewbaker, J. L. and Kwack, B. H. (1963). The essential role of calcium ion in pollen germination and pollen tube growth. American *Journal of Botany* 50: 859–65.

Bright, J., Simon, J., Hiscock, B. et al. (2009). Pollen generates nitric oxide and nitrite: a possible link to pollen-induced allergic responses. *Plant Physiology and Biochemistry* 47: 49–55.

Cai, G. and Cresti, M. (2010). Microtubule motors and pollen tube growth–still an open question. *Protoplasma*. 247: 131–43.

Carlisle, S., Bascom Jr., C. S., Hepler P. K. and Bezanilla, M. (2018). Interplay between ions, the cytoskeleton, and cell wall properties during tip growth. *Plant Physiology* 176: 28–40.

Castro, M. A. and Agullo, M. A. (1998). Anatomy of the stigma of *Vigna adenantha* (Leguminosae, Papilionoideae). *Bio cell* 22: 9–18.

Chae, K. and Lord, E. M. (2011). Pollen tube growth and guidance: roles of small, secreted proteins. *Annals of Botany* 108: 627–36.

Chapman, L. A. and Goring, D. R. (2010). Pollen–pistil interactions regulating successful fertilization in the Brassicaceae. *Journal of Experimental Botany* 61: 1987–99.

Chaudhary, A., Khanduri, P., Tandon, R. et al. (2014). Central cell degeneration leads to three-celled female gametophyte in *Zeylanidium lichenoides* Engl. (Podostemaceae). *South African Journal of Botany* 91: 99–106.

Chaudhary, A., Tofanelli, R. and Schneitz K. (2020). Plant Reproduction: Shaping the genome of plants. *eLife* 9: e54874.

Chavarría-Krauser, A. and Yejie, D. (2011). A model of plasma membrane flow and cytosis regulation in growing pollen tubes. *Journal of Theoretical Biology* 285: 10–24.

Chebli, Y., Kaneda, M., Zerzour, R. and Geitmann, A. (2012). The cell wall of the *Arabidopsis* pollen tube–spatial distribution, recycling, and network formation of polysaccharides. *Plant Physiology* 160: 1940–55.

Chebli, Y., Kroegerb, J. and Geitmanna, A. (2013). Transport logistics in pollen tubes. *Molecular Plant* 6:1037–1052.

Cheung, A. Y., Duan, Q. H., Costa, S. S. et al. (2008). The dynamic pollen tube cytoskeleton: live cell studies using actin binding and microtubule-binding reporter proteins. *Molecular Plant* 686–702.

Cheung, A. Y., Wang H. and Wu, H. M. (1995). A floral transmitting tissue-specific glycoprotein attracts pollen tubes and stimulates their growth. *Cell* 82: 383–93.

Coccussi, J. (1969). Orchid embryology: Pollen tetrads of *Epidendrum scutella* in the anther and on the stigma. *Planta* 84: 215–29.

Dafni, A. and Maués, M. M. (1998). A rapid and simple procedure to determine stigma receptivity. *Sexual Plant Reproduction* 11: 177–80.

de Graaf, B. H. J., Derksen, J. W. M. and Mariani C. (2001). Pollen and pistil in the progamic phase. *Sexual Plant Reproduction* 14: 41–55.

Dearnaley, J. D. W. and Daggard, G. A. (2001). Expression of a polygalacturonase enzyme in germinating pollen of Brassica napus. *Sexual Plant Reproduction* 13: 265–71.

Dettmer, J., Hong-Hermesdorf, A., Stierhof, Y. D. and Schumacher, K. (2006). Vacuolar HC-ATPase activity is required for endocytic and secretory trafficking in Arabidopsis. *The Plant Cell* 18: 715–30.

Dixit, R., Rizzo, C., Nasrallah, M. and Nasrallah, J. (2001). The *Brassica* MIP-MOD gene encodes a functional water channel that is expressed in the stigma epidermis. *Plant Molecular Biology* 45: 51–62.

Dresselhaus, T., Sprunck, S. and Wessel, G. M. (2016). Fertilization mechanisms in flowering plants. *Current Biology* 26: R125–R139.

Edlund, A. F., Swanson, R. and Preuss, D. (2004). Pollen and stigma structure and function: the role of diversity in pollination. *Plant Cell* 16: S84–S97.

Erbar, C. (2003). Pollen tube transmitting tissue: place of competition of male gametophytes. *International Journal of Plant Sciences* 164: S265–S277.

Escobar-Restrepo, J. M., Huck, N., Kessler, S. et al. (2007). The FERONIA receptor-like kinase mediates male–female interactions during pollen tube reception. *Science* 317: 656–60.

Fiebig, A., Kimport, R. and Preuss, D. (2004). Comparisons of pollen coat genes across Brassicaceae species reveal rapid evolution by repeat expansion and diversification. *Proceedings of the National Academy of Sciences USA* 101: 3286–91.

Franklin-Tong, V. E. (1999). Signalling and the modulation of pollen tube growth. *The Plant Cell* (online) 11: 727–38.

Ge L. L., Xie, C. T., Tian, H. Q. and Russell, S. D. (2009). Distribution of calcium in the stigma and style of tobacco during pollen germination and tube elongation. *Sexual Plant Reproduction* 22: 87–96.

Gotelli, M. M., Lattar, E. C., Zini, L. M. et al. (2017). Style morphology and pollen tube pathway. *Plant Reproduction* 30: 155–70.

Grebnev, G., Ntefidou, M. and Kost, B. (2017). Secretion and endocytosis in pollen tubes: models of tip growth in the spot light. *Frontiers in Plant Sciences* 154: 1–7.

Green, J. R. (1894). On the germination of the pollen grain and the nutrition of the pollen tube. *Annals of Botany* 8: 225–8.

Grossniklaus, U. (2017). Polyspermy produces tri-parental seeds in maize. *Current Biology* 27: R1300–R1302.

Guan, Y., Lu, J. Xu, J., et al. (2014). Two mitogen-activated protein kinases, MPK3 and MPK6, are required for funicular guidance of pollen tubes in *Arabidopsis*. *Plant Physiology* 165: 528–33.

Gunning, B. E. S. and Pate, J. S. (1969). 'Transfer cell'–plant cells with wall ingrowths specialized in relation to short distance transport of solutes-their occurrence, structure and development. *Protoplasma* 68: 107–33.

Hafidh, S., Fila, J. and Honys, D. (2016). Male gametophyte development and function in angiosperms: a general concept. *Plant Reproduction* 29: 31–51.

Hepler, P. K. and Winship, L. J. (2015). The pollen tube clear zone: clues to the mechanism of polarized growth. *Journal of Integrative Plant Biology* 57: 79–92.

Herrero, M. and Dickinson, H. G. (1979). Pollen–pistil incompatibility in Petunia hybrida: changes in the pistil following compatible and incompatible intra-specific crosses. *Journal of Cell Science* 36: 1–18.

Herrero, M. and Hormaza, J. l. (1996). Pistil strategies controlling pollen tube growth. *Sexual Plant Reproduction* 9: 343–7.

Heslop-Harrison, J. (1975a). Male gametophyte selection and the pollen–stigma interaction. In D. L. Mulcahy, ed., *Gamete Competition in Plants and Animals*. Amsterdam: North Holland Pub. Co., pp. 177–90.

Heslop-Harrison, J. (1975b). Incompatibility and the pollen–stigma interaction. *Annual Review of Plant Physiology* 26: 403–25.

Heslop-Harrison, Y. and Shivanna, K. R. (1977). The receptive surface of the angiosperm stigmas. *Annals of Botany* 41: 1233–58.

Heslop-Harrison, J. and Heslop-Harrison, Y. (1985). Surfaces and secretions in the pollen–pistil interaction: a brief review. *Journal of Cell Science (Suppl)* 2: 287–300.

Heslop-Harrison, J. (1987). Pollen germination and pollen tube growth. *International Review in Cytology* 107: 1–78.

Higashiyama, T. and Yang, W. (2017). Gametophytic pollen tube guidance: attractant peptides, gametic controls, and receptors. *Plant Physiology* 173: 112–21.

Hiscock, S. J. and Allen, A. M. (2008). Diverse cell signalling pathways regulate pollen–stigma interactions: the search for consensus. *New Phytologist* 179: 286–317.

Hiscock, S. J., Hoedemaekers, K., Friedman, W. E. and Dickinson, H. G. (2002). The stigma surface and pollen–stigma interactions in *Senecio squalidus* L. (Asteraceae) following cross (compatible) and self (incompatible) pollinations. *International Journal of Plant Sciences* 163: 1–16.

Holdaway-Clarke, T. L. and Hepler, P. K. (2003). Control of pollen tube growth: role of ion gradients and fluxes. *New Phytologist* 159: 539–63.

Hou, Y., Guo, X., Cyprys, P. et al. (2016). Maternal ENODLs are required for pollen tube reception in *Arabidopsis*. *Current Biology* 26: 2343–50.

Huang, W. J., Liu, H. K., McCormick, S. and Tang, W. H. (2014). Tomato pistil factor STIG1 promotes in vivo pollen tube growth by binding to phosphatidylinositol 3-phosphate and the extracellular domain of the pollen receptor kinase LePRK2. *The Plant Cell* 26: 2505–23.

Jayaprakash, P. (2018). Pollen Germination in vitro, Pollination in Plants. *Phatlane William Mokwala, Intech Open*, DOI: 10.5772/intechopen.75360. https://www.intechopen.com/books/pollination-in-plants/pollen–germination-in-vitro.

Jensen, W. A. (1964). Observations on the fusion of nuclei in plants. *Journal of Cell Biology* 23: 669–72.

Jensen, W. A. (1965a). The ultrastructure and histochemistry of the synergids of cotton. *American Journal of Botany* 52: 238–56.

Jensen, W. A. (1965b). The ultrastructure and composition of the egg and the central cell of cotton. *American Journal of Botany* 52: 781–97.

Jensen, W. A. (1973). Fertilization in flowering plants. *Bioscience* 23: 21–7.

Jin, X. F., Ye, Z. M., Amboka, G. M. et al. (2017). Stigma sensitivity and the duration of temporary closure are affected by pollinator identity in *Mazus miquelii* (Phrymaceae), a species with bilobed stigma. *Frontiers of Plant Science* 8: 783. doi: 10.3389/fpls.2017.00783.

Johnson, M. A, Harper, J. F. and Palanivelu, R. (2019). A fruitful journey: pollen tube navigation from germination to fertilization. *Annual Review of Plant Biology* 70: 809–37.

Juan, M. L. and Maria, H. (2017). Pollen tube access to the ovule is mediated by glycoprotein secretion on the obturator of apple (*Malus* × domestica, Borkh). *Annals of Botany* 119: 989–1000.

Ketelaar, T., Galway, M. E., Mulder, B. M. and Emons, A. M. (2008). Rates of exocytosis and endocytosis in Arabidopsis root hairs and pollen tubes. *Journal of Microscopy* 231: 265–73.

Khanduri, P., Chaudhary, A., Uniyal, P. L. and Tandon, R. (2014). Reproductive biology of *Willisia arekaliana* (Podostemaceae), a freshwater endemic species of India. *Aquatic Botany* 119: 57–65.

Kim, H. U, Chung, T. Y. and Kang, S. K. (1996). Characterization of anther-specific genes encoding a putative pectin esterase of Chinese cabbage. *Molecules and Cells* 6: 334–40.

Kim, S., Mollet, J., Dong, J. et al. (2003). Chemocyanin, a small, basic protein from the lily stigma, induces pollen tube chemotropism. *Proceedings of the National Academy of Sciences USA* 100: 16125–30.

Knox, R. B. and Heslop-Harrison, J. (1970). Pollen wall proteins: localization and enzymatic activity. *Journal of Cell Science* 6: 1–27.

Köhler, C., Scheid, O. M. and Erilova, A. (2010). The impact of the triploid block on the origin and evolution of polyploid plants. *Trends in Genetics* 26: 142–8.

Kost, B. (2008). Spatial control of Rho (Rac-Rop) signalling in tip-growing plant cells. *Trends in Cell Biology* 18: 119–27.

Lagera, A. J., Balinado, L. O., Baldomero, J. R. et al. (2017). Varying sugars and sugar concentrations influence in vitro pollen germination and pollen tube growth of *Cassia alata* L. *Journal of Young Investigators* 33: 42–5.

Lancelle, S. A. and Hepler, P. K. (1992). Ultrastructure of freeze-substituted pollen tubes of *Lilium longiflorum*. *Protoplasma* 167: 215–30.

Linskens, H. F. (1967). Pollen. *Handbuch der pflanzenphysiologie* XVIII: 368–406.

Liu, C., Shen, L. and Xiao, Y. et al. (2021). Pollen PCP-B peptides unlock a stigma peptide–receptor kinase gating mechanism for pollination. *Science* 372: 171–5.

Lora, J., Hormaza, J. I. and Herrero, M. (2016).The diversity of the pollen tube pathway in plants: toward an increasing control by the sporophyte. *Frontiers in Plant Science* 7: 107.

Lord, E. M. and Russel, S. D. (2002). The mechanisms of pollination and fertilisation in plants. *Annual Reviews in Cell Development and Biology* 18: 81–105.

Losada, J. M. and Herrero, M. (2017). Pollen tube access to the ovule is mediated by glycoprotein secretion on the obturator of apple (*Malus* × *domestica*, Borkh). *Annals of Botany* 119: 989–1000.

Mangla, Y., Tandon, R., Goel, S. and Raina, S. N. (2013). Structural organization of the gynoecium and pollen tube path in Himalayan sea buckthorn, *Hippophae rhamnoides* (Elaeagnaceae). *AoB PLANTS* 5: plt015; doi:10.1093/aobpla/plt015.

Mangla, Y. and Tandon, R. (2011). Insects facilitate wind pollination in pollen-limited *Cratevaadansonii* (Capparaceae). *Australian Journal of Botany* 59: 61–9.

Mascarenhas, J. P. and Lafountain, J. (1972). Protoplasmic streaming, cytochalasin B and growth of the pollen tube. *Tissue Cell* 4: 11–14.

Mayfield, J. A., Fiebig, A., Johnstone, S. E. and Preuss, D. (2001). Gene families from the Arabidopsis thaliana pollen coat proteome. *Science* 292: 2482–5.

Micheli, F. (2001). Pectin methylesterases: cell wall enzymes with important roles in plant physiology. *Trends in Plant Sciences* 6: 414–19.

Mizuta, Y. and Higashiyama, T. (2018). Chemical signalling for pollen tube guidance at a glance. *Journal of Cell Science* 131: 1–8.

Mollet, J. C., Park, S. Y., Nothnagel, E. A. and Lord, E. M. (2000). A lily stylar pectin is necessary for pollen tube adhesion to an in vitro stylar matrix. *Plant Cell* 12: 1737–50.

Muengkaew, R., Chaiprasart, P. and Wongsawad, P. (2016). Calcium-boron addition promotes pollen germination and fruit set of mango. *International Journal of Fruit Science* 17: 147–58.

Murphy, D. J. (2006). The extracellular pollen coat in members of the Brassicaceae: composition, biosynthesis, and functions in pollination. *Protoplasma* 228: 31–9.

Nawaschin, S. (1898). Resultateeiner Revision der Befruchtungsvorgängebei *Lilium martagon* and *Fritillaria tenella*. *Bulletin of Academy of Imperial Sciences St. Petersburg* 9: 377–82.

Newcombe, F. C. (1922). Significance of the behavior of sensitive stigmas. *American Journal of Botany* 9: 99–120.

Newcombe, F. C. (1924). Significance of the behavior of sensitive stigmas II. *American Journal of Botany* 11: 85–93.

Nguema-Ona, E., Coimbra, S., Vicer-Gebouin, M. et al. (2012). Arabinogalactan proteins in root and pollen-tube cells: distribution and functional aspects. *Annals of Botany* 110: 383–404.

Onelli, E. and Moscatelli, A. (2013). Endocytic pathways and recycling in growing pollen tubes. *Plants* 2: 211–29.

Podolyan, A., Luneva, O., Klimenko, E. and Breygina, M. (2021). Oxygen radicals and cytoplasm zoning in growing lily pollen tubes. *Plant Reproduction* 34: 103–15.

Raina, M., Kumar, R. and Kaul, V. (2017). Stigmatic limitations on reproductive success in a paleotropical tree: causes and consequences. *AoB PLANTS* 9: plx023. doi:10.1093/aobpla/plx023.

Russell, S. D. (1996). Attraction and transport of male gametes for fertilization. *Sexual Plant Reproduction* 9: 337–42.

Schoenaers, S., Balcerowicz, D., Costa, A. and Vissenberg, K. (2017). The Kinase ERULUS controls pollen tube targeting and growth in *Arabidopsis thaliana*. *Frontiers in Plant Sciences* 8: 1942. doi: 10.3389/fpls.2017.01942.

Scott, R. J., Armstrong, S. J., Doughty, J. and Spielman, M. (2008). Double fertilization in *Arabidopsis thaliana* involves a polyspermy block on the egg but not the central cell. *Molecular Plant* 1: 611–19.

Sehgal, A., Khurana, J. P., Sethi, M. et al. (2011). Occurrence of unique three-celled megagametophyte and single fertilization in an aquatic angiosperm- *Dalzellia zeylanica* (Podostemaceae-Tristichoideae). *Sexual Plant Reproduction* 24: 199–210.

Selinski, J. and Scheibe, R. (2014). Pollen tube growth: where does the energy come from? *Plant Signal and Behavior* 9: e977200.

Sever, K., Škvorc, Z., Saša Bogdan, S. et al. (2012). In vitro pollen germination and pollen tube growth differences among *Quercus robur* L. clones in response to meteorological conditions. *Grana* 51: 25–34.

Sharma, B. and Bhatla, S. C. (2013). Accumulation and scavenging of reactive oxygen species and nitric oxide correlate with stigma maturation and pollen–stigma interaction in sunflower. *Acta Physiologiae Plantarum* 35: 2777–87.

Shivanna, K. R. (1977). Pollen–stigma interaction–recognition acceptance and rejection. In Symposium on basic sciences and agriculture, Indian National Science Academy, New Delhi, pp. 53–61.

Shivanna, K. R. (2003). *Pollen Biology and Biotechnology*. Enfield: Science Publishers.

Shivanna, K. R. (2016). Fertilization in flowering plants. *Resonance* 21: 1007–18.

Shivanna, K. R. (2020a). The pistil: structure in relation to its function. In R. Tandon, K. R. Shivanna and M. Koul, eds., *Reproductive Ecology of Flowering Plants-Patterns and Processes*. Delhi: Springer-India.

Shivanna, K. R. (2020b). Pollen–pistil interaction and fertilization. In R. Tandon, K. R. Shivanna and M. Koul, eds., *Reproductive Ecology of Flowering Plants-Patterns and Processes*. Delhi: Springer-India.

Shivanna, K. R. and Rangaswamy, N. S. (1992). *Pollen Biology: A Laboratory Manual*. Springer-Verlag Berlin Heidelberg.

Smith, D. K., Jones, D. M., Lau, J. B. R. et al. (2018). A putative protein o-fucosyltransferase facilitates pollen tube penetration through the stigma–style interface. *Plant Physiology* 176: 2804–18.

Sogo, A., Noguchi, J., Jaffré, T. et al. (2004). Pollen–tube growth pattern and chalazogamy in *Casuarina equisetifolia* (Casuarinaceae). *Journal of Plant Research* 117: 37–46.

Sritongchuay, T., Bumrungsri, S., Meesawat, U. and Mazer, S. J. (2010). Stigma closure and re-opening in *Oroxylum indicum* (Bignoniaceae): causes and consequences. *American Journal of Botany* 97: 136–43.

Steer, M. W. and Steer, J. M. (1989). Pollen tube tip growth. *New Phytologist* 111: 323–58.

Steinhorst, L. and Kudla, J. (2013). Calcium–a central regulator of pollen germination and tube growth. *Biochimica et Biophysica Acta* 1833: 1573–81.

Stephan, O., Cottier, S. and Fahlen, S. et al. (2014). RISAP is a TGN-associated RAC5 effector regulating membrane traffic during polar cell growth in tobacco. *Plant Cell* 26: 4426–47.

Sterling, J. D., Quigley, H. F., Orellana, A. and Mohnen, D. (2001). The catalytic site of the pectin biosynthetic enzyme alpha-1,4-galacturonosyltransferase is located in the lumen of the Golgi. *Plant Physiology* 127: 360–71.

Swanson, R., Edlund, A. F. and Preuss, D. (2004). Species specificity in pollen–pistil interactions. *Annual Review of Genetics* 38: 793–818.

Tandon, R., Manohara, T. N., Nijalingappa, B. H. M. and Shivanna, K. R. (1999). Polyethylene glycol enhances in vitro germination and tube growth of oil palm pollen. Indian Journal of Experimental Biology 37: 169–72.

Taylor, L. P. and Hepler, P. K. (1997). Pollen germination and tube growth. *Annual Review of Plant Physiology and Plant Molecular Biology* 48: 461–91.

Trigo, S. G., Blanco-Touriñán, N., Thomas, A. et al. (2019). The Kinase ERULUS controls pollen tube targeting and growth in *Arabidopsis thaliana. Biorxiv preprint.* doi: https://doi.org/10.1101/428854.

Vikas, Singh V. and Tandon, R. (2012). Polyethylene glycol and polyamines promote pollen germination and tube growth in *Azadirachta indica* (Meliaceae). *The International Journal of Plant Reproductive Biology* 4: 17–23.

Williams, J. H. and Brown, C. D. (2018). Pollen has higher water content when dispersed in a tricellular state than in a bicellular state. *Acta Botanica Brasilica* 32: 454–61.

Williams J. H. (2008). Novelties of the flowering plant pollen tube underlie diversification of a key life history stage. *Proceedings of the National Academy of Sciences of the USA* 105: 11259–63.

Wilsen, K. L. and Hepler, P. K. (2007). Sperm delivery in flowering plants: the control of pollen tube growth. *BioScience* 57: 835–44.

Wu, Y. Z., Qiu, X., Du, S. and Erickson L. (1996). PO149, a new member of pollen pectate lyase-like gene family from alfalfa. *Plant Molecular Biology* 32: 1037–42.

Zafra, A., Rejón, J. D., Hiscock, S. J. and Alché, J. D. (2016). Patterns of RO accumulation in the stigmas of angiosperms and visions into their multi-functionality in plant reproduction. *Frontiers in Plant Science* 7: 1112.

Zheng, R. H., Shun, D. S., Xiao, H. and Tian, H. Q. (2019). Calcium: a critical factor in pollen germination and tube elongation. *International Journal of Molecular Sciences* 20: 420.

Zheng Y. Y., Lin, X. J., Liang, H. M. et al. (2018). The long journey of pollen tube in the pistil. *International Journal of Molecular Sciences* 19: 3529.

Zhou, Q. Y., Jin, X. B. and Fu, D. Z. (2004). Developmental morphology of obturator and micropyle and pathway of pollen tube growth in ovary in *Phellodendron amurense* (Rutaceae). *Acta Botanica Sinica* 46: 1434–42.

8

Self-Incompatibility

Self-incompatibility has evolved as a key mechanism to promote cross pollination in about 116 families of angiosperms.

8.1 Introduction

The pistil of a flower is exposed to all types of pollen grains in the atmosphere irrespective of whether they belong to the same species or not. However, mere landing of pollen on the stigma is not enough to effect fertilization. As we learnt in the last chapter, there are cellular interactions or cross talk that take place between the pollen and the pistil before successful fertilization. These specific interactions between pollen and pistil facilitate selection of the right type of pollen grains by the pistil and limit fertilization between incompatible gametes.

The inability of a functional male gamete and female gamete to fuse with each other and achieve fertilization is termed as sexual incompatibility. Sexual incompatibility may be interspecific or intraspecific. Following pollination, the ability of a pistil (or stigma) to reject pollen grains from other species is termed as **inter-specific incompatibility.** This type of incompatibility prevents the formation of inter-specific hybrids and maintains the identity of a biological species. The inter-specific incompatibility is controlled by several genes and is also referred to as heterogenic incompatibility. Interestingly, in nature there are several incidences where pistil carrying functional female gametes are unable to set fruits even when pollinated by viable and fertile self-pollen grains. Scientific investigation have established that the failure of fruit set in these plants is due to genetic factors which impose a physiological barrier to self-fertilization. This phenomenon of failure of a male gamete and a female gamete to achieve self-fertilization is termed as **intra-specific incompatibility** or more specifically, **self-incompatibility** (SI). In other words, self-incompatibility is the *inability of a fertile hermaphrodite plant to set seeds when self-pollinated.* The term self-incompatibility was first coined by Stout (1917); it allows flowering plants to avoid inbreeding and involves genetic mechanisms which prevent self-fertilization and promote out-crossing.

In a self-incompatible plant, whenever its own pollen grains reach stigma either pollen germination or pollen tube growth is terminated which results in failure of seed-set. Yet, there are incidences where self-pollen are able to germinate, and self-pollen tubes are even able to penetrate the ovules. In these cases either fertilization fails to occur, or if at all occurs, the zygote gets aborted after syngamy. This type of SI is called **Late Acting Self-incompatibility (LSI)**. The LSI is a marked reproductive feature of several angiosperm families, such as Bignoniaceae. SI is regarded as a mechanism that assures cross pollination and enhances fitness of progeny but at the same time it is also a major barrier that must be overcome in order to conduct many breeding experiments. Therefore, studies have been done to devise methods to overcome SI. This chapter examines aspects of SI, viz. classification, genetics, pollen–pistil interactions during self-incompatibility and methods to overcome such SI.

8.2 Self-incompatibility: Classification and Types

Self-incompatibility has been reported in a total of 116 families. Key examples of families where various types of SI are known include: most of the perennial grasses (Poaceae), legumes (Fabaceae), members of Brassicaceae, Asteraceae, Papaveraceae and Solanaceae. Self-incompatible systems in plants can be distinguished on the basis of morphology of plants as **Heteromorphic** and **Homomorphic systems**. As the name suggests, in heteromorphic SI, there is an occurrence of more than one morphological mating type within a species which can be distinguished on the basis of morphological differences in the length of pistil and stamens. Heteromorphic SI can be assessed without any breeding experiments as the morphological differences are quite apparent and are enough to distinguish between different mating types. For example, *Primula,* (a dimorphic species) exhibits two distinct mating types: thrum morph and pin morph. In thrum morph or 'short-styled' morph, the pistil has short style and stamens are long whereas in pin or 'long-styled' morph, the stamens are short and the pistil has long style (Fig. 8.1 A). *Lythrum* is a trimorphic, tristylous species where other than the long and short style morph, a 'mid-styled' morph also exists. In mid-style or M morph, length of style is in between the pistils of long and short style morph (Fig. 8.1 B). Some other examples of heteromorphic SI are *Oxalis, Turnera subulata, Pentanisia* sp., *Fagopyrum esculentum*, and *Eichhornia*. Heteromorphic SI is reported from 199 genera distributed in 25 families and 15 orders of angiosperms (Barrett 2019).

In homomorphic self-incompatible species, there can be several mating types within a species but all the mating types are morphologically similar. Therefore, breeding tests are required for recognition of different mating types. Homomorphic self-incompatibility has been demonstrated conclusively only in some families like Brassicaceae, Convolvulaceae, Asteraceae, Betulaceae, Caryophyllaceae, Poaceae and Oleaceae. Common examples of homomorphic self-incompatibility are *Petunia, Brassica*, and *Nicotiana*. Homomorphic SI can be further classified into two types based on genetic factors involved; **Sporophytic Self-incompatibility (SSI)** and **Gametophytic Self-incompatibility (GSI)**

Figure 8.1 **A.** *Primula* sp; a distylous system. **B.** *Lythrum salicaria*; a tristylous system. Levels of stigma and anthers are marked with an arrow and 'a' respectively. L-morph: long style morph, M-morph: mid style morph, S-morph: short style morph. **See Color Plates (page 490).**

Sporophytic self-incompatibility: This type of incompatibility is controlled by the genotype of the sporophytic tissue of the plant from which pollen is derived. In other words, incompatibility is controlled by the genotype of the anther (sporophyte). Examples: Brassicaceae, Asteraceae, Convolvulaceae, Betulaceae, Caryophyllaceae, Sterculiaceae, Polemoniaceae.

Gametophytic self-incompatibility: This type of incompatibility is determined by the genotype of the male gamete or the pollen grains. Examples: Plantaginaceae, Papaveraceae, Solanaceae, Poaceae.

8.3 Genetic Control of Self-incompatibility

8.3.1 Homomorphic Self-incompatibility

The first and the most plausible hypothesis for the genetic control of SI was postulated by East & Mangelsdorf (1925) explaining homomorphic SI in *Nicotiana*, popularly known as the **S-allele hypothesis**. According to this hypothesis, SI is regulated by a single locus 'S' and if pollen and pistil have a common S allele, pollen will be rejected by the pistil and rendered incompatible. Thus, this is also known as **Opposition S-alleles hypothesis**. The number of alleles for the S gene in homomorphic systems varies from species to species.

For example: as many as twelve S-alleles have been identified in *Prunus domestica* and *P. insititia* (Rosaceae) which can go up to as high as 40–60 as reported in *Oenothera organensis*, *Papaver rhoeas, Lolium perenne, Raphanus sativus* and *Brassica campestris*. Recently, a biallelic homomorphic system has been discovered in olives and other related genera. The two types of homomorphic SI systems can be distinguished on the basis of timing of expression of S-gene. In SSI, the S-alleles are activated before completion of meiosis in the microspore mother cell and products of both the alleles are distributed in all the pollen grains (Fig. 8.2 A). In GSI, there is delayed activation of S-alleles which occurs only after completion of meiosis. Thus, the products of one allele are present in two microspores, while the products of the other allele are present in the remaining two microspores of the tetrad (Fig. 8.2 B).

A. In sporophytic system

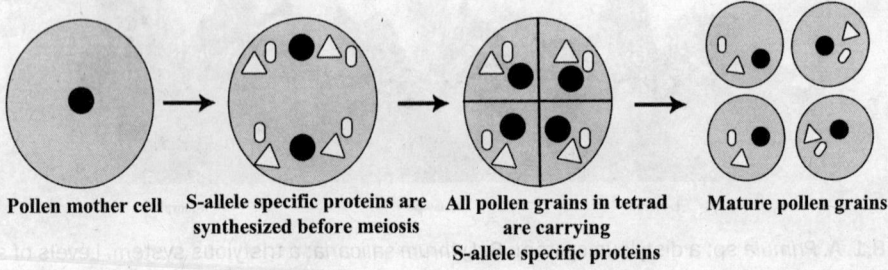

Pollen mother cell   S-allele specific proteins are   All pollen grains in tetrad   Mature pollen grains
                      synthesized before meiosis   are carrying
                                                        S-allele specific proteins

B. In Gametophytic system

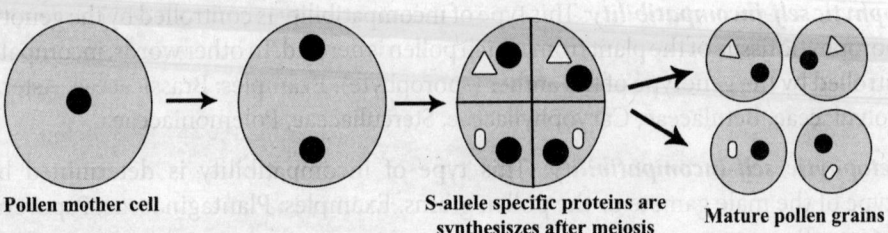

Pollen mother cell   S-allele specific proteins are   Mature pollen grains
                                                synthesiszes after meiosis

Figure 8.2 The accumulation of S-gene specific proteins (depicted with triangles and ovals) in pollen grains during microsporogenesis. (*Source*: Based on and modified after Shivanna 2002.)

8.3.1.1 *Gametophytic Self-incompatibility (GSI)*

In GSI systems, incompatibility phenotype of the pollen is determined by its own haploid genome such that the pollen grains that carry allele common to any of the alleles present in pistil are rejected. In GSI, S alleles are expressed co-dominantly in the pistil. Consider a plant carrying alleles S_1 and S_2. During meiosis, half the pollen grains will receive the S_1 allele and the other half will receive the S_2 allele. If the pistil of such a plant gets self-pollen grains (some with allele S_1 and others with alleles S_2) none of the pollen will be able to effect fertilization and pollen tube inhibition will occur in the style (Fig. 8.3 A). Now consider, the same plant gets pollinated with the pollen grains derived from a plant with S_1S_3 genotype. In that case only 50% of the pollen grains carrying the S_3 allele would be functional and will be able to bring about fertilization as allele S_3 is not common between the pollen and the pistil (Fig. 8.3 B). However, if the pistil with the S_1S_2 genotype receives pollen grains from a parent with S_3S_4 genotype, 100% pollen grains would lead to fertilization as none of the alleles are common between the two parents (Fig. 8.3 C).

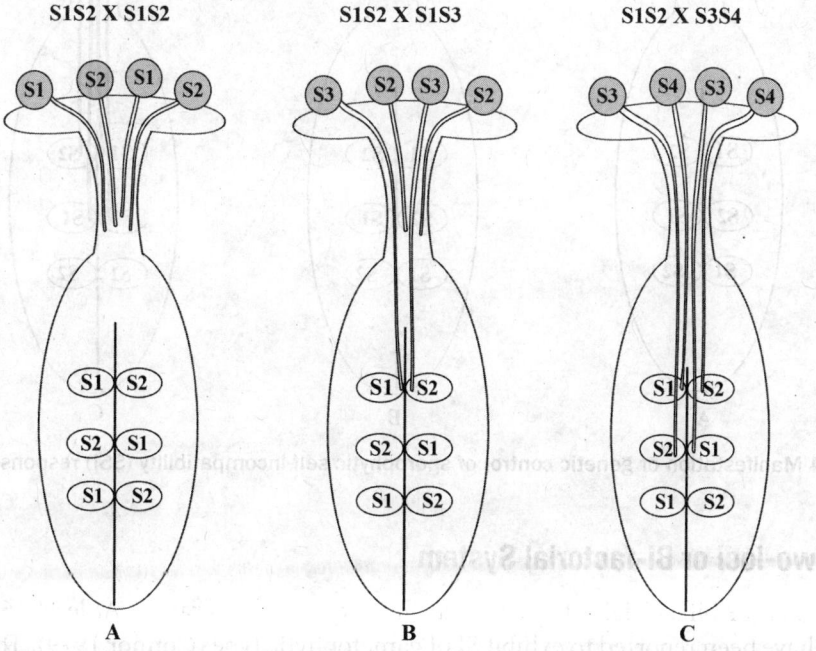

Figure 8.3 Manifestation of genetic control of gametophytic self-incompatibility. *Refer to text for details.*

8.3.1.2 *Sporophytic Self-incompatibility (SSI)*

In SSI systems, the incompatibility phenotype of pollen is determined by the diploid genome of the plant that produced it. In SSI there is a dominant interaction between S alleles (multiallelic), due to which all the pollen grains behave in the same manner, irrespective of which allele they carry. For example, a plant of genotype S_1S_2 would produce pollen grains with S_1 and S_2 allele but all the pollen grains would behave phenotypically either as S_1 or S_2; depending on dominance of either S_1 or S_2 in the plant from which pollen grains are derived (Shivanna 2002). Consider a condition where the S_1 allele is dominant in the parent plant with S_1S_2 genotype. The pollen grains with S_1 or S_2 would not be able to effect fertilization in pistils with genotype S_1S_2; S_1S_3; S_1S_4 and so on, because all the pollen grains will behave as S_1 (Fig. 8.4 A). Similarly, pollen grains derived from a plant with genotype S_2S_3 where S_2 is dominant over S_3 will not be able to fertilize a pistil with genotype S_1S_2 (Fig. 8.4 B). However, pollen grains of plant with genotype S_3S_4 (with either S_3 or S_4 dominant) will be able to carry effective fertilization within pistils with genotype S_1S_2. In other words, such pollen grains would be 100% compatible (Fig. 8.4 C).

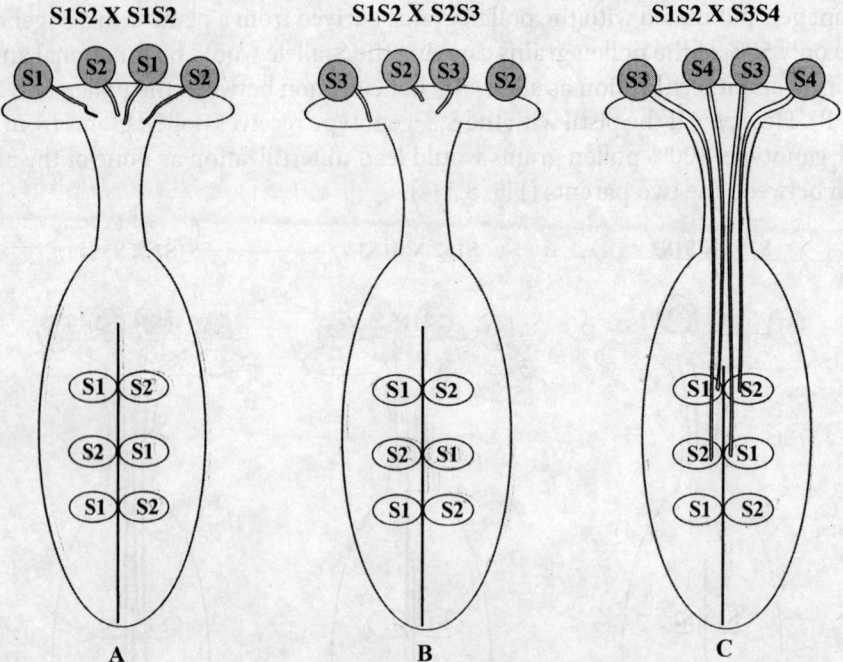

Figure 8.4 Manifestation of genetic control of sporophytic self-incompatibility (SSI) response.

8.3.2 Two-loci or Bi-factorial System

The two loci (or bifactorial) SI system was first explained in grasses. At least 16 genera of Poaceae have been reported to exhibit SI of gametophytic type (Connor 1979). The two loci controlling GSI in grasses are S and Z, both being polyallelic. Up to 17 S and 17 Z alleles have been shown to occur in a naturally occurring perennial ryegrass (*Lolium perenne*) population (Fearon et al. 1994). According to McCubbin & Dickinson (1997), there exists a co-operation between S and Z loci in the pollen though both perform their actions independently in the pistil. The rejection reaction occurs when the combination of S and Z alleles in the pollen is matched by one of the four possible combinations of S and Z alleles; in the diploid stigma. This suggests that the two-loci or the bi-factorial GSI system is more stringent than a unifactorial system.

8.3.3 Heteromorphic Self-incompatibility

Unlike homomorphic systems, several genes are believed to control heteromorphic SI. The genes controlling the flower morphology, and the incompatibility type are tightly linked loci; governed by the **S supergene**. Early clues for genetic control of distyly in *Primula* sp. were provided by Ernst (1928). In his study, he proposed the role of two genes: one for style length and one for stamen length. Later Brieger (1930) expanded it to four, and Stren (1930) and Ernst (1933) suggested that as many as six allelomorphic genes had to be present for heterostyly (based on Haldane 1933; Ernst 1955). JBS Haldane (1933) a population geneticist, deserves special mention for his thorough investigation of these allelomorphs in relation to features (pollen size, stigma type, exine pattern) of pin, thrum and intermediate morphs in *Primula*. In addition, he described two new allelomorphs for heterostyly. He was the first to provide the denotation of 'S' for dominant thrum morph. Dowrick (1956) proposed, that the S locus in *Primula* contains five genes (S supergene complex locus) for the following traits: *the style length gene* (G for short style, g for long style), *style incompatibility gene* (I^S for style incompatibility of short style, i^S for style incompatibility of long style), *pollen incompatibility gene* (I^P for pollen incompatibility of short anther, i^P for pollen incompatibility of long anther), *pollen size gene* (P for large pollen grain, p for small pollen grain) and *anther height gene* (A for long anther, and a for short anther). Based on supergene hypothesis, the S allele would consist of the GI^SI^PPA gene cluster (i.e., haplotype) and the s allele would consist of the gi^si^ppa haplotype. These haplotypes are mostly inherited without recombination. The supergene hypothesis for the distyly S-locus is now supported by genetic studies in *Turnera subulata* (Labone et al. 2010), *Fagopyrum esculentum* (Matsui & Yasui 2020). It is important to note that in heteromorphic systems, incompatibility in the pollen grains is determined sporophytically.

In dimorphic systems length of style is controlled by a single gene, with two alleles **S** and **s** (diallelic SI). The allele for short style; S (Thrum morph) is dominant over the allele for long style; s (Pin morph). Long style morphs are homozygous for recessive allele, i.e., ss while the short style morphs/plants are heterozygous or homozygous for dominant allele, i.e., Ss or SS respectively (Fig. 8.5; Barrett 2019).

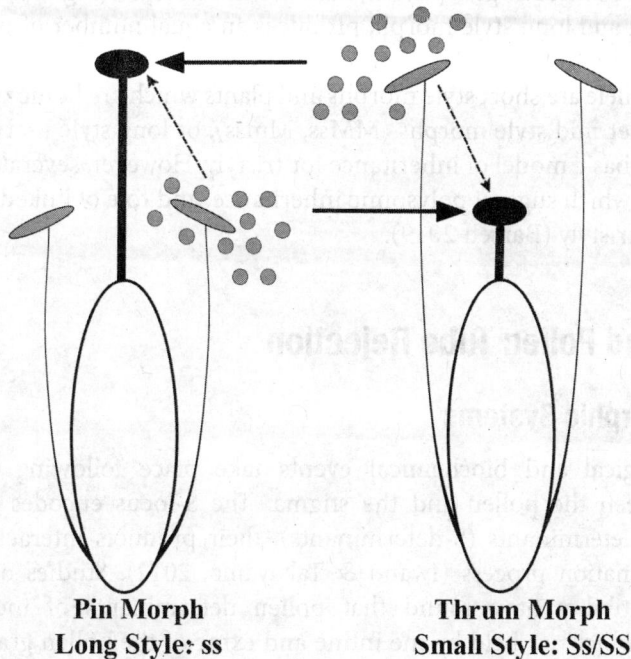

Pin Morph
Long Style: ss

Thrum Morph
Small Style: Ss/SS

Figure 8.5 Arrow with solid lines indicate the compatible crosses while broken line arrow lines show the incompatible cross in distyly.

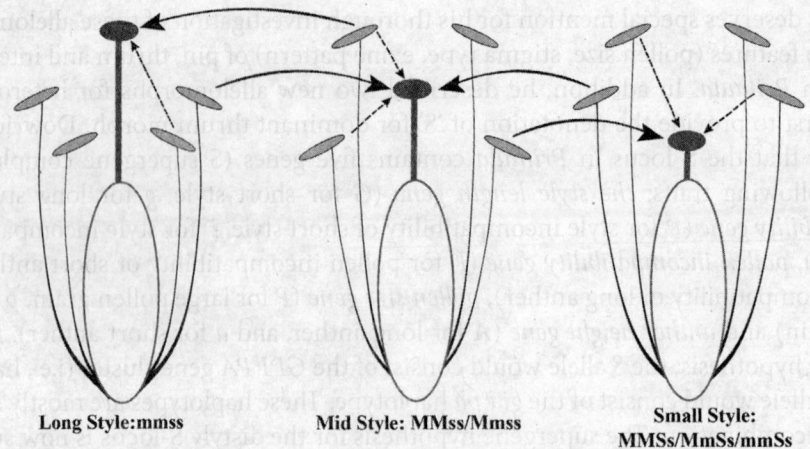

Long Style:mmss   **Mid Style: MMss/Mmss**   **Small Style:
MMSs/MmSs/mmSs**

Figure 8.6 Arrows with solid lines indicate the compatible crosses while broken line arrows show the incompatible cross among tristylous morphs.

In trimorphic systems, 'tristyly' is controlled by two different genes namely S and M, with two alleles each. In such systems S is epistatic over M. The long style morphs are homozygous recessive for both the genes (mmss), while mid style morphs are homozygous recessive for the S gene and heterozygous or homozygous for the alleles of the M gene (MMss/Mmss). Short style morphs are heterozygous for alleles of gene S and homozygous or heterozygous for alleles of gene M, i.e., MMSs/MmSs/mmSs (Fig. 8.6). A cross between short style morph and long style morph, produces an equal number of progenies of short style morph and long style morphs. Here, one may summarize that the plants with at-least one dominant S-allele are short style morphs and plants which are homozygous recessive at this locus are either mid style morphs (MMss, Mmss), or long style morphs (ssmm). This model serves as a basic model of inheritance for tristyly. However, several variations of this basic model exist which suggest polysomic inheritance, and role of linked, or unlinked loci in governance of tristyly (Barrett 2019).

8.4 Pollen and Pollen Tube Rejection

8.4.1 Homomorphic Systems

Several physiological and biochemical events take place following an incompatible interaction between the pollen and the stigma. The S-locus encodes both pollen and pistil specificity determinants (S-determinants), their products interact and trigger self or cross discrimination process (Iwano & Takayama, 2012). Studies of inter and intra specific incompatibility have found that pollen determinants of incompatibility are S-locus proteins which are held by the intine and exine of the pollen grains. The proteins incorporated in intine are contributed by pollen cytoplasm (gametophytic origin, in GSI) while proteins present in the exine are donated by tapetum (sporophytic origin, in SSI) during microsporogenesis (Fig. 8.7). These proteins are suggested to have hydrolytic activity. The stage at which recognition and rejection of self-pollen occurs is variable among the incompatibility systems. The release of intine held proteins needs hydration of pollen on the stigma. Hence, it takes longer for pollen to set-up recognition, and rejection in GSI. The exine held proteins of sporophytic origin are readily available after landing of pollen on the stigma, resulting in the faster and immediate rejection, or recognition reactions in SSI. The stages or sequential reactions for effective SI are as follows (Shivanna 2002):

i. Production of S-allele specific products in pistil and pollen.
ii. Interaction of S-allele specific products in pistil with those of pollen grains for recognition of the pollen.
iii. Inhibition of incompatible pollen.

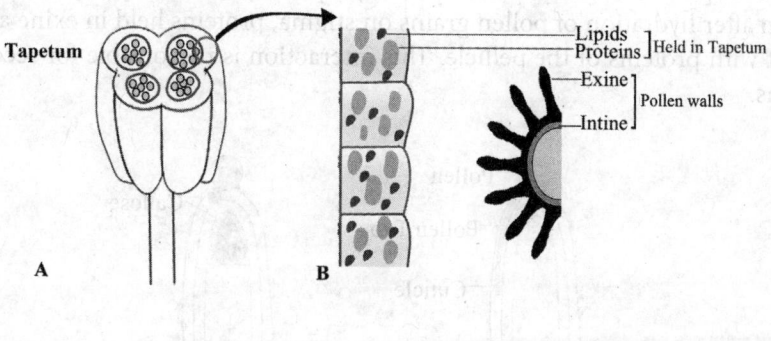

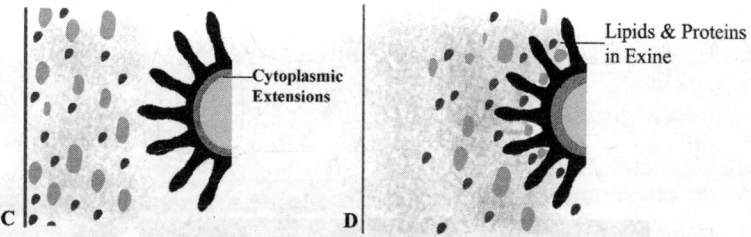

Figure 8.7 Diagrammatic representation of incorporation of proteins in SSI systems by tapetum (Sporophytic origin). Such exine-held proteins are readily available after hydration enabling speedy incompatibility reaction. (*Source:* Based on Shivanna 1979.)

8.4.1.1 *Sporophytic Self-incompatibility System*

Species where SSI exists, pollen grains are generally shed at the 3-cell stage, show a high rate of metabolism, and are short-lived. During SSI; the recognition, and rejection reactions are quite rapid and pollen tube growth is inhibited on the stigma itself. An extreme example is *Gaudinia fragilis* in which the incompatible response is observed within 30 seconds of pollen–stigma contact. SSI incompatible pollinations result in any one of the abnormal behaviors of pollen grains or pollen tubes listed below:

i.   Failure of pollen grain germination
ii.   Emergence of small protuberances that look like abortive pollen tubes
iii. Formation of malformed pollen tubes which are unable to penetrate the stigmatic tissue
iv.   Emergence of multiple pollen tubes
v.   Swollen pollen tube tip

The most characteristic feature of the rejection reaction of SSI is the development of callose plugs. The callose plug formation occurs at the point of contact of pollen grain with stigma, and also at the tip of the pollen tube (Fig. 8.8 A and B). However, the deposition of callose is a consequence, and not a cause, of the inhibition. The formation of callose plugs is a reliable histological and biochemical feature for the study of SI which can be easily assessed using decolorized aniline blue (Fig. 8.8 C). Another characteristic feature of SSI is the occurrence of dry types of stigmas that possess a hydrated layer *pellicle* over the lipidic cuticle layer (Fig. 8.8 D). Soon after hydration of pollen grains on stigma, proteins held in exine are released and interact with proteins of the pellicle. This interaction is responsible for recognition of pollen grains.

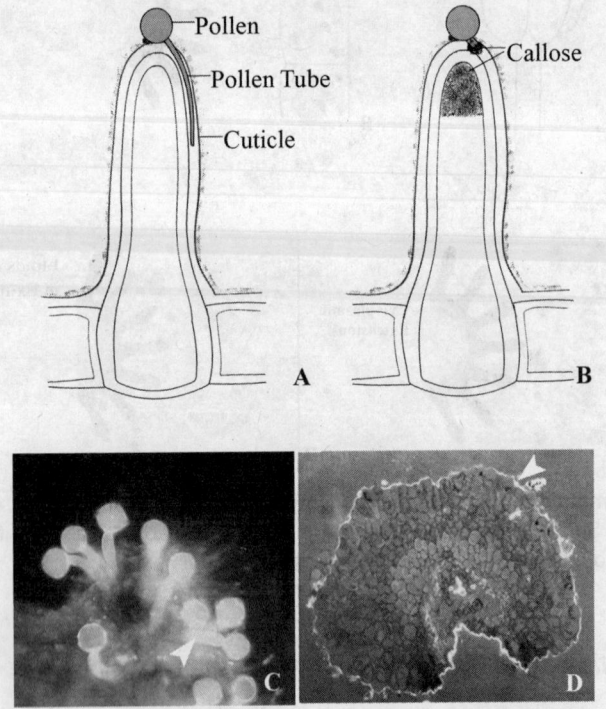

Figure 8.8 Pollen recognition and rejection in SSI. A. Pollen tube growth in case of cross-pollen. The pollen tube of cross pollen grain is able to penetrate the cuticle-pellicle layer of stigmatic papilla. **B.** Inhibition of self-pollen resulting in callose plug formation. **C** and **D.** Fluorescence micrographs showing a part of self-pollinated stigma in neem with high callose deposition (**C.** stained with decolorized aniline-blue) in pollen tubes (arrow head). **D.** Transverse section of dry stigma (*Hippophae rhamnoides*, stained with Auramine O) showing cuticle-pellicle layer (arrow head). **See Color Plates (page 490)**.

The pollen-pistil interaction eliciting a SI response is well elucidated in *Brassica*. In *Brassica*, molecular studies have identified three S locus specific proteins, namely: S-LOCUS RECEPTOR KINASE (SRK), S-LOCUS CYSTEINE RICH PROTEIN/S-LOCUS PROTEINS 11 (SCR/SP11), and the S-LOCUS GLYCOPROTEIN (SLG). In the rejection of self-pollen; pollen borne S locus protein 11 (SP11) interacts with stigmatic receptor kinase (SRK). This interaction initiates a signaling cascade within the stigmatic papilla cell, leading to rejection of self-pollen (Kakita et al. 2007; Kitashiba & Nasrallah 2014; refer to Box 8.2 for details). In *Arabis* and *Brassica*, it has been suggested that in case of compatible pollinations an enzyme 'cutinase' becomes activated after the recognition interaction which erodes the cuticle layer, thus allowing pollen tubes to penetrate the stigmatic tissue.

Box 8.1

Differences between SSI and GSI

SSI	GSI
• SSI is controlled by diploid genotype of sporophytic tissue of the plant from which pollen is derived. In other words, incompatibility is controlled by the genotype of anther.	• Incompatibility process is determined by the genotype of the male gametes or of the pollen grains.
• In SSI, there is a dominance interaction between S-alleles of pollen grains.	• In GSI, there is a co-dominance interaction between S-alleles of pollen grains.
• S allele specific proteins are synthesized before completion of meiosis.	• S-allele specific proteins are synthesized after completion of meiosis.
• Zone of pollen/pollen tube inhibition is stigma.	• Zone of pollen tube inhibition is style.
• Rejection reaction involves interaction between S-locus protein and stigmatic receptor kinase.	• Rejection reaction involves stylar RNases and Pollen S-locus F-box proteins.
• Stigma is dry type.	• Stigma is wet type.
• In species with SSI, pollen grains are generally shed at 3-cell stage.	• In species with GSI, pollen grains are generally shed at 2-cell stage.
• Pollen grains show high metabolic rate and are short lived.	• Pollen grains show low metabolic rate and are long lived.
• Examples: Brassicaceae, Asteraceae, Convolvulaceae, Caryophyllaceae.	• Examples: Plantaginaceae, Papaveraceae, Solanaceae, Poaceae.

8.4.1.2 *Gametophytic Self-incompatibility System*

Pollen grains of species showing GSI are usually shed at the 2-celled stage, exhibit slow metabolic rates, and are long-lived. Germination of pollen grains is not constrained by the stigma but pollen tube growth is inhibited in the style. Like SSI, callose plays a major role in pollen tube growth arrest in GSI as well. There is excessive deposition of callose in the tips of tubes which generally show swelling and/or bursting in the stylar region. Apart from callose, deposition of pectic material at the tip of the pollen tubes in the incompatible style has also been observed. In *Lycopersicon peruvianum*, pollen tubes have two layered cell walls, the outer layer is pecto-cellulosic made of loose fibrils and inner is homogenous containing callose. Pollen tubes in the incompatible styles also display similar wall structures during initial stages of growth. However, after traversing about two-third of the style, their inner wall gradually becomes thin and numerous particles accumulate in the tube cytoplasm. Eventually, the inner wall disappears and pollen tube bursts which indicates that the incompatibility reaction in GSI is an active process.

Usually, plants with GSI possess a wet type of stigma. Difference in the metabolism between self and cross-pollinated pistils has also been observed in *Nicotiana alata*. Styles with incompatible tubes show higher activity of Peroxidease-10, whereas, it is very low in style with compatible pollen tubes suggesting a probable role of Peroxidease-10 in rejection reactions.

At the molecular level GSI mechanisms result from an interaction between pistil and pollen products that are encoded by the multi-allelic self-incompatibility locus (S-locus). In Solanaceae and Rosaceae, S-allele specific proteins associated with GSI have been identified as Ribonucleases, referred to as stylar (S)-RNases. Pollen rejection in this system requires the interaction of (S)-RNase and multiple pollen SLFs (S-locus F-box proteins). These (S)-RNases inhibit the growth of any pollen tube bearing an S-allele that matches either of the stylar S-alleles via degradation of RNA. (Nowak et al. 2011, Williams et al. 2015; Vieira et al. 2019; refer to Box 8.2 for details).

Some species with GSI systems show typical features of plants with SSI, e.g., tricellular short-lived pollen grains, dry stigma as seen in Poaceae (Langridge & Baumann 2008), cessation of self-pollen tubes growth at stigma, e.g., *Oenothera* sp. (Dickinson & Lawson 1975), *Commelina* sp. (Owens 1981), *Boswelia serrata* (Sunnichan *et al.* 2005), and members of Papavaraceae. The pollen–pistil interaction at the stigma has been studied at molecular level in Papavaraceae, where GSI is based on programmed cell death (PCD). The pistil S-locus encodes a small protein (~15 kDa, called PrsS: *Papaver rhoeas* stigmatic S- protein) secreted by the pistil surface that binds to the self-pollen S-receptor (called PrpS: *Papaver rhoeas* pollen S); triggering a Ca^{2+} dependent signaling network, that results in pollen inhibition and programmed cell death (Eaves et al. 2014; Wilkins et al. 2014; refer to Box 8.2 for details).

Box 8.2

Molecular Mechanism of SI in Some Families

SI is controlled by single S-locus, with multiple alleles. These alleles are tightly linked and are inherited as a single segregating unit. The variants of this gene complex are termed as S-haplotypes. Each haplotype (such as S_1, S_2...S_n) encodes both male specific and female specific determinants (S-determinants, or products of S alleles). The male specific determinants are carried by pollen grains and female specific determinants are present in the pistil. The interaction between these S-determinants discriminate the self-pollen and manifest incompatibility reaction; if encoded by same haplotype (Fig. 8.9). The self/non-self-discrimination systems, can be classified into two types: 1. self-recognition and 2. non-self-recognition systems. The self-recognition system prevails in Brassicaceae and Papaveraceae, while the non-self-recognition system is found in Solanaceae, Rosaceae.

The first study to understand the molecules and biomolecules involved in SI was conducted by Nasrallah and associates (1986) in *Brassica oleracea*. They showed that each S-allele produces a specific protein and each S-allele specific protein is inheritable. Now it is well known that in the family Brassicaceae, self-incompatible *Brassica* sp., *Arabidposis lyrata* and *Arabidopsis halleri,* the incompatibility system involves S-locus derived proteins; S-locus receptor kinase (SRK), and S-locus protein 11 (SP11, or S-locus cysteine-rich protein, SCR). SRK and SP11 genes are tightly linked and inherited like a single Mendelian locus gene. SRK is a membrane-spanning kinase that is present in plasma membrane of stigmatic papillae cells (female-specific determinant) while SP11 is a small basic protein secreted from the anther tapetum and localized to the pollen coat (male-specific determinant). SP11 is a ligand of SRK, and a match between ligand and receptor leads to self-pollen recognition and initiates an incompatibility reaction.

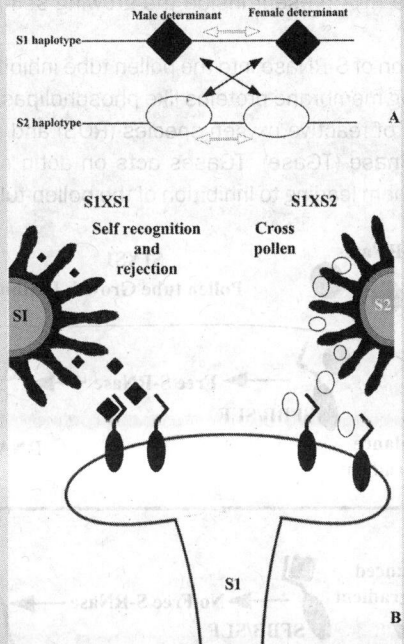

Figure 8.9 A. Interaction of S-haplotypes S1 and S2. The white double-sided arrows mark the self-incompatible reaction between male and female determinants of same haplotypes. The black double-sided arrows mark compatible reaction between haplotypes (S1 and S2). **B. Manifestation of SI by recognition of self-pollen and then rejection.** On the left side of the picture; the proteins carried by pollen (S1) act as ligands for receptors for stigma (S1) (female determinants). If there is a match between a ligand and a receptor; self-pollen grains get recognized and their germination is inhibited. Such manifestation of SI can be seen in family Brassicaceae. The other side (right hand) of the picture shows pollination by pollen carrying products of S2 haplotype thus, their male determinants are not recognized by stigma carrying S1 female determinants. Hence, pairing is compatible.

In *Papaver rhoeas* (Papavaraceae, GSI system), the S-allele female-determinant PrsS (*P. rhoeas* style S) is secreted by stigmatic papillae and is a highly polymorphic protein. However, the presence of a male-determinant on plasma membrane of pollen grain is still putative. After incompatible pollination/self-pollen landing on stigma, PrsS triggers a series of SI responses: (1) it increases the Ca^{2+} influx in the cytosol of pollen grain and (2) starts the depolymerization of the actin cytoskeleton. These two actions result in the cessation of pollen tube growth and programmed cell death of self-pollen grains on stigma.

In Solanaceae and Rosaceae, the female determinants S-allele products are S-RNases (~30 KD proteins) which possess ribonuclease activity. The male determinants are numerous F-box proteins known as SFBBs (S-locus F-Box Brothers) present in pollen. The female determinant is present in the extracellular matrix of style. In Solanaceae, recognition of non-self-pollen grains takes place. During cross-pollination SFBBs interact with S-RNase and degrade the S-RNase and therefore the pollen tube continues to grow. Whereas, during self-pollination, since self-pollen are not recognized, none of SFBBs interact with the self S-RNase. Such that, S-RNase exhibits cytotoxicity and degrades the RNA present inside the growing self-pollen tube, restricting their growth in the style (Fig. 8.10).

In Rosaceae, the penetration of S-RNase into the pollen tube inhibits its growth but also triggers a series of responses involving membrane proteins like phospholipases, intracellular variations of cytoplasmic Ca^{2+}, production of reactive oxygen species (ROS) and altered enzymatic activities, such as that of transglutaminase (TGase). TGases acts on actin and tubulin (cytoskeleton) of pollen tube and destabilizes them leading to inhibition of the pollen-tube growth (Duca et al. 2019).

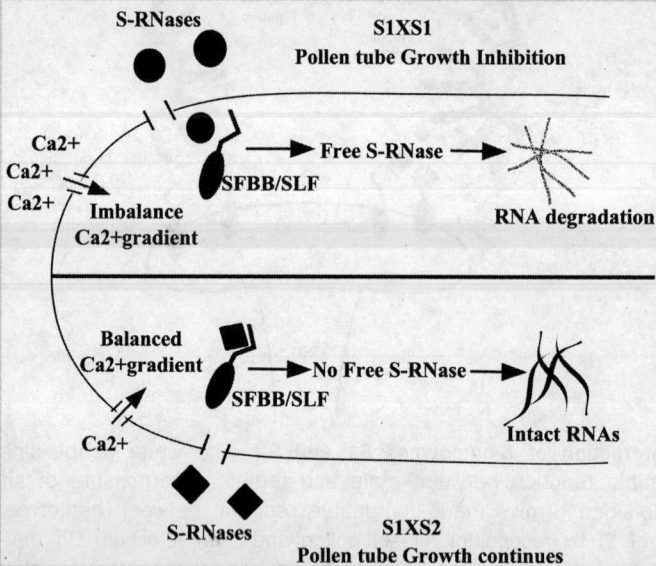

Figure 8.10 The pollen tube showing non-self-recognition system as found in Solanaceae, Rosaceae. In the case of self-pollen, male determinants (S1) SFBBs are not able to inactivate the S-RNase of S1 pistil (present in style). Hence, their self-pollen tube RNAs get digested and pollen tube growth ceases. While in cross-pollination where pistil is carrying S2 (or any other) female determinants, the S-RNases get deactivated by SFBBs/SLF. Thus, RNA in pollen tube remain intact and pollen tube continues to grow. In case of self-pollen tube, the Ca^{2+} gradient also gets disturbed and pollen tube wall synthesis also ceases.

Source: Charlesworth 2010; Iwano and Takayama 2012; Nasrallah 2019; Duca et al. 2019.

8.4.2 Heteromorphic Systems

The zone and mechanism of rejection reaction or pollen tube inhibition in cases of incompatible crosses (pin X pin **or** thrum X thrum) in heteromorphic systems are variable. In these systems, pollen wall-held proteins play no role in incompatible reactions and there is no callose formation in the stigmatic papillae after autogamy. In family Plumbaginaceae, inhibition occurs on the stigma surface in both morphs as the pollen tube fails to emerge or penetrate the stigmatic tissue. In *Fagopyrum* (Polygonaceae) and *Primula obconica* (Primulaceae), inhibition occurs inside the stigma in case of thrum X thrum pollination, while inhibition occurs in the style, in pin X pin cross. In *Linum grandiflorum*, pin X pin pollination results in failure of pollen grain adhesion and hydration; while in thrum X thrum pollination, pollen do germinate but pollen tubes get inhibited on the stigma itself. Likewise in short-styled flowers of *Psychotria* (Rubiaceae), the self-pollen tube growth is inhibited within the stigma while in long-styled *Psychotria* flowers, illegitimate pollen tube growth gets inhibited either within the stigma or at various levels in the style (Faivre 2002). In trimorphic *Lythrum*, the inhibition zone is also different among morphs. In short style morph, pollen tube growth is restricted at the base of stigma, while in mid and long style morphs, pollen tube growth stops in upper parts of the style. In *Primula vulgaris* there is no localized zone for pollen tube growth and inhibition.

As compared to the homomorphic SI, very little information is available in literature for the mechanism of inhibition in heteromorphic systems. In dimorphic *Linum grandiflorum*, it has been postulated that the rejection of incompatible pollen grains is a passive mechanism involving osmotic potential. Lewis (1949) observed a marked difference between the osmotic potential of style and pollen grains among two morphs. Osmotic potential of style in thrum morph was 1.0, and in pollen was 7.0; while in the style of pin morph it was 1.75, and in its pollen 4.0. In the compatible crosses (pin X thrum and thrum X pin), the ratio of osmotic potential of pollen:style was 4:1. In an incompatible cross of pin X pin morph this ratio was 5:2 and in an incompatible cross of thrum X thrum morph this ratio was 7:1. On the basis of these observations it was proposed that a 4:1 ratio of osmotic potential of style and pollen grains is optimal for pollen hydration, germination and tube growth. Deviation from this ratio leads to incompatibility. If the ratio is too high (in thrum X thrum morph it is 7:1) then pollen tubes will absorb water and burst and if the ratio is too low (pin X pin morph, 5:2) pollen hydration will be prevented. Later, Shivanna (2002) suggested that in *L. grandiflorum* intra-morph incompatibility manifests at three levels: pollen adhesion, pollen hydration and pollen tube growth in stigma. The first two levels show passive processes while the third operates as an active process. Similar observations are also known from *Primula*.

Box 8.3

Unique Cases of SI

1. Late Acting Self-incompatibility: This phenomenon was first described by Seavey & Bawa (1986). Most self-incompatibility systems primarily function in a pre-zygotic manner through failure of self-pollen germination and pollen tube growth in stigma or style. However, there are many species in which pollen grains germinate, self-pollen tubes grow successfully in the style and reach ovary and then, undergo self-incompatible inhibition. Such cases where inhibition of self-pollen tubes occur in the ovary either pre-zygotically or post-zygotically are said to show Late acting Self-incompatibility (LSI). LSI is present in many families like Bignoniaceae, Malvaceae, Apocynaceae, Anacardiaceae and mainly in tropical flora. The manifestation of LSI can be in one of the following ways: (1) self-pollen tubes may grow into the ovary but fail to penetrate the ovules; (2) self-pollen tubes even penetrate the ovules but their growth is arrested at micropyle/nucellus or syngamy fails to occur (even after the discharge of male gametes into the synergids), (3) zygote and endosperm form but embryogenesis fails. The first two (1 and 2) consequences of self-pollination can be grouped together and described as pre-zygotic LSI whilst the condition 3 is regarded as post-zygotic LSI. The various studies attempting to decipher underlying genetics of LSI suggest that LSI is governed by 'S' gene with two alleles (e.g., *Theobroma cacao, Asclepias exaltata, Ipomopsis tenuifolia*) (Gibbs 2014 and references therein).

2. Pseudo Self-compatibility and Self-fertility: Grasses in general exhibit a stringent two locus gametophytic SI system. However, several studies have shown the existence of selfed seeds per inflorescence in some grasses (approx. 5% of total seeds), viz. *Secale cereale, Lolium perenne, Panicum virgatum* and *Festuca pratensis*. This selfed seed production in grasses is referred to as pseudo self-compatibility. Usually, pseudo self-compatibility is seen after artificial self-pollination or under different environmental conditions, and was thought to be a non-genic effect. For example, in glass house experiments, *L. perenne* showed higher selfed seed set when temperature was increased from 30.7°C to 34°C at floral anthesis. Later, various studies produced similar results and it was documented that increase in this self-fertility could be achieved only when pollen grains are exposed to heat. The heat exposure to stigma or variable pollination conditions had a little or no effect. However, the genetic nature of pseudo-compatibility came to light with the studies on *Festuca pratensis* and other grasses that highlighted polygenic control and role of modifier genes other than S and Z locus. It was also proposed that this mechanism is independent from the SI genotypes.

Another interesting incidence is the presence of fully self-fertile plants within some allogamous grass species like *Secale cereale, Lolium perenne* and *Festuca pratensis*. The genetics conferring self-fertility to these variants is much more complex than self-incompatibility. For example, in *Phalaris coerulescens*, the segregation analysis studies revealed that at least three genes are present which may act independently for the manifestation of self-fertility. This self-fertility as a trait is dominant and epistatic over self-incompatibility. Recently, frequent coexistence of self-compatible and self-incompatible plants in *Ligustrum vulgare* (Oleaceae) with a diallelic genetic mechanism controlling self-incompatibility has also been reported (Do Canto et al. 2016 and references therein; Cauwer et al. 2020).

8.5 Methods to Overcome Self-incompatibility

Self-incompatibility (SI) is a unique reproductive mechanism adopted by flowering plants to promote outbreeding and has played a pivotal role in evolution and diversification of angiosperms. However, at the same time SI has been a limiting factor in plant improvement programs. Improvement of cultivated plants relies on the discovery and introduction of genes from wild species for agronomic traits followed by inbreeding of the F_1 hybrid cultivar. Main challenge to this approach is the occurrence of self-incompatibility in the cultivars preventing inbreeding of hybrids. Therefore, to implement efficient breeding programs, the overall understanding of the SI system and also the development of mechanisms to breakdown the SI is important. Some of the methods devised to overcome SI are given below:

8.5.1 Mixed Pollination

Techniques of mixed pollination are adopted for species where the pollen grains are unable to germinate on the stigma due to absence of certain exine proteins. The stigma in such species is dusted with compatible as well as incompatible pollen together. The compatible pollen grains are either inactivated by irradiation, or are killed by treating them with chemicals like methanol or sometimes even subjected to repeated freezing and thawing treatments so that they cannot do fertilization. Such pollen, however, still have the ability to germinate. Also, there is no change in their exine held proteins, which are released once the pollen lands on compatible stigma. These compatible pollen grains are known as the *mentor pollen* or the *recognition pollen*. The recognition proteins from the wall of mentor pollen help the incompatible pollen to germinate on the stigmas. Mentor pollen also provide a pollen factor or P-factor which interacts with S-factor from the stigma and renders the stigma accessible to incompatible pollen. Many a time, even after germinating on the stigma, the pollen tubes are inhibited in the stylar canal. In such cases, mentor pollen provide a pollen growth promoting substance which allows sustained growth of pollen tubes in the style. Mentor pollen is also known to provide substances critical for sustained growth of ovules, ovary and other fruit tissues, thus indirectly influencing fertilization events and seed maturation. This method has been successful in overcoming incompatibility in lily.

8.5.2 Bud Pollination

In *Petunia*, incompatibility manifests in the style and any self-pollination done at the time of anthesis is unsuccessful. However, it has been shown that if self-pollination is done two days before anthesis or at the bud stage it leads to formation of seeds. It is likely that the incompatibility factor is produced in the style at the time of flower maturity or flower opening. Hence, if the pollination is done at the bud stage, when the incompatibility factor is yet to be produced, the pollen tubes grow normally in the style and this leads to fertilization. This is usually effective in overcoming GSI.

8.5.3 Stub Pollination

Incompatibility in plants where rejection reaction occurs on the surface of the stigma (in case of SSI) can be overcome by removing the stigma and even some portion of the style, followed by self-pollination at the cut end of the style. This process has proved successful in overcoming incompatibility and is referred to as stub-pollination. In *Ipomea trichocarpa*, the pollen grains are unable to germinate on the stigma followed by self-pollination but if the pollen is dusted on the stub of the style, germination is normal.

8.5.4 Intra Ovarian Pollination

As the name indicates, pollination is done directly inside or within the ovary by introducing a pollen suspension. This method is employed to overcome both GSI and SSI and was first devised by Kanta & Maheswari (1963) in *Papaver somniferum*. In the procedure, the surface of the ovary is first sterilized with ethanol and two punctures are made opposite to each other in the ovary. Through one of the punctures, the pollen suspension is added to the ovarian cavity with the help of a hypodermic syringe. The second puncture or the opening allows air to escape. Once the pollination is accomplished, the punctures are sealed with petroleum jelly. Pollen suspension is prepared by mixing the pollen with distilled water and any other additive that might be useful for pollen germination of the species. The major drawback of this technique is that it is applicable only to plants with large ovaries like members of Papaveraceae where this method proved successful (Fig. 8.11).

Figure 8.11 Intra ovarian pollination in *Papaver* sp. A. Young ovary where punctures have been made **B-C.** Ovaries after six and nine days of pollination demonstrate swelling. Note the puncture (arrow) **D.** An ovary (cut opened) showing developing seeds (arrow) after intra ovarian pollination. (*Source:* Kanta 1960, *reproduced with permission*.)

8.5.5 *In vitro* Pollination and Fertilization

In this method, all stigmatic, stylar and ovarian tissues are completely removed and the pollen grains are directly dusted on the ovules, or ovules attached with placenta. These ovules are cultured in a nutrient medium which promotes pollen germination and subsequently fertilization. Under *in vitro* conditions, pollination done directly on ovules is called as *ovular pollination* while pollination on ovules attached with placenta is known as *placental pollination*. This technique has proved beneficial in *Petunia axillaris* where normal seed set was observed. The major disadvantage of this method is the need for elaborate tissue culture techniques and aseptic conditions.

8.5.6 Modification of Stigmatic Surface

In those species, where the surface of stigma proves a barrier to pollen germination, the stigma surface is modified using organic solvents such as hexane, or detergents like Triton X-100 or even by application of chemicals like Concanavalin-A. In *Brassica*, Concanavalin-A is applied to the surface of the stigma which binds and masks the pellicle, the layer responsible for incompatibility. Triton X-100 and hexane have similar mechanism of action. Similarly, in rye and *L. perenne* application of calcium channel blockers lanthanum chloride (La^{3+}), protein kinase inhibitor lavendustin A, and verapamil on stigma allowed incompatible pollen grains to grow their tubes down the style and to the ovary (Do Canto et al. 2016; Klaas et al. 2011).

Box 8.4

Differences between Intra Ovarian Pollination and *In vitro* Pollination & fertilization

Intra ovarian pollination	*In vitro* pollination & fertilization
• It is an *in vivo* technique.	• It is an *in vitro* technique.
• Pollen suspension in injected into ovarian cavity throaugh a pore or puncture using hypodermal syringe.	• Pollen grains are dusted near the ovules or placenta placed over culture medium.
• Easy to perform and cost effective.	• Elaborate procedure involved and costly.
• No need of tissue culture techniques.	• Needs tissue culture under aseptic conditions.
• This technique can be employed only for plants with large ovary and ovarian cavity.	• This method can be employed for even small sized ovary.

8.5.7 Heat Treatment of Style

Moderately high temperatures have also proved beneficial in overcoming self-incompatibility in certain plants. Styles of *Lilium longiflorum* subjected to a temperature of 50°C for 6 minutes prior to pollination, renders self-pollen compatible. Temperature above this, proves injurious to the pistil while lowering the temperature proves ineffective in overcoming incompatibility. It is suggested that increase in temperature either inactivates or selectively denatures the enzymes responsible for self-incompatibility. Possibly heat-sensitive inhibitors of pollen tube growth get inactivated. Temperature required for overcoming incompatibility is species specific, e.g., in *Lilium*, 50°C is effective whereas in rye 30°C is enough for promoting selfing.

8.5.8 Irradiation

X-ray irradiation of pollen or the style can induce temporary breakdown of incompatibility reactions, e.g., *Oenothera organensis* and *Prunus avium*. Irradiation of pollinia of *Dendrobium* with gamma rays has also been shown to be effective in overcoming self-incompatibility. Irradiation is suggested to either cause unstable cytological changes in the pollen or the style or induce mutation in the incompatibility gene.

8.5.9 Increased Level of Carbon Dioxide

The percentage of carbon dioxide in the atmosphere is 0.03. If this level is raised to 4–6% or 100 fold at a relative humidity of 100% for several hours prior to pollination, then pollinations which would otherwise have been self-incompatible under normal conditions, behave as compatible pollinations. This method has proved beneficial in breeding of *Brassica* species.

8.5.10 Parasexual Hybridization

Production of hybrids through fusion of isolated protoplasts of two different species is called *parasexual hybridization* or *somatic hybridization*. Parasexual hybridization helps to overcome the incompatibility encountered at stigma, style or ovary. The technique was first developed by E.C. Cocking in 1969. He demonstrated that naked cells or the protoplasts can be obtained through the enzymatic degradation of the cell wall. Bhojwani & Cocking (1972) demonstrated for the first time the possibility of fusion of protoplasts of microspores or young pollen. Microspore protoplasts fuse readily as they are rich in cytoplasm and this is very important in hybridization as they are haploid. Protoplast fusion can be achieved in three steps:

    i.   Isolation of protoplasts
    ii.   Fusion of isolated protoplasts to create hybrid protoplasts
    iii.   Culture of hybrid protoplasts to regenerate into whole plants

The somatic cells in plants are bound by a rigid cell wall made up of cellulose. Also, the adjacent cells in a tissue are cemented together by a pectin rich matrix. Thus, to obtain protoplasts from somatic cells, treatment of cells with a mixture of cellulase and pectinase enzymes is needed. Concentration of enzymes required for the procedure is dependent on the tissue used and varies from species to species. Osmotic fragility of the isolated protoplasts is most crucial in the entire sequence of events starting from enzyme solution to washing medium, fusion medium and culture medium. Commonly, a metabolically inert sugar like mannitol is used as an osmotic stabilizer. Freshly isolated protoplasts in the presence of suitable osmoticum appear spherical.

Once the protoplasts are isolated, the next step is their fusion which requires the help of agents called *fusogens*. One frequently used fusogen is Polyethylene glycol (PEG) because of its high efficiency of inducing heterokaryon formation. PEG has been used for fusion of protoplasts from diverse taxa such as soybean, corn, and pea, or even fusion of a plant cell with an animal cell. Equal quantities of dense suspension of the two types of protoplasts are mixed along with 15% PEG solution. After10–30 minutes, PEG solution is diluted with the medium in which it was originally prepared. Finally, PEG is washed off completely. After this the protoplasts become closely appressed to each other over major portion of their membranes. The cytoplasmic continuities expand and fuse to form broader connections. Eventually the protoplasts fuse completely and attain a spherical shape. Fusion of two dissimilar protoplasts leads to the formation of a heterokaryon. When subjected to culture conditions, the two nuclei of a hetrokaryon fuse to establish a true hybrid cell. In a suitable culture medium, the fused protoplasts synthesize a new wall around themselves and are reconstituted as cells. These cells divide and re-divide to form a callus which ultimately forms a plantlet.

8.5.11 Genome Editing

In recent years, focus has shifted to generating self-compatible lines of an otherwise self-incompatible species or cultivars using gene editing tools like Clustered Regularly Interspaced Short Palindromic Repeat (CRISPR). Through this method, genes related to incompatibility are made inoperative (gene knockout), e.g., studies in tomato and wild relatives of potato have demonstrated that loss of gene coding S-RNase prevents S-RNase ribonuclease activity leading to self-compatibility (Rodriguez et al. 2019).

Glossary

Distyly: Floral dimorphism where two floral types differ reciprocally in length of style and stamen.

Fertile pollen grains: Pollen grains which have capability to fertilize the ovules.

Haplotype: The term haplotype is derived from two words 'haploid' and 'genotype' and indicates the inheritance of the group of genes in a cluster from a single parent. Haplotype also includes a pair of genes present on one chromosome, or all of the genes on a chromosome that are inherited together from single parent.

Heteromorphic incompatibility: A type of self-incompatibility where two or three morphologically different mating types occur in a species that restrain self- or intra-morph mating due to different positions of style and stamen.

Homomorphic incompatibility: Type of self-incompatibility where several mating types occur in a species which are indistinguishable based on morphology but exhibit different genotypes.

Late acting self-incompatibility: A self-incompatibility system where rejection of self-pollen tubes occurs much later, either at ovule, or after formation of zygote (post-zygotic).

Pseudoself-compatibility: Seed formation through self-pollination in self-incompatible grasses.

Self-compatibility: Ability of hermaphrodite plants to set seed from self-pollination.

Self-incompatibility: Mechanism limiting seed set from self-pollination in fertile hermaphrodite plants.

Tristyly: Floral polymorphism that consists of three floral morphs differing in length of style and stamen.

Viable pollen grains: Pollen grains which have ability to transfer male gametes to the ovules are known as viable pollen grains.

Key Questions

Q8.1   Differentiate between:
   a. Homomorphic and Heteromorphic self-incompatibility
   b. Sporophytic and Gametophytic self-incompatibility
   c. Intra ovarian and *In vitro* pollination

Q8.2   Write short notes on:
   a. Genetic control of homomorphic self-incompatibility
   b. Genetic control of heteromorphic self-incompatibility
   c. Parasexual hybridization
   d. S-allele hypothesis
   e. Mixed pollination
   f. Stub pollination
   g. GSI in grasses

Q8.3   Fill in the blanks:
   a. The S-allele hypothesis for genetic control of homomorphic incompatibility was postulated by..........................
   b. Grasses are characterized by.......................... type of self-incompatibility.
   c. The Poaceae exhibit.................type of genetic control for gametophytic type of self-incompatibility.
   d. Rejection reaction occurs at the stigma surface in.....................type of self-incompatibility.
   e. The characteristic feature of rejection reaction of SSI is the development of ....
   f. In GSI, pollen tube growth is inhibited at the.................
   g. Intra ovarian pollination method was first developed by..................in *Papaver somniferum*.
   h. ....is an example of fusogen.
   i. The technique which involves the fusion of isolated protoplasts to produce hybrid is called ....
   j. The technique of parasexual hybridization was first shown by....................
   k. Thrum-morph flowers in distyly are characterized by presence of.............................

Q8.4   Give an example of species and family which exhibit the following:
   a. Trimorphic mating system (tristyly)
   b. Distyly or dimorphic mating system
   c. Pseudo self-compatibility
   d. Homomorphic self-incompatibility
   e. Heteromorphic self-incompatibility

Q8.5   Expand the following:
   a. SSI
   b. GSI
   c. CRISPER

Practicals

Exercise 8.1 To determine the interspecific incompatibility in species using semi *in vivo* pollination and decolorized aniline blue staining method

Theory

Manual pollination exercises are extensively used to decipher the mating/breeding system of a species. Manual pollination is usually performed on the flowers attached to the mother plants (*in vivo* method). Sometimes, *in vivo* manual pollination is difficult to perform especially in plants with small flowers. To overcome it, a variation called semi *in vivo* pollination can be executed, where pistils are dissected from the flower, implanted on a suitable medium and then pollinated. Agar with sucrose is most commonly used medium as it provides the osmoticum, the nutrition and support to the pistils to keep them alive for a longer duration. Manual pollination must be done with viable pollen grains on a receptive stigma. Thus, prior to manual pollination, pollen viability and stigmatic receptivity must be assessed.

Manually pollinated pistils can be used to study the behavior of pollen and pollen tube growth. This method can also be employed for studying interspecific incompatibility and self-incompatibility. In self-incompatible pollination, pollen fail to germinate due to accumulation of callose and even if pollen tube emerges it remains only as a small protuberance and aborts because it is plugged with callose. Thus, staining such pollinated stigmas with callose specific dye helps in determining incompatibility. One such dye is Aniline Blue; which in decolorized state (at pH-9.2), specifically stains callose. When observed under a fluorescence microscope, the callose fluoresces.

The stigma is an open system where pollen of various species (especially neighboring co-flowering species) and other particles fall without significant restrictions. The stigma and pistil are enabled and equipped with mechanisms to select the legitimate pollen grains and reject all illegitimate pollen grains. With these mechanisms even viable and fertile pollen grains of different species or genera are rejected. This ability of the pistil (or stigma) to reject pollen grains from other species or genera, is called inter-specific incompatibility or sexual incompatibility. The prevalence of such incompatibility helps in prevention of formation of inter-specific or inter-generic hybrids. In most of the cases of inter-specific pollination, the pollen grains fail to germinate and the pollen-stigma interface accumulates callose. Moreover, if pollen tube emerges it shows deformities and callose deposition.

Material Required

Flowers of two different species/genera (e.g., *Brassica, Catharanthus, Petunia, Crotolaria, Nicotiana*), small petri plate sets (35 mm or as available), glass vials/ screw cap bottle (30–50 ml), glass slides, coverslips, measuring cylinder, camel hair brush (with few hairs, 0 No.), a pair of forceps, needles, blade.

Chemicals

Agar-agar, Sucrose, Acetic Alcohol (1:3, v/v), 1N HCl, double distilled water.

Equipment

Weighing balance, Hot plate, Hot air oven (with thermostat), Fluorescence microscope.

Procedure

Step 1.Preparation of petriplate with Agar medium (100 ml)

1. Weigh 2 g agar and 10–15 g sucrose. Add the agar in ~50 ml of double distilled water. Heat the content over a hot plate, stirring it all the while, to prevent agar from charring. As the agar dissolves, add sucrose and mix well. Make up the final volume to 100 ml using double distilled water. (This medium can also be prepared in a microwave.)
2. Pour the medium in petriplates to make a thin layer (~5mm). Do not fill the petriplates to their brim.
3. Allow the agar medium to solidify. Cover the petriplates and store them at 4°C. Such plates can be stored for 4–5 days.

Note: if the species selected have large pistils, then wide-mouth glass bottles or plastic jars with lid may be used. Accordingly, the amount of agar medium in the jar should be increased so that the pistil base can be easily embedded.

Step 2

a. Pollen collection: The pollen grains can be simply collected by shaking the mature anthers over a butter paper sheet or in a clean glass petriplate. This can be done easily in species with dry pollen grains. For species with sticky pollen grains, it is always advisable to collect mature anther that can be forced to dehisce by keeping in an improvised humidity chamber for about 20–25 minutes. Then pollen grains can be directly collected using a camel hair brush. An alternative method for species with sticky pollen grains is to directly collect pollen grains from dehisced anthers on butter paper, using a camel hair brush. Pollen grains should be collected just before the start of exercise, to avoid loss of pollen viability. Collect the pollen grains of the selected species separately.

b. Pollen viability: Use some pollen grains of each species for assessing pollen viability by any test (TTC or FCR) as detailed earlier (exercise 4.6). Pollen samples showing about 50%–60% viability can be used for pollination.

c. Stigmatic receptivity: Identify the most receptive stage of stigma among the flower developmental stages (*see Exercise 7.2*) or if already known then directly collect 10–15 mature flowers of each species.

Step 3 Semi in vivo pollination

1. Take two petriplates with solid agar medium. Label them as 1 and 2.
2. Take flowers of any two species and carefully remove calyx, corolla and stamens. While doing so, care must be taken to not damage the pistil or touch the stigma.
3. Implant the pistils (n=20–30 or more) of one of the species in petriplate 1 and of the second species in petriplate 2.
4. Using a camel hair brush transfer the pollen grains of first species (petriplate 1) to stigmas of the other (petriplate 2) and vice versa.
5. Cover the petriplates and incubate them at room temperature. The total duration of incubation generally ranges between 24–72 hours.

Step 4 Fixation

1. Fix the manually pollinated pistils at intervals of 0 min, 30 min, 60 min, 2 hours, 4 hours, 8 hours and so on after pollination.
2. Take 10–15 ml of acetic alcohol in a screw cap bottle and dip the pistils in it for about 24 hours.
3. After fixation, the pistils can be stored in 70% ethanol until further processing.

Step 5 Tissue clearing and staining

1. Gently wash the fixed/stored pistils two or three times with double distilled water.
2. Transfer the washed pistils in another screw cap bottle containing 1N HCl and fasten the cap.
3. Incubate the bottle or vial in a hot air oven at 50–60°C for about 1–3 hrs. Check the extent of softening of pistil tissue after regular intervals. The tissue must not be over softened (*Note: Here one may increase the concentration of HCl to 5N and correspondingly decrease the duration of incubation*).
4. After appropriate incubation, wash the pistils twice carefully with double distilled water.
5. Transfer the washed pistils to a screw cap bottle containing decolorized aniline blue solution. Cover the bottle with aluminum foil and incubate the pistils at 4°C for 24 hrs.
6. Mount the pistil over a slide in aniline blue solution, place the coverslip and observe under the UV-fluorescence microscope (*Note: Slightly press the coverslip to spread the tissue of pistil for better view of the pollen tubes*).

Result

Only the sites of callose depositions will show fluorescence. Accumulation of callose indicates the inhibition or rejection of pollen grains and inter-specific incompatibility.

Exercise 8.2 To determine the type of homomorphic self-incompatibility (GSI or SSI) in a species by employing semi *in vivo* pollination and decolorized aniline blue staining method

Theory

In species with homomorphic self-incompatibility, the site of pollen or pollen tube inhibition is in the style for GSI systems or on the stigma in SSI systems. This feature can serve as a marker for establishing the type of self-incompatibility in a species. The semi *in vivo* pollination and aniline blue staining method is very useful for establishing the zone of inhibition of self-pollen grains and pollen tubes. Here, the localization of callose present in growing pollen tubes is an indication of the site of inhibition.

Suggested plant material: *Brassica* sp., *Petunia* sp., *Nicotiana* sp.

Material Required
As in Exercise 8.1

Procedure

Step 1
Follow the step 1 as in Exercise 8.1

Step 2
To study the type of self-incompatibility in a species, pollinate individual pistils with pollen of the same flower.

Step 3 Manual and semi in vivo pollination

1. Take petriplates with solid agar medium and label them.
2. Take flowers of the selected species and carefully remove the calyx, corolla and stamens of the flowers (*care must be taken not to damage pistils or touch stigmas*). Put the pistils (n=20–30 or more) in pteriplates and pollinate them immediately with self-pollen, using a camel hair brush.
3. Cover the petriplates and incubate them at room temperature for 24–72 hours. (*Note: The duration of incubation varies between species.*)
4. Repeat steps 3 and 4 for all the species collected.

Step 4 Fixation, Tissue clearing and Staining
Follow the steps 4 and 5 of Exercise 8.1
The fixation, tissue clearing and staining of pistils of different species must be done separately.

Observation and Result
If the pollen tube growth appears inhibited in the style, then it is likely that the self-incompatibility is of gametophytic type (GSI). If the pollen tubes and pollen show deposition of callose on the stigma surface, the self-incompatibility is possibly of sporophytic type (SSI).

Exercise 8.3 To determine the mating/breeding system of a species using *in vivo* manual pollination

In vivo manual pollinations are generally conducted to assess the breeding system of a species in its natural population. Before the start of this exercise, one must be acquainted with the developmental stages and floral biology of the species under investigation. For example, in a hermaphrodite flower one needs to emasculate the flower before dehiscence of anthers and stigma maturity. Moreover, knowledge of the stage and duration of anther dehiscence is a prerequisite for pollen collection.

Material Required
Camel hair brush (with few hairs, 0/1 No.), a pair of forceps, needles, blade, metallic tags, muslin cloth bags/butter paper bags, thread, butter paper.

Procedure

1. Consider two trees of a species in your locality and mark them as Tree A and Tree A*.
2. Mark mature flowers (n=30) on each tree using paint, metallic tags or paper tags.
3. Before the onset of anthesis in the marked flower, emasculate some flowers (n=20) on each tree.
4. Cover rest of the flowers (n=10) with muslin cloth bag/paper bag without emasculation (*for selfing* or *autogamy*).
5. At the time of stigma maturity in emasculated flowers (n=10), collect pollen from A and A* separately and perform manual cross pollination (A to A* and vice versa; *xenogamy*). After manual cross pollination cover the flowers.
6. In the remaining emasculated flowers (n=10), perform manual geitonogamous pollinations. Each flower of A and A*, is pollinated with pollen from some other flower of A and A* respectively. Bag the flowers after pollination.
7. Remove the bags after 2–3 days and observe the initiation of fruit formation in flowers of the different pollination treatments. Record the observations.
8. In the fruiting season, once again observe the number of fruits that appear in the marked flowers with different pollination treatments and compute percent fruit formation in each treatment.
9. Predict the breeding of species using the data (*see Tables 8.1 and 8.2*).

Precautions

1. Sufficient pollen load on stigma during manual pollination must be ensured through examination of stigma with hand lens.
2. While doing manual pollination do not touch or damage the stigma with the brush.
3. Pollen viability and fertility must be assessed before performing manual pollination.

Table 8.1 Sample table for the calculation of fruit set after different pollination treatments

Pollination treatment	Number of flowers given treatment (n)	Fruit initiation (after 2–3 days)	Number of fruits
Autogamy	10	no	zero
Geitonogamy	10	no	zero
Xenogamy	10	yes	7
Your Answer: Breeding system of the given species is.....................................			

Table 8.2 Sample table for computing fruit set after different pollination treatments

Pollination treatment	Number of flowers given treatment (n)	Fruit initiation (after 2–3 days)	Number of fruits
Autogamy	10	Yes	8
Geitonogamy	10	Yes	8
Xenogamy	10	no	zero
Your Answer: Breeding system of the given species is.....................................			

Exercise 8.4 To calculate Index of Self-incompatibility (ISI)

The existence of self-incompatibility is known from more than 100 plant families and approximately 40% of species (Igic et al. 2008). However, not all species fit into the categories of *Self-incompatible* (SI) and *Self-compatible* (SC). There are a number of species which have intermediate states between SI and SC. In other words, in some species fruit set occurs both by selfing and cross pollination (*also referred to as mixed mating system*). Nonetheless, the amount of fruit set varies between these two strategies and the variation can be depicted as the strength of SI in form of an index known as the Index of Self-Incompatibility (ISI) (Zapata & Arroyo 1978; Mangla & Tandon 2011). The ISI is calculated by dividing the percentage of fruit set resulting from self-pollination (whether autogamy or allogamy) to cross pollination (Zapata & Arroyo 1978). The ratio of '1' or more suggests that the species has a potential for reproduction by both self and cross means. The minimum possible value is '0' which indicates that the species is completely self-incompatible (Table 8.4, Zapata & Arroyo 1978). The calculation of ISI is important for crop breeding programmes as knowledge of the breeding system of a species is vital. To calculate ISI, the protocol for *in vivo* manual pollination detailed in Exercise 8.3 may be used. A sample table (Table 8.3), referring some suggested species and hypothetical examples, is provided to record the results.

Table 8.3 Sample table for calculation of ISI

Species	Self-pollination			Cross pollination			Ratio
	Number of flowers	Fruit set	Percentage fruit set (P_s)	Number of flowers	Fruit set	Percentage fruit set (P_c)	P_s/P_c
X	80	20	25	80	40	50	0.5
Y	44	0	0	52	33	63	0
Brassica							
Crotolaria							
Crateva adansonii							

The results of hypothetical examples when compared with values given in Table 8.4 indicate that species X is relatively self-incompatible, while species Y is completely self-incompatible.

Table 8.4 Value of ISI and its inference

Index of Self-Incompatibility	State
0	Complete Self-Incompatibility
< 0.2	Severe Self-Incompatibility
0.2 ≤ 1	Relative Self-Incompatibility or Relative Self-Compatibility
≥ 1	Self-Compatibility

Exercise 8.5 To analyze the breeding system of a species by calculating its Pollen:Ovule ratio

Pollen production and ovule production are highly variable among flowering plants. Pollen production is variously related to modes of pollination such as species exhibiting anemophily produce large numbers of pollen grains as the chances of reaching conspecific stigma are relatively low. In contrast, species with self-pollination and entomophilous modes produce a comparatively smaller number of pollen grains. Similar generalization can also be made between ovule numbers and modes of pollination. For example, anemophilous species mostly tend to produce one ovule per ovary while entomophilous and self-pollinating species produce large number of ovules. An investigation by Cruden (1977) indicated that the pollen to ovule ratio (i.e., the number of pollen grains per ovule) reflects the breeding system of a species. For any species, the pollen-ovule ratio may be determined by dividing the total number of pollen grains produced per flower by the number of ovules per flower. Cruden (1977) highlighted that a strong relation prevails between the mode of pollen transfer and its pollen: ovule ratio (P/O). For instance, cleistogamous flowers have a lower P/O ratio than autogamous flowers while xenogamous flowers have a higher P/O ratio than autogamous flowers. Although the P/O is an indicator of breeding system of a species; proper manual pollination experiments are mandatory to confirm this.

Procedure

1. To estimate pollen production and ovule count in a flower refer to Exercises 4.7 and 5.6 respectively.
2. Calculate the Pollen:Ovule ratio:

P/O= Number of pollen grains in a flower/Number of ovules in a flower

3. Predict the breeding system of species based on values given in Table 8.5 below (*based on Cruden 1977; Shivanna & Tandon 2014*).

Table 8.5 Breeding systems based on pollen:ovule ratio (Cruden 1977)

Pollen:ovule ratio	Breeding system
2.7–5.4	Cleistogamy
8.1–39	Obligate autogamy
31.9–396	Facultative autogamy
244.7–2,588	Facultative xenogamy
2,108–195,525	Obligate xenogamy

Bibliography

Abdallah, D., Baraket, G., Perez, V. et al. (2019). Analysis of self-incompatibility and genetic diversity in diploid and hexaploid plum genotypes. *Frontiers of Plant Sciences*10: 896. doi: 10.3389/fpls.2019.00896.

Barrett, S. C. H. (2019). 'A most complex marriage arrangement': recent advances on heterostyly and unresolved questions. *New Phytologist* 224: 1051–67.

Bartoš, M., Janeček, Š., Janečková P. et al. (2020). Self-compatibility and autonomous selfing of plants in meadow communities. *Plant Biology* 22: 120–8.

Besnard, G., Cheptou, P-O. and Debbaoui, M. et al. (2020). Paternity tests support a diallelic self-incompatibility system in a wild olive (*Olea europaea* subsp. *laperrinei*, Oleaceae). *Ecology and Evolution* 10: 1876–88.

Bhojwani, S. S. and Bhatnagar, S. P. (eds.). (2008). *The Embryology of Angiosperms*. Noida, India: Vikas Publishing House Private Limited.

Bhojwani, S. S. and Cocking, E. C. (1972). Isolation of protoplasts from pollen tetrads. *Nature New Biology* 239: 29–30.

Brieger, F. (1930). *Selbststerilit/tt und Kreuzungssterilit/it imPflanzenreich und Tierreich. Monogr. Ges, gebiet d. Phys. d. Pflanzen u. Tiere*. J. SPRINGER Verlag, Berlin 1930, 2z, 395 S. m. 118 Abb. (not seen in original).

Brom, T., Vincent, C. and Sylvain, B. (2020). Breakdown of gametophytic self-incompatibility in subdivided populations. *Evolution* 10. doi:1111/evo.13897.

De Cauwer, I., Vernet, P., Billiard, S. et al. (2020). Widespread coexistence of self-compatible and self-incompatible phenotypes in a diallelic self-incompatibility system in *Ligustrum vulgare* (Oleaceae). *Cold Spring Harbor Laboratory*. https://doi.org/10.1101/2020.03.26.009399.

Charlesworth, D. (2010). Self-incompatibility. *F1000 Biology Reports* 2: 68.

Connor, H. E. (1979). Breeding systems in the grasses: a survey. *New Zealand Journal of Botany* 17: 547–74.

Cruden, R. W. (1977). Pollen–ovule ratios: a conservative indicator of breeding systems in flowering plants. *Evolution* 31: 32–46.

Cruden, R. W. (2000). Pollen grains: why so many? *Plant Systematics and Evolution* 222: 143–65.

De Nettancourt, D. (1977). *Incompatibility in Angiosperms*. Berlin: Springer Verlag.

de Vos Jurriaan, M., Hughes, C. E., Schneeweiss, G. M, et al. (2014). Heterostyly accelerates diversification via reduced extinction in primroses. *Proceedings of the Royal Society: Biological Sciences* 281: 20140075.

Dickinson, H. G. and Lawson, J. (1975). Pollen tube growth in the stigma of *Oenothera organensis* following compatible and incompatible intraspecific pollinations. *Proceedings of the Royal Society of London B* 188: 325–44.

Do Canto, J., Studer, B. and Lubberstedt, T. (2016). Overcoming self-incompatibility in grasses: a pathway to hybrid breeding. *Theoretical and Applied Genetics* 129: 1815–29.

Dowrick, V. P. J. (1956). Heterostyly and homostyly in *Primula obconica*. *Heredity* 10: 219–36.

Duca, S. D., Aloisi, I., Parrott, L. and Cai, G. (2019). Cytoskeleton, transglutaminase and gametophytic self-incompatibility in the Malinae (Rosaceae). *International Journal of Molecular Sciences* 20: 1–11.

East, E. M. and Mangelsdorf, A. J. (1925). A new interpretation of the hereditary behavior of self-sterile plants. *Proceedings of National Academy of Sciences USA* 11: 166–71.

Eaves, D. J., Flores-Ortiz, C., Haque, T. et al. (2014). Self-incompatibility in *Papaver*: advances in integrating the signalling network. *Biochemical Society Transactions* 42: 370–6.

Ernst, A. (1928). ZurVererbung der morphologischen Heterostyliemerkmale. *Ber. dtsch. bot. & S*. 46: 573–88 (not seen in original).

Ernst, A. (1955). Self-fertility in monomorphic Primulas. *Genetica* 27: 391–448.

Faivre, A. E. (2002).Variation in pollen tube inhibition sites within and among three heterostylous species of Rubiaceae. *International Journal of Plant Sciences*163: 783–94.

Fearon, C., Cornish, M., Hayward, M. et al. (1994). Self-incompatibility in ryegrass. X. number and frequency of alleles in a natural population of *Lolium perenne* L. *Heredity* 73: 254–61.

Gibbs, P. E. (2014). Late-acting self-incompatibility – the pariah breeding system in flowering plants. *New Phytologist* 203: 717–34.

Goring, D. R. (2000). The search for components of the self-incompatibility signalling pathway(s) in *Brassica napus*. *Annals of Botany* 85: 171–9.

Haldane, J. B. S. (1933). Two new allelomorphs for heterostyly in *Primula*. *American Naturalist* 67: 559–60.

Igić, B., Lande, R. and Kohn J. R. (2008). Loss of self-incompatibility and its evolutionary consequences. *International Journalof Plant Sciences* 169: 93–104.

Iwano, M. and Takayama, S. (2012). Self/non-self discrimination in angiosperm self-incompatibility. *Current Opinion in Plant Biology* 15: 78–83.

Kakita, M., Shimosato, H., Murase, K. et al. (2007). Direct interaction between S-locus receptor kinase and M-locus protein kinase involved in *Brassica* self-incompatibility signaling. *Plant Biotechnology* 24: 185–90.

Kanta, K. (1960). Intra ovarian pollination in *Papaver rhoeas* L. *Nature* 188: 683–4.

Kao, T. H. and McCubbin, A. F. (1996). How flowering plants discriminate between self and non-self pollen to prevent inbreeding. *Proceedings of National Academy of Sciences* 93: 12059–65.

Kitashiba, H. and Nasrallah, J. B. (2014). Self-incompatibility in Brassicaceae crops: lessons for interspecific incompatibility. *Breeding Science* 64: 23–37.

Klaas, M., Yang, B., Bosch, M. et al. (2011). Progress towards elucidating the mechanisms of self-incompatibility in the grasses: further insights from studies in *Lolium*. *Annals of Botany* 108: 677–85.

Labonne, J., Tamari, F. and Shore, J. (2010). Characterization of X-ray-generated floral mutants carrying deletions at the S-locus of distylous *Turnera subulata*. *Heredity* 105: 235–43.

Langridge, P. and Baumann, U. (2008). Self-Incompatibility in the grasses. In V. E. Franklin-Tong, ed., *Self-Incompatibility in Flowering Plants-Evolution, Diversity, and Mechanisms*. Berlin Heidelberg: Springer-Verlag.

Lewis, D. (1949). Incompatibility in flowering plants. *Biological Reviews* 24: 472–96.

Mangla, Y. and Gupta, C. K. (2015). Self-incompatibility. Lesson developed as e-resource in open education under section 'Botany'. Institute of lifelong Learning (ILLL), University of Delhi. ISSN:978-93-85611-91-9.

Mangla, Y. and Tandon, R. (2011). Insects facilitate wind pollination in pollen-limited *Cratevaadansonii* (Capparaceae). *Australian Journal of Botany* 59: 61–9.

Matsui, K. and Yasui, Y. (2020). Buckwheat heteromorphic self-incompatibility: genetics, genomics and application to breeding. *Breeding Science* 70: 32–8.

McCubbin, A. and Dickinson, H. (1997). Self-incompatibility. In *Pollen Biotechnology for Crop Production and Improvement*. Cambridge: Cambridge University Press, pp. 199–217.

Megumi Iwano, M. and Takayama, S. (2012). Self/non-selfdiscrimination in angiosperm self-incompatibility. *Current Opinion in Plant Biology* 15: 78–83.

Nasrallah, J. B. (2019). Self-incompatibility in the Brassicaceae: regulation and mechanism of self-recognition. *Current Topics in Developmental Biology* 131: 435–52.

Nowak, M. D., Davis, A. P., Anthony, F. and Yoder, A. D. (2011). Expression and trans-specific polymorphism of self-incompatibility *RNases* in *coffea* (*Rubiaceae*). *PLoS One* 6:e21019. doi: 10.1371/journal.pone.0021019.

Owens, S. J. (1981). Self-incompatibility in the Commelinaceae. *Annals of Botany* 47: 567–81.

Power, J. B. and Cocking, E. C. (1971). Fusion of plant protoplasts. *Science Progress* 59: 181–98.

Raduski, A. R., Elizabeth, B. H. and Igić, B. (2012). The expression of self-incompatibility in angiosperms is bimodal. *Evolution* 66: 1275–83.

Rodriguez-Leal, D., Xu, C., Kwon, C. T, et al. (2019). Evolution of buffering in a genetic circuit controlling plant stem cell proliferation. *Nature Genetics* 51: 786–92.

Seavey, S. R. and Bawa, K. S. (1986). Late-acting self-incompatibility in angiosperms. *Botanical Review* 52: 195–219.

Shivanna, K. R (ed). (2002). Self-incompatibility. In *Pollen Biology and Biotechnology*, pp. 140–66.

Shivanna, K. R. and Tandon, R. (2014). *Reproductive Ecology of Flowering Plants:A Manual*. Springer India.

Shivanna, K. R. (1989). Recognition and rejection phenomena during pollen–pistil interaction. *Proceedings of Indian Academy of Sciences* 88: 115–41.

Sunnichan, V. G., Mohan Ram, H. Y. and Shivanna, K. R. (2005). Reproductive biology of *Boswellia serrata*, the source of salai guggul, an important gum-resin. *Botanical Journal of the Linnean Society* 147: 73–82.

Takayama, S. and Isogai, A. (2005). Self-incompatibility in plants. *Annual Review of Plant Biology* 56(1): 467–89.

Tandon, R. (2011). Reproductive biology of *Azadirachta indica* (Meliaceae), a medicinal tree species from arid zones. *Plant Species Biology* 26: 116–23.

Vieira, J., Rocha, S., Vázquez, N. et al. (2019). Predicting specificities under the non-self gametophytic self-incompatibility recognition model. *Frontiers in Plant Sciences* 10: 879. doi: 10.3389/fpls.2019.00879.

Wilkins, K. A., Poulter, N. S. and Franklin-Tong, V. E. (2014). Taking one for the team: self-recognition and cell suicide in pollen. *Journal of Experimental Botany* 65: 1331–42.

Williams, J. S., Wu, L., Li, S. et al. (2015). Insight into *S-RNase*-based self-incompatibility in *Petunia*: recent findings and future directions. *Frontiers in Plant Science* 5:41. doi: 10.3389/fpls.2015.00041.

Zapata, T. R. and Arroyo, M. T. K. (1978). Plant reproductive ecology of a secondary deciduous tropical forest in Venezuela. *Biotropica* 10: 221–30.

9

Endosperm

Different forms of endosperm fulfill the various requirements of mankind.

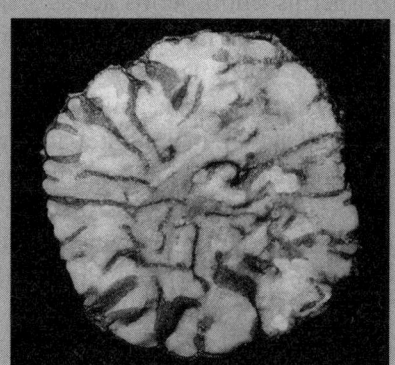

9.1 Introduction

Strasburger's work in *Monotropa* identified the embryo as the product of fertilization of egg cell by the one of the male gametes. The mystery around the fate of the second male gamete discharged by the pollen tube was resolved through the discovery of double fertilization by Nawaschin (1898). It is a common knowledge now, that one of the male gametes undergoes fusion with the nucleus of the egg cell and the other fuses with the two polar nuclei of the central cell. The fusion of three nuclei in the latter is known as triple fusion which was also referred to as **vegetative fertilization** by Strasburger in 1900. Triple fusion results in a triploid nucleus known as **Primary Endosperm Nucleus (PEN)**, which divides and forms the endosperm. Thus, double fertilization initiates development of embryo and endosperm. Discovery of triple fusion led to questions like *what is endosperm and what role does it play?* The term endosperm means 'with-in the seed,' i.e., a tissue that develops inside a seed. A plethora of studies have established that endosperm is the nutritive tissue for a growing embryo inside a seed. The two tissues are closely connected in their growth within a seed, reflecting the importance of the embryo-endosperm relationship. Recent investigations show that failure of endosperm formation leads to the abortion of the developing embryo which establishes that embryo development is regulated by endosperm.

Formation of the PEN is a well-organized event which is preceded by several ultrastructural changes in the central cell. The PEN follows different developmental pathways forming the basis for classifying the types of endosperm. The PEN and the endosperm cells are mostly triploid but the ploidy level may vary with the type of female gametophyte from which a central cell develops. In most angiosperm families, the endosperm is short-lived and the developing embryos consume the endosperm completely before germination. This leaves mature seeds without any endosperm and such type of seeds are known as

non-endospermous or ex-albuminous seeds, e.g., *Cucurbita*, pea, and beans. In other angiosperms, endosperms act as a storage tissue and persist in mature seeds. Such seeds where endosperm is present at maturity are known as **endospermous or albuminous seeds**, e.g., cereals, coconut, and castor bean. In such seeds endosperm serves as a reservoir of nutrients during seed germination. The endosperm of albuminous seeds is a major source of oils and starch, making it a an important food and industrial raw material. More than 60% of human nutritional requirements are fulfilled by endosperm alone, especially, the endosperm of cereals. Here, it is worth mentioning that cereals have a specialized type of endosperm in their seeds which has been investigated intensively.

Presence of endosperm during the development of seeds is a characteristic feature of angiosperms. However, there are some exceptions to this, like members of families Orchidaceae, Podostemaceae, and Trapaceae. Plants in these families are characterized by lack of endosperm in their developing seeds which is compensated by other specialized tissues that provide nutrition to the growing embryos.

This chapter presents an account of ultrastructure profiles of PEN, different types of endosperm development, endosperm in cereals, embryo-endosperm relationship and developmental biology of endosperm in detail.

9.2 Ultrastructural Changes in the Central Cell and Formation of Primary Endosperm Nucleus

Central cell is the largest cell of an embryo sac and it usually possesses two large, flattened, and haploid nuclei, known as the polar nuclei. It is a highly vacuolated cell at maturity, containing a prominent central vacuole and a peripheral cytoplasm. Unlike other cells of embryo sac, central cell is completely surrounded by cell wall and a plasma membrane. Some plasmodesmatal connections also exist in the common walls between the central cell and other cells (i.e., antipodals and egg cell) of the embryo sac.

In most angiosperms, the two polar nuclei of embryo sac fuse together and form a secondary nucleus before fertilization. During the process of fusion, the nuclear membrane of the two nuclei evaginate to make contact, followed by the formation of nuclear bridges between their nucleoplasm which enlarge and complete the process of fusion (Fig. 9.1). Often, small segments of cytoplasm remain entrapped in the nuclear bridges which are later excluded (Fig. 9.2). A big vacuole occupies the chalazal end of the central cell and the secondary nucleus lies in the portion of central cell cytoplasm which is just below the egg cell. Newly formed secondary nucleus initially becomes rich in RNA and protein content. By this time, the central cell becomes densely cytoplasmic and the smooth and rough ERs get arranged around the secondary nucleus. Also, ribosomes, Golgi-bodies, well-developed plastids, mitochondria and other organelles aggregate in the cytoplasm of central cell. These changes indicate that the central cell is metabolically active and the secondary nucleus is ready for fusion with sperm nucleus.

During fertilization, the secondary nucleus fuses with the sperm nucleus to give rise to a triploid Primary Endosperm Nucleus (PEN). At this stage, the fertilized central cell is also known as *Primary Endosperm Cell* (PEC), as it has its own cell wall and PEN remains confined within it. The PEN is formed before the formation of zygote and remains present just below the egg cell (Fig. 9.1). Sperm cytoplasm plays no role in the formation of PEC. However, the membrane of PEN is contributed by secondary nucleus as well as by male nucleus. The cytoplasm of PEC is rich in cell organelles such as endoplasmic reticulum, Golgi bodies, mitochondria and ribosomes reflecting the ontogeny of PEC from a metabolically active central cell. So, the features of the newly formed PEC do not differ much in ultrastructure from those of the central cell except for the presence of abundant protein bodies, starch granules and polysaccharides in PEC. The PEN/PEC divides rapidly soon after its formation eventually forming the endosperm.

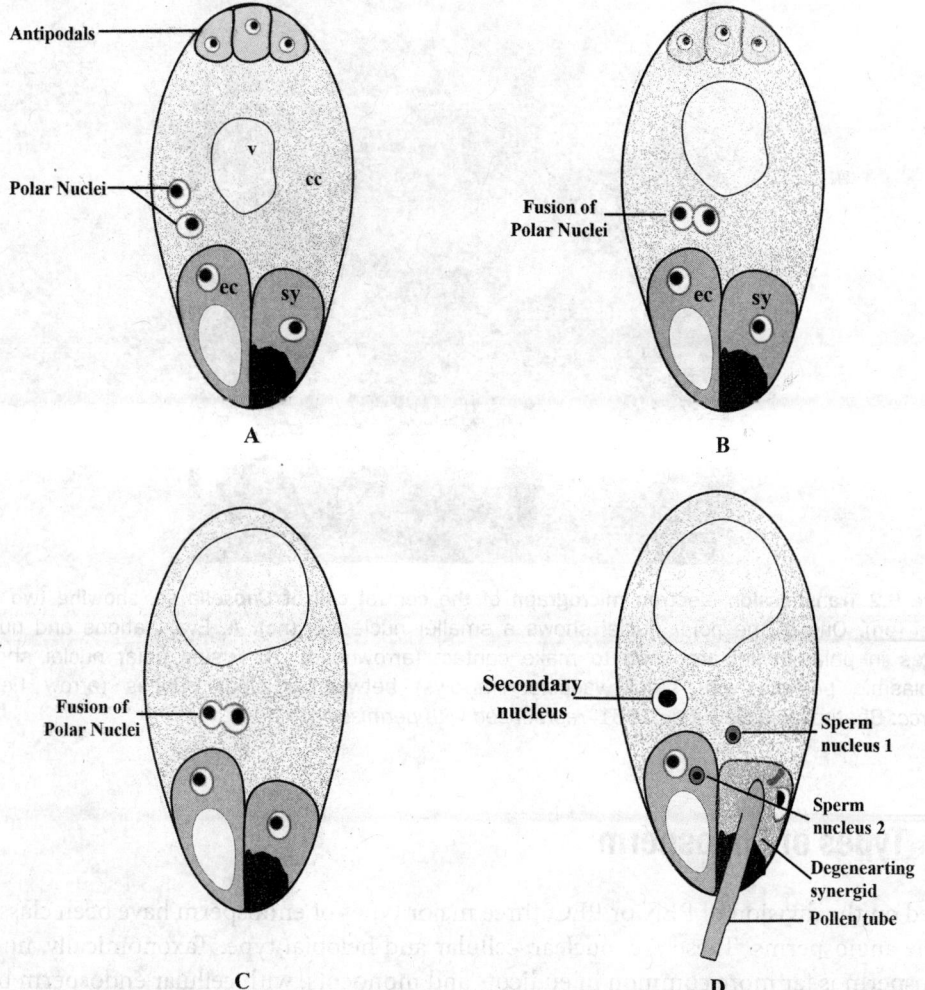

Figure 9.1 Depiction of the formation of secondary nucleus before the arrival of the male gametes.
A. Mature embryo sac. B–C. Polar nuclei draw closer until their nuclear membranes (through evagination) make contact, and connections between nucleoplasm occur forming nuclear bridges. D. Embryo sac with secondary nucleus. **See Color Plates (page 491).**

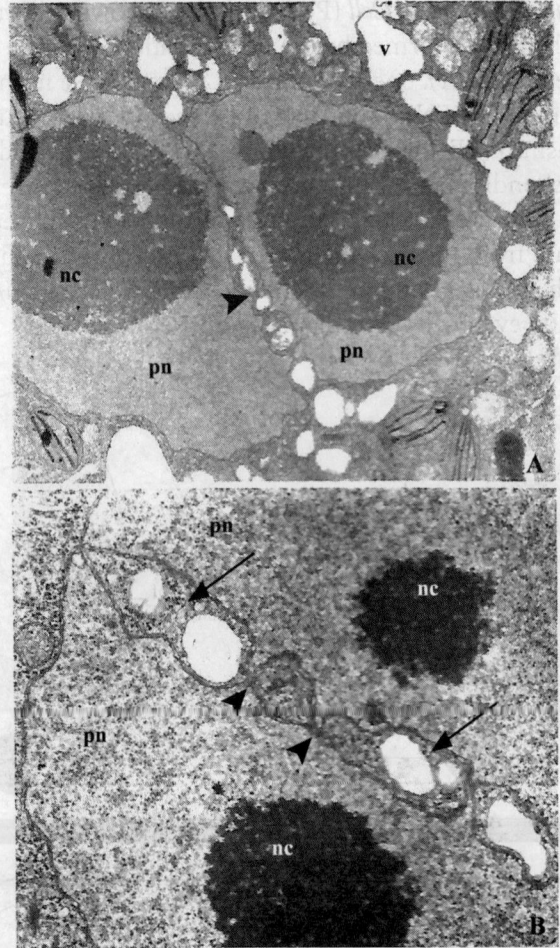

Figure 9.2 Transmission electron micrograph of the central cell of *Capsella* sp. showing two polar nuclei (pn). One of the polar nuclei shows a smaller nucleolus (**nc**). **A.** Evaginations and nuclear bridges in polar nuclei are seen to make contact (arrowhead). **B.** Fusing polar nuclei showing cytoplasmic pockets with small vacuoles (arrows) between nuclear bridges (arrow heads). (*Source*: Bhatnagar & Sawhney 1981, *reproduced with permission*.)

9.3 Types of Endosperm

Based on the division of PEN or PEC, three major types of endosperm have been classified in the angiosperms. These are: nuclear, cellular and helobial type. Taxonomically, nuclear endosperm is far more common in eudicots and monocots, with cellular endosperm being frequent among basal angiosperms and dicots. The helobial type is mostly restricted to monocots. However, this generalization has exceptions in different taxonomic groups.

Box 9.1

Genomic Imprinting in Endosperm

In most plants, endosperm is a triploid tissue with two identical copies of the maternal genome and one copy of the paternal genome. Thus, unlike other cells and tissues of a plant, the ratio of maternal to paternal genomes in the endosperm is 2 (maternal): 1 (paternal), which if expressed, can cause genomic imbalance and influence the development of the endosperm and the seed (Li & Dickinson 2010). An excess dose of maternal genome can lead to precocious cellularization and reduced endosperm growth. To avoid this imbalance, flowering plants undergo genomic imprinting in the endosperm. Genomic imprinting is the phenomenon in which a set of genes is expressed according to their parent of origin (Huh et al. 2007).

In different cells of any particular organism, only a certain number of genes express at a given time according to the requirement of a cell. Rest of the genes do not express or are silenced. Gene silencing is a mechanism by which cells shut down large numbers of genes when their expression is not required. In the endosperm, the paternal alleles are silenced whereas the maternal alleles remain active resulting in mono-parental imprinted expression.

The silenced genes are often associated with having DNA methylation which is also a major pathway involved in the regulation of endosperm gene imprinting. It is mediated by METHYLTRANSFERASE1 (MET1), which maintains DNA methylation of CpG sites (*Cytosine* and *Guanine rich sites*) of a silenced gene. For the gene to express, the methylation of DNA has to be removed. In plants, parental genomic imprinting occurs during gametogenesis which leads to differential expression of the two parental alleles. It has been demonstrated in *Arabidopsis* that MET1 maintains CpG methylation silencing marks on the paternal alleles of imprinted genes in sperm cells. On the other hand, the methylation marks are removed from maternal alleles in central cell, for genes to express. After fertilization, the active maternal alleles and inactive paternal alleles are inherited. Hence, the endosperm inherits silenced paternal alleles (p) and two copies of active maternal alleles (m), resulting in mono-parental imprinted expression in the endosperm (Kinoshita et al. 2004; Berger & Chaudhury 2009; Gehring et al. 2009; Jullien & Berger 2009). Such a mechanism applies to most maternally expressed imprinted genes which are silenced in the sperm cells.

9.3.1 Nuclear Endosperm

Nuclear endosperm is the most common type of endosperm development. After fertilization, the PEN divides repeatedly without cell wall formation, resulting in a characteristic coenocytic-stage (free nuclear stage) of PEC. Thus, numerous nuclei remain suspended in PEC and this coenocytic stage is also referred to as the *syncytial phase* (Fig. 9.3 A–C). The divisions in PEN may be synchronous or asynchronous. This is followed by cellularization which takes place after a specific number of divisions in each species, e.g., in *Arabidopsis*, cellularization starts after eight rounds of divisions. To begin *cellularization phase*, vacuole in PEC enlarges and all the nuclei migrate towards the periphery of the cell (Fig. 9.3 D). After this, the cell walls get laid down in a centripetal manner which is initiated in the micropylar region. Once cellularization is complete, the peripheral cells of the endosperm undergo synchronous cell divisions to form five–six layers of cells. With the onset of cellularization, the central vacuole retracts and completely disappears as the entire endosperm becomes cellular (Fig. 9.3 E–H). Nuclear endosperm is quite common in families like Cucurbitaceae, Leguminosae, Brassicaceae, Phytolaccaceae, Primulaceae, Proteaceae, Sapindaceae, Poaceae, Arecaceae, Cyperaceae, and Limnanthaceae. Many authors (Newcomb 1978; Lersten 2004) find the term 'nuclear endosperm' to be inappropriate to describe such type of development of endopserm. They reason that the term *nuclear endosperm* should apply only to noncellular endosperm; and not to the type of endosperm where the term 'nuclear' suggests just the first phase of growth. According to them, this two-step development pattern of endosperm may be more appropriately labelled as the **coenocytic/multicellular type**.

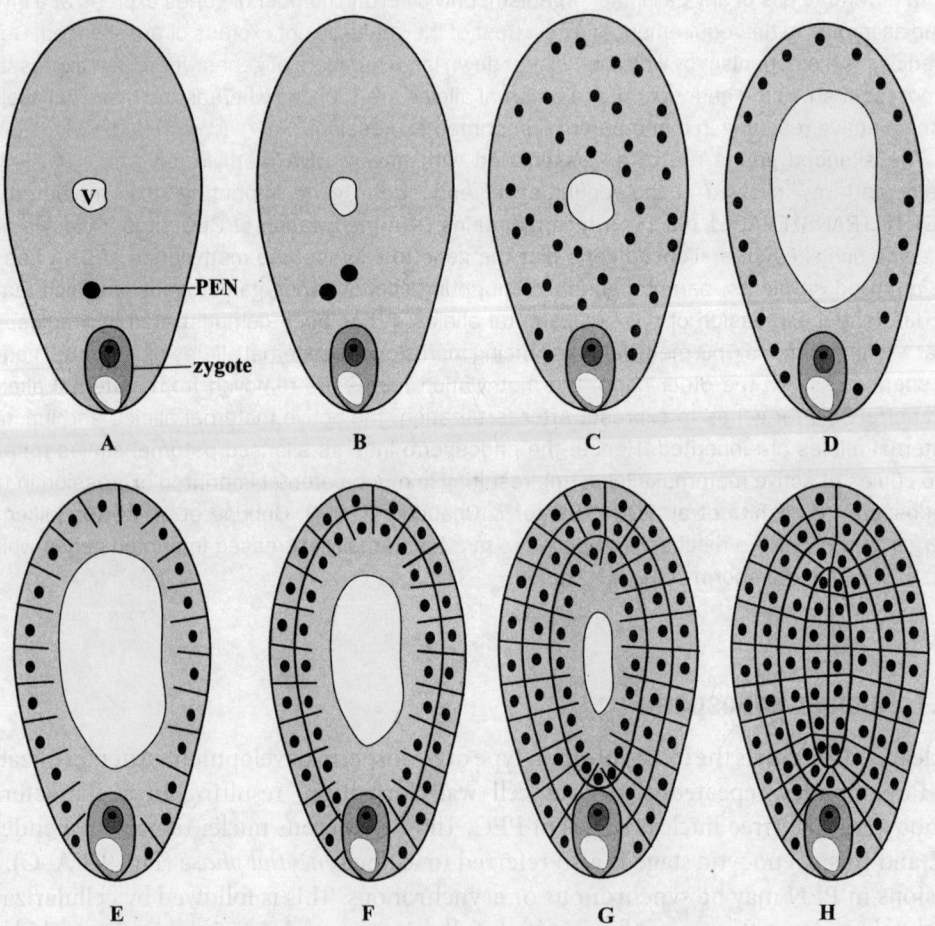

Figure 9.3 Diagrammatic representation of free nuclear endosperm and its cellularization. A. PEN: Primary endosperm nucleus. B. First division in PEN. C. Syncytial phase/Coenocytic stage D. Migration of nuclei towards periphery due to appearance of vacuole in the centre. E.–H. Cellularization in centripetal manner due to radial microtubular system (for details see Fig. 9.15), cell division in peripheral layers and gradual disappearance of the vacuole. (*Source:* Adapted and modified after Olsen 2004 *with permission.*)

A well-documented example of true nuclear endosperm is *Arachis hypogaea* (the common peanut, family Leguminosae) (Prakash 1960). In peanut, the coenocytic condition persists till the development of cotyledons during embryogenesis. The developing cotyledons absorb nutrients from the coenocyte which shows a corresponding decrease in its content and volume. Very late in the embryo development, a central vacuole appears in the coenocytic PEC. This leads to cellularization, resulting in two thin layers of peripheral endosperm cells which may persist till seed maturity (Fig. 9.4).

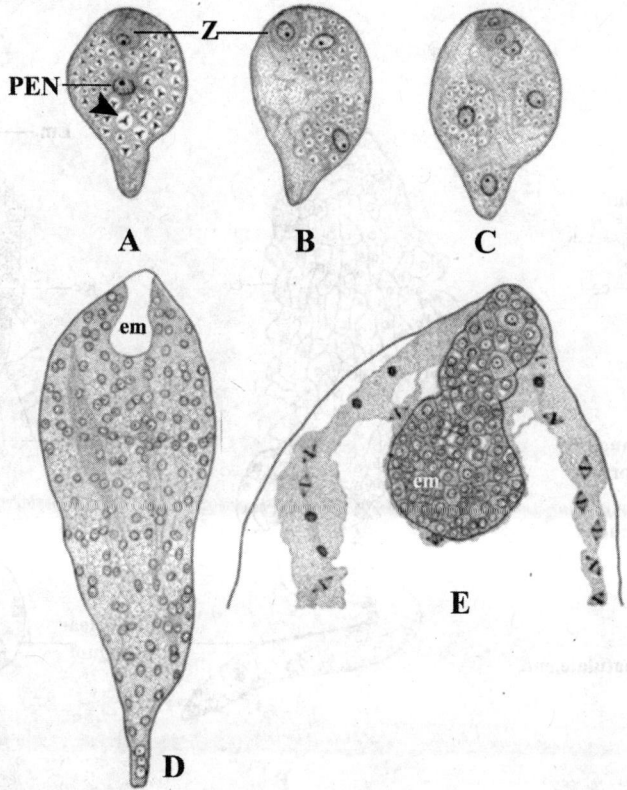

Figure 9.4 Free nuclear or coenocytic type of endosperm development in *Arachis hypogea*. A. Zygote **(Z)** at the top and primary endosperm nucleus **(PEN)** in the centre. Arrow head indicates presence of starch grains. **B.** First division in PEN and 2-nucleate PEC. **C.** 4-nucleate PEC with starch still present; pro-embryo at top. **D.** Coenocyte with pro-embryo at globular stage. **E.** Late globular stage of embryo **(em)**: the endosperm nuclei can be seen dividing and are in free nuclear state. (*Source*: Prakash 1960, *reproduced with permission.*)

Nuclear endosperm in coconut (family Arecaceae) is distinct and a suitable classroom example. In coconut, after fertilization (when the fruit is about 50 mm long), the PEN undergoes a number of divisions and the embryo sac cavity gets filled with liquid syncytium (also referred as coconut milk) where nuclei of variable sizes reside. As the fruit increases in size (~100 mm), numerous large spherical vesicles appear which enclose free nuclei and behave like cells. The number of nuclei per cell varies. These free cells gradually settle on the periphery and form a white gelatinous layer. As these cells undergo mitotic divisions, the gelatinous mass increases in thickness. This white gelatinous layer is the cellularized endosperm and commonly known as coconut meat. The coconut water in a mature coconut is devoid of any nuclei and cells. *Areca catechu* also shows a similar sort of endosperm formation. Here, the embryo sac cavity is small and after cellularization of free nuclei it gets filled compactly with cells. So, the seed becomes extremely hard. It is clear from all the above examples that irrespective of the type of development, endosperm eventually becomes cellular sooner or later. However, the extent of cellularization of endosperm varies with species.

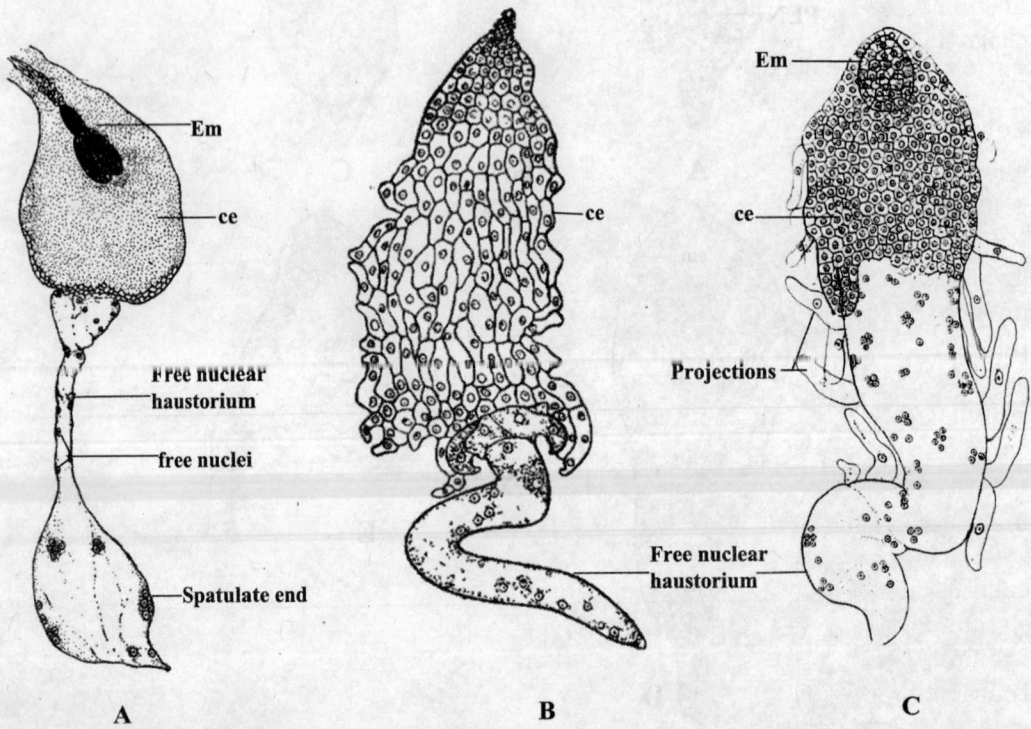

Figure 9.5 Diagram of endosperm showing morphological variation in the haustoria. ce: cellularized endosperm; **Em**: embryo. A. *Crotolaria*, **B.** *Grevillea robusta*, **C.** *Lomatia polymorpha* (*Source*: A. Rau 1951; B. Kaushik 1938; C. Vankata Rao 1963, *reproduced with permission.*)

A feature commonly seen in species with nuclear endosperm is a haustorium developing from the acellularized portion, e.g., Cucurbitaceae, Fabaceae; however, some families like Phytolaccaceae, Primulaceae, and Boraginaceae lack chalazal haustorium. The structure of haustorium varies in its assembly among species. The variation in cellularization within endosperm is well studied in members of family Cucurbitaceae. Here, after the syncytium stage, only the micropylar region of the endosperm becomes cellular and forms *endosperm proper* while its chalazal portion remains free nuclear. Soon after fertilization, the chalazal portion of the embryo sac extends downwards in a tube-like extension. The free nuclei of endosperm migrate into the tube and form a coenocytic elongated haustorium. Among cucurbits, *Echinocystis lobata* has the longest haustorium (16 mm). In *Cucumis melo,* and *C. Sativus* (Fig. 9.5 A) the tip of haustoria becomes spatulate. The coenocytic haustorium remains free nuclear throughout its life-span in most of the cucurbits except *Citrullus, Coccinia indica,* and *Momordica.* In these species, multinucleate chambers or uninucleate cells develop in haustorium.

The size and shape of free nuclear haustorium also varies among legumes. In *Crotolaria* only the micropylar portion of coenocytic endosperm becomes cellular, while the chalazal portion remains coenocytic and develops haustorium. There is a constricted haustorium in *Mimosa pudica*, while in *Cassia* it is bulbous and in *Calliandra,* it is coiled. Likewise, in Proteaceae members – *Grevillea robusta* and *Lomatia polymorpha*, a well-developed chalazal haustorium is present with the micropylar region being cellular (Fig. 9.5 C and D).

9.3.2 Cellular Endosperm

The cellular endosperm is characterized by cellular development where karyokinesis is followed by cytokinesis. Thus, the endosperm is ab *initio* cellular. *Drimys winteri* is a typical example of a cellular endosperm. The first division in the PEC is mostly transverse giving rise to two unequal chambers: larger micropylar and smaller chalazal. The transverse divisions in the chalazal chamber form two unequal sized cells; of which the cell towards the extreme chalazal end degenerates. The remaining chalazal, and micropylar cells divide and form the cellular endosperm (Fig. 9.6) (Bhandari & Venkataraman 1968).

Cellular endosperm is commonly present among the members of families Acanthaceae, Scrophulariaceae, Bignoniaceae, Lentibulariaceae, and Campanulaceae. Presence of the chalazal, or micropylar, or both chalazal and micropylar haustoria is very common in cellular endosperm. The haustorial behavior is apparent by its hypertrophoid nature, with large nuclei, dense cytoplasm, and perhaps branching. The haustoria in the cellular endosperm are more varied than in the nuclear endosperm in terms of their ontogeny and structure.

Family Acanthaceae is characterized by the initial asymmetric division in PEC. Further contribution of so formed cells towards formation of the haustorium is variable among members. In some members, the haustorium is altogether absent (e.g., *Acanthus*). In others like *Thunbergia,* first transverse division in PEC forms two chambers of which the micropylar chamber develops into a haustorium (Fig. 9.7). In some species like, *Ruellia, Eranthemum,* and *Barleria* (Fig. 9.7); both chalazal and micropylar haustoria are present. In these the first division in PEC is transverse and unequal as a rule, leading to a smaller cell at micropylar (upper) end and a larger cell (or basal cell) at chalazal end. The upper smaller cell divides further transversely to form a chamber in the centre and a cell at the micropylar end. The cell in the centre turns into the endosperm proper. The micropylar cell and basal cell may show free nuclear divisions, or may be ab *initio* cellular, and form the multinucleate or multicellular haustorium. The number of cells/nuclei in a haustorium varies among different taxa (Mohan Ram & Wadhi 1964).

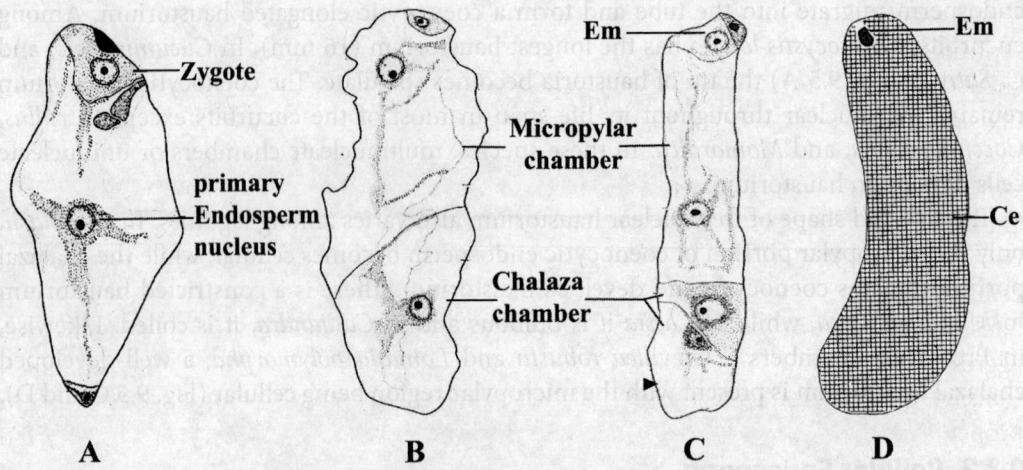

A   B   C   D

Figure 9.6 Cellular endosperm in *Drimys winteri*. A. Embryo sac with zygote and Primary endosperm nucleus migrates towards the centre of PEC. **B–D.** After first division in PEC two unequal chambers (micropylar and chalazal) can be recognized. The subsequent division in chalazal chamber is also unequal and lowermost cell gets degenerated (arrowhead). The smaller cell derived alone contributes in the formation of cellular endosperm (ce). **Em:** Embryo. (*Source:* Bhandari & Venkataraman 1968, *reproduced with permission.*)

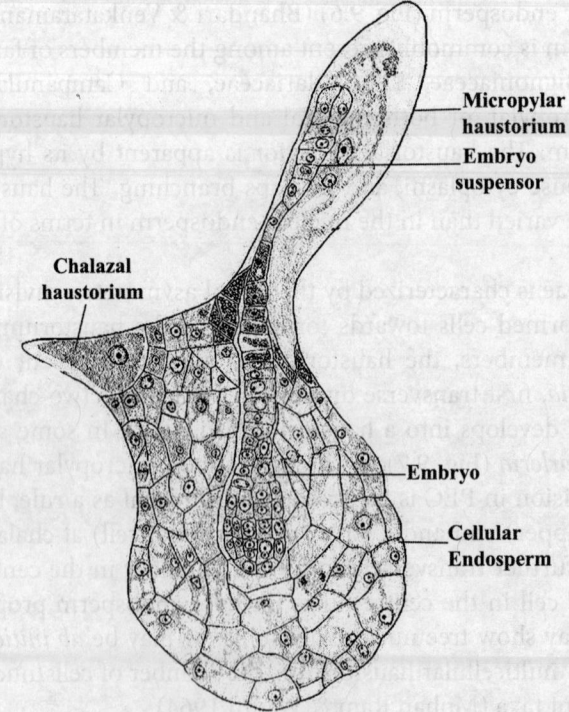

Figure 9.7 Endosperm in *Barleria cristata* (Acanthaceae) showing both chalazal and micropylar haustoria. (*Source:* Mohan Ram & Wadhi 1964, *reproduced with permission.*)

Aggressive and branched micropylar haustoria are present in *Impatiens roylei* and *Hydrocera triflora*. In *Impatiens roylei,* the first division of PEC gives rise to two chambers: smaller micropylar and larger chalazal. The smaller cell undergoes a series of transverse divisions and the uppermost cell of the tier forms the haustorium. The other cells contribute to endosperm (Maheshwari 1950).

The chalazal haustoria can be seen in *Magnolia, Iodina rhombifolia, Comandra umbellata,* and *Cansjera rheedii*. In *Magnolia obovata,* the first transverse division in PEC gives rise to two chambers of equal size. The divisions in the micropylar chamber are fast and form the cellular endosperm. The chalazal cell divides once and elongates into a tail-like 2-celled haustorium (Fig. 9.8). The haustoria in *Iodina rhombifoila,* and *Cansjera rheedii* are more aggressive, much branched hypertrophied structure, and exhibit further proliferations. In *Cansjera rheedii,* after the first division in PEC; the micropylar chamber forms the endosperm proper. The chalazal cell forms the aggressive haustorium which grows until it reaches to the base of the ovary (Fig. 9.9 A–B). However, in *I. rhombifoila* the chalazal haustorium originates from the embryo sac even before fertilization. The chalazal end of embryo sac extends and forms a caecum like structure which penetrates into the placenta. After the first transverse division in PEC, the chalazal cell nucleus becomes hypertrophied and enters into caecum. The micropylar chamber forms the endosperm proper (Fig. 9.9 C–D).

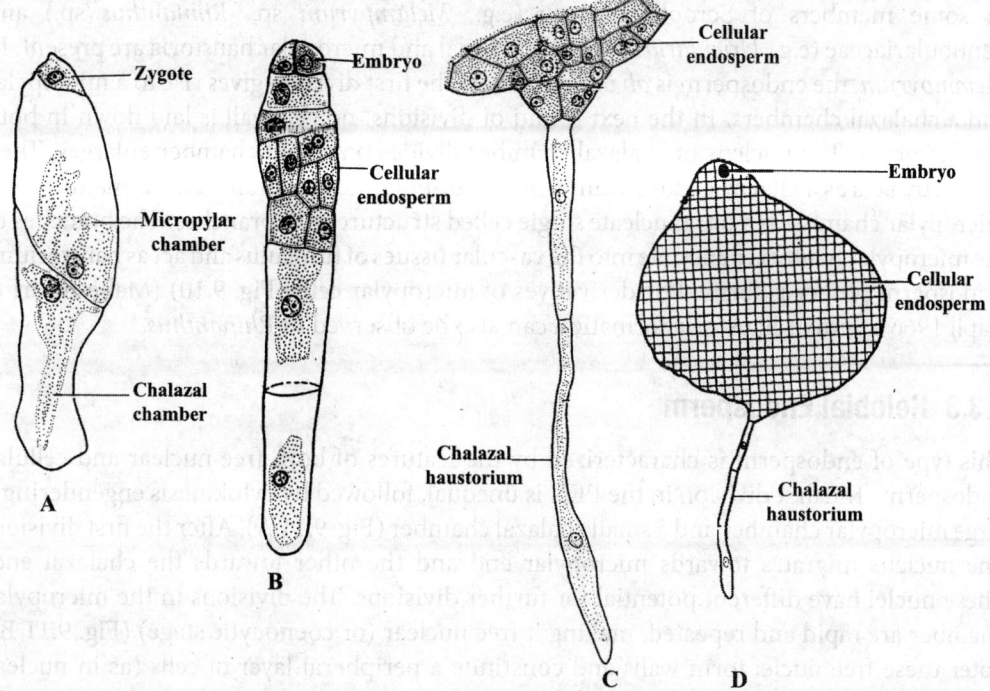

Figure 9.8 Endosperm and its chalazal haustorium development in *Magnolia obovata*. **A.** First division in primary endosperm cell forms two chambers – a micropylar and a chalazal. **B–D.** The cellularized endosperm is formed by micropylar chamber. Chalazal chamber (cell) divides only once and gives rise to a two celled tail like haustorium. **D.** Mature cellularized endosperm. (*Source:* After Kapil & Bhandari 1964, *reproduced with permission.*)

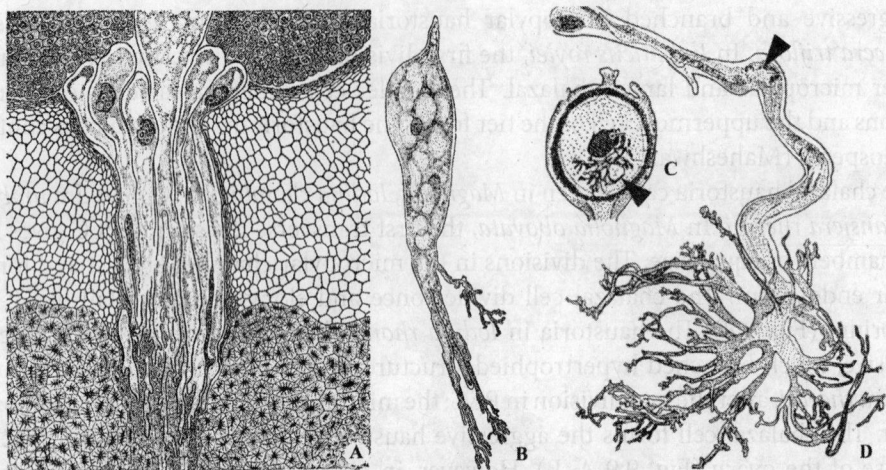

Figure 9.9 Endosperm with aggressive haustorium. A–B. *Cansjera rheedii* A. A longitudinal section (LS) of lower part of fruit showing the primary and secondary haustoria. B. Dissected primary haustorium. **C–D.** *Iodina rhombifoila.* C. LS of fruit showing aggressive haustorium (arrowhead). D. Enlarged view of haustorium with numerous branches and hypertrophied nucleus (arrowhead). (*Source:* A–B: After Swamy 1960, C–D Bhatnagar & Sabharwal 1969, *reproduced with permission.*)

In some members of Scrophulariaceae (e.g., *Melampyrum* sp., *Rhinanthus* sp.) and Lentibulariaceae (e.g., *Uricularia* sp.); both chalazal and micropylar haustoria are present. In *Melampyrum*, the endosperm is *ab initio* cellular. The first division gives rise to a micropylar and a chalazal chambers. In the next round of divisions, no cell wall is laid down in both the chambers. The nucleus of chalazal chamber divides once and chamber enlarges. Then it directly acts as a chalazal haustorium which is bi-nucleate and branched. At maturity, the micropylar chamber is a tetranucleate single celled structure with branches. The branches of the micropylar hautoria penetrate into the vascular tissues of funiculus and act as haustorium. Endosperm proper is formed by derivatives of micropylar cells (Fig. 9.10) (Maheshwari & Kapil 1966). Similar haustoria formation can also be observed in *Rhinanthus*.

9.3.3 Helobial Endosperm

This type of endosperm is characterized by the features of both free nuclear and cellular endosperm. The first division in the PEN is unequal, followed by cytokinesis engendering a large micropylar chamber and a small chalazal chamber (Fig. 9.11 A). After the first division, one nucleus migrates towards micropylar end and the other towards the chalazal end. These nuclei have different potential for further divisions. The divisions in the micropylar chamber are rapid and repeated, making it free nuclear (or coenocytic stage) (Fig. 9.11 B). Later these free nuclei form walls and constitute a peripheral layer of cells (as in nuclear endosperm). After the completion of cell wall formation, centripetal growth occurs and results in an enormous endosperm (e.g., *Asphodelus* sp.; Figure 9.11 C; Eunus 1952). The nucleus of the chalazal chamber either does not divide or if it divides, it remains coenocytic and finally degrades (as in *Eremurus* sp.). Rarely the chalazal chamber becomes cellularized, e.g., in *Trillium undulatum*. In some incidences such as in *Enalusa coroides, Asphodelus* sp. the nucleus in the chalazal chamber becomes hypertrophied and later degenerates.

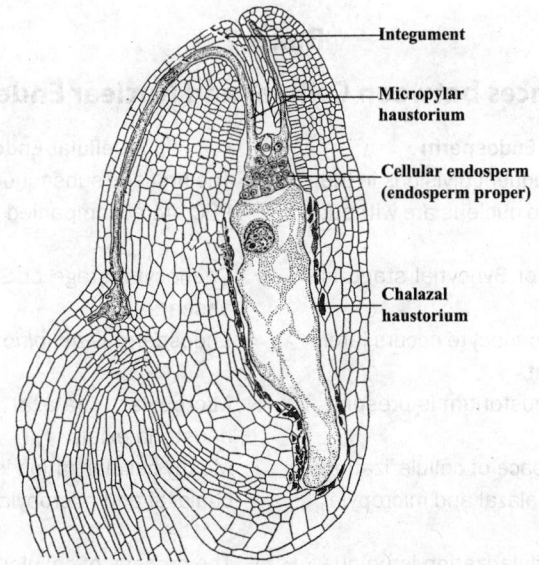

Figure 9.10 Longitudinal section of young seed showing chalazal and micropylar haustoria of endosperm in *Melampyrum*. (*Source*: Maheshwari & Kapil 1966, *reproduced with permission*.)

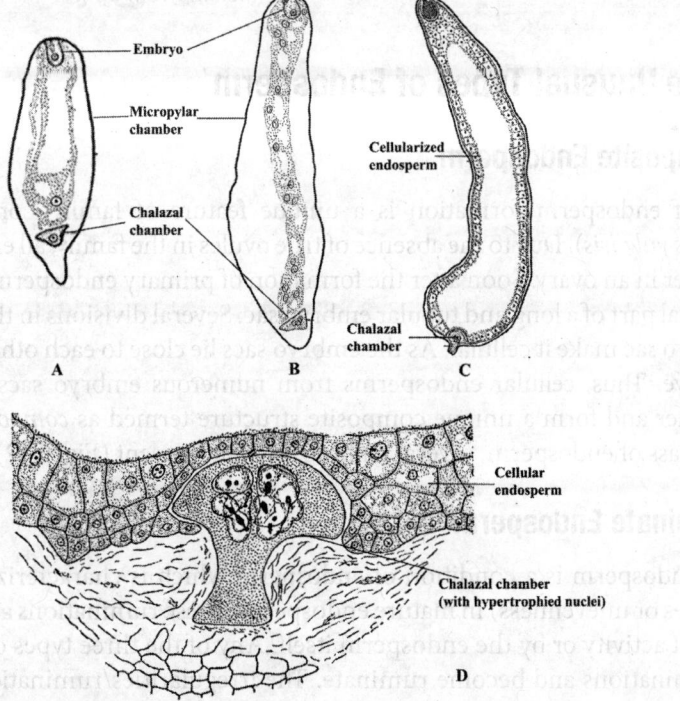

Figure 9.11 Development of helobial endosperm in *Asphodelus tenuifolius*. (*Source*: https://learning. uonbi.ac.ke/courses/SBT403/scormPackages/path_2/lecture_5_endosperm_ development.html.)

Box 9.2

Differences between Cellular and Nuclear Endosperm

Nuclear Endosperm	Cellular Endosperm
• The first and subsequent divisions in the primary endosperm nucleus are without cytokinesis.	• The first and subsequent divisions in the PEC are accompanied by cytokinesis.
• Coenocytic stage or Syncytial stage is present.	• Coenocytic stage or Syncytial stage is absent.
• Cellularization in coenocyte occurs later in the development.	• Endosperm is *ab initio* cellular.
• Mostly chalazal haustorium is present.	• Micropylar, chalazal or both type of haustoria exist.
• The distinction in pace of cellularization can be made at chalazal and micropylar poles.	• Mostly cellularization is evenly paced at chalazal and micropylar poles.
• The process of cellularization is unique due to the presence of radial microtubular system and inactive phragmoplast.	• The process of cellularization is simple.
• Examples: Arecaceae (*Cocos nucifera*), Cucurbitaceae (*Cucumis* sp.)	• Examples: Acanthaceae (*Thunbergia* sp., *Ruellia* sp.), Scrophulariaceae (*Melampyrum* sp.).

9.4 Some Unusual Types of Endosperm

9.4.1 Composite Endosperm

This type of endosperm formation is a unique feature of family Loranthaceae (except *Struthanthus vulgaris*). Due to the absence of true ovules in the family, all embryo sacs remain close together in an ovary. Soon after the formation of primary endosperm nucleus, it moves to the chalazal part of a long and tubular embryo sac. Several divisions in the chalazal portion of the embryo sac make it cellular. As the embryo sacs lie close to each other, their separating walls dissolve. Thus, cellular endosperms from numerous embryo sacs which eventually come together and form a unique composite structure termed as *composite endosperm*. In this fused mass of endosperm, several embryos remain present (Fig. 9.12).

9.4.2 Ruminate Endosperm

Ruminate endosperm is a condition of endosperm which is characterized by rumination (irregularities or unevenness) in mature endosperm. These ruminations are caused by either the seed coat activity or by the endosperm itself. Any of the three types of endosperm may develop ruminations and become ruminate. The irregularities/rumination caused by seed coat may be due to (a) an excessive in-growth or invaginations in the seed coat as observed in members of Annonaceae, Vitaceae, Aristolochiaceae; or (b) unequal radial elongation of the cells of one or more layers of seed coat, e.g., *Passiflora*, *Coccoloba uvifera*, *Diospyros*, *Myristica fragrans* (Fig. 9.13).

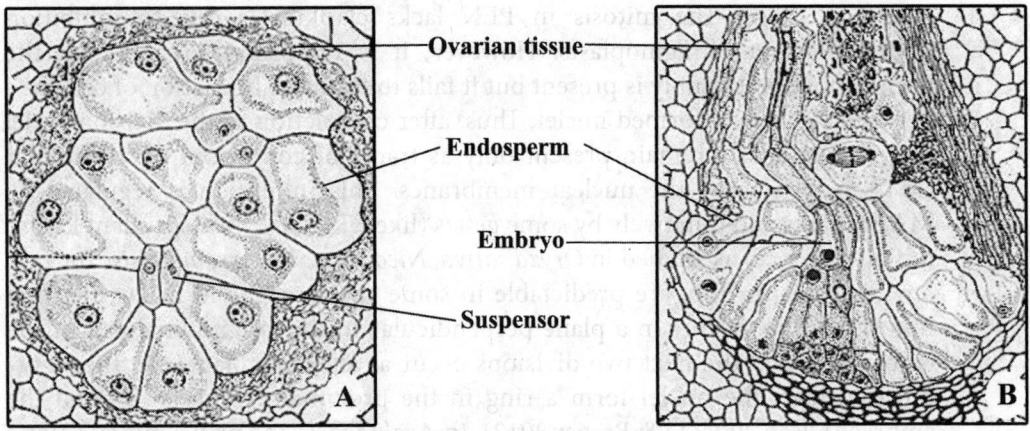

Figure 9.12 Composite endosperm in Loranthaceae members. A. *Amyema* sp. B. *Tolypanthus* sp. Note that the suspensor is biseriate. (*Source*: Dixit 1958, *reproduced with permission*.)

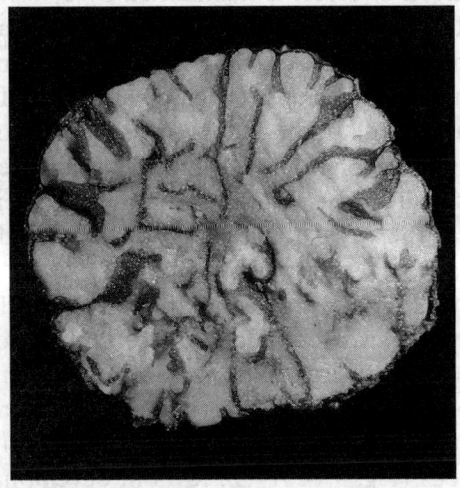

Figure 9.13 Rumination in endosperm of nutmeg (*Myristica fragrans*).

9.5 Cellularization of Nuclear Endosperm

The process of cellularization in nuclear endosperm is a variation of conventional cytokinesis among angiosperms. The notable differences are the cytoskeletal arrangement and alveoli formation (Olsen 2001; Li & Berger 2012). During the development of free nuclear endosperm, the following four phases can be observed: 1. Syncytium stage; due to inhibition of phragmoplasts functioning. 2. Arrest of mitotic division in nuclei. 3. Arrangement of cytoskeleton as a radial micro-tubular system (RMS) and alveoli formation. 4 Mitosis and cell layer development. The description of these phases is as follows:

1. ***Synctium Stage***: The mitosis in PEN lacks cytokinesis due to inhibition of functions of phragmoplasts. However, it is important to note that the phragmoplasts assembly is present but it fails to facilitate formation of cell plate between two newly formed nuclei. Thus, after completion of the division cycle, the phragmoplasts remain present only as traces of condensed micro-tubular (MT) material over the nuclear membranes. This inhibition is regulated by MAP kinases and putatively by some genes (like *KRP3*, cyclin-dependent kinase inhibitor in rice) as studied in *Oryza sativa*, *Nicotiana*, and *Arabidopsis*. The first few divisions in PEN are predictable in some species. Like in maize, the first mitosis in PEN occurs in a plane perpendicular to the longitudinal axis of the embryo sac and the next two divisions occur at alternate planes. At the eight-nuclear stage, the nuclei form a ring in the proximal cytoplasm around the embryo (Olsen 2004; Li & Berger 2012). In *Arabidopsis*, the nuclei migrate along the micropylar-chalazal pole.

2. ***Arrest of Mitotic Divisions***: After a period of syncytial divisions and nuclear migration, cell division arrests for some time before the process of cellularization initiates. This period of arrest is hallmarked by cytoskeletal rearrangement for cellularization. The duration of this cessation period varies among species, e.g., it is 2–3 days in maize and *Arabidopsis* and 6 days in barley. The number of nuclei at which mitotic divisions stop also varies in different species, suggesting a purely genetical control. In the endosperm coenocyte of maize and *Arabidopsis*, cellularization and cytoskeletal rearrangement initiates after eight rounds of mitotic divisions while in *Helianthus*, it commences even when less than ten free nuclei are present. The arrest of mitosis is attributed to the RETINOBLASTOMA proteins (Rb) and several transcription factors.

3. ***RMS and Alveoli Formation***: Before the cellularization, enlargement of the vacuole pushes the nuclei to the boundary of central cell coenocyte. This movement of nuclei is followed by the subdivision of the endosperm coenocyte into cellular compartments or endosperm cellularization. It is initiated by the formation of a Radial Microtubule System (RMS) originating from the membrane of each endosperm nucleus (Fig. 9.14 A–C). RMS in developing nuclear endosperm is well observed in various species (wheat, *Arabidopsis*, *Phaseolus*). At this stage each nucleus remains surrounded by a mass of cytoplasm and its boundary is delimited by a dense cortical array of cytoskeleton. Such a unit is known as a *nucleo-cytoplasmic domain* (NCD, olsen 2001; 2004). RMSs from sister nuclei extend and initially overlap, followed by appearance of cytoplasmic phragmoplasts that deposit anticlinal cell walls. These walls surround each nucleus from all the sides except at the side facing the central vacuole (Fig. 9.14 D). Such structure is known as alveoli, or a tube-like walled structure. The cells may have hexagonal or pentagonal shaped boundaries. For the formation of cell plates between sister nuclei, a complex mechanism based on vesicles derived from the endoplasmic reticulum has been proposed.

The next step is the extension of the alveoli. The alveolus extends toward the central vacuole (i.e., in a centripetal manner). The microtubular arrays (MT) (in barley and rice) undergo dramatic reshaping, assuming a tree-like shape. These tree-like MTs clasp the elongated nucleus to the central cell wall as well as send a canopy of microtubules up towards the central vacuole near the upper end of the alveolus (Olsen 2004; Fig. 9.14 E). Role of *Formins*, as actin polymerization-regulating factors has been described for this RMS arrangement and endosperm cellularization.

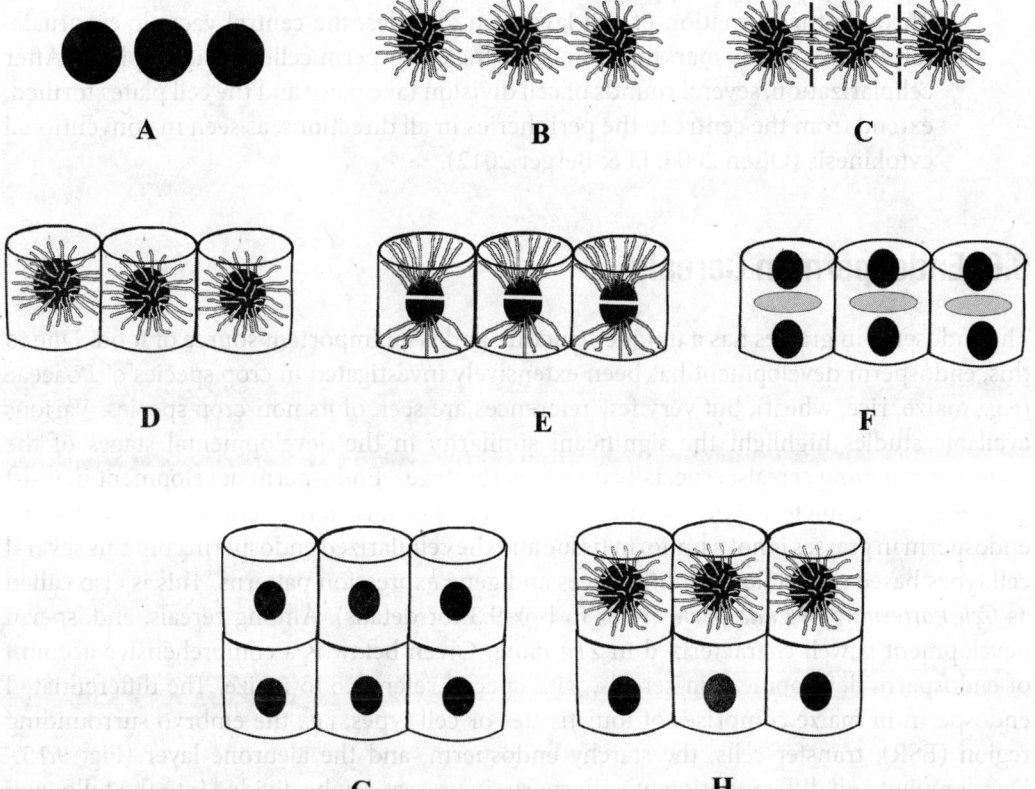

Figure 9.14 Diagrammatic representation of cellularization in free-nuclear endosperm of cereals and *Arabidopsis*. **A.** Free nuclei of syncytium/coenocyte lie along the central cell wall due to appearance of a large vacuole. **B. Radial microtubule systems (RMS)** originate from each nucleus, extend, and then approach each other. **C.** Appearance of cytoplasmic phragmoplasts at the zones between two opposing RMS of sister nuclei. Later cell plate formation occurs here, represented as solid black lines. **D. Alveoli formation:** It occurs due to wall formation by cytoplasmic phragmoplasts which surround each nucleus (except at centripetal wall towards vacuole). **E. Alveoli extension:** Nuclei remain anchored to the central cell wall at one end due to microtubule array and extend towards another side (towards vacuole). The periclinal nuclear division occurs in the alveoli. **F.** Microtubular array forms adventitious phragmoplasts and deposit a cell wall perpendicular to the former central cell wall. This closes former alveoli and generates a second layer of alveoli **(G–H)**. (*Source:* Based on Olsen 2004, *reproduced with permission.*)

4. ***Mitosis and Cell Layer Development:*** The completion of alveoli formation is followed by synchronous rounds of mitosis in all alveolar nuclei. The plane of division is periclinal (i.e., oriented in parallel to the former central cell wall) (Fig. 9.14 F–H). The arrangement of cytoskeleton remains the same as it was in the syncytium. However, by this time the phragmoplast becomes functional and deposits a cell wall perpendicular to the former central cell wall (of sister nuclei). The formation of these periclinal cell walls results in a layer of complete cells on the periphery of the developing endosperm (or coenocyte), and an inner layer of alveoli with their openings towards the central vacuole (Fig. 9.14 G–H). The same process is subsequently repeated several times towards the central vacuole. Through the formation of cell layers on all sides, the central vacuole eventually diminishes in size marking the end of the endosperm cellularization phase. After cellularization, several rounds of cell division take place and the cell plates formed; extend from the centre to the peripheries in all directions, as seen in conventional cytokinesis (Olsen 2004; Li & Berger 2012).

9.6 Endosperm in Cereals

The endoserm in grasses has a unique structure and is an important source of food. Due to this, endosperm development has been extensively investigated in crop species of Poaceae (e.g., maize, rice, wheat), but very few references are seen of its non-crop species. Various available studies highlight the significant similarity in the developmental stages of the endosperm among cereals; especially during early stages. Endosperm development in most grasses is of the nuclear type and the mature endosperm is starchy and dry. However, the endosperm in grasses is not a uniform tissue and the cellularized endosperm contains several cell types based on cytological differences and gene expression patterns. This is also called as *Cell Patterning in Endosperm* (refer to Box 9.3 for details). Among cereals, endosperm development is well characterized in *Zea mays*. Given below is a comprehensive account of endosperm development in cereals, with special reference to maize. The differentiated endosperm in maize comprises of four tissues or cell types, i.e., the embryo surrounding region (ESR), transfer cells, the starchy endosperm, and the aleurone layer (Fig. 9.15). Development and differentiation of endosperm in grasses can be divided into the following phases: 1. Early development including double fertilization and syncytium formation, 2. Cellularization, 3. Differentiation of cell types, i.e., transfer cells, aleurone, starchy endosperm and embryo surrounding cell. It also includes, periods of mitosis and endo-reduplication, accumulation of reserve food compounds and maturation, 4. Programmed cell death (PCD), dormancy, and desiccation (Sabelli & Larkins 2009; Li & Berger 2012).

1. ***Early development including double fertilization and syncytium formation:*** PEN divides immediately and rapidly after triple fusion in grasses. The divisions in PEN are free nuclear (without cytokinesis) and synchronous leading to the formation of a syncytium. In maize, nuclear divisions in the syncytium generate up to 512 nuclei while in *Triticum* sp. and *Hordeum* sp., more than 2,000 syncytial nuclei

may be counted. These nuclei migrate towards the chalazal region of the embryo sac due of enlargement of the central vacuole, and get distributed at the periphery of the primary endosperm cell.

2. **Cellularization**: Like other species with nuclear type of endosperm, cellularization in cereals is through cytoskeletal arrangement and alveoli formation. In maize, after the first layer of cell formation on periphery, the cell wall formation occurs centripetally followed by synchronous and periclinal divisions, until the central cell cavity is completely filled with cells.

3. **Differentiation**:

a. **The Embryo Surrounding Region (ESR)**: The embryo surrounding region is represented by the cells that cover the region of the endosperm in which the embryo develops (Fig. 9.14 and 9.15). In maize, these cells are recognizable by their dense cytoplasmic contents, rich in small vacuoles, and with a complex membrane system. They play an important role in embryo nutrition by supplying the sugars, primarily through an apoplastic route. ESR also establishes a physical barrier between the embryo and the endosperm during seed development. The mechanism of differentiation of ESR is lesser known. However, in maize and barley at coenocyte stage, the cellularization starts around the embryo. This is attributed to a cytoplasmic arrangement of phragmoplasts.

b. **Transfer cells**: Several cell layers of the endosperm, near the placenta stop dividing and differentiate as transfer cells (TCs). These cells are very elongated with extensive cell wall invaginations and increased plasma membrane surface. TCs are responsible for the intake of maternally provided nutrients which are required for starch synthesis in the endosperm.

Transfer cells facilitate sugar and amino acid uptake by the endosperm. TCs are noticeable in maize, less distinguishable in wheat (*Triticum aestivum*) and barley (*Hordeum vulgare*), and are barely seen in rice. Studies have suggested high presence of aquaporins, sugar transporters, amino acid transporters, and ion channels in barley and maize endosperm TCs.

c. **Aleurone**: The aleurone layer covers the entire limit of the endosperm except for the region of transfer cells. The number of cell layers in aleurone varies from single (as in maize) to three (as in barley) to several layers (varieties of rice). This layer(s) is derived by the peripheral daughter cells of the syncytium or alveolar nuclei (*first layer of alveoli*). In maize, the aleurone layer consists of an estimated 250,000 cells, whereas barley has around 100,000 aleurone cells. The aleurone cells are cuboidal in section. These cells show marked presence of well-developed endoplasmic reticulum, a large number of mitochondria, dense and granular cytoplasm, and small vacuoles with inclusion bodies. The granular appearance is due to the presence of many aleurone grains. These cells may contain anthocyanins, which are responsible for the colorful grains of maize. The aleurone grains are basically storage structures that remain surrounded by lipid droplets. These grains have lytic vacuoles and protein storage vacuoles in them. Both kinds of vacuoles may contain two major types of inclusion bodies: *globoid bodies,* and *protein-carbohydrate bodies*. The globoid bodies contain a crystalline matrix of phytin,

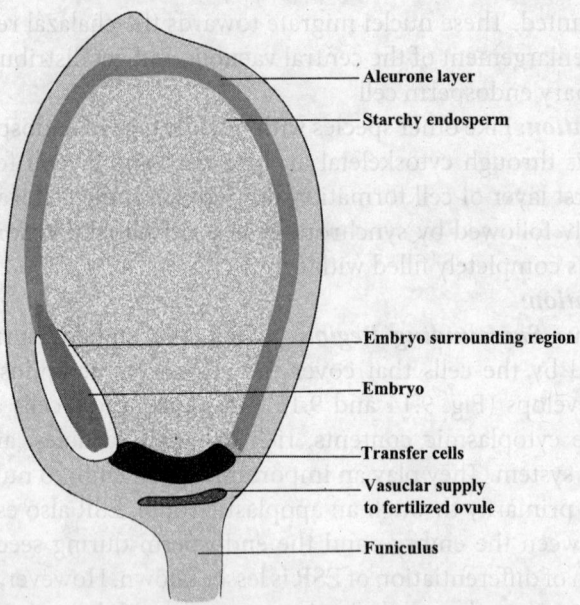

Figure 9.15 Diagram of longitudinal section of young maize kernel showing cellularized endosperm containing several cell types: the embryo surrounding region, transfer cells, the starchy endosperm, and the aleurone layer (Fig. 9.14). (*Source*: Based on Olsen 2001.) (*Refer to Box 9.3 and Fig. 9.16 for additional details.*)

protein and lipid. Accumulation of storage proteins in discrete cellular bodies is an adaptation that probably avoids their exposure to degrading enzymes (Sabelli & Larkins 2009). The aleurone is the only 'live' tissue in a mature endosperm, having a specific developmental program which protects it from desiccation (Hoecker et al. 1995).

d. **Starchy endosperm:** The central portion of the endosperm is composed of starchy endosperm cells which are rich in starch and gluten proteins. The cells of starchy endosperm differentiate into (i) two to three layers of protein rich and starch poor cells present below the aleurone layer called sub-aleurone cells, and (ii) the central starch rich cells of endosperm. In wheat an additional layer of prismatic cells between the sub-aleurone cells and central cells differentiates (Evers & Millar 2002). Starch grain accumulation starts soon after cellularization in the grasses. The transition from the cell division or the cellularization phase into the storage phase of endosperm development is accompanied by extensive reprogramming of gene expression patterns. During this phase genes encoding the enzymes of starch biosynthesis (ADP-glucose pyrophosphorylase (AGPase), Starch Synthase (SS), Starch-Branching (BE) and Starch-Debranching (DBE) enzymes) are expressed and accumulation of starch and other storage compounds occurs. Starch accumulation phase is associated with high metabolic activity and is energy limited (Sabelli & Larkins 2009). Starch is synthesized from sucrose after it is converted into glucose. Thus, starch accumulation is regulated by sucrose and the induction of sucrose synthase enzyme. High activity of invertase is also seen during the early formative phase in maize and sorghum. Invertase is required for cleavage of sucrose into glucose and fructose; to be transported into the endosperm by hexose transporters. Among the phytohormones, ABA signalling has also been indicated in the onset of starch biosynthesis.

4. *Maturation and desiccation:* The last stage of endosperm development is characterized by *Programmed Cell Death* (PCD). PCD facilitates nutrient hydrolysis and uptake by the embryo at germination. In maize, the PCD initially starts in starchy endosperm, while in wheat it is a random process. Irrespective of its initiation, PCD occurs in all cell types of endosperm except the aleurone layer. In grasses, a range of proteases possibly play a role in PCD. It is also governed by increased levels of ethylene and ABA signaling.

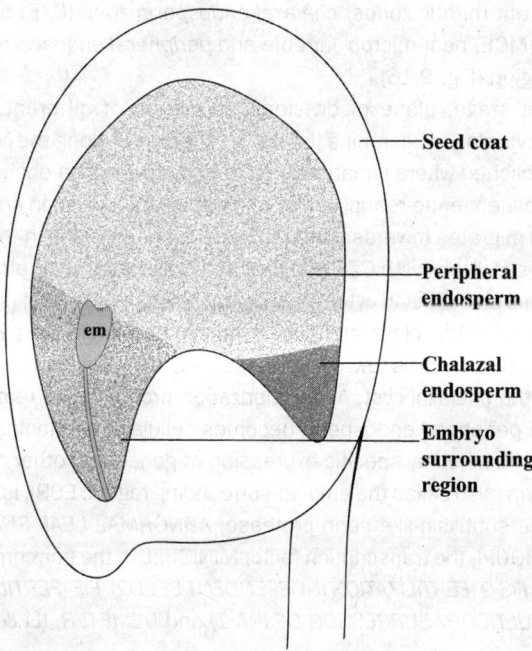

Figure 9.16 Diagram of a fertilized ovule of *Arabidopsis* showing differentiated zones in endosperm. (*Source*: Based on Li & Berger 2012.)

Box 9.3

Patterning in Endosperm

Ample evidences show that endosperm is not a formless and uniform tissue. It is rather quite apparent that cellularized endosperm comprises several cell types based on expression patterns and cytological differences. These cell types are arranged along the axis running from the anterior micropylar pole to the posterior chalazal pole. Other cell types are arranged along the radial symmetry. These two axes of patterning are conserved amongst flowering plants. A short description of these axes and cell types is as follows:

1. Polar Organization: An antero-posterior axis or micropylar-chalazal axis (MC axis) already exists in the embryo sac which is retained in endosperm as well. In many species (*A. thaliana* and maize) this axis becomes curved due to curvature of ovule, such that the micropyle and the chalaza become positioned side by side. The first two divisions in PEN take place along the MC axis of the endosperm. The third division is perpendicular to the MC axis. This is followed by establishment of different mitotic zones: *chalazal endosperm zone* (CZE) near the chalazal pole, *micropylar endosperm* (MCE) near micropylar pole and *peripheral endosperm* in the central part or embryo surrounding region (Fig. 9.16).

During the syncytial stage, plane of divisions, formation of different mitotic domains and organization of nucleo cytoplasmic domains (NCDs) is differential along the MC axis. The CZE is the first domain to be established where variations in mitotic activities are observed. The 2-4 nuclei at chalazal pole show prominent endo-reduplication followed by their division and finally form a cyst (of indefinite ploidy), which migrates towards chalaza. The nuclear divisions in the micropylar pole cells no longer take place in synchrony with CZE and are faster. Cellularization also initiates early in this region. The three domains also differ in cytological organization of the cytoskeleton. In CZE, posterior-directed migration of NCDs takes place and each syncytial division in the peripheral endosperm is followed by gradual migration of NCDs towards the cyst. Theses NCDs fuse to generate multinucleate nodules and fuse with the chalazal cyst. As cellularization proceeds, all remaining NCDs fuse into the cyst, and the entire peripheral endosperm becomes cellular. The identity of the three domains along the MC axis is established by specific expression of genes and other markers. In *A. thaliana*, the micropylar endosperm (also called the embryo-surrounding region, ESR), is marked by the sucrose transporter AtSUC5, the subtilisin-like serin protease, *ABNORMAL LEAF SHAPE 1* gene, the Basic helix loop helix factor ZHOUPI, the transcription factor MINISEED3, the polycomb group (PRC2) genes, including *MEA (MEDEA), FIS 2 (FERTILIZATION INDEPENDENT SEED2), FIE (FERTILIZATION INDEPENDENT ENDOSPERM), MSI1 (MULTICOPY SUPRESSOR OF IRA 1)* and *DEMETER*, (Li & Berger 2012).

2. Radial organization: Radial organization is more apparent in cereals and has been studied in *A. thaliana* among dicots. Radial organization is characterized by the differentiation of the outer most layer of endosperm after cellularization and subsequent cell divisions. In cereals, the outer endosperm cell layer undergoes a specific series of events leading to its differentiation into a specialized layer, the aleurone layer. In maize, the *DEFECTIVE ENDOSPERM KERNEL 1* (*DEK1*), encoding a cysteine proteinase (Lid et al. 2002), and *CRINKLY 4* (*CR4*) encoding a leucine-rich repeat kinase, are important for aleurone differentiation. The function of these genes is controlled or regulated by a gene *SUPERNUMERARY ALEURONE LAYER 1* (SAL1). Recent phylogenetic analysis suggests that Dek1 has different functions in dicots and monocots. In *A. thaliana DEK1* may have a general role in cell differentiation (outer most layer of endosperm) but in monocots *DEK1* functions for aleurone differentiation. However, it is undetermined whether the aleurone layer in the cereal endosperm is equivalent to the outer endosperm layer in *Arabidopsis*. The successive divisions in outer cell layer fill the embryo sac cavity in radial fashion and form starchy endosperm. The differentiation of transfer cells is also a part of radial organization (Li & Berger 2012).

Box 9.4

Endosperm: Reservoir of Nutrients

Persistent endosperm in various species is a reservoir of many biomolecules including starch, proteins and lipids. These serve as a source of nutrition for the humans and livestock. The synthesis and accumulation of starch and proteins in cereal endosperm is most significant, as the cereals are the vital source of carbohydrates and proteins. The morphology of starch grains plays an important role in grain digestibility and industrial applications. Within family Poaceae, two types of starch grains are present. The compound granules which develop as 'multi-granules' within one amyloplast, (A-type) are small and polyhedral and are quite common type in *Oryza sativa* (rice) and *Avena sativa* (oat). The simple granules grow as only one large granule in one amyloplast; and are typical of most species of subfamily Panicoideae (B-type). The tribe Triticeae appears to be unique in showing a so-called bimodal form of starch accumulation, with both large A-type, and small B-type granules. Differences in the structure of A-type and B-type starch grain lead to variations in swelling, gelatinization and pasting properties of the flour.

In general, two major types of seed storage proteins occur in all the seed-forming species: the globulins and certain albumins. The storage proteins in cereals endosperm are quite variable and responsible for the cohesive and visco-elastic properties of the dough made from endosperm flour. These properties are essential for making bread, pasta and other baking stuff. The chief storage proteins in cereals endosperm are prolamins (highly hydrophobic and soluble in alcoholic solutions or denaturing solvents) and globulins (soluble in saline solutions). Additional proteins like α-amylase inhibitor, urease, and lectins too accumulate in minute quantities.

Two basic types of prolamins are present in the cereal family: (1) those found in Triticeae, which are closely related and comprise monomeric gliadins and polymeric glutenins in wheat, hordein in barley, and secalins in rye, and (2) those found in Panicoideae, which include zeins of maize and the related proteins in sorghum (kafirins), millet. Prolamins are rich in proline and glutamine and are generally deficient in essential amino acids lysine and tryptophan (charged amino acids). High amount of prolamins are present (~50% to 60% of total endosperm proteins) in *Hordeum vulgare* (barley), *Pennisetum* (millet), *Secale* (rye), *Sorghum bicolor* (sorghum), *Triticum* (wheat), and *Zea mays* (maize). In contrast, only 5% to 10% of endosperm proteins in *Oryza sativa* (rice) and *Avena sativa* (oat) are prolamins and the bulk of protein is made up of globulins.

Oilseed crops are among the most precise examples of species storing large amounts of oil in the endosperm. Species storing oil in the endosperm are *Ricinus communis* (castor bean), *Elaeis guineensis* (oil palm), *Papaver somniferum* (opium poppy) and *Euphorbia* which store up to 50% oil to the seed dry weight. These reserves are packaged into discrete organelles known as lipid bodies. Oil is accumulated in spherosomes, or oleosomes, and usually consists of triacylglycerols surrounded by a half-unit phospholipid membrane and few proteins. Many seeds also store complex polysaccharides such as mannans and xyloglucans in the endosperm cell walls, e.g., *Phoenix dactylifera* (oil palm), and *Phytelephas macrocarpa* (Ivory nut) have mannans as the major storage carbohydrate (near 79% of the endosperm dry weight). The impregnation of the cell walls with these compounds makes endosperm very hard.

Source: Lopes & Larkins 1993; Sabelli & Larkins 2009.

9.7 Hormonal Regulation of Endosperm

The endosperm is a known source of phytohormones, many of which have been characterized. It is also known that these phytohormones indirectly influence the embryo and seed growth. But the exact role which they are playing in endosperm itself is still not understood.

- **Cytokinins:** The abundance of cytokinin in early endosperm development was recognized nearly half a century ago. However, its origin and functions in this tissue are still unclear. Cytokinins have been well identified in the maize (Zeatin, a natural cytokinin) and coconut endosperm. Use of liquid syncytial endosperm of *Cocos nucifera* (Coconut) in *in vitro*/tissue cultures is known to increase the cell division and somatic embryo differentiation. In *Arabidopsis*, the expression of cytokinin-biosynthetic genes has been detected in the chalazal part of endosperm (or posterior pole), suggesting a localized production of cytokinin. The high level of cytokinin activity found in early developing seeds indicates that cytokinins could be involved in the growth of the seed and also control seed mass and yield. However, multiple mutant analyses suggest bigger seeds are produced by a combination of mutations that block cytokinin signaling. Mutants affected in cytokinin signaling also display many other defects like impaired ovule development, and lesser number of ovules eventually leading to reduction in the number of seeds per silique. Thus, increase in seed size in such mutants can be attributed to more maternal reserves allocated per seed due to a smaller number of ovules as compared to wild types.
- **Abscisic acid (ABA):** Near the seed maturation stage in *Arabidopsis*, functional analysis of abscisic acid (ABA)-biosynthetic genes have shown that the endosperm also contributes to the production of ABA along with embryo, which together help in inducing seed dormancy. This is also highlighted that ABA operates along with the dynamic balance of gibberellic acid.
- **Auxins and other hormones:** The role of auxins has been investigated in maize endosperm where it controls cell division. The production of auxins and their transporters has been localized in the maize ESR and TCs. The studies also indicate a potential gradient of auxin activity along the MC axis in the endosperm. Recently, Batista et al. (2019) showed that increased auxin biosynthesis in the *Arabidopsis* endosperm prevents its cellularization and thus affects seed development. Reducing auxin biosynthesis and signaling re-establishes endosperm cellularization and seed viability. Auxin efflux from the endosperm has also been shown to be necessary for the post-fertilization differentiation of maternal tissues, which is necessary for efficient resource provision to the developing endosperm. Other than auxins, gibberellins along with the ethylene signaling pathway is required for proper differentiation of barley endosperm TCs.

9.8 Functions of Endosperm

Recent molecular studies have brought to light many more functions of endosperm other than provision of nutrition to growing embryo. Endosperm is also involved in regulation of embryo development and seed growth highlighting the interdependence and importance of embryo-endosperm relationship. The detailed functions of endosperm are as follows:

9.8.1 Nutrition of Embryo

1. As the embryo sac has little nutrition for the developing embryo, it is the endosperm that stores and provides nutrition to the embryo up to its maturity. The position of the endosperm between the embryo and maternal tissue chalaza, clearly suggests that endosperm plays a role in maternal nutrient transfer to the embryo.

2. In cereals, the starchy endosperm contains food reserves which are used during seed germination and seedling establishment. Whereas in dicot seeds, the cellular endosperm is used as a nutrient source for embryo growth during seed development. *Arabidopsis* mutants that are defective in endosperm cellularization produce small dry seeds due to the lack of nutrients during embryonic growth (Sørensen et al. 2002).

3. In the absence of endosperm in families like Podostemaceae, structures like haustorium and nucellar plasmodium provide nourishment to the embryo. Occurrence of these structures also indicates that a key function of the endosperm is providing nutrition to embryo.

9.8.2 Regulation of Embryo Development

In most angiosperms, the zygote divides only after the formation of endosperm. Studies have shown that if endosperm aborts, the embryo growth also ceases, suggesting endosperm to be regulating the development of embryo. Some of the ways in which endosperm controls the development of embryo are listed below:

1. One important role that the endosperm plays during embryogenesis is formation of a continuous cuticular layer between the developing embryo and the endosperm. This cuticle layer is required for preventing fusion of the embryo surface with surrounding endosperm tissues during seed development and also organ fusion after germination (Jeffree 2006). Such a cuticle also prevents seedlings dessication during germination. The region of the endosperm that directly surrounds the developing embryo (the embryo surrounding region or ESR) expresses several genes involved in cuticle formation: for example, the *ABNORMAL LEAF SHAPE1* (*ALE1*) is involved in cuticle formation in embryos of *Arabidopsis* and *Zea mays*. *ALE1* codes for a subtilisin-like serine protease and this is expressed predominantly in the ESR. *ale1* mutants produce embryos that adhere to the endosperm during development and germinating seedlings are extremely prone to desiccation (Tanaka et al. 2001).

2. The transcription factor ZHOUPI (ZOU, also called Retarded Growth of Embryo1 or RGE1), is also seen exclusively in the endosperm of developing seeds (e.g., in *Zea mays*). After fertilization, ZOU is initially present uniformly in the endosperm, subsequently becoming restricted to the ESR together with *ALE1*. *zou* mutant embryos show severe defects in cuticle formation and in epidermal cell adhesion (Yang et al. 2008). This observation supports the theory that signaling from the endosperm is essential for embryogenesis.

3. A study by Wu et al. (2013) revealed that embryo-endosperm are not just limited by cuticle, but, an additional glycoprotein-rich sheath is also present at the embryo–endosperm interface (e.g., in *Solanum* sp., *Arabidopsis*). This is suggested to be necessary for the embryo to separate from gliding into neighboring endosperm tissues. The material that constitutes the sheath is deposited on the embryo surface towards the outside of cuticle. It is derived from the endosperm, and its production depends upon ZOU. Thus, indicating a complex molecular relationship between embryo and endosperm.

4. The endosperm is largely an ephemeral tissue that breaks down almost completely during seed development providing physical space (along with nutrients) to growing embryos. In *zou* mutants of *Arabidopsis*, lack of endosperm breakdown leads to a constraint on embryo expansion and dramatically reduces the final size of embryo.

5. Recent advances in seed biology have also shown that the endosperm is capable of sensing environmental signals to regulate the growth of the embryo.

9.8.3 Seed Development

1. Auxin efflux from the endosperm has been shown to be necessary for the post-fertilization differentiation and proliferation of maternal tissues, i.e., integuments to form seed coat. Thus, weak endosperm development may prevent proper seed coat elongation.

2. The growth of the seed (both in monocots and dicots) is coupled with the rapid growth of the endosperm. Thus, the endosperm regulates seed size by controlling the seed volume. Some genes (viz. *APETALA2* and *AINTEGUMENTA* in *A. thaliana*) and transcription factors, express synchronously in embryo and endosperm. Such studies have highlighted that these genes control endosperm growth which in turn affects the seed size. Another pathway through which endosperm controls seed size involves sucrose metabolism. The sucrose entering the seed is converted into hexose that accumulates in the endosperm, causing a rapid increase in endosperm volume. The hydrolysis of sucrose to hexose probably also contributes to a high-water potential leading to water uptake by the endosperm. Thus, leading to an overall increase in the seed size. In maize, cell wall invertases INCW1 and INCW2 have been characterized in TCs which ensure the cleavage of sucrose delivered from maternal tissues in the apoplast. Mutation in these invertases causes a reduction in endosperm and seed size.

3. Some studies suggest that a coordination between embryo-endosperm is necessary for seed growth and seed size. For example, in *A. thaliana,* gene *CLAVATA3/ Embryo Surrounding Region-Related 8 (CLE8)* is expressed in endosperm which in turn regulates the expression of gene *Wuschel-Like Homeo Box 8 (WOX8)* of embryo. Thus, together endosperm gene *CLE8* and embryo gene *WOX8* form a signaling module that helps in seed growth and controls overall seed size.

9.8.4 Seed Germination

1. The role of the endosperm in seed germination is well known in cereals. The analysis and prevalence of abscisic acid (ABA)-biosynthetic genes and gibberellic acids (GAs) show that the endosperm also helps in induction of dormancy in embryo and seed. The embryo synthesizes and secretes gibberellin into the aleurone layer of the endosperm which induces the synthesis of α-amylase in the aleurone layer. During germination, α-amylase produced by aleurone cells helps in remobilization of the reserves of the starchy endosperm. The hydrolyzed starch and proteins are then utilized during seed germination and seedling establishment.
2. In dicots like *Arabidopsis*, the remnants of endosperm in the form of a single layer of living cells is present at seed maturity. This endospermic tissue plays an active role in the regulation of seed germination. These endospermic cells accumulate lipids in the form of triacylglycerols (TAGs) which are catabolized into sucrose through gluconeogenesis during and after germination (Penfield et al. 2004).
3. Transcriptome analysis of endosperm has also provided new insights into the regulatory functions of the endosperm during seed germination. For example: Gibberellin-responsive element (GARE) is required for gibberellin-induced transcription in the aleurone layer during germination in cereals. In *Arabidopsis*, the transcription factors ABI3, FUSCA3 (FUS3) and LEAFY COTYLEDON2 (LEC2) express both in the embryo and endosperm and are regulators of seed maturation and germination (Kagaya et al. 2005; Mendoza et al. 2005).

9.9 Nutritive Tissues Other Than Endosperm

9.9.1 Pseudoembryo Sac/Nucellar Plasmodium

Absence of double fertilization and consequently endosperm is a unique feature of family Podostemaceae. In the absence of endosperm, a novel tissue develops in the ovules to provide nutrition to the growing embryo. Close to fertilization, the nucellar cells present below the embryo sac start disintegrating and form a cavity below the embryo sac where the cytoplasm and nuclei create a plasmodial mass. This plasmodial mass is known as *nucellar plasmodium* or *pseudo-embryo sac* which eventually gets consumed by the growing embryo (Fig. 9.17).

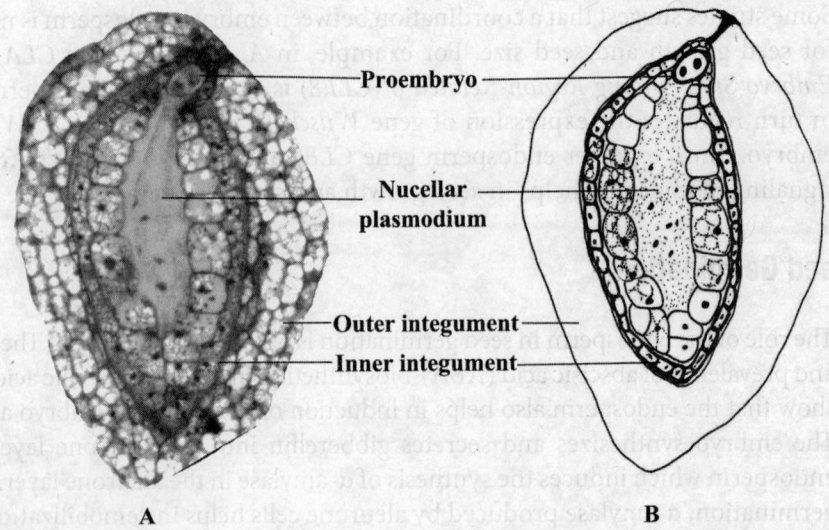

Figure 9.17 **A.** Longitudinal section of fertilized ovule of *Willisia arekaliana* (Podostemaceae) showing pseudo-embryo sac or nucellar plasmodium. **B.** Sketch of fertlized ovule with pseudo-embryo sac or nucellar plasmodium.

9.9.2 Perisperm

In most angiosperms, the nucellus is partially or completely consumed during megagametophyte development. However, in some families like Pipearceae, Nymphaeaceae, Zingerberaceae, Amaranthaceae, Chenopodiaceae, Cannaceae nucellus does not disintegrate and is persistent in the mature seed. Such a persistent nucellus in seed is known as Perisperm. In such seeds, nutrient acquisition, storage, and mobilization is done by perisperm and endosperm only serves as an interface between the embryo and the perisperm. Some examples of species showing perisperm are sugar beet, quinoa (*Chenopodium quinoa*), black pepper (*Piper nigrum*), *Agave,* and *Coffea arabica.* The term *perisperm* usually implies a massive, nutrient rich, non-vascularized tissue (chalazal and nucellar origin) present in fertilized ovules. However, in *Trithuria* sp. development of perisperm is precocious and a massive storage tissue is present in non-fertilized ovules which persists till seed matures (Rudall 1997; Rudall et al. 2008).

In *Piper nigrum* (Piperaceae), the major part of mature seed is perisperm which is also the economically important part of the seed. The remnants of endosperm surrounding the embryo occupy very small portion, situated laterally at the terminal part of the seed. Development of perisperm begins with rapid divisions in the nucellar cells below the embryo sac. These growing cells undergo fusion resulting in large composite cells, rich in cytoplasmic contents, oil globules and starch. In the mature seeds, the perisperm is distinguishable into an outer protein-rich zone and an inner starch-filled zone (Fig. 9.18). In *Chenopodium quinoa* (Chenopodiaceae), the endosperm in the mature seed is only 1–2 cell layers thick and is present as a remnant only. The perisperm is a spheroidal central mass of dead, thin-walled cells filled with angular starch grains (Prego et al. 1998).

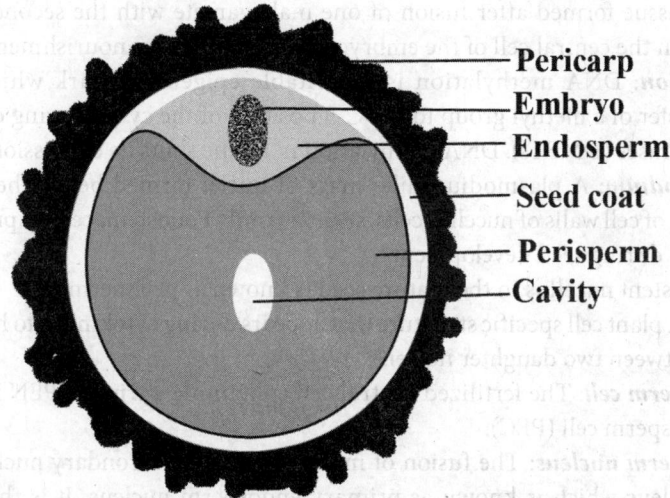

- Pericarp
- Embryo
- Endosperm
- Seed coat
- Perisperm
- Cavity

Figure 9.18 Diagram of a section of *Piper nigrum* fruit showing different parts. Note: perisperm occupies the major volume and contributes maximum to the seed mass.

9.9.3 Chalazosperm

Chalazosperm is a starch-rich, loosely packed tissue derived from the chalaza of the ovule, seen in mature seeds of *Cyanastrum* and *Canna*. In these plants, the endosperm is consumed during embryo development and chalazosperm accumulates starch and persists in mature seed as storage tissue.

Glossary

Cellular endosperm: Type of endosperm where divisions (karyokinesis) in primary endosperm nucleus are followed by wall formation.

Chalazosperm: An embryo nourishing tissue formed by divisions in chalazal nucellar cells after fertilization.

Coenocyte or synctium: A cell with several nuclei which remain floating in the common cytoplasmic mass. A coenocyte is formed when karyokinesis is not followed by cytokinesis.

Composite endosperm: Due to absence of true ovules in Loranthaceae members, the endosperm from many embryo sacs come together and fuse into what is called as composite endosperm.

CpG sites: The regions of DNA where a cytosine nucleotide is followed by a guanine nucleotide in the linear sequence of bases along its 5' → 3' direction, is known as CpG sites. When CpG sites occur with high frequency in genomic regions then these are called CpG islands (or CG islands).

Endosperm patterning: Pattern of gene expression and cytological differences seen in different cell types of a cellularized endosperm.

Endosperm: A tissue formed after fusion of one male gamete with the secondary nucleus (or polar nuclei) in the central cell of the embryo sac. It is meant for nourishment of the embryo.

DNA Methylation: DNA methylation is a heritable epigenetic mark which involves the covalent transfer of a methyl group to the C-5 position of the cytosine ring of DNA by DNA methyltransferases (DMETs). DNA methylation in a gene shuts its expression.

Nucellar plasmodium: A plasmodium like mass of nuclei formed below the embryo sac by disintegration of cell walls of nucellar cells, seen in family Podostemaceae. It provides nutrition to the embryo during seed development.

Perisperm: Persistent nucellus in the mature seed is known as perisperm.

Phragmoplast: A plant cell specific structure that appears during cytokinesis to help in formation of cell wall between two daughter nuclei.

Primary endosperm cell: The fertilized central cell containing a triploid PEN is also known as primary endosperm cell (PEC).

Primary endosperm nucleus: The fusion of male gamete with secondary nucleus gives rise to a triploid nucleus which is known as primary endosperm nucleus. It is the progenitor for endosperm.

Secondary nucleus: In most of the plant species the polar nuclei of central cell fuse before arrival of male gametes and form the secondary nucleus.

Key Questions

Q9.1   Match Column A with Column B:

Column A		Column B	
a.	Aggressive haustorium	I.	*Asphodelus* sp.
b.	Pseudo-embryo sac	II.	Loranthaceae
c.	Composite endosperm	III.	*Zea mays*
d.	Embryo surrounding region	IV.	Free nuclear endosperm
e.	Helobial endosperm	V.	Podostemaceae
f.	Coenocyte	VI.	*Iodina rhombifolia*
g.	Chalazal endosperm	VII.	*Piper nigrum*
h.	Perisperm	VIII.	*Arabidopsis*
i.	Free nuclear endosperm	IX.	*Antigogon*/ Palms
j.	Ruminate endosperm	X.	*Cocus nucifera*

Q9.2   Enumerate the differences between cellular and free nuclear endosperm.

Q9.3   Write notes on:
   a. Endosperm haustorium in Acanthaceae
   b. Physiological role of endosperm
   c. Nutritional role of endosperm
   d. Composite endosperm

e. Unique embryological features of family Loranthaceae
f. Nucellar Plasmodium
g. Endosperm in cereals
h. Endosperm patterning

Q9.4 Endosperm fulfills the nutritional or dietary requirements of mankind. Justify the statement with suitable examples.

Q9.5 Discuss the unique embryological features of family Podostemaceae with suitable diagrams.

Q9.6 Elaborate upon the distribution and development of the helobial type of endosperm with the help of an example and illustrations.

Q9.7 'Endosperm is a formless tissue'. The statement is true or false. Justify your answer.

Q9.8 Find the odd one out and justify your answer:
a. Perisperm, Cellular endosperm, Primary endosperm cell, Primary endosperm nucleus
b. Embryo surrounding region, Aleurone layer, Coenocyte, Transfer cells
c. Synctium, Alevoli, RMS, Cellular endosperm

Q9.9 Enumerate the functions of endosperm.

Practicals

Exercise 9.1 To dissect and study free-nuclear type of endosperm from developing seeds

Suggested Materials: Young seeds of Cucumber (*Cucumis sativus*); pods of *Senna occidentalis*.

Procedure

Cucumis sativus

1. Place a young (appx. 4–6 mm long) seed on a clean glass slide and add a drop of acetocarmine.
2. Place the slide on the stage of dissection microscope.
3. Insert a needle into the broad chalazal end of the seed (up to half the depth) and cut through the upper seed tissues by moving the needle towards the pointed micropylar end. This will open the seed into two longitudinal halves.
4. Gently scrap both halves with fine camel hair brush/needle as one of the halves contains the endosperm. The endosperm will slide in the stain.
   (*Alternately, the seed can be cut in half longitudinally, that is, parallel to the broader plane of the seed; with a blade.*)
5. Remove the extra seed tissues and place a coverslip on the endosperm.

Senna occidentalis

1. Place a seed (2–3 mm long) on glass slide and add a drop of acetocarmine stain.
2. Carefully peel off the black/brown seed coat from around the endosperm with the help of needles under a dissection microscope.

3. Split the seed into two halves and gently scrap both halves with fine camel hair brush/needle as one of the halves contains the endosperm. The endosperm will slide in the stain.
4. Place a coverslip on the endosperm.

Observations

Both in *Cucumis sativus* and *Senna occidentalis* the upper part of the endosperm (micropylar) is cellularized and lower part is a free-nuclear haustorium. In *Cucumis*, the haustorium is long, tail-like, narrow and with a spatulated end. In *Senna,* the haustorium is broad with finger-like projections at the end (Fig. 9.19).

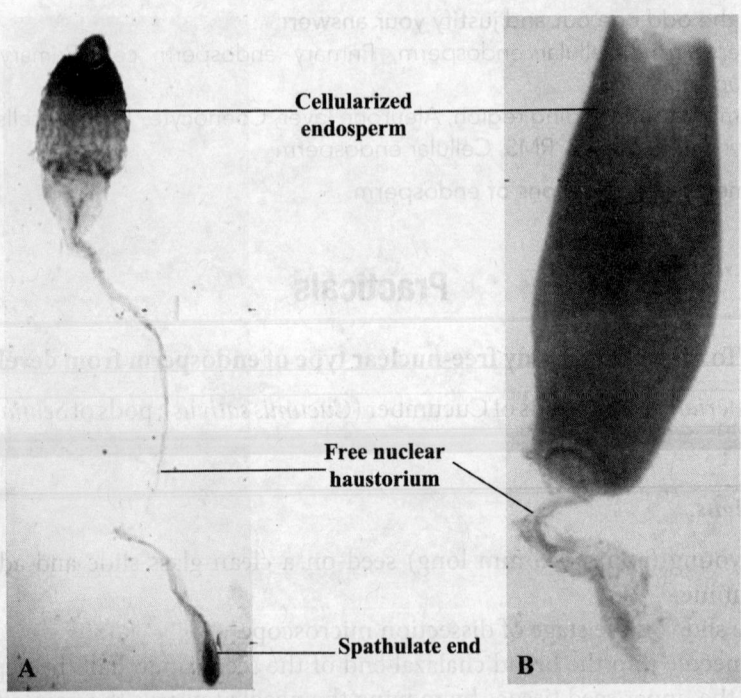

Figure 9.19 Whole mount of dissected endosperm A. *Cucumis sativus.* **B.** *Senna occidentalis.* **See Color Plates (page 491).** (*Source:* Courtesy of Dr P. Chitralekha.)

Bibliography

Arekal, G. D. (1963). Embryological studies in canadian representatives of the tribe Rhinantheae, Scrophulariaceae. *Canadian Journal of Botany* 41: 267–302.

Batista, R. A., Figueiredo, D. D., Santos-González, J. and Köhler, C. (2019). Auxin regulates endosperm cellularization in *Arabidopsis. Genes and Development* 33: 1–11.

Berger, F. and Chaudhury, A. (2009). Parental memories shape seeds. *Trends in Plant Science* 14: 550–6.

Bhandari, N. N. and Venkataraman, R. (1968). Embryology of *Drimys winteri. Journal of Arnold Arboretum* 49: 509–24.

Bhatnagar, S. P. and Sabharwal, G. (1966). Female gametophyte and endosperm of *Iodina rhombifolia* Hook. and Arn. *Phytomorphology* 16: 588–91.

Bhatnagar, S. P. and Sawhney, V. (1981). Endosperm–its morphology, ultrastructure and histochemistry. *International Review of Cytology* 73: 55–102.

Cabej, N. R. (2019). Plant Epigenetics. In *Epigenetic Principles of Evolution*. Cambridge, MA: Academic Press, pp: 733–81. doi:10.1016/b978-0-12-814067-3.00015-6.

Chopra, R. N. and Sachar, R. C. (1963). Endosperm. In P. Maheshwari, ed., *Recent Advances in the Embryology of Angiosperms*. Delhi: International Society of Plant Morphologists.

Chourey, P. S., Jain, M., Li, Q. B. and Carlson, S. J. (2006). Genetic control of cell wall invertases in developing endosperm of maize. *Planta* 223: 159–67.

Cossegal, M., Vernoud, V., Depege, N. and Rogowsky, P. M. (2007). The embryo surrounding region. *Plant Cell Monographs* 8: 57–71.

Dixit, S. N (1958). Morphological and embryological studies in the Loranthaceae. VIII. *Tolypanthus* (Bl) Bl considerably and migrates from the tip of the embryo sac. *Phytomorphology* 11: 335–45.

Eunus, A. M. (1952). Contributions to the embryology of Liliaceae. III. Embryogeny and development of the seed of *Asphodelus tenuifolius* Cav. *Lloydia* 15: 149–57.

Evers, T. and Millar, S. (2002). Cereal grain structure and development: some implications for quality. *Journal of Cereal Science* 36: 261–84.

Figueiredo, D. D., Batista, R. A., Roszak, P. J. et al. (2016). Auxin production in the endosperm drives seed coat development in *Arabidopsis*. *eLife* 5: pii: e20542.

Gehring, M., Bubb, K. L. and Henikoff, S. (2009). Extensive demethylation of repetitive elements during seed development underlies gene imprinting. *Science* 324: 1447–51.

Hoecker, U., Vasil, I. K. and McCarty, D. R. (1995). Integrated control of seed maturation and germination programs by activator and repressor functions of Viviparous-1 of maize. *Genes and Development* 9: 2459–69.

Huh, J. H., Bauer, M. J., Hsieh, T. F. and Fischer, R. (2007). Endosperm gene imprinting and seed development. *Current Opinion in Genetics and Development* 17: 480–5.

Ingram, G. C. (2020). Family plot: the impact of the endosperm and other extra-embryonic seed tissues on angiosperm zygotic embryogenesis. *F1000 Research*. doi.org/10.12688/f1000research.21527.1.

Jeffree, C. E. (2006). The fine structure of the plant cuticle. In M. Riederer and C. Muller, ed., *The Biology of Plant Cuticle*. Oxford: Blackwell, pp. 11–125.

Johri, B. M. and Kapil, R. N. (1953). Contribution to the morphology and life history of *Acalypha indica* L. *Phytomorphology* 3: 137–51.

Jullien, P. E. and Berger, F. (2009). Gamete-specific epigenetic mechanisms shape genomic imprinting. *Current Opinion in Plant Biology* 12: 637–42.

Kagaya, Y., Toyoshima, R., Okuda, R. et al. (2005). LEAFY COTYLEDON1 controls seed storage protein genes through its regulation of *FUSCA3* and *ABSCISIC ACID INSENSITIVE3*. *Plant Cell and Physiology* 46: 399–406.

Kapil, R. N. and Bhandari, N. N. (1964). Morphology and embryology of *Magnolia* Dill. Ex Linn. *Procedings of the Indian National Science Academy B.* 30: 245–62.

Kaushik, S. B. (1938). The endosperm in *Grevillea robusta* Cunn. *Current Science* 7: 332.

Kinoshita, T., Miura, A., Choi, Y. et al. (2004). One-way control of FWA imprinting in *Arabidopsis* endosperm by DNA methylation. *Science* 303: 521–3.

Kuriachen, P. M. and Dave, Y. (1989). Anatomical and histochemical changes in berries of *Piper nigrum* L. *Journal of Plant Biology* 32: 11–21.

Lersten, N. R. (2004). *Flowering plant embryology*. Oxford: Blackwell, Publishing Ltd.

Li, J. and Berger, F. (2012). Endosperm: food for humankind and fodder for scientific discoveries. *New Phytologist* 195: 290–305.

Li, N. and Dickinson, H. G. (2010). Balance between maternal and paternal alleles sets the timing of resource accumulation in the maize endosperm. *Proceedings of the Royal Society. B, Biological sciences* 277: 3–10.

Lid, S. E., Gruis, D., Jung, R. et al. (2002). The defective kernel 1 (dek1) gene required for aleurone cell development in the endosperm of maize grains encodes a membrane protein of the calpain gene superfamily. *Proceedings of the National Academy of Sciences USA* 99: 5460–5.

Liu, Y., Yan, Z., Chen, N. et al. (2010). Development and function of central cell in angiosperm female gametophyte. *Genesis* 48: 466–78.

Lopes, M. A. and Larkins, B. A. (1993). Endosperm origin, development, and function. *Plant Cell* 5: 1383–99.

Maheshwari, P. and Kapil, R. N. (1966). Some Indian contributions to the embryology of angiosperms. *Phytomorphology* 16: 239–91.

Maheshwari, P. (1950). *An Introduction to the Embryology of Angiosperms*. New York: McGraw–Hill.

Mendoza, M. S., Dubreucq, B., Miquel, M. et al. (2005). LEAFY COTYLEDON 2 activation is sufficient to trigger the accumulation of oil and seed specific mRNAs in *Arabidopsis* leaves. *FEBS Letters* 579: 4666–70

Miyawaki, K., Matsumoto-Kitano, M. and Kakimoto, T. (2004). Expression of cytokinin biosynthetic isopentenyltransferase genes in Arabidopsis: tissue specificity and regulation by auxin, cytokinin, and nitrate. *Plant Journal* 37: 128–38.

Mohan Ram, H. Y. and Wadhi, M. (1964). Endosperm in Acanthaceae. *Phytomorphology* 13: 82–91.

Nawaschin, S. (1898). Resultateeiner Revision der Befruchtungsvorgängebei *Lilium martagon* and *Fritillaria tenella*. *Bulletin of Academy of Imperial Sciences St. Petersburg* 9: 377–82.

Newcomb, W. (1978). The development of cells in the coenocytic endosperm of the African blood lily *Haemanthus katherinae*. *Canadian Journal of Botany* 56: 483–501.

Olsen, O. A. (2001). Endosperm development: cellularization and cell fate specification. *Annual Review of Plant Physiology and Plant Molecular Biology* 52: 233–67.

Olsen, O. A. (2004). Nuclear endosperm development in cereals and *Arabidopsis thaliana*. *Plant Cell* 16 (Suppl): S214–S227.

Penfield, S., Rylott, E. L., Gilday, A. D. et al. (2004). Reserve mobilization in the *Arabidopsis* endosperm fuels hypocotyl elongation in the dark, is independent of abscisic acid, and requires PHOSPHOENOLPYRUVATE CARBOXYKINASE1. *Plant Cell* 16: 2705–18.

Prakash, S. (1960). The endosperm of *Arachis hypogaea*Linn. *Phytomorphology*10: 60–4.

Prego, I., Maldonado, S. and Otegui, M. (1998). Seed structure and localization of reserves in *Chenopodium quinoa*. *Annals of Botany* 82: 481–8.

Rau, M. S. (1951). The endosperm in *Crotolaria*. *Current Science* 3: 73.

Rudall, P. J., Remizowa, M. V., Beer, A. S. et al. (2008). Comparative ovule and megagametophyte development in Hydatellaceae and water lilies reveal a mosaic of features among the earliest angiosperms. *Annals of Botany* 101: 941–56.

Rudall, P. J. (1997). The nucellus and chalaza in monocotyledons: structure and systematics. *Botanical Review* 63: 140–84.

Sabelli, P. A. and Larkins, B. A. (2009). The development of endosperm in grasses. *Plant Physiology* 149: 14–26.

Sørensen, M. B., Mayer, U., Lukowitz, W. et al. (2002). Cellularization in the endosperm of *Arabidopsis thaliana* is coupled to mitosis and shares multiple components with cytokinesis. *Development* 129: 5567–76.

Swamy, B. G. L. (1960). Contributions to the embryology of *Cansjera rheedii*. *Phytomorphology* 10: 397–409.

Tanaka, H., Onouchi, H., Kondo, M. et al. (2001). A subtilisin-like serine protease is required for epidermal surface formation in *Arabidopsis* embryos and juvenile plants. *Development* 128: 4681–9.

Vijayaraghavan, M. R. and Prabhakar, K. (1984). The endosperm. In B. M. Johri, ed., *Embryology of Angiosperms*. Springer-Verlag. pp 319–70.

Wu, C. C., Diggle, P. K. and Friedman, W. E. (2013). Kin recognition within a seed and the effect of genetic relatedness of an endosperm to its compatriot embryo on maize seed development. *Proceedings of the National Academy of Science USA* 110: 2217–22.

Yang, S., Johnston, N., Talideh, E. et al. (2008). The endosperm-specific ZHOUP1 gene of *Arabidopsis thaliana* regulates endosperm breakdown and embryonic epidermal development. *Development* 135: 3501–9.

10

Zygotic Embryogenesis

The organized divisions and specifications of cells during embryogenesis is called embryo patterning which results in a well-organized embryo from single celled structure the 'zygote'.

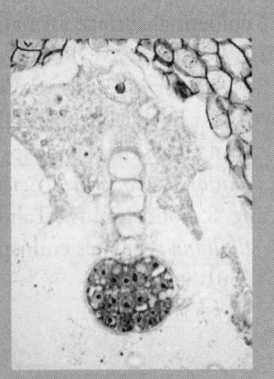

10.1 Introduction

Zygote is a unique cell from which the life cycle of angiosperms begins. It is a product of fertilization between the sperm and the egg cell. Zygote undergoes organized divisions and cell-specifications to give rise to an embryo, a young sporophyte. The process of development of an embryo from a zygote is known as **embryogenesis**. A particular pattern, form, and polarity exists during the development of embryo which is the outcome of several cellular, molecular and genetic mechanisms. These programmed changes enable the embryo to form the future sporophyte as a unique entity. The present chapter deals with the embryogenesis in angiosperms, encompassing embryo patterning, genetics and physiology involved. Embryos can also develop from somatic cells under *in vitro* conditions and resulting embryos are known as somatic embryos and the process as somatic embryogenesis. In this chapter, the term embryogenesis will be used for only zygotic embryogenesis.

10.2 Structure of the Embryo

Typical mature embryos in both monocots and dicots are similar in the basic design as they share similar embryogeny, at least up to a particular stage (i.e., octant, as will be discussed in Section 10.4); after which ontogenetic differences appear between them. A typical dicot embryo comprises of an embryonal axis with two cotyledons attached to it laterally. The part of embryonal axis above the level of cotyledons is known as the epicotyl; which terminates into the embryonic shoot (also called plumule). The part of embryonal axis below the level of cotyledons is known as the hypocotyl; which gives rise to an embryonic root (also called radicle) (Fig. 10.1 A). A typical monocot embryo differs from a dicot embryo in having only one cotyledon attached to the embryonal axis (Fig. 10.1 B).

The embryos in cereals like *Zea mays*, *Triticum* sp., *Oryza* sp. need special mention because of additional embryonic organs associated with them namely, scutellum, coleoptile and coleorhiza (Fig. 10.1 C). Scutellum is thought to be a seed leaf or a single massive cotyledon in cereals, which fully covers the embryo. Lateral to the scutellum, a short embryonal axis is attached which is divided into an epicotyl *(above the level of scutellum)* and a hypocotyl *(below the level of scutellum)*. The epicotyl comprises of several young leaves covered by a sheath called coleoptile. The hypocotyl consists of root apical meristem covered by coleorhiza, which develops a small outgrowth on one side, called the epiblast. In many plant species, the growing embryo is connected to the maternal tissue via a file of cells, known as the suspensor (Maheshwari, 1950).

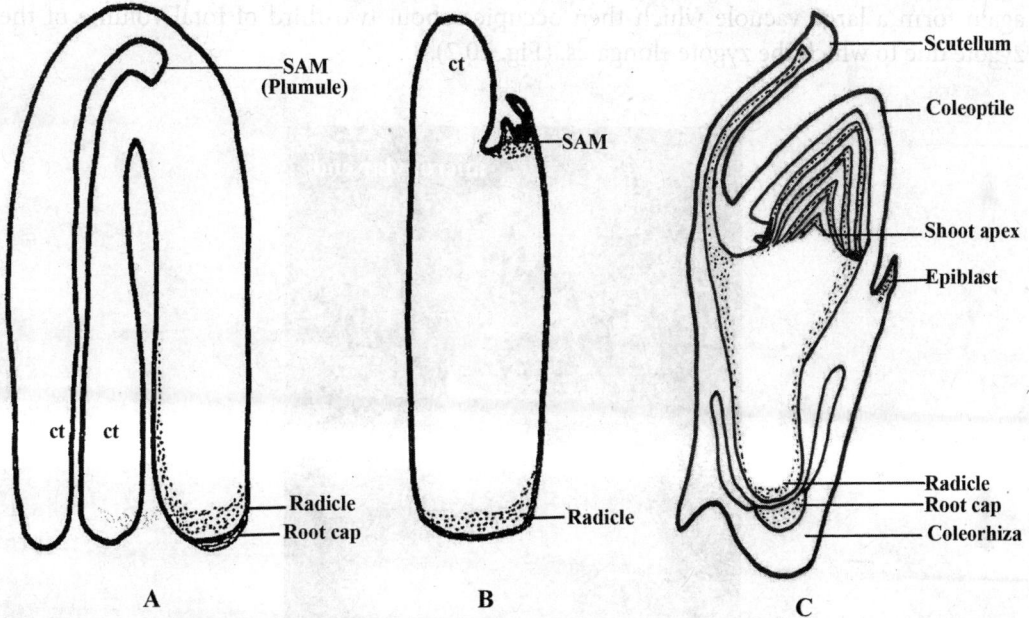

Figure 10.1 Diagrammatic representation of different parts of embryos A. A typical dicotyledonous embryo (*Capsella bursa-pastoris*). **B.** A monocotyledonous embryo (*Hydrilla* sp.). **C.** A grass embryo. ct: cotyledon, SAM: shoot apical meristem. (*Source for 10.1 B.* Maheshwari & Johri 1950, *reproduced with permission from JIBS.*)

10.3 Egg to Zygote: Post-Fertilization Changes, Ultrastructure and Polarity

In plants, the zygote is an undifferentiated cell at its inception which soon undergoes organized and progressive post-fertilization changes in the cytoplasm and organelles. Immediately after its formation, the zygote undergoes dormancy. The dormancy period varies greatly between species and may be of a few hours or may extend up to several days. For example, in *Capsella* sp. the dormancy period is about 24 hours, while in *Hippophae rhamnoides* it is around 50 days. The dormancy period may be linked to the type of endosperm present in the species. Usually, species with cellular endosperm undergo shorter dormancy periods while species with nuclear endosperm have a longer dormancy period (Raghvan 2000).

The dormancy period is necessary for some very apparent post-fertilization changes to occur, like a shift in the volume of zygote. In certain species the volume of zygote decreases significantly like in *Gossypium hirsutum*, *Hibiscus* sp., *Lagerstroemia speciose*, *Brassica campestris*, *Capsella* sp., while the zygote enlarges in species like *Datura stramonium*, *Nicotiana tabacum*, *Arabidopsis*, *Cypripedium insigne*, *Hordeum vulgare*. In *Arabidopsis* and *Niotiana tabacum*, the change in volume of zygote is found to be associated with the vacuolization. Soon after fertilization, the large micropylar vacuole of the fertilized egg is replaced by numerous small vesicle-like vacuoles. Gradually these small vacuoles fuse and again form a large vacuole which then occupies about two-third of total volume of the zygote due to which the zygote elongates. (Fig. 10.2).

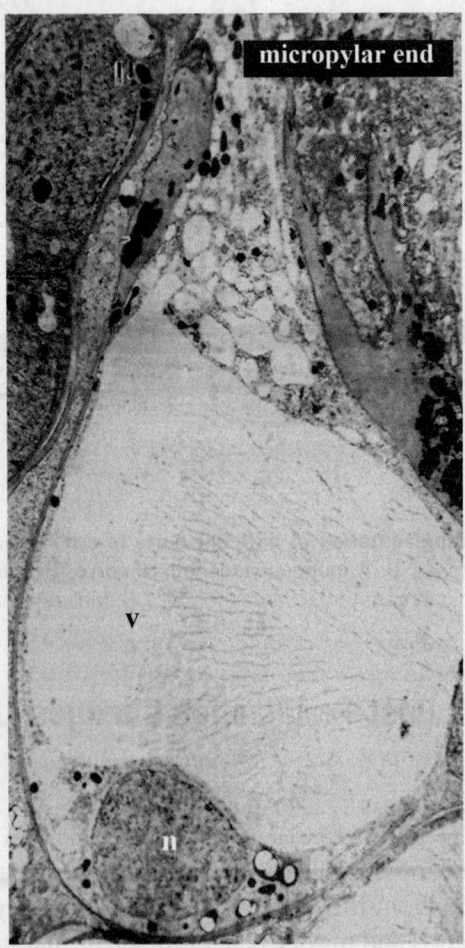

Figure 10.2 Transmission electron micrograph of an elongated zygote (*Nicotiana tabacum*) exhibiting apical-basal polarity. n: nucleus, v: vacuole.(Source: Huang et al. 1993, *reproduced with permission*.)

One of the first patterning events after fertilization is the formation of an apical-basal axis of the zygote and the attainment of polarity. A zygote is anchored to the micropylar end of the ovule at one point and hangs in the embryo sac cavity at the other end. The polarity of the egg cell from which a zygote develops is transiently lost after fertilization. However, during the dormancy period, zygote elongates and reacquires the polarity by positioning its nucleus towards the chalazal end and its large vacuole towards the micropylar end e.g., *Arabidopsis*. The total number of ER, ribosomes, starch grains in the plastids, cristae in mitochondria, dictyosomes activity and cytoplasmic content also increases towards the chalazal end (e.g., *Zea mays, Capsella* sp., *Linum* sp., *Quercus*). Cytoplasmic organelles also show movement from micropylar end to chalazal end of the zygote and accumulate around the nucleus (e.g., in Cotton). A considerable rearrangement of microtubules from random to transverse orientation may also be seen during the elongation of the zygote.

The establishment of polarity in the ultrastructural profile of the zygote results in the formation of two poles: the basal and the apical. The basal pole lies at the micropylar end of the embryo sac and mostly possesses a large vacuole. The apical pole is present towards the chalazal end of embryo sac and has a large number of cell organelles including nucleus (Fig. 10.2). This polarity lays the foundation for future divisions and differentiation of the embryo.

The increased activity of ribosomes and dictyosomes is indicative of high metabolic activity of fertilized egg cell. In some species like *Capsella* and *Nicotiana*, high concentrations of mRNA have been observed at the chalazal end of the zygote. The higher activity of secretory cell organelles (ER, dictyosomes, polysomes) is also seen during this time and this is necessary for the wall synthesis around the zygote as the unfertilized egg does not possess a cell wall towards its chalazal end. Initially, a callose wall is laid only at the chalazal end of the zygote. Soon, a wall completely surrounds the zygote, blocking the plasmodesmata and insulating the newly formed cell. The insulation is beneficial as it protects the zygote from the neighboring sporophytic cells (Natesh & Rau 1984; Raghvan 2000).

10.4 Embryogeny: From Zygote to Multi-cellular Embryo

Embryogenesis is the process of formation of a mature multi-cellular embryo from a single celled zygote. Attainment of polarity in zygote, makes it all set for its first division. In the majority of angiosperms, the first division in the zygote is invariably asymmetric and transverse leading to formation of two unequal sized cells: a smaller *apical cell* (**ap/ac**), (also referred to as a terminal cell) and a larger *basal cell* (**bc**). Generally, the basal cell gives rise to a suspensor or haustorium, and the apical cell forms the organo-genetic parts of the embryo. The other differences in the apical and basal cells are summarized in Box 10.1. Certain plant families deviate from this generalization. For example, in Poaceae the first division in zygote is oblique, where as in Loranthaceae, the first division is vertical.

Box 10.1

Differences between an Apical Cell and a Basal Cell of Pro-embryo

Basal Cell	Apical Cell
• Present towards the micropylar end of embryo sac.	• Present towards the chalazal end of embryo sac.
• Larger in size.	• Smaller in size.
• Possesses a large vacuole.	• Generally devoid of vacuoles; if present they are very small and inconspicuous.
• Cytoplasm is sparse.	• Cytoplasm is dense.
• Devoid of organelles except nucleus and vacuole.	• Rich in organelles.
• Generally, forms a suspensor and root apex.	• Forms organo-genetic parts of embryo like cotyledons, Shoot Apical Meristem (SAM).
• The subsequent divisions are transverse.	• The subsequent divisions can be either transverse or longitudinal.

10.4.1 Types of Embryo Development

Different types of embryo development are seen in angiosperms. These types are designated on the basis of the contribution of apical and basal cells in the formation of a mature embryo, and the plane of division in the apical cell. Together, these decide embryo patterning. While division in the basal cell is mostly transverse, plane of division in the apical cell varies, based on which there are two major groups of embryogeny in angiosperms. It is important to note that a particular type of embryogeny is not restricted to a particular family. Different genera and even a species may exhibit different types of embryogeny within a family, e.g., *Anemone rivularis* shows two types of embryogeny (both Solanad and Crucifer types). Moreover, there are species like *Paeonia, Drimys winteri* which do not fit in any of these classes.

The following types of embryogeny are prevalent in the angiosperms (Fig. 10.3):

Group 1. When the plane of division in the apical cell is longitudinal

I. The basal and apical cell both contribute to the formation of embryo

a. Asterad type: The embryo is generated from the derivatives of both basal and apical cell. For example, in *Lactuca sativa*, the derivatives of apical cells contribute to the formation of cotyledons and the plumule, while those of basal cell form roots, root cap, hypocotyl and suspensor. The asterad type of embryogeny and its variations can be seen in families Asteraceae, Lamiaceae, Polygonaceae, Urticaceae, Geraniaceae, and Poaceae.

II. The derivatives of basal cell may contribute to the formation of embryo

b. Onagrad type (or Crucifer type): The mature embryo is generally derived from the apical cell only. The basal cell contributes towards the formation of a suspensor and an organo-genetic part of embryo called hypophysis; as seen in Onagraceae, Brassicaceae, Lythraceae, Ranunculaceae, Rutaceae, Scorphulariaceae, Lilliaceae, and Juncaceae

Group 2. When the plane of division in the apical cell is transverse

I. The basal and apical cell both contribute to the formation of embryo

a. Chenopodiad type: Here, the embryo arises from the division products of both apical and basal cells. Usually, the hypocotyl of mature embryo is derived from the basal cell while the epicotyl arises from the derivatives of the apical cell. Such embryogeny is prevalent in Chenopodiaceae, Amaranthaceae, Polymoniaceae, and Boraginaceae.

II. The basal cell does not contribute to the formation of embryo but forms a suspensor

b. Solanad Type: It is characterized by the formation of embryo primarily by the apical cell. The divisions in the basal cell give rise to a suspensor. Such embryo development is seen in families Solanaceae, Papaveraceae, Linaceae, and Rubiaceae.

c. Caryophyllad type: The embryo is formed exclusively by the apical cell. The basal cell does not divide any further. However, if a suspensor is present, then it is derived from the basal cell and is unicellular. This type of embryogeny can be seen in Caryophyllaceae, Crassulaceae, Holoragaceae, Portulacaceae, Alistamaceae, and Arecaceae.

Group 3. When the plane of division in the zygote is longitudinal

a. Piperad type: This type of embryogeny is the characteristic of families like Piperaceae, and Loranthaceae, where the first division in the zygote is vertical. No apical and basal cells are formed, and instead, two elongated cells parallel to each other are cut off to form the embryo.

10.4.2 Early Embryogenesis: Octant Configurations of Pro-embryo

During embryogenesis, all the involved cells acquire specific structure and function. The organized divisions and specifications of cells during embryogenesis is called *embryo patterning*. However, the processes that generate different cell types from the zygote remain largely unknown. The pattern formation starts with the divisions in apical and basal cell (*ac*, and *bc*, respectively) of the 2-celled pro-embryo which gives rise to the tetrad stage of pro-embryo. First division (and subsequent divisions) in the basal cell (*bc*) is transverse and two cells (*ci* and *m*) remain arranged in single tier at the 4-celled pro-embryo stage (Fig. 10.4). The basal cell will continue to divide symmetrically and in the transverse direction forming the suspensor. However, the apical cell may undergo divisions in either of the two planes; transverse and longitudinal. Hence, two configurations are possible in subsequent 4-celled pro-embryo stages which are linear (Fig. 10.4 A) and T-shaped (Fig. 10.4 B–C). In the linear tetrad, two daughter cells (*I* and *I'*) of *ac* undergo two vertical divisions at right angles to each other and form an octant (eight-cell stage) with two tiers of 4 cells each (Fig. 10.4 A). Similarly, the T-shaped pro-embryo may give rise to an octant with two tiers of 4 cells each (*I* and *I'*) after two rounds of division (one longitudinal and one transverse)

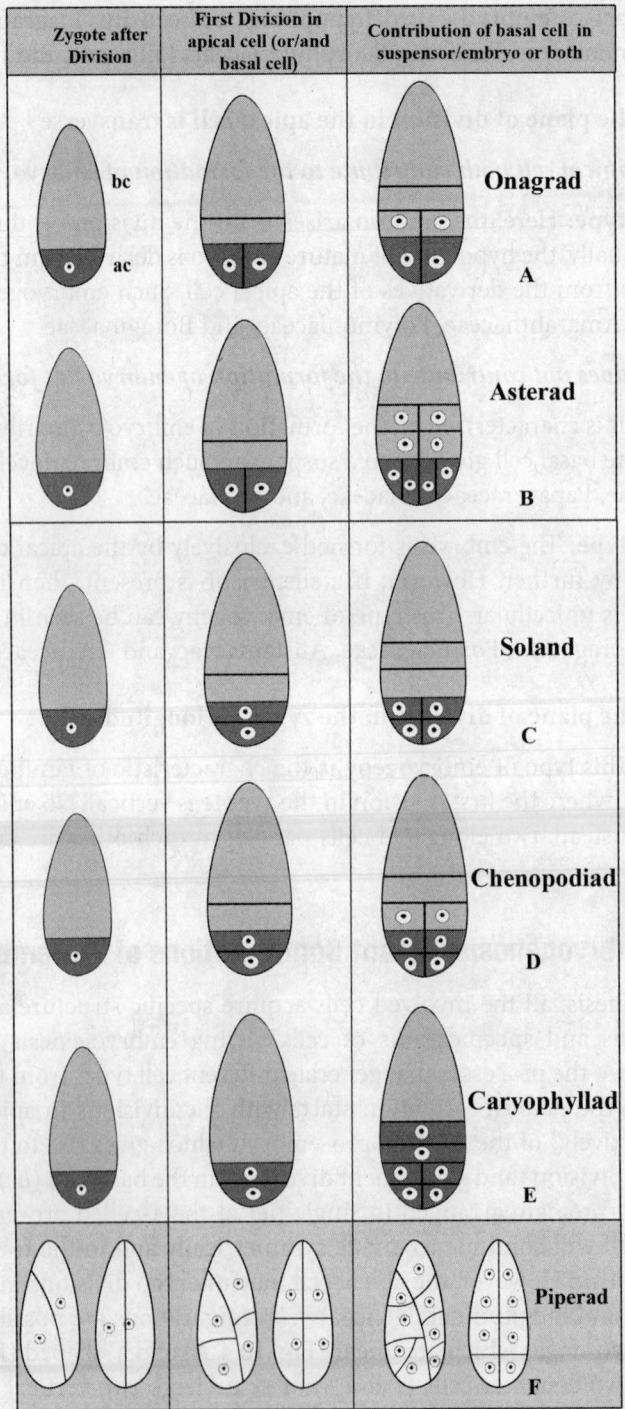

Figure 10.3 Types of embryogeny in angiosperms. The large basal cell (bc) and its derivatives are light grey, while the smaller apical cell (ac) and its derivatives are dark grey. (*Source*: After Maheshwari 1950.)

(Fig. 10.4 B). In an octant formed from the T-shaped pro-embryo, a third type of pattern is also seen where all the eight cells are in one tier (*q*), i.e., an axial quadrant (4 cells) surrounded by four peripheral cells (Fig. 10.4 C). Both the types of octants (superimposed type: Fig. 10.4 A–B, and single tier type: Fig.10.4 C) occur in monocots as well as in dicotyledons. Thus, at the eight-cell (octant) stage, pro-embryo consists of an upper and a lower tier of four cells each and a suspensor. These three domains remain fixed and go on to form the organo-genetic parts of mature embryo.

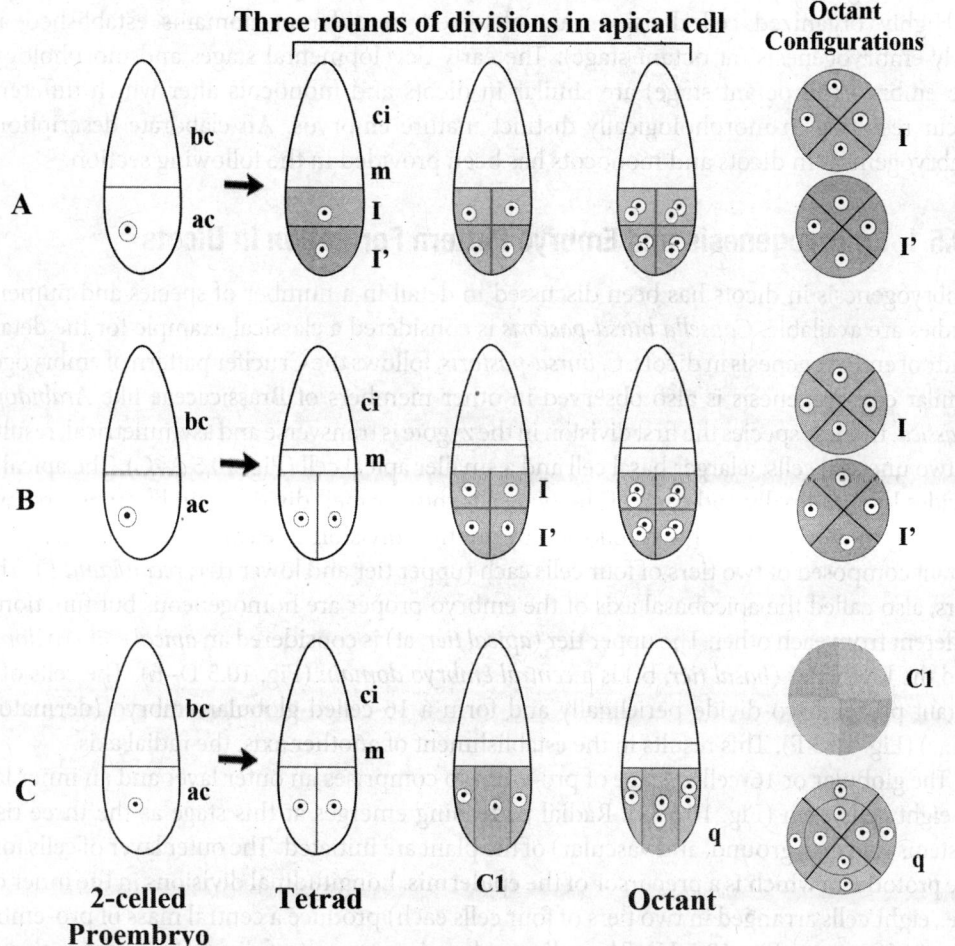

Figure 10.4 Embryo patterning till octant stage. **A–B.** Formation of an octant with two superimposed tiers of *I* and *I'* from linear tetrad (A) and from T-shape tetrad (B). **C.** Octant from T-shape pro-embryo where eight cells are in one tier '*q*' (an axial quadrant remained surrounded by four peripheral cell). C1: transverse section of pro-embryo at 4-celled stage.
Note: First plane of division in apical cell of pro-embryo is marked by black line. Second plane of division is marked by red line whereas third plane of division is noticeable by nuclei with red boundaries. Octant configurations: as observed in transverse section of embryo proper. **See Color Plates (page 492).** (*Source:* Based on Natesh & Rau 1984.)

10.5 Embryogenesis and Embryo Patterning

Embryogenesis is the development and differentiation of the future sporophyte with the aims to: 1. Establish the basic body plan of future sporophyte (organogenesis) through patterning. Radial patterning produces three tissue systems (epidermis, ground tissue and vascular system), and the axial patterning establishes the apical-basal (shoot-root) axis. 2. Form the meristematic tissue for postembryonic elaboration of the body structure (leaves, roots, flowers). 3. Develop an accessible food reserve (in form of cotyledons) for the germinating embryo until it becomes autotrophic.

Highly organized cell divisions are observed in different domains established after early embryogenesis (at octant stage). The early developmental stages and morphology of pro-embryo (till octant stage) are similar in dicots and monocots after which differences occur resulting in morphologically distinct mature embryos. An elaborate description of embryogenesis in dicots and monocots has been provided in the following section.

10.5.1 Embryogenesis and Embryo Pattern Formation in Dicots

Embryogenesis in dicots has been discussed in detail in a number of species and numerous studies are available. *Capsella bursa-pastoris* is considered a classical example for the detailed study of embryogenesis in dicots. *C. bursa-pastoris*, follows the Crucifer pattern of embryogeny. Similar embryogenesis is also observed in other members of Brassicaceae like *Arabidopsis, Brassica*. In these species the first division in the zygote is transverse and asymmetrical, resulting in two unequal cells: a larger basal cell and a smaller apical cell (Fig. 10.5 A–C). The apical cell divides longitudinally and each of the two cells thus formed divide again by transverse walls resulting in a quadrant. This is followed by another division in each of the cells, yielding an octant composed of two tiers of four cells each (upper tier and lower tier; *recall I and I'*). These tiers, also called the apicobasal axis of the embryo proper are homogeneous but functionally different from each other. The upper tier (*apical tier*, at) is considered an *apical embryo domain* and the lower tier (*basal tier*, bt) is a *central embryo domain* (Fig. 10.5 D–E). The cells of the octant pro-embryo divide periclinally and form a 16-celled globular embryo (dermatogen stage) (Fig. 10.5 F). This results in the establishment of another axis, the radial axis.

The globular or 16-celled stage of pro-embryo comprises an outer layer and an inner layer of eight cells each (Fig. 10.5 G). Radial patterning emerges at this stage as the three tissue systems (dermal, ground, and vascular) of the plant are initiated. The outer layer of cells forms the protoderm which is a precursor of the epidermis. Longitudinal divisions in the inner core (i.e., eight cells arranged in two tiers of four cells each) produce a central mass of pro-embryo (Transition stage, Fig. 10.5 H). The cells produced as a result of divisions in upper tier give rise to the precursors of the shoot meristem and a middle layer of cells which eventually generates cotyledons (lateral organs of the embryo with a leaf like structure). Divisions in the lower tier produce precursor cells of vascular tissue and a middle layer of cells that constitute the future ground tissue (cortex and pith) of stem and root (heart-shape stage, Fig. 10.5 I).

Notably, at the transition stage of embryogenesis, the radially symmetrical apical region of embryo becomes bilaterally symmetrical due to the lateral expansion of the distal poles present at the apical domain (epicotyl). The laterally expanded poles form the early

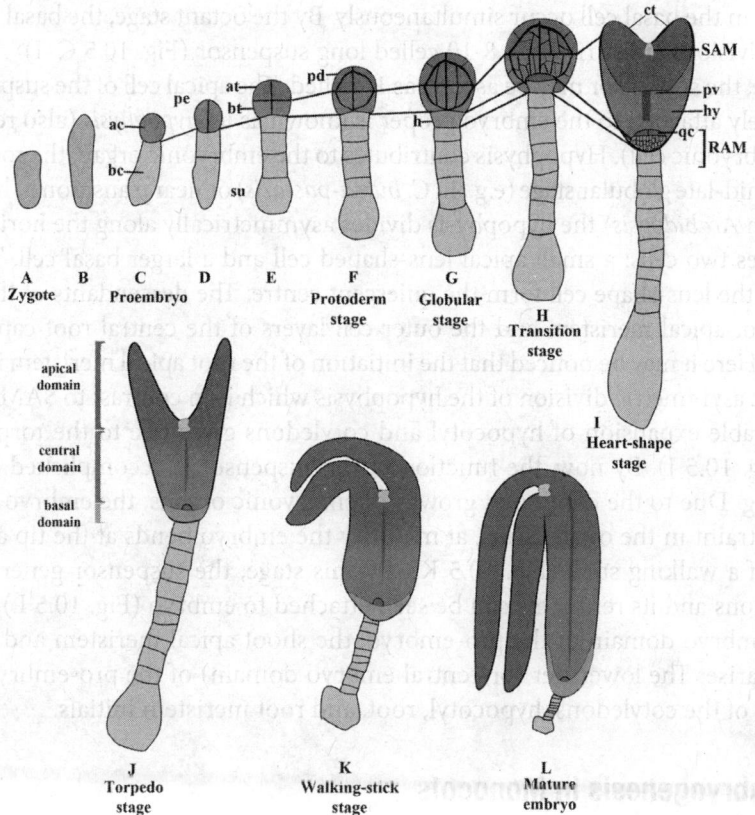

Figure 10.5 Diagrammatic representation of embryogeny and embryo patterning in dicots (*Capsella bursa-pastoris, Arabidopsis*). Zygote (**A**) after elongation (**B**) undergoes asymmetric division, forming apical cell (**ac**) and basal cell (**bc**) (**C**). The first division in ac is longitudinal (**D**). At the octant stage two tiers (lower tiers: *apical embryo domain* (**at**) and *central embryo domain (or basal tier;* **bt**) are recognized. These contribute differently to embryo patterning (**E–I**). Close to reaching the globular stage of embryo formation, the hypophysis (**h**) gets distinguished from the suspensor (**G**). Closer to transition stage (or late globular stage, **H**) the hypophysis cuts-off a lens shape cell which forms the quiescent centre in root apex; the other derivatives form the parts of root apical meristem (**RAM, I**). It is important to note that by the time a heart-shaped stage is achieved, different organo-genetic parts of mature embryo can be seen. Embryo maturation (**J–K**): torpedo-shape (**J**), Walking stick stage (**K**), mature embryo (**L**). **ct**: cotyledons, **SAM**: shoot apical meristem, **pv**: provascular system, **hy**: hypocotyl. Black line connecting different stages suggests an organo-genetic differentiation of different parts of embryo from the respective tier. **See Color Plates (page 492)**. (*Source*: Adapted and modified after Peris et al. 2010, *with permission*.)

cotyledons (heart-shaped stage). The shoot apex is differentiated from the cells present between the cotyledons. Here, it is important to remember that the shoot apical meristem (SAM) cannot be recognized by stereotypic cell divisions during the transition. Several molecular and physiological studies have established that genes meant for formation of meristematic cells and polar transport of auxin get activated by the mid-globular stage (refer to Box 10.2 for details). After the transition to heart-shape stage, the hypocotyl is derived by divisions and differentiation of the cells present in the lower tier.

Divisions in the basal cell occur simultaneously. By the octant stage, the basal cell through transverse divisions, gives rise to a 8-10 celled long suspensor (Fig. 10.5 C–I). At the heart-shaped stage, the suspensor may be as long as 16 celled. The apical cell of the suspensor which is immediately attached to the embryo proper is known as its *hypophysis* (also recognized as an extra-embryonic cell). Hypophysis contributes to the embryonic organ, the root apex (Fig. 10.5 G). At mid-late globular stage (e.g., in *C. bursa-pastoris*) or near transition to heart-shaped stage (e.g., in *Arabidopsis*) the hypophysis divides asymmetrically along the horizontal plane and generates two cells: a small apical lens-shaped cell and a larger basal cell. Two vertical divisions in the lens-shape cell form the quiescent centre. The descendants of the larger cell form the root apical meristem and the outer cell layers of the central root cap (columella, epidermis). Here it may be noticed that the initiation of the root apical meristem is marked by a stereotypic asymmetric division of the hypophysis which is in contrast to SAM.

Considerable expansion of hypocotyl and cotyledons gives rise to the torpedo-shaped embryo (Fig. 10.5 J). By now, the function of the suspensor is accomplished and it starts degenerating. Due to the continued growth of embryonic organs, the embryo experiences spatial constraint in the ovule. Thus, at maturity the embryo bends at the tip and assumes the shape of a walking stick (Fig. 10.5 K). By this stage, the suspensor generally has lost all its functions and its remnants can be seen attached to embryo (Fig. 10.5 L). Thus, from the apical embryo domain of the pro-embryo, the shoot apical meristem and most of the cotyledons arise. The lower tier (or central embryo domain) of the pro-embryo forms the abaxial part of the cotyledons, hypocotyl, root, and root meristem initials.

10.5.2 Embryogenesis in Monocots

The chief difference between the mature embryos of monocots and dicots is in the number of cotyledons; one in monocots and two in dicots. The nature of the single cotyledon of the monocots was variously questioned by earlier workers. Detailed investigations by Swamy & Lakshmanan 1962, and Lakshmanan 1972; on the ontogeny of monocot embryos in different families of angiosperms provided the answers. Their extensive studies established that the epicotyl in monocot embryos is a terminal structure. Its position appears lateral due to rapid growth of cotyledon and slow growth of epicotyl. Lakshmanan (1972) and others have demonstrated that the epicotyl and the cotyledon arise from the same terminal tier.

Ontogenically, monocot embryo differs from dicot embryo in the contribution of cells of the terminal tier (or quadrant, *recall I'*) in the formation of cotyledon. In dicots, the two opposite cells of terminal tier equally contribute to the formation of cotyledons, while in monocots, the number of cells of the tier that give rise to the cotyledon varies. For example, in family Philydraceae all the four cells of a tier form the cotyledon, while in families Pontederiaceae, Iridaceae and Sparganiaceae, three cells participate in the formation of a cotyledon. In families Hydrocharitaceae, Amaryllidaceae and Potamogetonaceae only two adjacent cells of a tier form the cotyledon (Natesh & Rau 1984).

Very little is known till date about the embryo patterning in monocots (except in grasses: maize, rice and wheat). Here, a general account of development of a monocotyledonous embryo is discussed with special emphasis on *Najas lacerata,* and *Halophila ovata. Cocos nucifera* (coconut) represents a different type of cotyledon and shoot apex ontogeny among monocots, hence a brief account of that is also provided.

Development of a monocotyledon embryo in Najas lacerata: The first division in the zygote takes place by a transverse wall engendering two daughter cells–a smaller apical cell (*ac*) and a larger basal cell (*bc*) (Fig. 10.6 A). The latter (*bc*) does not divide any further and eventually matures into a single-celled suspensor haustorium. The *ac* again segments in transverse plane, forming cells '*c*', and, '*d*'. Further, a transverse division in cell *d* results in an upper '*m*' and a lower '*ci*' cell (Fig. 10.6 B). The cell *ci*, finally builds a filamentous row of four cells by transverse divisions, designated by symbols *n, o, h,* and *s*. The derivatives of the '*n*' and '*o*' contribute to the embryo proper, while the '*h*' and '*s*' persist as such between the embryo and the suspensor (Fig. 10.6 H–S).

The terminal cell *c* first divides by a vertical wall and the resulting two daughter cells again divide in the same plane but at right angles to the first division. Thus, a tier of four cells comes into being. Periclinal divisions in this four-celled tier; results in an octant tier '*q*' (four axial cells surrounded by four peripheral cells). The peripheral cells form a uniform layer of protoderm, by laying down anticlinal walls and also envelop the inner derivatives of *q*. At this stage the embryo is slightly globular in shape. Three of the four axial cells show pronounced divisions and the fourth remains without divisions or undergoes a few divisions. The unequal rate of division among the axial cells disrupts the symmetry resulting in the embryo being humped or notched. The notched part (overgrown part) of the embryo represents the cotyledon. The slow dividing axial cell, laterally gives rise to the beginnings of epicotyl (Fig.10.6 G–I).

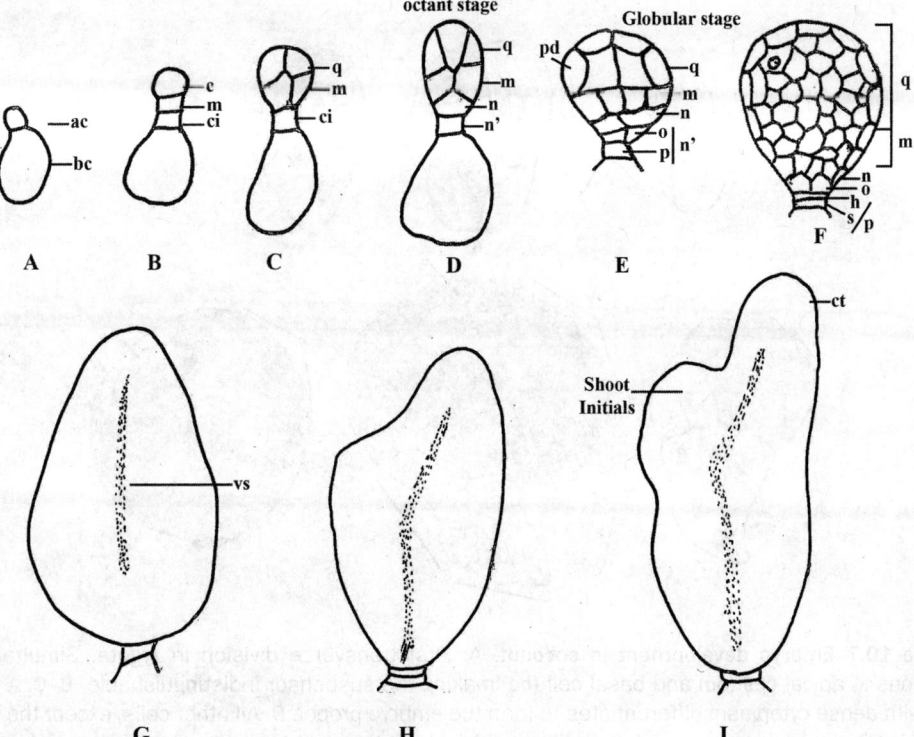

Figure 10.6 Different stages of embryogeny in a monocot (*Najas lacerata*). (*Source*: Swamy & Lakshmanan 1962, *reproduced with permission from JIBS*.)

Soon after the differentiation of protoderm, the divisions in cells '*m*', '*n*' and '*o*' are initiated. The inner derivatives (axial cell) divide in diverse planes. The cells in the core region (majorly derivatives of '*m*' and '*n*' only; a few from '*q*') correspond to the 'plerome initials' (vascular system). The next layer differentiates as the 'periblem initials' (cortex). In general, the cells derived from the pro-embryonal tiers *m* and *n* show enlargement as compared to the daughter cells of tier '*q*'. Here a distinction can be made between the derivatives of these groups and it may be noted that ontogenetically, the cotyledon and the epicotyl are distinct terminal structures. Later, the derivatives of '*n*' participate in the organization of the radicle.

Development of embryo in Cocos nucifera: The zygote in coconut divides transversely producing a typical small apical cell and large basal cell. Both apical and basal cells divide actively, and initially, the suspensor cannot be distinguished from the embryo proper. As embryogeny progresses, one of the derivatives of the apical cell becomes more densely cytoplasmic than the rest of the cells of the embryo proper (including basal cells). This distinctive cell continues to divide while the other cells fail to do so. Eventually, the progeny of this cell forms the entire globular embryo. All other cells, whether derived from the apical or basal cell form the suspensor. The single cotyledon initiates from one flank of globular pro-embryo and later becomes a crescent-shaped ridge. The centrally located apical domain of globular pro-embryo forms the shoot tip and leaf primordia (epicotyl or plumule). The cotyledon grows profusely, curves and covers the whole embryo (Fig. 10.7). This profuse growth of cotyledon is distinctive among monocots and is restricted to palms.

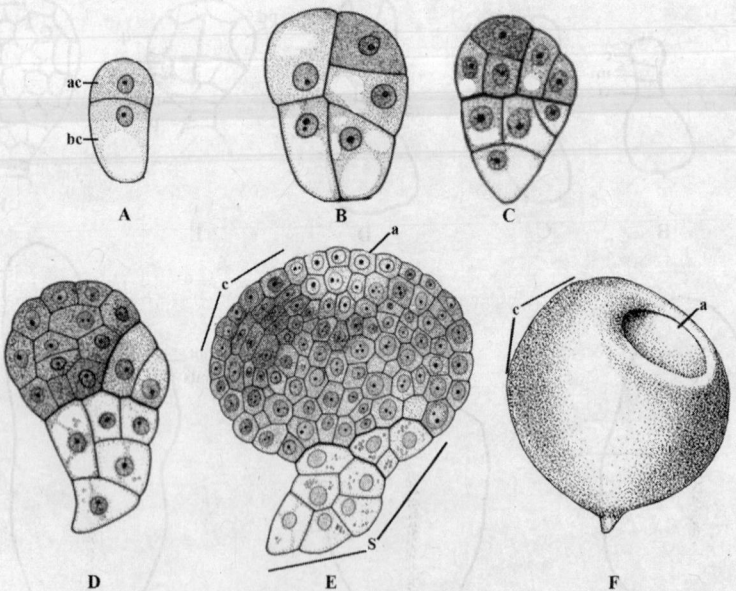

Figure 10.7 Embryo development in coconut. A. First transverse division in zygote. Simultaneous divisions in apical cell (**ac**) and basal cell (**bc**) making the suspensor indistinguishable. **B–C.** A single cell with dense cytoplasm differentiates to form the embryo proper. **D.** All other cells, except the apical cell contribute to suspensor (**s**) formation; which is distinguishable with progressive embryogeny. **E.** Initiation of single cotyledon (**c**) and apical domain (**a**) from a flank of the globular pro-embryo. **F.** 3-D diagrammatic view of development of early stage of profuse cotyledon (**c**) encircling apical domain (**a**). (*Source*: Adapted and modified from Haccius & Philip 1979, *with permission*.)

Box 10.2

Hormonal Regulation of Embryo Patterning in Flowering Plants

The embryo patterning in flowering plants depends on appropriate regulation of the critical morphogenetic processes during embryo development. Early experiments by Liu et al. (1993) on *Brassica rapa* using three auxin transport inhibitors (9-hydroxyfluorene-9-carboxylic acid, trans-cinnamic acid, and 2,3,5-triiodobenzoic acid) provided novel insights on the role of auxin in acquisition of radial to bilateral symmetry in early embryogenesis. The mutant embryos generated in the study were with ring-like cotyledons and additional shoots. This study also established polar transport of auxin during embryogenesis. Since then, an ever-increasing body of evidence concluded that it is not just the presence of auxin which is essential for embryo pattern formation; but all processes including auxin activity – biosynthesis and transport, auxin perception by its receptor, and auxin response are crucial. Various genes and proteins have been characterized for each aspect of auxin regulation related to embryogenesis. A concise account of it is provided here.

   Studies based on analysis of *Arabidopsis* mutants, identified several genes of auxin efflux carriers, viz. *PIN1*, *PIN3*, *PIN4* and *PIN7* and for auxin influx *AUX1*, *LAX1* (*LIKE-AUX1*) and *LAX2*. These efflux carriers are required for the differential auxin distribution during embryogenesis (Fig. 10.8), and *AUX1*, *LAX1* and *LAX2* play a role in both shoot and root-pole formation, in association with PIN efflux carriers (Robert et al. 2015). The individual PIN genes act superfluously, and a mutation in a single PIN gene does not hamper embryogenesis. Mutation studies have clearly suggested that only *pin1 pin3 pin4 pin7* quadruple mutants are strongly defective in the overall establishment of apical-basal polarity (reviewed in Petrášek & Friml 2009). Other auxin transporters like ABCB1 and ABCB19 (*ATP-Binding Cassette subfamily B transporters of the multidrug resistance phosphoglycoprotein*) also play a role in embryogenesis, but not in the embryo patterning. They only help to maintain the auxin distribution in the outer layers of the embryo. At the pro-embryo stages; ABCB1 transporters are localized in all suspensor cells and pro-embryonal cells, while localization of ABCB19 is restricted to the suspensor-forming cells or derivatives of basal cell (Petrášek & Friml 2009; Robert et al. 2015).

   After the first anticlinal division of a zygote, auxin accumulation increases in the apical cell. The differential distribution of auxin is a result of the activity of PIN7 transporters; which are

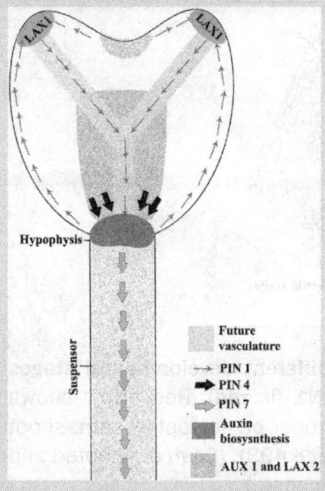

Figure 10.8 Diagrammatic representation of auxin transport. Auxin efflux and influx carriers during heart-shape stage of embryogenesis have been shown. The auxin synthesis occurs at shoot apical meristem and in the suspensor; which are situated at two opposite poles. Auxin is transported by PIN1 and LAX1 (localized in protoderm) from the suspensor towards the cotyledons tip (red arrows). Auxin is transported by efflux via PIN1 & PIN4. The AUX1and LAX2 transporters shows influx via the shoot apical meristem to the root meristem (hypophysis). PIN7 (localized in suspensor) transports the auxin away from the embryo. **See Color Plates (page 493).** (*Source:* Based on Petrášek & Friml 2009; Locascio et al. 2014; Robert et al. 2015.)

localized on the apical side of adjacent suspensor cells (in plasma membrane). At this stage, PIN1 presumably mediates the uniform distribution of auxin between cells of the growing pro-embryo. Later, during the early globular stage, PIN1 gradually re-localizes towards the uppermost suspensor cell, the hypophysis. Simultaneously, the polarity of PIN7 shifts from apical to basal in the cells of the suspensor. This is the central moment in the setting of the basal end of the apical-basal embryo axis occurring at an early globular stage. These coordinated PIN7 polarity rearrangements, at later stages (at heart stage) are also aided by the action of PIN4. The PIN7 polarity rearrangements lead to a shift and show apical-to-basal flow of auxin; and hence the auxin accumulation in the hypophysis. Here the role of AUX1and LAX2 transporters has also been demonstrated. This pattern of auxin distribution and response are crucial for the specification of the hypophysis as the precursor of the root meristem. At the transition stage (globular to heart stage) auxin maxima are created by PIN1 activity in the epidermis; in the inner cells of cotyledon primordia. However, PIN1 mediates basipetal auxin transport towards the root pole too. Thus, the bilateral symmetry is established and cotyledon primordia emerge (Weijers & Jurgens 2005; Petrášek & Friml 2009; Robert et al. 2015).

Among grasses, ZmPIN1-mediated auxin transport during early stages of maize embryogenesis is known. The ZmPIN1 protein is localized in embryo plasma membranes (Figure 10.9a–d). After the first division of the zygote, several cell divisions in different planes lead to the formation of the small embryo and the larger suspensor (pro-embryo stage). At this stage, auxins accumulate in the endosperm above the embryo but not in the embryo itself, and ZmPIN1 is localized at the cell boundaries of the undifferentiated pro-embryo core, without any polarity (Fig. 10.9 B). Later, adaxial/abaxial polarity is established by the outgrowth of the scutellum at the abaxial side of the embryo (transition stage, Fig. 10.9 C). ZmPIN1 is polarly localized in the apical anticlinal membranes, marking the provascular cells of the differentiating scutellum and an auxin flux toward the tip of the single maize cotyledon (late transition stage). Here, a switch from the apical to basal membrane localization of ZmPIN1 proteins characterizes the coleoptile stage and the establishment of an auxin flux from both the differentiated scutellum and the shoot apical meristem (SAM) that is responsible for the differentiation of embryonic roots follows (Fig. 10.9 D). The final stage, in which auxin polar transport is involved during embryogenesis, is the formation of leaf primordia, suggested by *Zm*PIN1 localization in the subapical region of the meristem (Locascio et al. 2014).

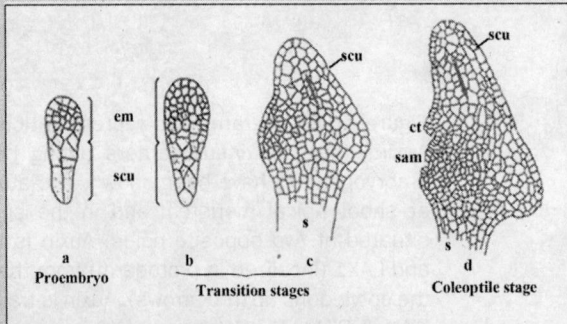

Figure 10.9 Median-longitudinal sections of maize embryos in different developmental stages (pro-embryo, transition and coleoptile) showing location of ZMPIN1 (in red). Red arrow shows (in stage c and d) auxin polar efflux mediated by PIN. **em:** embryo proper, **ct:** coleoptile, **sam:** shoot apical meristem, **scu:** scutellum, **s:** suspensor. **See Color Plates (page 493).** (*Source:* Adapted and modified after Locascio et al. 2014, published under CC BY 4.0 License.)

Box 10.3

Differences between Embryogeny and Embryos of Dicots and Monocots

Dicot Embryogeny and Embryo	Monocot Embryogeny and Embryo
• In dicots, the two opposite cells of terminal tier of octant pro-embryo contribute to the formation of cotyledons.	• In monocots, a variable number of cells of the octant pro-embryo give rise to the cotyledon.
• Embryo development can be divided in different stages, viz. globular, heart, torpedo and walking stick.	• Only globular stage is seen.
• During transition from globular to heart-shape, bilateral symmetry appears due to equal growth of cotyledons.	• There is rapid growth of only one cotyledon.
• Epicotyl and hypocotyl both are terminal structures.	• In monocot embryos, epicotyl is a terminal and hypocotyl is a lateral structure.
• Mature embryo has two cotyledons.	• Mature embryo has a single cotyledon.

10.5.3 Embryogenesis in Poaceae

Among the monocotyledons, the grasses (Poaceae members) exhibit a peculiar development and structure of a mature embryo. The chief features of the grass embryo are the presence of scutellum (an absorptive organ), coleorhiza and coleoptile (the cap-like structures sheathing the radicle and plumule respectively), and the epiblast (a flap-like outgrowth on the side of coleorhiza). These entities can be seen in a median-longitudinal section of grass embryo in wheat, maize, rice (Figs.10.1 C and 10.10 F). The development of these peculiar structures and grass embryo are discussed with reference to *Triticum aestivum* (wheat).

In wheat, the first division of the zygote is transverse-oblique, or oblique and asymmetrical. This division produces small apical cell and a large basal cell. Both the cells divide further by oblique walls and form T-shaped 4-celled pro-embryo (Fig. 10.10 A–B). The repeated oblique divisions of these cells generate the globular pro-embryo. Octant configuration in wheat embryogeny corresponds to cells in one tier, i.e., '*q*'. One or two cells derived from basal cell form the suspensor. Following the divisions, at 16–32 celled stage, the globular embryo becomes club-shaped and the organogenesis begins (Fig. 10.10 C–D). At this stage the scutellum appears as an ambiguous elevation in the apical-lateral region of pro-embryo and is the first organ of grass embryo ontogenetically (Fig. 10.10 E). By the time scutellum appears, the opposite side of the embryo also expands by divisions and this marks the differentiation of shoot apex (coleoptile) and leaf primordia. The differentiation of the radicle is the last event in embryogenesis of grasses. It differentiates endogenously in the central zone of embryo (Batygina 1978).

Oblique divisions in zygote and pro-embryo are specific to wheat. However, not all members of Poaceae may undergo oblique divisions. For instance, in *Eragrostis unioloides* the zygote and pro-embryo divide by perfect transverse divisions as in dicots and in other monocots. Here, the daughter cells of the basal cell also participate in the formation of mature embryo along with suspensor and embryogeny being of the asterad type.

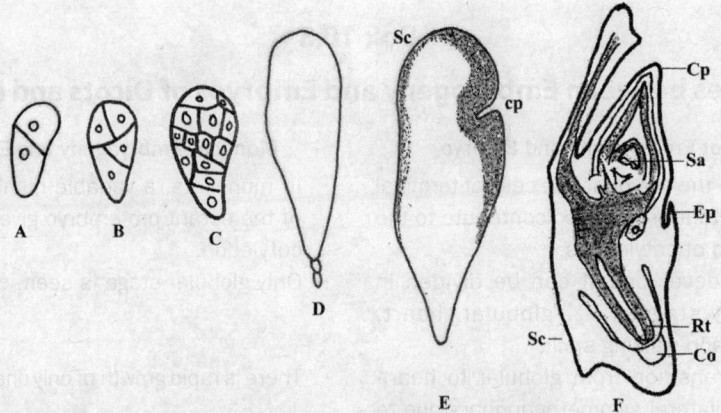

Figure 10.10 Diagrammatic representation of embryogenesis in Wheat. A. First oblique division in zygote. **B.** Further oblique divisions in basal and apical cell. **C.** Globular embryo. **D.** club shaped embryo. **E.** Differentiation of coleoptile (**Cp**) and scutellum (**Sc**). **F.** Mature embryo. **Sa:** shoot apical meristem; **Ep:** epiblast; **Rt:** root; **Co:** coleorhiza. (*Source:* Redrawn from Batygina 1978, *with permission from INSA.*)

Box 10.4

Differences between Zygote and Embryo

Zygote	Embryo
· Zygote is the product of fusion between egg cell and sperm cell nuclei.	· Embryo is a structure that develops from zygote.
· Divisions in the zygote lead to the formation of an embryo.	· Embryo is the earliest stage of a sporophyte or a plant.
· Zygote has apical-basal polarity.	· Embryo has both apical-basal polarity and radial symmetry.
· Zygote is unicellular.	· Embryo is multicellular.

10.6 Unusual Features of Embryo and Embryogenesis in Some Angiosperms

There are several angiosperms which exhibit deviation from the regular pattern of embryogeny or peculiarity in the structure of embryo, viz. irregular shape, lack of apical-basal polarity or organs in mature embryo (e.g., Ranunculaceae, Pandanaceae, Papavaraceae, Piperaceae, Orobanchaceae, Orchidaceae). Some may show late differentiation as in *Drimys winteri*, which differentiates after the 20-celled stage of pro-embryo and does not follow any particular type embryogeny. The description of some classical examples is given in the present section.

10.6.1 Embryo Development in *Paeonia*

Yakovlev & Yoffe (1957, 1961) described a unique type of embryogeny in some species of *Paeonia* which was further established by Cave et al. (1961) and others. Though *Paeonia* is a dicot, its embryo development is quite similar to that of gymnosperms. The first and subsequent mitotic divisions in the zygote are not accompanied with cell wall formation (cytokinesis), hence are free-nuclear and eventually reach a coenocytic stage. Thus, the pro-embryo is a large cell with numerous free-nuclei (Fig. 10.11 A). As the development proceeds, a large vacuole appears in the center and pushes the nuclei towards the periphery (Fig. 10.11 B). Afterwards, the cellularization of the free-nuclei starts and several embryo primordia are formed from the marginal cells (Fig. 10.11 C). Finally, only one of them develops into a mature dicotyledonous embryo (Fig. 10.11 D). Since, the development of embryo in *Paeonia* does not correspond to any of the existing angiosperms, it was proposed that this unique kind of embryo development be included as a *Paeoniad type* of embryogeny (Yakovlev 1983).

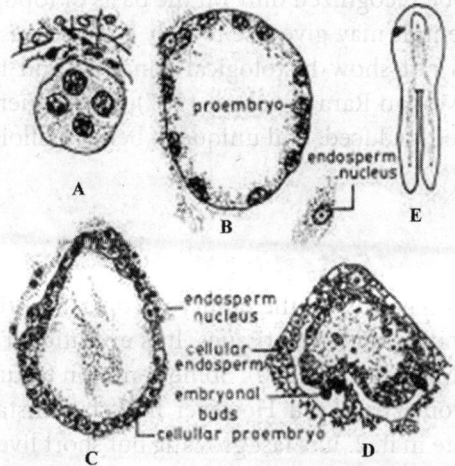

Figure 10.11 Embryo development in *Paeonia* sp. **A–B.** Coenocytic stage of embryo development. **C.** Centripetal cellularization stage of pro-embryo. **D.** Embryonal buds and note the cellular endosperm. **E.** Differentiated dicotyledonous embryo. (*Source: Johri & Ambegaokar 1984, reproduced with permission from INSA.*)

10.6.2 Reduced Embryos

In a number of angiosperm taxa the mature embryo is devoid of organs, such embryos are known as *reduced embryos*. Some of the examples are:

- Simplest reduced embryo reported so far is from *Monotropa uniflora* where seeds are shed at the 2-cell stage of embryo (Fig. 10.12 A). Further growth is intra-seminal (inside the seed). Similar examples can be observed in members of Papavaraceae and Oleaceae.

- Orchids and Orobanchaceae: In these families, the embryo remains undifferentiated and lacks true plumule, cotyledons and radicles. In Orchidaceae, the mature embryo is an ovoidal-globose mass of undifferentiated cells (Fig. 10.12 B). Here, polar orientation of the embryo can be marked by its position in fertilized ovule. The cells of embryo towards the chalazal side are smaller and apparently mark the radicle pole while the opposite end depicts the plumular pole. During germination, only the plumular pole becomes active and forms the whole plant. Likewise, mature embryo in *Orobanche* is globose. During natural seed germination, only the radicle pole becomes active and generates the whole plant. No activity can be seen in plumular pole.
- *Utricularia*: In most *Utricularia* species, only initial histogenesis occurs during the embryogenesis. The epidermis, the meristematic tissue at the apex, and the storage parenchyma are formed but no radicle pole differentiates. Due to this unusual organogenesis in the embryos, provascular tissue is absent in most *Utricularia* species.
- Podostemaceae: Family Podostemaceae is unique in possessing highly simplified embryos which lack bipolarity of a usual embryo. The polar ends, i.e., hypocotyl and epicotyl can be recognized only on the basis of topography. Plumule is either absent or if present, it may give rise to only a few leaves between the cotyledons. The radicle does not show histological zonation and true roots are lacking in mature plants. (Mohan Ram & Sehgal 1997). Consequently, the adult plant body is highly simplified, reduced, and unique in being thalloid.

10.7 Suspensor

Suspensor is a file of cells that pushes the embryo proper into the endosperm cavity and connects it to the surrounding maternal tissues. It is evolutionarily conserved but diverse in size and morphology (Maheshwari 1950). Some common features of a suspensor are the following: 1. It develops from a basal cell. However, in certain instances, daughter cells from the apical cell also integrate in it. 2. It is fast growing but short lived. Generally, a suspensor exists during transition from the globular to the heart-shaped stage of the pro-embryo. 3. It has an absorptive behavior and plays a role in the nutrition of embryo.

10.7.1 Structure

Suspensor exhibits considerable variation in its shape, size, and number and ploidy of cells. It may be highly reduced in size e.g., *Bryonia*, *Euphorbia* or may even be absent in some taxa like *Lycopis, Tilia, Viola,* and *Penaea*. *Phaseolus coccineus* is a unique example where number of cells in a suspensor can be as high as ~200 at the heart-shaped stage of embryo. It attains its maximum size by cotyledonary stage. A long filamentous suspensor is also a feature of Brassicaceae, where the number of cells may reach up to 20.

In the families where endosperm is absent (e.g., Orchidaceae, Podostemaceae), a well-defined suspensor exists. Orchidaceae shows considerable variability in the structure of suspensor which may be single celled (sac like, or tubular) or single-tier filamentous, or even multi-cellular and clustered (Fig. 10.13 A). Existence of suspensor haustoria is also seen in families like Podostemaceae, Crassulaceae, Tropaeolaceae, Rubiaceae, and

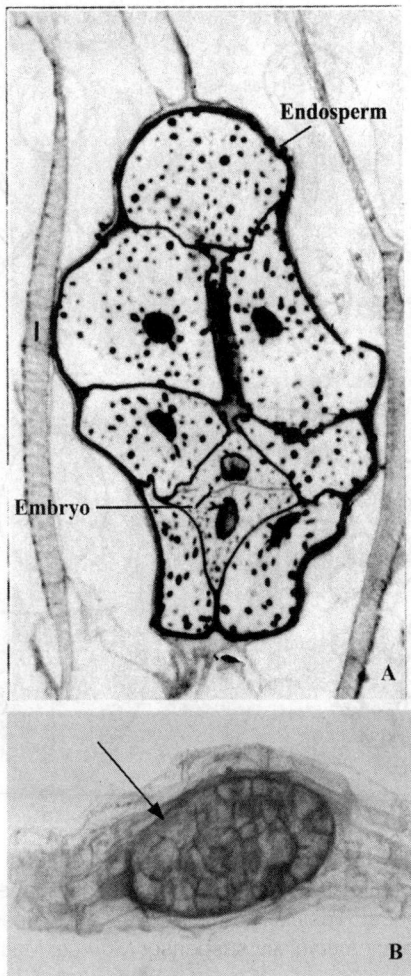

Figure 10.12 A. Reduced embryo of *Monotropa*. **B.** An orchid seed (*Paphiopedilum armeniacum*) with embryo (arrow). (*Source*: A: Olson 1980, *reproduced with permission*; B: Xu et al. 2020, published under CC BY 4.0 License.)

Fabaceae. In *Polyplerum stylosum* (*syn: Dicraea stylosa*; Podostemaceae), the nucleus of the basal cell becomes highly hypertrophied and the basal cell grows haustorial branches that penetrate the different parts of ovules (Fig. 10.13 B). In *Rubia cordifolia* (Rubiaceae), the suspensor is initially filamentous but later, some of cells of the micropylar region throw out lateral protrusions and swellings. These cells are designated as vesicular cells (Fig. 10.13 C). In Crassulaceae (e.g., *Sedum* sp.), the apical cell forms the 4-celled suspensor. The basal cell enlarges and its nucleus becomes hypertrophied followed by a longitudinal division. These newly formed cells form extensive 2-celled haustorium (Fig. 10.13 D) and cover the micropylar region. In some cases, these two cells show development of haustorial branches which penetrate raphe or other parts of ovule (e.g., in *Sedum ochroleucum*). In *C. bursa-pastoris*, near the globular stage of embryo, a terminal basal cell enlarges and becomes vacuolated. It subsequently develops into a haustorium.

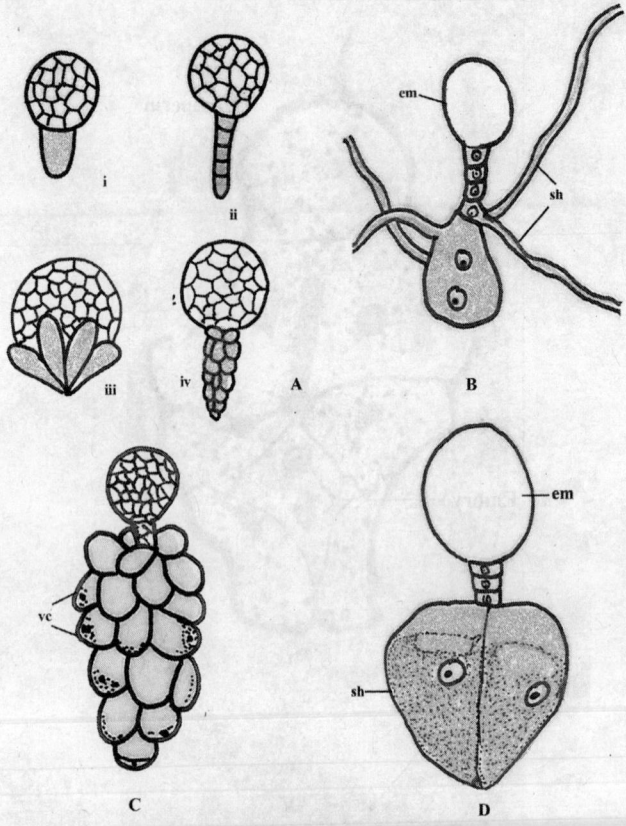

Figure 10.13 Structure variability in suspensor. A. Orchids. Suspensor can vary from a single cell structure to a multicellular structure. **B.** Haustorial branches in suspensor of *Polypleurum stylosum* (Podostemaceae). **C.** Rubiaceaae: Vesicle-like cells (**vc**) of suspensor. **D.** 2-celled suspensor haustorium in Crassulaceae. **em:** embryo; **sh:** suspensor (*Source:* Masand & Kapil 1966, *reproduced with permission.*)

10.7.2 Ultrastructure

The chief function of the suspensor is to provide nutrition and regulate the development of embryo. Ultrastructural studies made in *C. bursa-pastoris, Phaseolus* sp. *Stellaria media* (Yeung & Clutter 1979; Schulz & Jensen 1968; 1969) and *Arabidopsis thaliana* have been instrumental in determining the role of suspensor in nutrition of embryo.

In general, suspensor cells acquire ultrastructural specialization by the heart-shaped stage of pro-embryo (*C. bursa-pastoris*) or globular-stage of pro-embryo (*Phaseolus* sp.). The first step towards the specialization is the initiation of wall invaginations or wall infolds which is a widespread phenomenon and has been observed in numerous species, e.g., *Phaseolus* sp., *C. bursa-pastoris, Diplotaxis* sp., *Vigna* sp., *Alyssum* sp., *Brassica rapa, Glycine max, Medicago* sp., *Alisma lanceolatum, Ipomea purpurea, Trapaeolum majus.* The wall projections can appear either in outer or in inner walls of suspensor cell and is a common feature shared by basal cell (or haustorial cell) and other cells of suspensor. Even in *Stellaria media*, where the basal cell is undivided and forms a single cell suspensor, massive wall labyrinths develop on the inner surface.

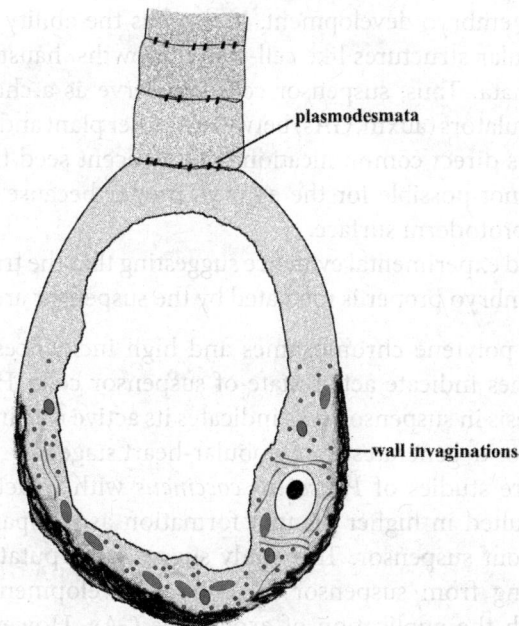

Figure 10.14 Generalized diagrammatic representation of ultrastructure of the haustorial cell of the suspensor. Note the wall ingrowths and that most of the cell organelles are concentrated at micropylar region of cell. The haustoria and other cells of suspensor remain connected with numerous plasmodesmata.

As a common attribute of suspensor, wall ingrowths go together with increase in number of other cell organelles like ribosomes, plastids, mitochondria, dictyosomes, microbodies, plastids and smooth ERs. This increase in cell organelles is followed by their arrangement in rafts parallel to the wall invaginations, especially of mitochondria, ERs and dictyosomes. These features support the transfer-cell like activity of haustorial cell. Even an increasing gradient of cell organelles (towards the chalazal end) may occur in suspensor (Fig. 10.14). In *C. bursa-pastoris* increasing gradation of ribosomes prevails in suspensor towards chalazal end of embryo while in other species there is decreasing gradient in plastids complexity and the number of microbodies. This ultrastructure profile of suspensor and basal cells is in contrast to cells of the embryo proper. The cells of suspensor also remain connected to each other by numerous plasmodesmata present at their transverse walls which help in transport of nutrient to embryo. However, the cellular processes by which the suspensor transfers substances to the embryo proper still needs to be determined.

10.7.3 Role of Suspensor

Abundant evidences suggest that suspensor plays role in embryo development by providing nutrition to the embryo. The position and structure of the suspensor also provides clues for its possible role(s) during embryogenesis. It is believed to push the embryo proper into the endosperm cavity and connects the embryo proper to surrounding maternal and endosperm tissues. The suspensor thus, serves as a conduit for nutrients and growth regulators required for embryo development. It also has the ability to transfer molecules due to presence of cellular structures like cell-wall ingrowths, haustorial outgrowths, and numerous plasmodesmata. Thus, suspensor cells can serve as a channel for exchange of nutrient and growth regulators (auxin, GAs) between mother plant and embryo. Additionally, the suspensor facilitates direct communication with adjacent seed tissues (e.g., seed coat, endosperm), which is not possible for the embryo proper because it is surrounded by a cuticle that covers the protoderm surface.

The direct studies and experimental evidence suggesting that the transfer of nutrients and growth factors to the embryo proper is mediated by the suspensor are summarized below:

- Prevalence of polytene chromosomes and high incidences of endo-duplication in chromosomes indicate active state of suspensor cells. High rate of RNA and protein synthesis in suspensor cells indicates its active role in early embryogenesis, as the suspensor degenerates after globular-heart stage.

- *In vitro* culture studies of *Phaseolus coccineus* with intact embryos along with suspensor resulted in higher plantlet formation as compared to the cultures of embryos without suspensor. This study suggests the putative role of gibberellic acid originating from suspensor in embryo development by substituting the suspensor with the application of exogenous GAs. However, recent transgenic and molecular studies confirm that presence of GAs and their biosynthesis is not a common feature of suspensor seen in all the species.

- Experiments with labeled metabolites, such as sucrose (e.g., *Phaseolus coccineus*) and polyamine (in *Arabidopsis thaliana*; *Phaseolus coccineus*), in developing seeds demonstrated the direct movement of molecules from the suspensor to the embryo proper in an energy dependent manner.

- Recent studies using transgenic approach have shown that in *Arabidopsis*, the suspensor at globular-stage is connected symplastically to the embryo proper. The visualization of symplastic movement of green fluorescent proteins (GFPs) from the suspensor to the embryo proper provides evidence for this.

- Genes such as the *Arabidopsis* 'AtSUC3'; a sucrose transporter gene involved in transferring molecules, gets up-regulated in the suspensor.

- Auxin is transported from the suspensor to the embryo proper in *Arabidopsis* embryos by PIN7 auxin efflux carrier protein, specifically localized in the suspensor. Also PIN7-mediated auxin transport is essential for embryo development (refer to Box 10.1 for details).

The functional potential of the suspensor goes beyond providing nutrition to the developing embryo. In several species, suspensor cells have been shown to have the potential to generate a new embryo (Lakshmanan & Ambegaokar, 1984, Yeung & Meinke, 1993). Investigations of suspensor mutants (*sus1*, *sus2*, and *raspberry1*) of *Arabidopsis* also provide genetic evidence that the suspensor has the capacity to develop embryo-like structures. Suspensor-derived embryogenesis in *Arabidopsis* can occur when the embryo proper is damaged or experimentally ablated (Liu et al. 2015). Hence, suspensor cells are thought of as a dormant backup system which can be activated if embryo proper is damaged.

10.8 Embryo Nourishment

Developing embryo remains *in-milieu* of sporophytic cells, suspensor, endosperm and embryo sac. Understanding the modes of embryo nutrition is an interesting aspect of embryogenesis. The nutritive role of the suspensor and endosperm in embryo nourishment is well established but the influence of maternal sporophytic cells on embryo nutrition is less well known. The incidences of embryo sac haustoria in some species suggest that sporophytic cells may nurture the growing embryo. The growing embryo sac leads to the degeneration of the nucellus and surrounding cells. The content of degenerated cells gets incorporated in the endosperm which could be thought of as indirect nutrition of embryo by sporophytic cells via endosperm. Similarly, information pertaining to embryo sac driven nutrition is limited to a few species. For instance, in *Euphorbia* sp., *Glycine max, Arabidopsis, Butomus*, cotton wall projections develop after fertilization at the micropylar end of embryo sac similar to transfer cells. In the experiments with labelled photosynthates (e.g., *Glycine max*), accumulation of radioactivity at hypostase and near these projections suggests that two-way nutritional supply to embryo is operational in the embryo sac.

The endosperm is the tissue which has a direct role in embryo nutrition. Numerous studies have also demonstrated the precise embryo-endosperm relationship in the context of nutrition and developmental regulation (*see Chapter 9 for details*).

Embryo-Endosperm Relationship: The outer wall ingrowths of suspensor cells (e.g., *C. bursa-pastoris, Diplotaxis* sp., *Vigna* sp., *Alyssum* sp., *Brassica rapa, Glycine max*) or from the embryo sac wall (e.g., *Pisum* sp., cotton) are structural adaptations for nutrient absorption from the endosperm and have been variously studied. Irrespective of their origin these wall projections remain in contact with endosperm. The widespread presence of these wall labyrinths in the suspensor suggest the transfer like activity of these cells and support the *Solute Absorption Theory* for embryo-endosperm nutrition relation postulated by Schulz & Jensen (1969). According to this theory, these proliferations increase the surface area and support the short-distance transport of the solutes/metabolites from endosperm to the embryo. The plasmodesmata perforating the suspensor-embryo walls serve as a duct for this transport. Also, the concentration of various organelles in the vicinity of wall ingrowths has been ascribed to the increase in the active transport of solutes across plasma membrane.

Soon after the degeneration of suspensor, direct absorption of endosperm by the growing embryo has been also observed. This is evident by the presence of an apparent zone of degenerated endosperm cells around the cotyledons (as in dicots) or near the tip of scutellum and coleoptile (in grasses) depicting the direct utilization of endosperm. It has also been postulated that enzymes might be playing their role in the lysis of endosperm, e.g., in *Nicotiana tabacum* during seed development when cation dependent phosphatase activity increases significantly. The lysed cells of endosperm can directly serve as a source of metabolites, and as precursors of RNA and DNA. More insights on embryo-endosperm relationship are given in chapter 9 (*see sections 9.81 and 9.8.2*).

Box 10.5

Genetic Regulation of Zygotic Embryogenesis in Dicots

The characterization of three regions (or domains) in the dicot embryo (see *Fig. 10.5*), which drive the concept of embryo patterning, is the result of a plethora of studies aimed at genetic mutant analysis in *Arabidopsis thaliana*. A comprehensive list of some genes involved and their function in embryogenesis is given below:

Gene	Organism	Function
ARABIDOPSIS RESPONSE REGULATOR (ARR7 and ARR15)	*Arabidopsis*	Regulate the asymmetric division of hypophysis cell and its derivatives. The lens-shaped upper daughter cell establishes the organizing center of the RAM, known as the quiescent center
BODENLOS (BDL)	*Arabidopsis*	Apical-basal embryonic pattern formation
CLAVATA3 (CLV3)	*Arabidopsis*	Shoot apical meristem
CUP SHAPED COTYLEDON 1 (CUC 1)	*Arabidopsis*	Delineate the boundaries of organ primordia especially cotyledons)
CYSTATIN (NtCYS)	*Nicotiana tabacum*	Prevention of precocious cell death of basal cell lineage
EMBRYO SURROUNDING FACTOR 1 (ESF1)	*Arabidopsis*	Embry patterning and suspensor formation
EMBRYONIC FACTOR 1 AND 19 (FAC1 and FAC 19)	*Arabidopsis*	Initiation of zygotic division
FASS (FS)	*Arabidopsis*	Cell division plane and morphogenesis
GAMETE CELL DEFECTIVE 1 (GCD1)	*Arabidopsis*	Gamete maturation and initiation of zygote division
GNOM and GROUNDED (GN and GRD)	*Arabidopsis*	Zygote elongation, asymmetric division and basal cell fate determination
MERISTEM LAYER 1 and PROTODERM FACTOR 2 (AtML1 and PDF2)	*Arabidopsis*	Protoderm formation
MONOPTEROS (MP)	*Arabidopsis*	Apical-basal embryonic pattern formation
PIN-FORMED 7 (PIN7)	*Arabidopsis*	Establishment of apical-basal auxin gradients
SHORT SUSPENSOR (SSP)	*Arabidopsis*	Zygote elongation/zygote asymmetric division
TARGET OF MONOPTEROS (TMO)	*Arabidopsis*	Specification of the vascular precursors
WRKY DNA-Binding Protein 2 (WRKY2)	*Arabidopsis*	Polarization of zygote
WUSCHEL (WUS)	*Arabidopsis*	Regulates size of the shoot meristem with CLV3

WUSCHEL RELATED HOMEOBOX PROTEIN 2 and 8 (WOX2 and WOX8)	Arabidopsis	Apical-basal axis formation
WOX1, WOX3 and WOX5	Arabidopsis	Initiation of SAM
WOX8 and WOX9	Arabidopsis	Epidermal, vascular and ground tissue
YODA (YDA, A MAPKK kinase gene)	Arabidopsis	Basal cell fate determination
ZYGOTE ARREST 1 and ZYGOTE STAGE ZEUS 1 (ZYG1 and ZEU1)	Arabidopsis	Initiation of zygotic division

Source: Modified after Zhao et al. 2017.

Glossary

Coleoptile: The protective sheath covering the epicotyl or young shoot tip in a grass embryo.

Coleorhiza: The protective sheath covering the hypocotyl or young root tip in a grass embryo.

Embryo: Embryo is the resultant product of divisions in zygote which encompasses all the organs for future sporophyte.

Embryo patterning: The organized divisions in zygote and functional specifications of cells during embryogenesis, is called embryo patterning.

Embryogeny or zygotic Embryogenesis: The process of development of embryo from a zygote is known as embryogeny or zygotic embryogenesis.

Hypocotyl: The part of embryonal axis below the level of cotyledons is known as hypocotyl; it terminates into an embryonic root (radicle).

Hypophysis: A derivative of basal cell found at the interface between the suspensor and the embryo proper and contributes to formation of root apex.

Pro-embryo: Embryo from two-celled stage up to late globular stage is called pro-embryo.

Reduced embryos: In certain plant species (like *Monotropa*) and families (like Orchidaceae), the embryo does not show differentiation of either radicle or plumule or sometimes both. Such embryos which are an undifferentiated mass of cells are commonly referred to as reduced embryos.

Suspensor: A structure connecting the embryo to the mother plant, formed as a result of a series of transverse divisions in the basal cell of pro-embryo.

Zygote: The product of fusion between egg cell and sperm cell is referred to as zygote.

Key Questions

Q10.1 Fill in the blanks:
  a. The first division in the zygote of grasses is accompanied with.....................wall formation.
  b. Solute absorption theory was put forward by..................

c. The formation of embryo from zygote is known as..............
d. Embryo development in monocots and dicots is similar till..................stage.
e. ....is an example of species with reduced embryos.
f.   Undifferentiated embryos are present in the family members of.............
g. In .... the embryogeny is similar to gymnosperms.
h. ....is thought to be the single massive cotyledon in cereals which covers the embryo.
i.   The first division in zygote is ....in Loranthaceae.

Q10.2 Differentiate between:
a. Zygote and Embryo
b. Apical Cell and Basal cell
c. Dicot embryo and Monocot embryo
d. Solanad and Asterad type of embryogeny

Q10.3 Elaborate on embryo-endosperm relationship.

Q10.4 Discuss how the embryogeny in *Paeonia* is similar to gymnosperms.

Q10.5 Write notes on:
a. Ultrastructure of suspensor
b. Role of suspensor
c. Octant configurations of pro-embryo
d. Embryogenesis in grasses

Q10.6 What are 'reduced embryos' in angiosperms. Discuss with suitable examples and diagrams.

Q10.7 The zygote is a polar structure. Justify the statement with reasons.

Practicals

Exercise 10.1 To study zygotic embryogenesis in dicots using micrographs

Capsella bursa-pastoris, *Brassica* are suitable classroom examples to study dicot embryogenesis. The details of dicot embryogenesis are provided in Section 10.5.1. Identifying features for each stage are enumerated and supplemented with photographs of embryogenesis in *Brassica* sp. (Fig. 10.15). To study developmental details; sections of resin embedded fertilized ovules at different time intervals are cut using microtome.

Embryogenesis in *Brassica* sp. (Fig. 10.15)

A. Longitudinal section of ovule showing first transverse division in the zygote. The first division generates a large cell, i.e., basal cell (bc) towards the micropylar end and small apical cell (ac) towards the chalazal end of ovule. The apical cell is characterized by dense cytoplasm and a prominent nucleus. In the photograph, the free nuclei that can be seen are free-nuclear endosperm (Fig. 10.15 A).

B. The micrograph shows the first longitudinal division in the apical cell (arrow). Below it a suspensor can be seen. This is formed by multiple transverse divisions in the basal cell (Fig. 10.15 B).

C. The longitudinal section of fertilized ovule shows the pro-embryo at the octant stage. Protoderm (pd) precursors have been determined. This stage is also referred as early globular stage. Endosperm is in free nuclear stage (Fig. 10.15 C).

D. The pro-embryo is at globular stage with well-developed suspensor. Here, an early indication of differentiation of shoot apical meristem can be seen as a depression (Fig. 10.15 D; arrowhead).

E. The longitudinal section of a fertilized ovule shows the pro-embryo at late globular to early heart-shape stage. Epidermis is well differentiated. A lens-shaped cell, i.e., hypophysis (h) can also be seen adjacent to the pro-embryo. The hypophysis cuts-off a lens-shaped cell which forms the quiescent center in root apex and other derivatives form the parts of root apical meristem (Fig. 10.15 E).

F. The longitudinal section of developing seed (or fertilized ovule) shows pro-embryo at heart-shaped stage of embryogenesis. The two young cotyledons are distinguishable as protuberances. A very long suspensor is present through which the embryo is attached to the mother plant. Endosperm, in which the embryo is embedded, is cellularized (Fig. 10.15 F).

G. Cross section of the young seed in which embryo is at early cotyledon/torpedo stage. Two cotyledons are distinguishable. The shoot apical meristem is present between the cotyledons (Fig. 10.15 F; inset).

Exercise 10.2 To dissect young embryo from developing seed

The embryo sac resides within an ovule, which upon fertilization becomes the home for developing embryos. The micropylar end of the embryo sac generally remains present at the pointed end of the ovule and can be studied by dissecting out the ovule and observing it under the microscope. When fertilized ovules are dissected at different stages of seed development or time intervals after fertilization, the embryos with suspensor can be seen without sectioning.

Suggested material
Developing fruits (fresh or fixed) of *Calendula*, *Lobularia maritima* (sweet alyssum), *Tropaeolum majus*. Collect the young fruits of sweet alyssum, heads of *Calendula* at different time intervals. The material can be fixed in FAA (Formalin: Acetic acid: 70% Ethanol; 5:5:90 v/v).

Requirements
Acetocarmine stain (1%), microslides, coverslips, a pair of forceps, needles, dissection and compound microscope.

The photograph shows the first longitudinal division in the apical cell (arrow). Below, the suspensor can be seen. This is formed by multiple transverse divisions in the basal

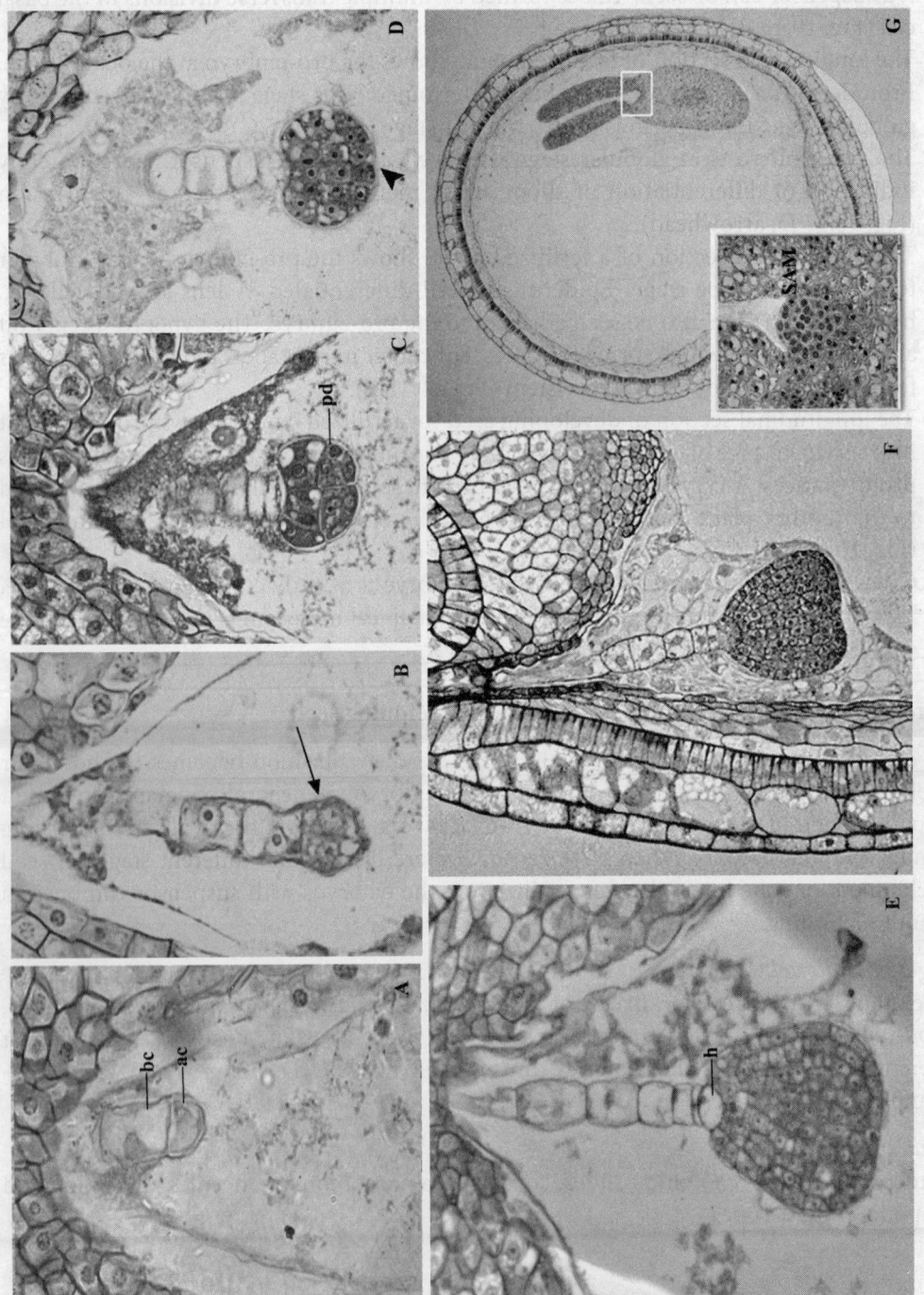

Figure 10.15 Embryogenesis in *Brassica* sp. See Color Plates (page 494).

Procedure

1. For *Calendula* take a fertilized head. Note that the point where ovules are attached to head is the micropylar side. Carefully remove few fertilized ovules using forceps.
2. Place an ovule on the glass slide with in a drop of acetocarmine and place it under the dissection microscope.
3. Carefully insert needle into micropylar end and slowly move it till the chalazal end. This will help is removing the fruit walls.
4. You may observe the stained embryo within translucent cell layers of the integuments covering the embryo sac.
5. Gently slit the integuments and embryo sac longitudinally and a gentle jerk will release the embryo with suspensor into the acetocarmine stain.
6. Remove extra tissue. Gently place a coverslip over dissected pro-embryo and observe under the compound microscope (low magnification).

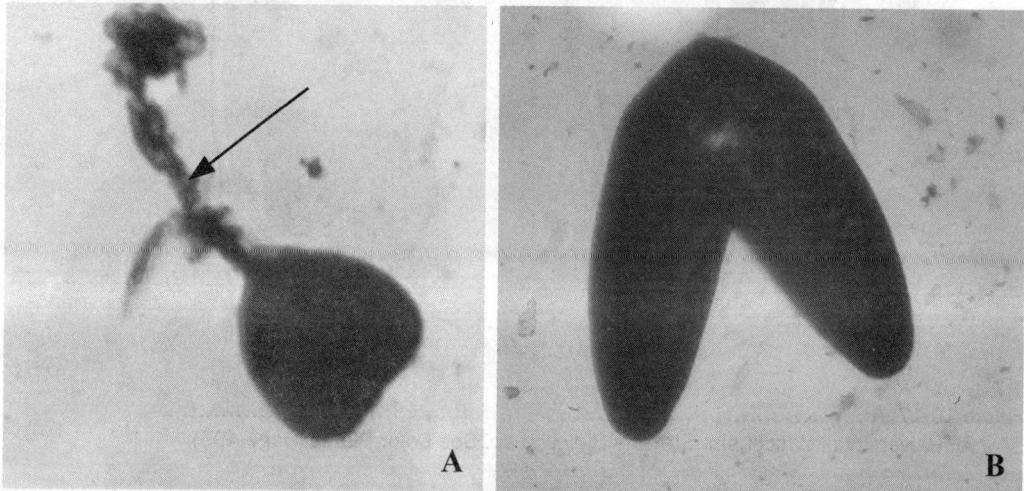

Figure 10.16 Dissected embryos of *Calendula officinalis*. A. Heart-shape pro-embryo with suspensor (arrow) **B.** Torpedo stage pro-embryo.

Observations

The developing pro-embryo(s) with suspensor can be observed. The stage might vary with development (Fig. 10.16).

Precautions

1. First remove the fruit walls gently and do not press the fertilized ovule as it would damage the embryo and suspensor.
2. After removing fruit walls give a gentle jerk to ovules, otherwise embryo will be damaged or suspensor will break.
3. Do not tap coverslip placed over dissected embryos.

Exercise 10.3 Identify section A and B. Label the different parts and enumerate their features.

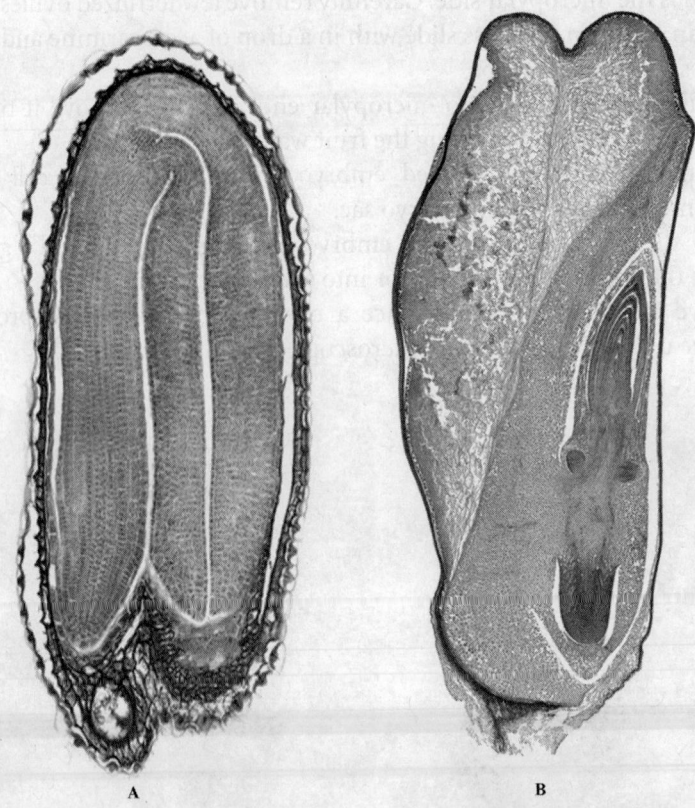

A   B

Figure 10.17 A..............................   **B**.....................................
Source: University of Wisconsin, Stevens Biology Lab. **See Color Plates (page 495).**

Bibliography

Armenta-Medina, A. C. and Gillmor, C. S. (2019). Genetic, molecular and parent-of-origin regulation of early embryogenesis in flowering plants. *Current Topics in Developmental Biology* 131: 497–543.

Barthlott, W., Große-Veldmann, B. and Korotkova, N. (2014). Orchid seed diversity: a scanning electron microscopy survey. *Botanic Garden and Botanical Museum Berlin-Dahlem*. Berlin: Englera.

Bartosz, J., Płachno, B. J and Świątek, P. (2010). Unusual embryo structure in viviparous *Utricularia nelumbifolia*, with remarks on embryo evolution in genus *Utricularia*. *Protoplasma* 239: 69–80.

Batygina, T. B. (1978). Embryology of wheat. *Proceedings of National Academy of Sciences B (India)*. 44: 13–29.

Batygina, T. B. (1969). On the possibility of separation of a new type of embryogenesis in angiopserms. *Review Biology* 32: 335–41.

Bhandari, N. N. and Venkataraman, R. (1968). Embryology of *Drimys winteri*. *Journal of the Arnold Arboretum*. 49: 509–24.

Bhojwani, S. S., Bhatnagar, S. P. and Dantu, P. K. (2015). *The Embryology of Angiosperms*. Noida: Vikas Publishing House Private Limited.

Cave, M. S., Arnott, H. J. and Cook, S. A. (1961). Embryogeny in the California Peonies with reference to their taxonomic position. *American Journal of Botany* 48: 397–404.

Czapik, R. and Izmailow, R. (2001). Zygotic embryogenesis: structural Aspects. In S. S. Bhojwani and Y. H. Soh, eds., *Current Trends in the Embryology of Angiosperms*. Philadelphia: Kluwer Academic Pub.

Deshpande, P. K. (1976). Development of embryo and endosperm in *Eragrostis unioloides* (Poaceae). *Plant Systematics and Evolution* 125: 253–9.

Forestan, C. and Varotto, S. (2010). PIN1 auxin efflux carriers localization studies in *Zea mays*. *Plant Signaling and Behavior* 5: 436–9.

Haccius, B. and Philip, V. J.(1979). Embryo development in *Cocos nucifera* L.: a critical contribution to a general understanding of palm embryogenesis. *Plant Systematics and Evolution* 132: 91–106.

Huang, B. Q., Strout, G. W. and Russell, S. D. (1993). Fertilization in *Nicotiana tabacum*: ultrastructural organization of propane-jet-frozen embryo sacs in vivo. *Planta* 191: 256–64.

Johri, B. M. and Ambegaokar, K. B. (1984). Some unusual features in the embryology of angiosperms. *Proceedings of the National Academy of Sciences, Plant Sciences* 93: 413–27.

Kawashima, T. and Goldberg, R. B. (2010). The suspensor: not just suspending the embryo. *Trends in Plant Science* 15: 23–30.

Lakshmanan, K. K. (1972). Monocot Embryo. *Vistas in Plant Science*. In T. M. Varghese and R. K. Grover, eds., *Intl Bio Sci*, Hissar, pp. 61–110 (not seen in original).

Lersten, N. R. (2004). *Flowering Plant Embryology*. Blackwell Publishing.

Liu, C. M., Zhi-hong, X. U. and Chuab, N. H. (1993). Auxin polar transport is essential for the establishment of bilateral symmetry during early plant embryogenesis. *The Plant Cell* 5: 621–30.

Locascio, A., Villanova, I. R., Bernardi, J. and Varotto, S. (2014). Current perspectives on the hormonal control of seed development in Arabidopsis and maize: a focus on auxin. *Frontiers in Plant Sciences* 5: 1–22.

Maheswari, P. and Johri, B. M. (1950). The occurrence of persistent pollen tubes in *Hydrilla, Ottelia* and *Boerhaavia,* together with a discussion of the possible significance of this phenomenon in the life-history of Angiosperms. *Journal of Indian Botanical Society* 29: 47–51.

Maheswari, P. (1950). *An Introduction to the Embryology of Angiosperms*. New Delhi: Tata McGraw-Hill publishing Company.

Masand, P. and Kapil, R. N. (1966). Nutrition of the embryo sac and embryo – a morphological approach. *Phytomorphology* 16: 158–75.

Mohan Ram, H. Y. and Sehgal, A. (1997). In vitro studies on developmental morphology of Indian Podostemaceae. *Aquatic Botany* 57: 97–132.

Moller, B. and Weijers, D. (2009). Auxin control of embryo patterning. *Perspectives in Biology* 1: 1–13.

Murgai, P. (1959). The development of the embryo in *Paeonia – a* reinvestigation. *Phytomorphology* 9: 275–7.

Natesh, S. and Rau, M. A. (1984). Embryo. In B. M. Johri, ed., *Embryology of Angiosperms*. Berlin: Springer-Verlag, pp. 377–443.

Olson, R. (1980). Seed morphology of *Monotropa uniflora* L. (Ericaceae). *American Journal of Botany* 67: 968–74.

Peris, C. I. L., Eike, H., Rademacher, E. H. and Weijers, D. (2010). Green beginnings – pattern formation in the early plant embryo. *Current Topics in Developmental Biology* 91: 1–27.

Petrášek, J. and Friml, J. (2009). Auxin transport routes in plant development. *Development* 136: 2675–88.

Raghvan, V. (1997). *Molecular Embryology of Flowering Plants*. New York: Cambridge University Press.

Raghvan, V. (2000). *Developmental Biology of Flowering Plants*. New York: Springer.

Robert, H. S., Grunewald, W., Sauer, M. et al. (2015). Plant embryogenesis requires AUX/LAX-mediated auxin influx. *Development* 142: 702–11.

Schulz, P. and Jensen, W. A. (1968). *Capsella* embryogenesis: the egg, zygote, and young embryo. *American Journal of Botany* 55: 807–19.

Schulz, P. and Jensen, W. A. (1969). *Capsella* embryogenesis: the suspensor and the basal cell. *Protoplasma* 67: 139–63.

Swamy, B. G. L. and Lakshmanan, K. K. (1962). Contributions to the embryology of the Najadaceae. *Journal of Indian Botanical Society* 42: 247–66.

Swamy, B. G. L. and Lakshmanan, K. K. (1962). The origin of epicotylary meristem and cotyledon in *Halophila Ovata* Gaudich. *Annals of Botany* 26: 243–9.

Swamy, B. G. L. and Parameswaran, N. (1962) On the origin of cotyledon and epicotyl in *Potamogeton Indicus*. Österreichische Botanische Zeitschrift 109: 344–9.

Schrick, K. and Laux, T. (2001). Zygotic embryogenesis: developmental genetics (*The Formation of an Embryo from a Fertilized Egg*). In S. S. Bhjwani and W. Y. Soh, eds., *Current Trends in the Embryology of Angiosperms*. Dordrecht: Kluwer Academic Publishers, pp. 249–78.

Xu, X., Fang, L., Li, L. et al. (2020). Abscisic acid inhibits asymbiotic germination of immature seeds of *Paphiopedilum armeniacum*. *International Journal of Molecular Science* 21: 9561. doi:10.3390/ijms21249561.

Yakovlev, M. S. and Yoffe, M. D. (1957). On some peculiar features in the embryogeny of *Paeonia* L. *Phytomorphology* 7: 74–82.

Yakovlev, M. S. and Yoffe, M. D. (1961). Further studies of the new type of embryogenesis in angiosperms, (in Russian). *Bot. Zhurn.* 46: 1402–21 (not seen in original).

Yakovlev, M. S. (1983). Paeoniaceae. In M. S. Yakovlev, ed., *Comparative Embryology of Flowering Plants, Phytolaccaceae – Thymelaeaceae* (in Russian). Leningrad: Nauka, pp. 70–7 (not seen in original).

Yeung, E. C. and Clutter, M. E. (1979). Embryogeny of *Phaseolus coccineus*: the ultrastructure and development of the suspensor. *Canadian Journal of Botany* 57: 120–36.

Zhao, P., Begcy, K., Dresselhaus, T. and Sun, M. (2017). Does early embryogenesis in eudicots and monocots involve the same mechanism and molecular players? *Plant Physiology* 173: 130–42.

11

Polyembryony and Apomixis

Apomixis is a form of asexual reproduction through seeds.

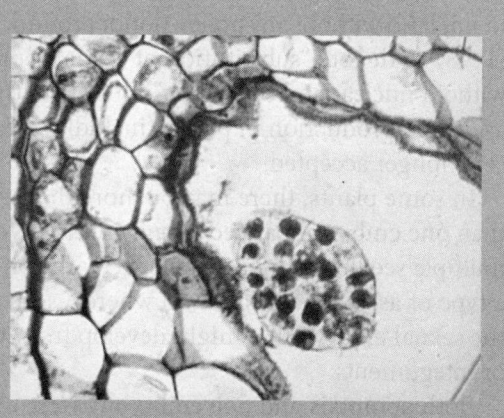

11.1 Introduction

Reproduction is the ultimate goal of every life-form on earth. Accordingly, flowering plants have evolved diverse and versatile strategies to ensure their reproductive success. Broadly, reproduction in higher plants can be divided into two types: sexual and asexual reproduction. Sexual reproduction in vascular plants is complex wherein, multicellular haploid and diploid generations alternate. The diploid sporophyte undergoes meiosis to produce haploid gametes which undergo fusion or syngamy to give rise to seed, the next sporophytic generation. On the other hand, asexual reproduction in plants occurs when a plant produces offspring without meiosis and syngamy. New individuals produced through asexual reproduction are genetically identical to the mother plant. Both sexual and asexual reproduction, have distinct advantages for natural plant populations. Sexual reproduction introduces genetic variability in a population and thus increases the adaptability of species to changing environments. By contrast, asexual reproduction eliminates the cost and the complexity associated with biparental sexual reproduction, and also fixes the genotype of mother plant as offsprings produced are clonal.

When vascular plants reproduce asexually, new individuals may be produced from somatic cells or somatic structures (*vegetative reproduction*) or through seeds that are produced without fertilization (*apomixis or agamospermy*). Vegetative reproduction occurs through propagules like bulbils, suckers, and tubers, which are generated from vegetative parts of a plant. Apomixis (away from mixing), is the formation of an embryo and seed from an unreduced gametophyte or sporophyte. Thus, apomixis leads to the formation of a seed without the processes of meiosis (apomeiosis), and fertilization (nuclear fusion). The discovery of apomixis in higher plants is attributed to the observation of a solitary female plant of an Australian species *Alchornea ilicifolia* (syn. *Caelebogyne ilicifolia*) by

Smith (1841). This female tree would constantly form seeds at the Royal Botanic Gardens in England without any pollen donor around. The term apomixis was introduced by Winkler (1908) to denote "substitution of sexual reproduction by an asexual reproduction process without nuclear and cell fusion". This led to the use of term apomixis to describe all forms of asexual reproduction in plants (including vegetative reproduction), but this generalization is no longer accepted.

In some plants, there may be more than one embryo in a seed. The occurrence of more than one embryo in a seed is known as **polyembryony** and may result in the emergence of multiple seedlings from a single seed. The phenomenon of polyembryony too is considered a type of asexual reproduction where a seed may contain additional embryo(s) other than the sexual embryo that might develop from the sporophytic cells of an ovule, viz. nucellus, or integuments.

Both apomixis and polyembryony were considered to be of rare occurrence among the angiosperms. However, recent studies have confirmed their frequent prevalence among flowering plants. Hojsgaard et al. (2014) found the presence of apomixis in at least in 80 families (12%) and 326 genera (1.8%) of flowering plants. It is also well represented among basal angiosperms (Amborelleales, Nymphaeales and Austrobaileyeales), monocotyledons and eudicots. Certain angiospermous families like Rosaceae, Asteraceae, Fabaceae, Poaceae, and Orchidaceae are well known for presence of a high number of apomictic genera. Similarly, polyembryony also shows wide occurrence in 255 genera belonging to 153 families of angiosperms (Carman 1997). Polyembryony has been reported in many cultivated plants, most well-known being almond, citrus, mango, soybean, apple, grape and olive cultivars. In this chapter, the phenomena of apomixis and polyembryony have been discussed in detail. Both apomixis and polyembryony are botanical curiosities that are advantageous to plant species. Research has proven both to have several practical applications, which are summarized in this chapter.

Box 11.1

Differences between Vegetative Reproduction and Agamospermy

Vegetative Reproduction	Agamospermy
• Vegetative reproduction is the reproduction through propagules generated from vegetative structures, viz. stem, root or leaf of a plant.	• Agamospermy is the reproduction through seeds that are produced without fertilization.
• Propagules for reproduction are bulbils, suckers, tubers.	• Propagule for reproduction are seeds.
• Reproductive organs may be sterile.	• Reproductive organs are fertile and functional.
• Seen in all plant groups including lower plants, gymnosperms and angiosperms.	• Seen mostly in seed forming plant groups, i.e., gymnosperms and angiosperms.

11.2 Polyembryony

Polyembryony can be defined as the simultaneous occurrence of two or more embryos in a seed. This phenomenon was discovered by Leeuwenhoek in 1719, who observed the formation of two plantlets from one citrus seed. The additional embryos may develop from various maternal tissues of the ovule or through zygote. Strasburger (1878) reported the formation of embryos from nucellar cells in *Citrus* and called the phenomenon **nucellar** or **adventive polyembryony**. This was followed by several studies which showed development of embryo from cells of integuments and also from the gametophytic cells of embryo sac other than the egg cell. The multiple embryos in an ovule may also develop by cleavage of zygote or proembryo (often referred as *sexual twinning*). Incidences of polyembryony due to presence of multiple embryo sacs are also quite common in several angiosperms. Similarly, participation of haploid cells of embryo sac like synergids and antipodals in embryo formation is also seen. Different ways in which the extra embryo/s develop other than the sexually produced embryo are discussed in the present section.

11.2.1 Cleavage of Zygote or Proembryos

The cleavage of zygote, or sexually produced embryo is the simplest method for the development of multiple embryos. Cleavage polyembryony occurs usually at very early embryonal stages of the zygote or the pro-embryo. The resulting embryos are genetically identical to each other. Cleavage polyembryony, though more common in gymnosperms, is also found sporadically in angiosperm families, viz. Poaceae, Orchidaceae and Rutaceae. New embryos can originate from either the individual cells of the embryo proper or from the suspensor cells (suspensory polyembryony, e.g., *Stachyurus chinensis*, *Zygophyllum fabago*; Fig. 11.1 A–B). These new embryos, after developing their own axes (shoot apex–root apex), become fully independent. However, in members of Poaceae, the emerging embryos are not independent as they share the suspensor, parts of scutellum, and radicle with each other. As a result, polymeric structures with multiple shoot meristems and a common root-pole are formed (Erdelská & Vidovencová 1994).

Jeffrey (1895) described cleavage polyembryony for the first time in *Erythronium americanum*, wherein soon after fertilization, the zygote undergoes repeated divisions to form an undifferentiated embryonic mass of cells. Outgrowths arise from this embryonic mass, proliferate and give rise to two or three (rarely four) independent embryos. Cleavage polyembryony is also commonly seen in family Orchidaceae (*Spathoglottis* sp., *Eulophia* sp., *Cypripedium japonicum*). Swamy (1943) observed many variations in the development of embryos in *Eulophia epidendraea* (Fig. 11.1 C–E), notably:

1. The repeated divisions in the zygote result in an irregular mass of cells. Later, multiple embryos develop from its chalazal end (Fig. 11.1 C).
2. The filamentous embryo becomes branched and each branch forms an embryo (Fig. 11.1 D)
3. The pro-embryo gives out small buds or outgrowths or splits and each of these may function as an embryo (note: em1 and em2 in Fig. 11.1 E).

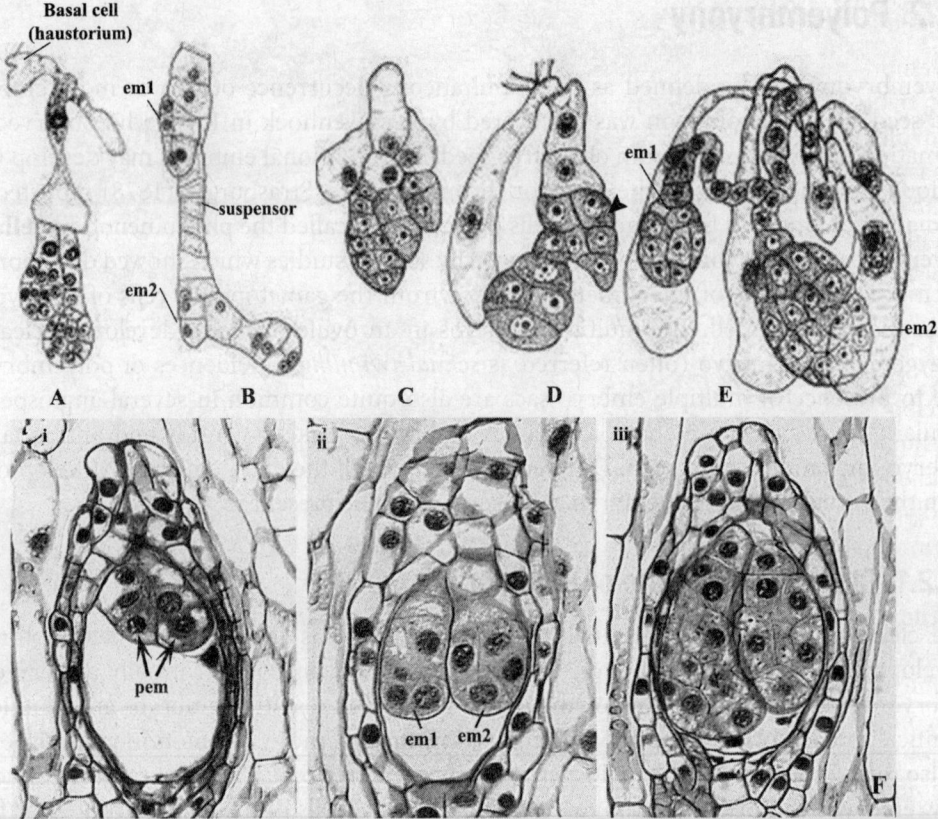

Figure 11.1 A. Cleavage polyembryony in *Stachyurus chinensis*. **B.** Embryonal buds (**em1** and **em2**) arising from the suspensor in *Zygophyllum fabago*. **C–E.** Development of multiple embryos in *Eulophia epidendraea*. **F.** Longitudinal section of fertilized ovules of *Cypripedium japonicum* (**i–iii**) showing embryo development due to cleavage of pro-embryo (pem). (*Source*: A–E: Lakshmanan and Ambegaokar 1984, *reproduced with permission*; F. Ghimire et al. 2020, published under CC BY License).

11.2.2 Formation of Embryos from Cells of Embryo Sac other than Egg Cell

In an embryo sac, the egg cell is destined to undergo fertilization and form a zygotic embryo. However, besides the egg cell, other constituent cells of an embryo sac like synergids, central cell, and antipodals also have the ability to form an embryo. Out of the three, synergids are the most common source of multiple embryos. Synergids can get fertilized by a sperm cell (e.g., *Poa, Aristolochia*) giving rise to diploid embryos. The formation of diploid embryo from a synergid can be the result of the entry of more than one pollen tube into the embryo sac, or due to presence of extra male gametes in the same pollen tube. The synergids also carry the potential to develop into haploid embryos without fertilization as seen in *Argemone mexicana* and *Phaseolus vulgaris*. Though embryos derived from antipodal cells are rare, they have been observed in species such as *Paspalum, Ulmus and Sedum*. Generally, such embryos degenerate early in their development and do not mature.

11.2.3 Presence of More than One Embryo Sac within Same Ovule

Polyembryonate conditions may also arise due to the occurrence of multiple embryo sacs in an ovule. Additional embryo sac(s) may develop from the somatic tissue (nucellus) of the ovule or from the derivatives of the same megaspore mother cell, and/or two or more megaspore mother cells may differentiate from the archesporial cells (*this phenomenon is known as apospory, a type of gametophytic apomixis*). In such cases, extra embryos develop from fertilization of egg cells of multiple embryo sacs. The incidences of multiple embryo sacs are frequent in families Poaceae (*Poa*, *Cenchrus*) and Asteraceae.

11.2.4 Formation of Embryos by Sporophytic Tissue of Ovule

Development of embryos from the diploid sporophytic cells of the ovule (integument or nucellus) is called **adventitious embryony**. It is the most common form of polyembryony and the embryos arising from the sporophytic cells are called **adventive embryos**. This type of polyembryony is also considered a type of *sporophytic apomixis*. Adventive embryony is commonly seen in *Citrus*, Mango, *Commiphora wightii* (Burseraceae), members of family Orchidaceae, Buxaceae, Cactaceae, Myrtaceae and Bignoniaceae. Embryos derived from nucellus are far more common than those obtained from the integuments. To distinguish the origin of adventive embryo(s), terms such as nucellar embryony (embryos arising from the nucellus) and integumentary embryony (embryos arising from the cells of integument) are commonly used. Naumova (1981) suggested that type of adventive embryony is often associated with the type of ovule. In crassinucellate ovules, **nucellar embryony** is predominant, e.g., *Citrus*, *Opuntia* and in tenuinucellate ovules, **integumentary embryony** is more common, e.g., *Euonymus*.

About 70% of the reported studies of polyembryony mention nucellar embryony as the cause of additional embryos. *Citrus* and Mango (*Mangifera* sp.) are the classical and well documented examples of this type of polyembryony. In the developing seeds of *Citrus*, two to ten embryos (and many times up to 13) are of frequent occurrences. Thus, at the time of germination, multiple seedlings can be observed coming out from a single seed.

Development of adventive embryos: Some of the diploid sporophytic cells of nucellus and integuments have the potential to form embryos. Such cells can be distinguished by their dense cytoplasm, rich in cell organelles like ribosomes, endoplasmic reticulum and plastids. In *Citrus* and *Handroanthus serratifolius* (Fig. 11.2 and 11.3), such cells gradually develop a thick cell wall blocking their cytoplasmic connections with the surrounding cells, and are hence called Adventive Embryo Initial Cells (AEICs) or Adventitious Embryo Precursor cells (AEPs; Alves et al. 2016). The AEICs soon start active divisions and this results in the formation of a small group of cells which protrude into the embryo sac and eventually develop into mature embryos. Mostly, the zygotic embryo also develops at the same time; however, there are several features of adventive embryos which distinguish them from sexual ones. The adventive embryos lack a suspensor while the sexual embryos have a suspensor anchored at the micropylar end of the embryo sac. Also, adventive embryos are always in lateral position whereas the sexual embryos always remain upright being attached to the micropylar end. Adventive embryos can be present at different developmental stages

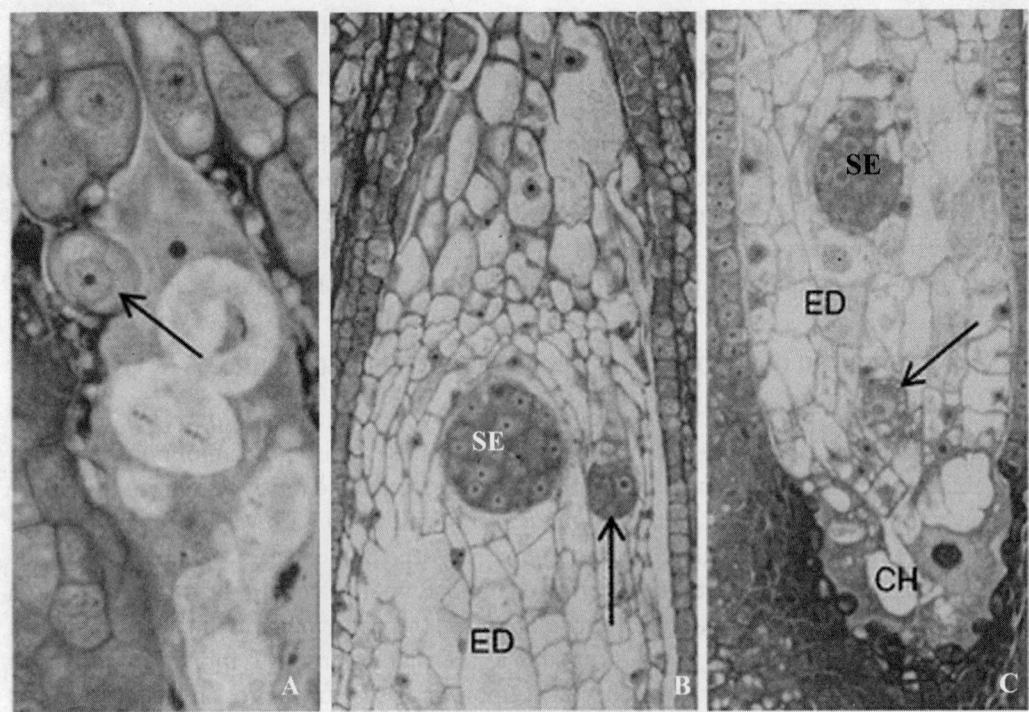

Figure 11.2 Adventitious embryos/polyembryony in *Handroanthus serratifolius*. **A.** Differentiated adventitious embryo precursor cell (or AEP; arrow) in the integument towards the micropylar region. **B–C.** Adventitious embryo (arrow) along with sexual embryo (**SE**) at micropylar end and at chalazal end (arrow, C). **CH:** chalazal haustorium; **ED:** cellular endosperm. **See Color Plates (page 496).** (*Source:* Alves et al. 2016, *reproduced with permission*.)

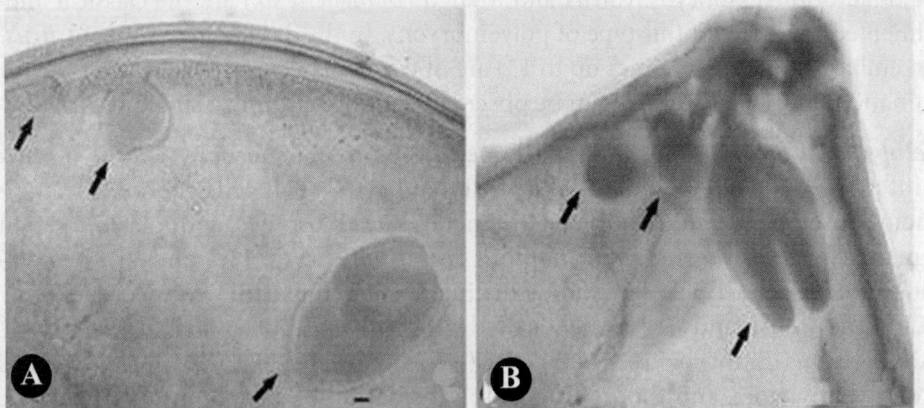

Figure 11.3 Incidences of nucellar polyembryony in *Commiphora wightii*. **A–B.** Multiple embryos originating from nucellus can be seen which are at different developmental stages. A heart-shape embryo is apparent in B. **See Color Plates (page 496).** (*Source:* Geetha et al. 2013, *reproduced with permission*.)

in an ovule as their growth and differentiation are not synchronous (e.g., in *Commiphora wightii*; Fig. 11.3). They even lack a definite shape while sexual embryos pass through several developmental stages each with a characteristic shape.

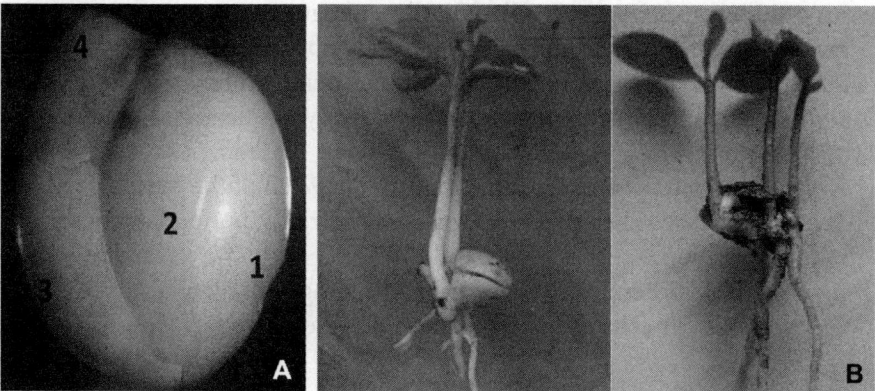

Figure 11.4 A. Mature seed of *Citrus* sp. showing polyembryony (4 embryos). **B.** Polyembryonate seed at germination showing emergence of multiple seedlings. **See Color Plates (page 496).** (*Source:* Kishore et al. 2012, *reproduced with permission.*)

Adventive embryony may or may not require pollination and fertilization for stimulation (Maheshwari 1952). In *Nigritella nigra* neither pollination nor fertilization are required for formation of adventive embryos. In most other plants, either stimulus of pollination, or pollination followed by fertilization, is necessary for the development of adventive embryos. In *Citrus*, it has been observed that poor growth or absence of endosperm, results in the reduction in the growth of adventive embryos. However, the growth of sexual embryo is not affected.

11.3 Classification of Polyembryony

Based on the source and genetic constitution of embryos, different workers have proposed different classifications for polyembryony. Typically, polyembryony is a naturally occurring phenomenon and is thus always spontaneous. However, many scientists have attempted the induction of multiple embryos under *in vitro* conditions using chemicals, X-ray irradiated pollen grains and through tissue culture practices (somatic embryogenesis). Such type of polyembryony is known as the induced type of polyembryony.

- Spontaneous polyembryony has been further classified by Ernst (1910) into two classes; on the basis of origin of supernumerary embryos:
  i. *True polyembryony:* When two or more embryos in an embryo sac originate either through cleavage of zygote/embryo or from any other cell of embryo sac (synergid, antipodal); or from sporophytic cells (nucellus, integuments) of the ovule. E.g., *Citrus*, mango.
  ii. *False polyembryony:* Occurrence of multiple embryos due to the presence of multiple embryo sacs in an ovule. E.g., *Poa, Cenchrus.*
- In 1969, Bouman and Boesewinkel categorized spontaneous polyembryony into four categories, based on the origin of the supernumerary or additional embryos:
  i. Due to cleavage of zygote or proembryos (i.e., from a new sporophytic cell).
  ii. Due to the presence of more than one embryo sac within the same ovule (from the egg cells) or any other cell of embryo sac.

    iii. From the sporophytic cells (nucellus, integuments).
    iv. From the male gametophyte (rare and disputed).
- Based on the genetic constitution of the originating cell of supernumerary/
  additional embryos spontaneous polyembryony has been classified into two types
  by Yakovlev (1967):
    i.   Gametophytic: embryos develop from any cell of the female gametophyte or
         embryo sac other than the egg cell (synergids, antipodals), with or without
         fertilization.
    ii. Sporophytic: embryos develop from cleavage of zygote or embryo or from the
         sporophytic cells of an ovule (nucellus, integuments).

11.4 Causes and Inheritance of Polyembryony

Although many attempts have been made so far to explain the causes of polyembryony; still
there is a huge gap in our understanding of the phenomenon. Haberlandt (1921; 1922) proposed
the **necro-hormone theory** as plausible cause of polyembryony. According to the theory, the
degenerating cells of the nucellus act as a source of stimulus for the surrounding cells to divide
and form adventive embryos. However, the theory could not be validated in subsequent studies
because adventive embryos could not be induced by damaging nucellar cells.

Ganeshaiah et al. (1991) proposed a model for the occurrence of polyembryony in *Citrus*
sp. in the context of **parent–offspring conflict theory**. According to this theory, a plant
produces numerous offsprings (seeds) in one reproductive event. During the development,
each offspring attempts to solicit maximum resources from the mother and act as a 'selfish
offspring'. This contest for resources between offspring leads to reduction in the brood size
(i.e., *number of seeds in an ovary*). On the other hand, the mother (maternal parent) puts
in effort to equally distribute resources among all offspring and consequently increase its
brood size. This causes a conflict between the mother and the offspring over the number
of seeds to be produced or developed. A plausible explanation provided for polyembryony
is that the mother enhances its brood size by producing additional embryos within a seed
rather than producing more seeds. Thus, it was suggested that the polyembryony is favored
by the maternal parent. A recent study by Woo et al. (2019), revealed maternal regulation of
polyembryony along with suppression of zygotic embryo. In this study, the offspring derived
from several genetic crosses using polyembryonic citrus cultivars as a female parent, showed
enhanced polyembryonic traits and resulted in poor occurrence of zygotic individuals.

Phytohormonal regulation of polyembryony has also been shown in some plants.
In *Citrus*, presence of volatile and non-volatile embryogenic inhibitors leads to a
monoembryonate condition. On the other hand, the ovules of polyembryonate species
lack such volatile and non-volatile embryogenic inhibitors. The non-volatile components
of the inhibitors are auxins, ABA and GA3. Ethanol and ethylene are the important volatile
inhibitors produced by the ovule of monoembryonate species of citrus (*Citrus medica*).
Presence of ethylene receptor proteins in nucellar cells has also been suggested. Such
receptors are thought to bind with the ethylene molecules and repress the development of
nucellar embryos.

Earlier studies on **genetic regulation** of polyembryony have suggested that the polyembryony in *Citrus* and mango is under the control of a gene with two alleles; 'P' and 'p'. The polyembryonic condition is determined by a heterozygous state of alleles (Pp), while monoembryonic condition (in citrus species) is determined by homozygous recessive (pp) condition (Parlevliet & Cameron 1959; Aron et al. 1988). It has been proposed by Esen & Soost (1977), that in monoembryonic species, these recessive alleles may synthesize a potent inhibitor of embryogenesis. However, in a cross between monoembryonic and polyembryonic parents, and between polyembryonic parents; a variable ratio of polyembryonic offspring has been recorded. The variabilities encountered suggest that apart from the products of P gene, some modifier or duplicate genes also play a minor role in defining the polyembryonate or monoembryonate condition. Recently Wang et al. (2017) identified a gene *CitRWP* in *Citrus* whose expression is specifically higher in polyembryonic cultivars and is assumed to be the key candidate gene controlling polyembryony. To identify whether a particular seedling is zygotic or nucellar in origin, several genetic markers based on RAPD, SSRs have been identified in mango and *Citrus*. These genetic markers have enormous implications in the breeding programs of these two crop species. However, these markers do not give much information about the inheritance pattern of polyembryony except its maternal regulation.

11.5 Practical Applications and Significance of Polyembryony

Earlier, polyembryony was thought to be a developmental abnormality. But research on various aspects of polyembryony made it clear that it is a phenomenon with possible practical applications and might be very useful in the field of cytogenetics and plant breeding.

    i.   Nucellar polyembryony has a great role in horticulture and crop breeding because it helps in producing genetically uniform seedlings of the desired parental type for better clones of shoot and rootstock.

    ii.   Polyembryony helps in the large-scale propagation of desired genotype which turns out better than the cuttings.

    iii.   Repeated vegetative propagation often leads to loss of vigor which is not seen in nucellar seedlings.

    iv.   The nucellar embryos are free from diseases and *in vitro* induction of nucellar embryony is the only practical method to raise virus free clones in *Citrus*.

    v.   Embryos arising from the gametic cells are haploid in nature. The haploids are very useful for cytogenetical studies and crop improvement.

    vi.   Homozygous diploids may be raised from haploids which are very useful for trait study and transfer during crosses.

    vii.   Adventive embryos can be used for producing artificial seeds.

11.6 Apomixis

Apomixis may be defined as an asexual mode of reproduction in which the ovule develops into a seed without undergoing meiosis and fertilization. Thus, apomictic plants produce seeds which are of identical genotype as the maternal parent. This makes apomixis, one of the most convenient forms of clonal propagation mediated by seed. Most apomicts occur in families Asteraceae, Rosaceae and Poaceae. Many studies have suggested that apomixis occurs more commonly in polyploids than in diploids. Also, apomixis is not very common among commercially important agricultural crops and trees with few exceptions like *Citrus*, *Pennisetum*, almond and apple.

Apomictic plants may also undergo sexual reproduction for seed set. Nearly 80% angiosperm species exhibit incidences of apomixis along with sexual reproduction in numerous identifiable combinations. The plants where seed set is achieved through both sexual reproduction and apomixis are known as facultative apomicts and the phenomenon is known as facultative apomixis (e.g., *Hieracium*, *Ranunculus kuepferi*, *Poa*, *Hippophae rhamnoides*, *Cortaderia*). Plants where seed set is possible only through apomictic pathways and there is complete absence of sexual reproduction are known as obligate apomicts. A very few incidences of obligate apomixis are known to exist as in *Pennisetum ciliare*, and pentaploid *Paspalum dilatatum*.

Crosses between sexual and apomictic types, in facultative apomictic species usually give rise to progenies that exhibit a range of mating behavior. Similar results were also observed by Mendel during his experiments with *Hieracium*. Mendel attempted to replicate his experiments of garden pea in *Hieracium* not knowing that latter was an apomict. In the crosses (both self- and cross-) he obtained maternal type plants along with hybrids. Also, the sister hybrids obtained in the F1 generation, were not uniform. This was in contrast to his first law of dominance. Additionally, he found that the progenies were uniform in F2 which was not concordant with his second law of independent assortment.

11.7 Types of Apomixis

Sexual reproduction in angiosperms (also termed as amphimixis), occurs through a series of events leading to the formation of a fertile and genetically unique seed. The events occurring in the pistil include; differentiation of archesporial cell from nucellus, differentiation of megaspore mother cell from archesporial cell, mega-sporogenesis, mega-gametogenesis, double fertilization, embryo-endosperm development; ultimately leading to the formation of a seed. Apomictic pathways produce fully mature and viable embryo in the seed by completely eliminating some of the events in the series. Nonetheless, embryo development through apomixis temporally mimics sexual reproduction. Thus, apomixis and sexual reproduction represent alternative strategies of reproduction that are developmentally related (Curtis & Grossniklaus 2008). Apomictic species have the following three characteristic features: 1. Generation of a cell capable of forming an embryo without undergoing meiosis (apomeiosis), i.e., the female gamete remaining unreduced; 2. Parthenogenesis, i.e., the spontaneous development of the embryo independent of fertilization; (despite being functional; male and female gametes never fuse) and 3. The potential to produce an endosperm autonomously or derived from fertilization (Koltunow 1993; Carman 1997; Bicknell & Koltunow 2004). Apomictic embryo development can happen through different pathways and based on that, apomixis can be of two types: sporophytic apomixis (or adventitious embryony) and gametophytic apomixis.

11.7.1 Sporophytic Apomixis

This is also referred to as adventive embryony which has already been discussed in section 11.2.4. In sporophytic or adventitious embryony, the embryos arise spontaneously from the sporophytic cells of an ovule like the nucellus and the integuments. The adventitious embryos generally develop late in the temporal sequence of ovule maturation (Fig.11.3 and Fig. 11.5). Here, sexually derived embryos either degenerate or compete with the apomictic embryo. Sporophytic apomixis differs from vegetative reproduction in formation of an embryo, and retaining the seed habit.

11.7.2 Gametophytic Apomixis

Gametophytic apomixis is characterized by the formation of unreduced single or multiple embryo sacs in an ovule and development of embryo from an unreduced egg cell without fertilization, i.e., parthenogenesis. The unreduced embryo sac(s) may be derived from unreduced MMC which circumvents meiosis or from the initials differentiated from the somatic cells of the ovule (generally nucellus). Based on this, gametophytic apomixis can be further divided into two types; **Diplospory** and **Apospory**. Further subdivisions of gametophytic mechanisms can be made based on features related to the involvement or avoidance of the different phases of meiosis, the number of mitotic divisions, and the final organization of the embryo sac (Fig. 11.5 and Fig. 11.6) (Crane 2001; Bhat et al. 2005).

11.7.2.1 *Diplospory*

In diplospory, the megaspore mother cell (MMC) or a cell at a position similar to that of the megasporocyte in an ovule forms the unreduced embryo sac. Diplospory is further divided into two types: **meiotic diplospory** and **mitotic diplospory**. In meiotic diplospory, a megaspore mother cell differentiates from the nucellus and undergoes meiosis. However, meiosis is inhibited at a particular stage leading to the formation of a restitution nucleus that eventually undergoes mitosis to form an unreduced embryo sac. In mitotic diplospory, the megaspore mother cell does not enter meiosis and instead undergoes mitosis to form an unreduced embryo sac (Nogler 1984).

Till date seven different types of diplosporous embryo sac developments have been reported and named after the genera in which they have been first studied (Nogler 1984; Crane 2001; Bhat et. al. 2005). In these types (except the Eragrostis-type), the final organization of female gametophyte is a typical 7-celled, 8-nucleate Polygonum-type (Fig. 11.6).

Meiotic Diplospory

    i.   **Taraxacum type:** The MMC enters the meiotic prophase but homologous chromosomes do not show any pairing due to precocious asynapsis. Thus, univalents remain scattered over the spindle at metaphase I and a restitution nucleus is formed after the first meiotic division. The MMC subsequently divides mitotically to form a dyad with somatic (2n) chromosome number. Generally, the micropylar cell of the dyad degenerates and chalazal cell undergoes three

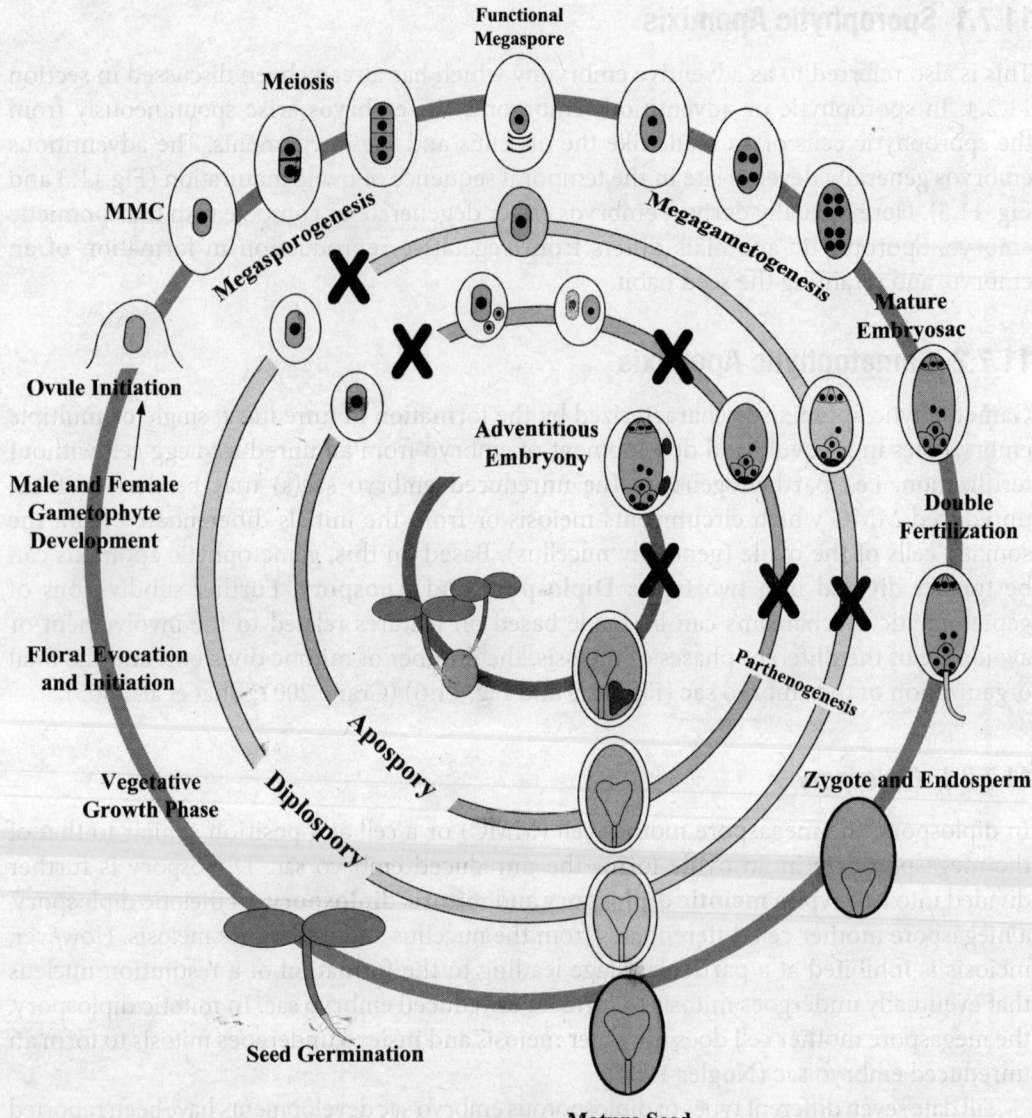

Figure 11.5 Schematic diagram comparing key events during sexual reproduction and different types of apomixis. *Diplospory:* Megasporogenesis is hampered due to formation of restitution nuclei or premeiotic chromosome doubling. Hence, unreduced megaspore gives rise to unreduced egg cell. *Apospory:* Aposporous initials appear from nucellus and sexual MMC degenerates. Later aposporous initial gives rise to unreduced female gametophyte (may be multiple in the same ovule). ***Adventitious embryony (sporophytic apomixis):*** Adventitious embryo precursor cells differentiate either from nucellus or integuments, which give rise to additional/supernumerary embryos in a mature seed. At germination, multiple seedlings may be seen. **See Color Plates (page 497).** (*Source:* Adapted and modified after Bicknell & Koltunow 2004.)

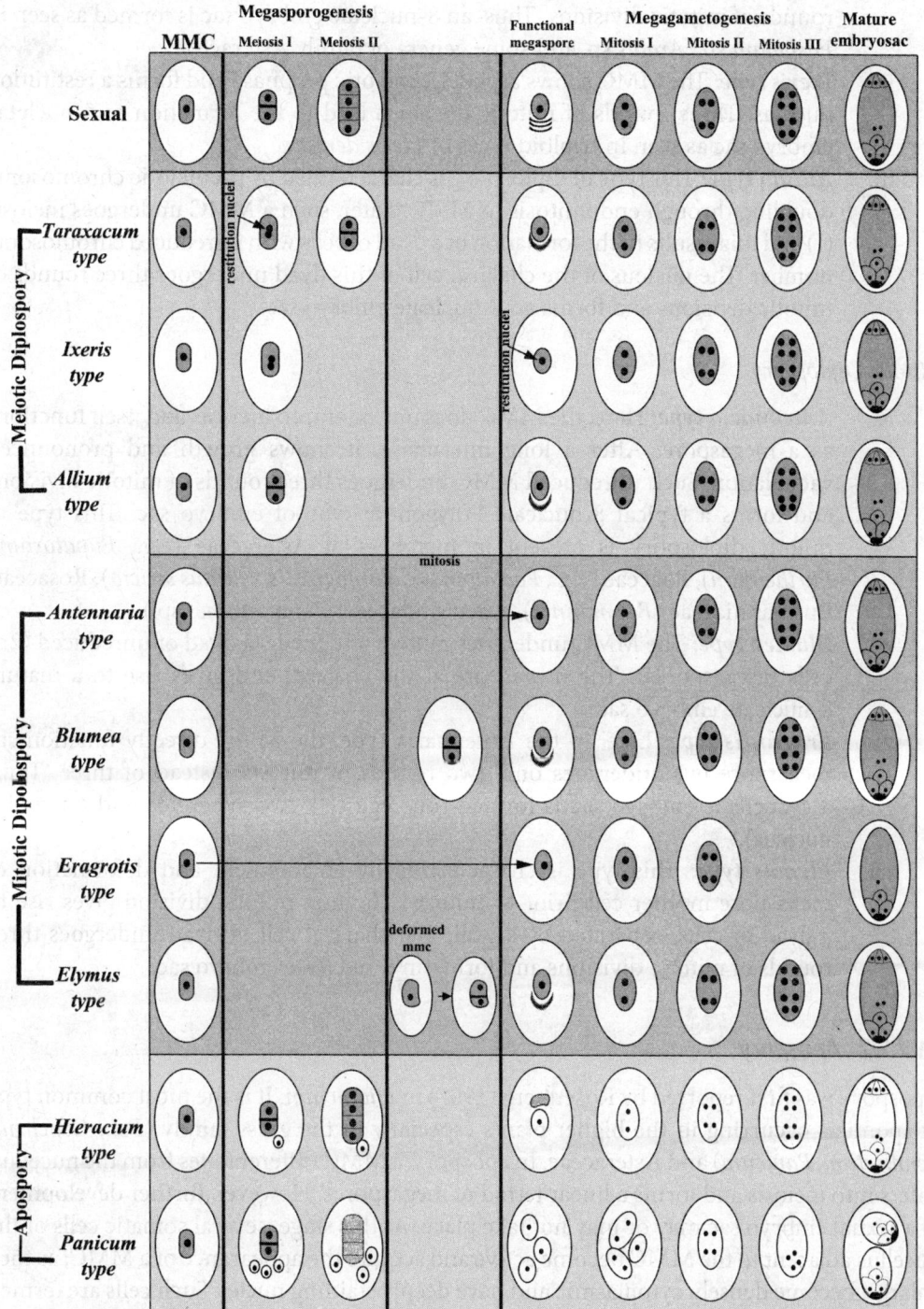

Figure 11.6 Diagrammatic scheme of embryo sac development in different types of gametophytic apomixis. **MMC:** megaspore mother cell. The micropyle is towards the bottom of page. (*Source:* Adapted and modified after Bhat et al. 2005.)

rounds of mitotic divisions. Thus, an 8-nucletae embryo sac is formed as seen in *Paspalum* sp., *Arabis* sp. and some genera of family Asteraceae.

ii.   *Ixeris type:* The MMC shows asyndetic meiotic prophase and forms a restitution nucleus. Three rounds of mitotic divisions lead to the formation of 8-nucletae embryo sac as seen in triploid races of *Ixeris dentata*.

iii. *Allium type:* This type of diplospory is characterized by premeiotic chromosome doubling through endomitosis in MMC. Later, such a MMC undergoes meiosis (I) and this results in the formation of a dyad of cells with unreduced chromosome number. The nucleus of the chalazal cell of this dyad undergoes three rounds of mitotic divisions and forms an 8-nucleate embryo sac.

Mitotic Diplospory

iv.   *Antennaria type:* Here, the MMC does not enter into meiosis and itself functions as a megaspore. After a long interphase, it shows growth and pronounced vacuolation. Such unreduced MMC undergoes three rounds of mitotic divisions and forms a typical 8-nucleate Polygonum-type of embryo sac. This type of mitotic diplospory is present in members of Asteraceae (e.g., *Eupatorium, Parthenium*), Poaceae (e.g., *Poa alpina, Calamogrostis, Nardus stricta*), Rosaceae, Burmanniaceae (*Burmannia*), Amaryllidaceae (*Zephyranthes* sp.).

v.   *Blumea type:* The MMC undergoes mitosis and forms a dyad of unreduced (2n) cells (megaspores). The megaspore at the chalazal end gives rise to a mature 8-nucleate embryo sac.

vi.   *Eragrostis type:* Like, in the Antennaria type, the MMC directly functions as megaspore but undergoes only two rounds of mitosis instead of three. Thus, a 4-nucletae embryo sac is formed (one egg cell, two synergids and one polar nucleus).

vii. *Elymus type:* This type is characterized by enlargement and deformation of megaspore mother cell prior to mitosis. The first mitotic division gives rise to a dyad of cells, separated by a wall. The chalazal cell of dyad undergoes three rounds of mitotic divisions and forms an 8-nucleate embryo sac.

11.7.2.2 *Apospory*

Apospory was first reported by Rosenberg (1907) in *Hieracium*. It is the most common type of apomixis occurring in the higher plants especially in the grass family (*Poa, Cenchrus, Pennisetum, Panicum*) and Asteraceae. In apospory, a MMC differentiates from the nucellus, enters into meiosis and forms a linear tetrad of megaspores. However, further development of a sexual embryo sac may or may not take place. At this stage, several somatic cells of the nucellus adjacent to the MMC become active and acquire the appearance of a MMC; as they enlarge, become densely cytoplasmic and have deeply staining nuclei. Such cells are termed as aposporous initials which divide and form unreduced embryo sacs. Thus, in contrast to diplospory, where the megaspore mother cell is the direct progenitor of the unreduced embryo sac, aposporous embryo sacs develop from nucellar cells which differentiate after megaspore mother cell differentiation. In many incidences; sexual or reduced embryo sacs may also attain maturity along with the aposporous embryo sacs. However, the development of the sexual embryo sac is often terminated at the megaspore mother cell or the megaspore stage. Also, aposporous embryo sacs develop faster than the sexual embryo sacs because they skip meiotic division. In apospory, the embryos develop parthenogenetically but endosperm formation may need pollination and fertilization. Apospory is of following two types (Fig. 11.6):

i. **Hieracium type:** In this type of apospory, several cells of nucellus adjacent to the chalazal end of sexual MMC become active and behave as aposporous initials. In certain incidences of Hieracium-type only single aposporous initial develops. Such aposporous initials divide mitotically and form an unreduced embryo sac which is 8-nucletae Polygonum-type. During the development of such an aposporous embryo sac, the sexual embryo sac may persist or degenerate. If an unreduced embryo sac persists, then in a mature ovule both reduced and unreduced embryo sacs will coexist.

ii. **Panicum type:** This type of apospory is characterized by appearance of several aposporous initials in the nucellus along with the MMC in an ovule. Generally, MMC degenerates and multiple embryo sacs develop from the unreduced aposporous initials. Here, a mature embryo sac is 4-nucletae (one egg cell, two synergids and one polar nucleus) with the absence of antipodals and second polar nucleus. The development of multiple aposporous initials may be asynchronous. Thus, multiple embryo sacs in an ovule may be seen at different developmental stages (Fig. 11.7). Generally, only one embryo sac gives rise to an embryo in a mature seed.

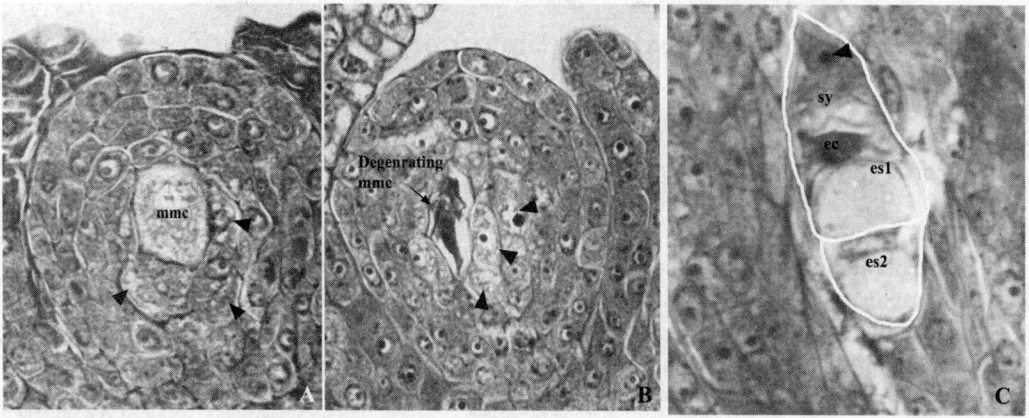

Figure 11.7 Longitudinal sections of ovules showing incidences of apospory in *Hippophae rhamnoides*. **A.** Arrowheads indicate the development of several aposporous initials along with megaspore mother cell (mmc). **B.** Degeneration of mmc. **C.** Presence of two embryo sacs (es1 and es2) in an ovule. In es1, a synergid (**sy**) and egg cell (**ec**) can be seen.

Box 11.2

Differences between Sporophytic Apomixis and Gametophytic Apomixis

Sporophytic Apomixis	Gametophytic Apomixis
• Sporophytic apomixis, or adventitious embryony is the development of embryo directly from sporophytic tissue of an ovule.	• Gametophytic apomixis is the development of embryo from the unreduced embryo sac in an ovule.
• Embryos are derived from the unreduced sporophytic cells of ovular tissues, i.e., nucellus, integuments.	• Embryos are derived from the unreduced egg cell.
• Embryo sac formation is not involved.	• Unreduced MMC gives rise to an embryo sac.
• In a seed, multiple embryos reside and give rise to multiple seedlings.	• In a seed, usually single embryo develops giving rise to a single seedling.
• Meiosis is completely absent.	• MMC may enter meiosis as is seen in some species.
• Usually, pollination stimulus and endosperm formation is required for the development of adventitious embryo.	• Pollination stimulus may or may not be required for embryo formation. Even endosperm formation may be autonomous.
• Commonly present in diploid taxa.	• Commonly associated with polyploids.
• Most common in *Citrus* sp., (Rutaceae), Mango (Anacardiaceae).	• Most common in Asteraceae, Poaceae, Rosaceae members.

Box 11.3

Differences between Diplospory and Apospory

Diplospory	Apospory
• Diplospory is development of unreduced embryo sac from MMC without meiosis.	• Apospory is the development of unreduced embryo sac from somatic cells of nucellus other than the MMC, i.e., aposporous initials.
• MMC may or may not enter meiosis.	• Aposporous initials do not enter meiosis.
• Formation of restitution nucleus may take place	• No such stage is seen in apospory.
• Invariably single embryo sac is present in an ovule.	• Single or multiple embryo sacs might be present in an ovule.
• Mostly mature embryo sac has typical *Polygonum* type of arrangement (except Eragrostis-type).	• Mature embryo sac may be Polygonum-type (in Hieracium-type) or may be 4-celled (as in Panicum-type, with antipodals being absent).
• No reduced embryo sacs are present in the ovule.	• Reduced embryo sacs may coexist with unreduced ones.
• More frequent in Asteraceae.	• More frequent in Poaceae.

11.8 Role of Pollination in Embryo and Endosperm Development among Apomicts

There are many apomictic species where apomictic seed formation requires the stimulus of pollination for endosperm development. Apomicts where endosperm formation takes place after the fusion of polar nuclei with sperm cell are called *pseudogamous apomicts*. Thus, in pseudogamous apomicts, pollination with fertile and viable pollen grains is mandatory for endosperm development. The members of family Rosaceae, *Ranunculus,* and *Panicum* are typically pseudogamous apomicts. However, in some apomictic species, endosperm development is autonomous, i.e., not dependent on the pollination and fertilization, and such apomicts are known as *non-pseudogamous apomicts* (e.g., *Hieracium, Antennaria*). Non-pseudogamous or autonomous apomicts such as *Hieracium* spp. can generate viable seeds in the complete absence of pollination. Similarly, in members of Poaceae, the embryo development can be observed prior to pollination. However, in some cases of non-pseudogamous apomicts, pollination might be required for stimulating and triggering embryo development.

11.9 Genetics of Apomixis

Studies analysing the progeny of crosses between apomictic and sexual forms have shown that apomixis is a genetically determined mechanism of reproduction. These studies suggest that apomixis is inherited as a dominant trait. The nucellar polyembryony (sporophytic apomixis) is controlled by a single dominant locus as discussed in section 11.4. Most of the studies investigating inheritance of gametophytic apomixis are limited to plants exhibiting apospory, viz. *Pennisetum* sp., *Panicum, Ranunculus.* Studies show that apospory is probably controlled by a single dominant gene locus (Asker & Jerling 1990). Savidan (1982) postulated that a supergene 'A' is responsible for apospory and parthenogenesis therein. Similarly, the other forms of gametophytic apomixis are controlled either by a single dominant Mendelian gene (as studied in many aposporous grasses, Savidan 1982; Nogler 1984), or by a small number of closely linked loci. However, for diplospory and parthenogenesis two unlinked loci have been recognized in Asteraceae. Studies involving sexual mutant analysis have also revealed that loci segregating with apomixis exhibit suppressed recombination. The identification of rare recombinants, or the genetic dissection of large genomic loci associated with apomixis revealed that three key events of apomixis, i.e., apomeiosis, parthenogenesis, and fertilization-independent endosperm formation, are each controlled by independent loci. Recent studies have established that apomictic and sexual pathways share a common network of genes functioning during the development of seed. Some of the genes involved are *DYAD, SWI1, Elongate1, SERK, ARG, MiMe*sets, *AGO, DMT, BBM, FIE, MEA, DME* (refer to Box 11.4 and 11.5 for details). Mutants of these genes/genomic regions exhibit apomixis components, which otherwise are involved in essential functions during megasporogenesis, meiosis initiation and progression, mega-gametogenesis, embryogenesis and endosperm development (Kaushal et al. 2019). Recent reports suggest that the apomixis is also epigenetically regulated (Hand & Koltunow 2014; Kaushal et al. 2019) (refer to Box 11.6 for details).

Box 11.4

Genetic Loci Linked to Apomixis

Direct identification of the genes and genomic determinants of apomixis have not seen much success. However, apomixis linked loci have been identified in some species. The genetic mapping studies suggest that genomic location of loci involved in the initiation of apomixis are few hundred kilobases long. In *Hieracium praealtum*, a member of Asteraceae, apomixis has been shown to be controlled independently by two unlinked loci: LOSS OF APOMEIOSIS (LOA) and LOSS OF PARTHENOGENESIS (LOP). LOA is essential for apomeiosis and suppression of all sexually derived megaspores, the activity of LOP in female gametophyte is required for autonomous development of both embryo and endosperm. Deletion of LOA or LOP results in partial reversion to sexual reproduction which suggests that apomixis is superimposed on a sexual pathway. Another locus associated with apomixis is Apomixis Linked Locus (*APOLLO*) in *Boechera*. *APOLLO* transcripts are down-regulated in sexual ovules when they enter meiosis and are up-regulated in apomeiotic ovules at the same stage of development.

In several species, the suppression of recombination around apomixis loci is found to be associated with heterochromatic or highly repetitive genomic regions. In *Pennisetum squamulatum*, the apospory-specific genomic region (ASGR) is located at the telomeric region of chromosomes containing abundant repetitive sequences and transposons. ASGR in *Pennisetum squamulatum* is found to be in the hemizygous condition (*i.e., single allele in diploid genome*) and in the chromosomal region containing LOA. Structurally ASGR of *P. squamulatum* is similar to genetic loci related to apospory in *Hieracium* spp. These structural similarities suggest that this locus might be necessary for function and maintenance of the apomictic trait. In *Paspalum* sp., the hemizygous apomictic controlling locus (ACL) also shows suppression of recombination and has undergone large-scale rearrangements due to transposable elements.

Inheritance of diplospory as single dominant locus has been described for *Erigeron anuus* and for *Taraxacum officinale*. In *Taraxacum*, two unlinked dominant loci control diplospory (DIPLOSPOROUS, DIP) and (PARTHENOGENESIS, PAR). Through transgenic mutant analysis of *Arabidopsis thaliana*, three genes have been proposed for apomeiosis. These genes which were found to be related with aberrant meiosis and formation of restitution nuclei in diplospory are: SWITCH1 (*SWI1*)/*DYAD* causing defects in female meiosis, CYCLIN A12 (*CYCA12*), also known as *TARDY ASYNCHRONOUS MEIOSIS* (*TAM*), forming unreduced female gametes due to failure of meiosis II. *CYCA12* codes for a cyclin A protein necessary for meiotic cell cycle progression, and *OMISSION OF SECOND DIVISION1* (*OSD1*) inhibits meiosis II. *PsASGR-BABYBOOM-like* (*PsASGR-BBML*) is the first gene responsible for parthenogenesis in the egg cell from ASGR to be isolated from the natural apomictic plants of *P. squamulatum*. This gene expresses in apomictically derived egg cells before double fertilization, and its transgenic activity in sexual individuals of tetraploid pearl millet lines induce the production of viable maternal haploids.

Polycomb-group (PcG) chromatin modelling complex, in particular, the Polycomb Repressive Complex 2 (PRC2) is known to play a role in suppressing seed development in the absence of fertilization. The FIS-PRC2 complex consists of several genes, viz. *MEDEA* (*MEA*), *FIS2*, *FERTILIZATION-INDEPENDENT ENDOSPERM* (*FIE*), and *MULTICOPY SUPPRESSOR OF IRA1* (*MSI1*). Loss of function/mutation in any of these genes results in fertilization-independent endosperm development and, in the case of *msi1* mutants, parthenogenetic embryo initiation was observed although viable seeds are not formed. The down-regulation of FIS–PRC2 genes investigated in autonomous apomict *Hieracium*, does not result in fertilization-independent seed development in sexual plants, although it is required for fertilization-independent embryo and endosperm development in apomicts (Rodrigues et al. 2008). This study highlighted that FIS–PRC2 complex and *HMSI1*, though expressed in apomictic and sexual ovaries within *Hieracium*, is unlikely to be linked to the LOP locus. The authors suggested that Hieracium *MSI1* (*HMSI1*) may be involved in fertilization-independent seed development downstream of LOP activity. Such studies demonstrate that though sexual developmental mutants identify candidate genes that initiate apomixis-like events, the natural apomictic pathway may be somewhat more complex with additional or alternative regulators.

Source: Koltunow et al. 2011; Kaushal et al. 2019; León-Martínez & Vielle-Calzada 2019.

11.10 Potential Applications of Apomixis

Apomixis offers huge potential for crop improvement and breeding. Generally, the hybrid seeds are known to lose their vigor after few generations, creating problems for farmers and crop breeders. In such cases, apomixis appears as an important trait which could lead to the formation of genetically uniform progenies or true-breeding hybrids through seeds. Thus, apomixis if once introduced in the crop may help in retaining the hybrid vigor in successive generation due to absence of genetic mixing and exchange. An apomictic pathway also obviates the need of pollinators and pollination. Hence, unavailability of pollinators may not lead to loss of yield in cross-pollinated species. Apomixis is also thought to be advantageous for rapid generation and multiplication of superior genotypes through seeds. It also reduces the cost and time involved in the conventional breeding programs. Moreover, apomixis provides reproductive assurance by overcoming the self-incompatibility barrier. Propagules of vegetative reproduction are prone to infection by pathogens, especially viruses. Virus free seeds produced via apomixis can be an alternative because such seeds carry much fewer plant viruses.

Scientists have been able to introduce the apomictic mode of reproduction in sexual crop plants through: a) hybridization or wide crosses with wild relatives of the apomicts, b) mutation breeding, and c) genetic transformation techniques. Hybridization or wide cross methodology to transfer apomixis through inter-specific hybridization rely on the availability of wild relatives. It is not necessary that for every targeted crop wild relatives with apomixis are available for crossing. So, first screening the reproductive pathway in wild relative or closely associated species needs to be done, but this is very tedious. Even if such species are available several generations of crosses are required. Thus, this method is very difficult, laborious and time consuming. The other two methods depend on direct transformation technologies and tissue culture, and hence, are preferred more.

Box 11.5

Genes Associated with Apomixis Traits

Some of the genes which demonstrate a phenotype inherent in apomicts are following:

Gene	Biochemical nature of the protein	Function
DYAD/SWITCH	Contains phospholipase C, phosphatidylinositol-specific domain. Involved in sister chromatid cohesion and meiotic chromosome organization during meiosis.	Loss of functional mutants form dyads instead of tetrads
UVI4-LIKE (OSD1)	Controls entry into the second meiotic division. Negatively regulates the anaphase-promoting complex/cyclosome	Mutants demonstrate diplospory
SPO11	Encodes DNA topo-isomerase VIA subunit, required for meiotic recombination	Involved in chromatin organization and promoting diplospory
REC8	Encodes a RAD21-like gene essential for meiosis Involved in homologous chromosome segregation, double-strand break repair, meiotic chromosome condensation, meiotic sister chromatid cohesion	Loss of functional mutants demonstrate diplospory phenotype
ARGONAUTE (AGO4, AGO6, AGO8, AGO5)	Encode an RNA slicer that selectively recruits microRNAs and siRNAs	Loss of functional mutants reveal apospory.
AGO9		Dipolospory
RDR6	Encodes RNA-dependent RNA polymerase	Loss of functional mutants produce ectopic megaspore mother cell
SGS3 (SUPPRESSOR OF GENE SILENCING 3)	Encodes for coiled domains suggesting oligomerization and a potential zinc finger domain	Phenotype of apospory
ARP6 (ACTIN-RELATED PROTEIN6)	Putative component of a chromatin-remodeling complex and histone modification in *Arabidopsis*	Mutants show defects in fertility and shorter flowering times. Loss of functional mutants demonstrate apospory
BBOOM (BABY BOOM)	Encodes an AP2-domain containing protein similar to ANT. AP2/ERF family of transcription factors	Induces parthenogenesis
LEC1	CCAT box-binding HAP3 subunit	A regulator of embryo development. Also involved in somatic embryogenesis.

LEC2	B3 family of transcription factors	Involved in somatic embryo-genesis
WUS	Homeobox protein required to keep the stem cells in an undifferentiated state	Induces somatic embryo formation
MSI1	Encodes a protein that functions in chromatin assembly as part of the CAF1 and FIE complex	Mutants exhibit parthenogenetic development of embryos and autonomous endosperm development
MEA	Component of PRC2, SET domain protein	Autonomous endosperm development. Mutations in locus cause embryo lethality
FIS2	DNA-binding transcription factor A negative regulator of seed development in the absence of pollination	Mutants show autonomous endosperm development, genomic imprinting
FIE	Encodes a protein Involved in chromatin silencing, histone methylation	Regulation of endosperm development, genetic imprinting
DEMETER (DME)	Encodes a DNA glycosylase DEMETER (DME)	Responsible for endosperm regulation mediated by the *MEDEA (MEA)*
APOLLO (Apomixis linked locus)	Aspartate glutamate Aspartate aspartate histidine exonuclease	Has both 'apoalleles' and 'sexalleles', apomictic *Boechera* sp. are heterozygous for the *APOLLO* gene, sexual genotypes are homozygous for sex alleles
HAPPY (Hypericum APOSPORY, HpARI)	Homologous to the *Arabidopsis thaliana* gene ARI7 encoding the ring finger protein ARIADNE7	Associated with apospory
AutE/AED (Autonomous Endosperm) (*Hieracium* subgenus *Pilosella*)	Loci of LOP and LOA region	Associated with apospory and autonomous endosperm formation
PsORC3a (*Paspalum simplex* Origin Recognition Complex)	Multiprotein complex which controls DNA replication and cell differentiation in eukaryotes	Specific for apomictic genotypes and aposporous initials
DMC1	Participation in meiosis in *Arabidopsis*	Diplospory
AGO 104	*AGO 104* and *AGO9* are homologous genes.	Their defects can produce a phenotype similar to diplospory in *Tripsacum*

Apostart	Speculated to participate in formation of 2n eggs (diplospory)	Diplospory in *Medicago sativa*
MSP1 (Multiple Sporocyte)	Controls early sporogenic development	Adventitious embryony
ORC (ORIGIN RECOGNITION COMPLEX)	Multiprotein complex which controls DNA replication and cell differentiation in eukaryotes	Adventitious embryony and apospory (*Paspalum simplex*)

Source: Adapted and modified after Brukhin & Baskar 2019.

Box 11.6

Epigenetic Control of Apomixis

Epigenetics refers to the heritable changes in gene expression pattern or phenotype of a gene without any alterations in its DNA sequence. The different epigenetic processes that change the way a gene works include DNA methylation, acetylation or chromatin modification, phosphorylation and ubiquitylation. Out of these, the best known epigenetic process, is DNA methylation. It is the addition or removal of a methyl group (CH_3), predominantly where cytosine bases occur consecutively. Methylation is associated with blocking of gene expression and demethylation does the opposite. Another process bringing about epigenetic regulation is chromatin modification. It involves addition of acetyl groups to chromatin, thus, altering chromatin structure to influence gene expression. In general, tightly folded chromatin does not display gene expression while it takes place in open chromatin.

The genetic bases of apomixis is still not clear. Despite much work, only a few genes responsible for apomixis are known. Rather, there is an increasing evidence that apomixis is controlled epigenetically. Various studies have shown that apomixis is a result of reversible down-regulation of genes involved in sexual pathways through epigenetic silencing. For example: Mutant analysis in *Arabidopsis* ovules have shown that mutants of *ARGONAUTE9* (*AGO9*) gene which is involved in the RNA directed DNA-methylation pathway, show apospory or diplospory (Olmedo-Monfil et al. 2010). Further support for epigenetic regulation of apomixis was provided by comparative analysis of sexual development in maize with the apomixis process in its wild relative, the apomictic *Tripsacum* (Garcia-Aguilar et al. 2010). The results show difference in expression of chromatin remodelling enzymes (e.g., CMT3 and DRM2) between apomicts and sexual individuals. Koltunow et al. (2011) while studying apomixis control in *Hieracium*, also indicated that apomixis is superimposed epigenetically over sexuality by the silencing action of two independent loci, LOA and LOP. In diplosporous *Eragrostis curvula*, an increased apomixis expression was associated with an increment in methylation of cytosine involving transposable elements (Zappacosta et al. 2014).

Through artificial phenotype reversion in a natural apomictic, *Paspalum simplex,* a significant repression of parthenogenesis was achieved by Podio et al. (2014). These results suggest that factors controlling repression of parthenogenesis might be inactivated in apomictic *Paspalum* by DNA methylation. Evidences for epigenetic control of apomixis are also known from *Cenchrus ciliaris*, where it was shown that certain regions of genome are hyper-methylated in obligate sexual plants compared to that in the obligate apomictic plants (Kumar & Bhat 2014).

All these studies, establish that during apomixis major genes involved in sexual pathway are down regulated epigenetically. However, it is still not known that suppression of sexual reproduction is the only driving force for establishment of apomixis.

Box 11.7

Apomixis: Polyploidy, Hybridization and Distribution

The gametophytic apomixis is a frequently occurring mode of reproduction in polyploid taxa. The sexual members of the same or closely related species are very commonly diploids. The reason(s) for this close association remains unclear. Recent studies have highlighted that polyploidy does not appear to be an absolute requirement for the expression of the apomixis though it may enhance the expression of apomixis in many systems. Gametophytic apomicts may form populations composed of individuals with different ploidy levels due to occasional fertilization of an unreduced egg with the sperm cell to give rise to polyploid plants. Another reason for common occurrence of gametophytic apomixis in polyploids could be the abnormal meiosis, where the chances of formation of fertile gametes are very low. This inability to form fertile gametes is thought to be an evolutionary force in polyploids to acquire apomixis. Such a population with individuals of varying ploidy levels is referred to as an *agamic complex*. As the sporophytic apomicts do not depend on the formation of unreduced gametes, they do not constitute agamic complexes.

Another important aspect of gametophytic apomixis is hybridization. Studies have recommended that hybridization does not induce apomixis *de novo* (e.g., *Ranunculus* sp.). The first generation apomictic hybrids (F1s) often show poor fertility due to meiotic aberrations. Thus, the chances of formation of functional and reduced female gamete through meiosis are low or negligible. Such observations (e.g., *Ranunculus*) led to the hypothesis that formation of unreduced gamete could provide a plausible mechanism to generate functional female gametes *(i.e., which ensure formation of embryo through parthenogenesis)* in apomictic hybrids. However, in an evolutionary context polyploidization of hybrids may also lead to meiotic stability through homologous pairing of the duplicated chromosomes.

Current bio-geographical records show that apomicts have wider distributions than their sexual relatives. Moreover, they occupy regions that were affected by the glacial period or with harsh, stressful environmental conditions. This can be due to the reproductive assurance which apomixis provides to plants and that enables them to form new populations starting with even a single individual. Thus, apomixis provides a double benefit to plants. Primarily, it allows for founder population events and spread of a species' population after seed dispersal. Secondly, apomixis creates clonal offspring and hence it multiplies genotypes which possibly have better adaptability. It is also suggested that sometimes avoidance of sexual reproduction is essential to prevent the breakup of the favorable gene combinations built by earlier selection, as the sexual reproduction cannot always bring favorable alleles together via gene reshuffling (recombination). This results in selection of a mechanism which avoids sexual reproduction and also has an advantage of clonal multiplication. So that such species have the potential of covering vast areas with highly adapted genotypes.

Source: Baker 1955; Mangla et al. 2015; Albertini et al. 2019; Kaushal et al. 2019.

11.11 Methods to Screen and Study Apomixis

- **Emasculation and Bagging Experiments:** This is the easiest and the most important method to elucidate the presence of apomixis in plants. Here, the flowers are emasculated before maturation and bagged using paper bags or muslin cloth. If there is fruit set even after emasculation and bagging, the plant is most likely an apomict. For such screening methods, a large number of flowers have to be emasculated and bagged.

- **Cyto-histological Screening:** Cyto-histological analysis of embryo sac development is a commonly used method to study apomixis. This is achieved through: histological sectioning, ovule clearing and fluorescence staining. For histological sectioning, ovules at different developmental stages are fixed in FAA (90ml 70% ethanol: 5ml formaldehyde: 5ml acetic acid) followed by dehydration in alcohol series and embedding in wax or resin. Such embedded material is used for sectioning using microtome and sections can be observed under a bright field microscope. Thus, the development of embryo sac in ovules of different development stages can be observed. For ovule clearing, fixed ovules are treated with a mixture of chloral hydrate, methyl salicylate and observed under the phase contrast microscope or differential interference contrast microscope. The cleared ovules can also be stained with Feulgen and observed under confocal microscopy using optical sectioning.

  It is well established that the aposporous initials, and diplosporous MMCs lack callose walls around them making them different from the sexual MMCs where presence of a callose wall is a key feature. Another method involves fluorescence microscopy where callose can be stained with decolorized aniline blue which fluoresces under the UV light (epi-fluorescence microscope). The absence of fluorescence around the MMCs indicates their apomictic nature.

- **Flow Cytometric Seed Screening (FCSS):** The ploidy of embryo and endosperm can be estimated using FCSS method as developed by Matzk et al. (2000). After estimating ploidy of embryo and endosperm (and the somatic cells) the reproductive pathways (apomictic or sexual) can be elucidated for a particular species.

- **Molecular Techniques:** With the availability of molecular markers like AFLP or RAPD, one can study the genetic variation at the DNA level. Apomictic progenies are mostly genetically uniform and identical to their mother plant. So, if variations are absent while analysing the parent plant with the progeny, then one can infer the presence of apomixis. Molecular methods are more advantageous than other methods but are tedious and need elaborate experimental set-ups.

Glossary

Adventive embryony: Development of embryos directly from the sporophytic tissue of the ovule, i.e., integument or nucellus.

Apomeiosis: Avoidance or failure of meiosis during the development of an embryo sac.

Apomixis: Asexual reproduction through seeds such that progeny is genetically identical to the maternal plant. E.g., *Taraxacum*, *Hieracium*, *Rubus*.

Apospory: Development of unreduced embryo sacs from nucellar cells other than the MMC. These cells are known as aposporous initial cells.

Asexual embryo formation: Embryo formation from an unreduced female gamete through parthenogenesis.

Asyndesis: Abnormal or suppressed stages of meiosis which may lead to abortion of meiosis.

Autonomous apomixis: Incidences of apomixis where both embryo and endosperm development take place without pollination and fertilization.

Diplospory: A gametophytic apomictic pathway where a diploid embryo sac develops from the megaspore mother cell.

Gametophytic apomixis: A type of apomixis in which egg cell of a well-developed embryo sac produces an embryo parthenogenetically.

Hemizygous: When only one allele is present instead of two in a diploid cell, the allele is called hemizygous and the condition is known as hemizygosity.

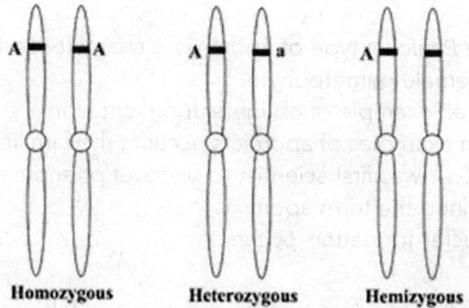

Parthenogenesis: Development of an egg cell into an embryo without fertilization.

Polyembryony: Development of more than one embryo in a seed such that more than one seedling emerges from the seed at the time of germination.

Pseudogamous apomixis: Incidences of apomixis where the embryo develops parthenogenetically (asexually without fertilization of egg cell), but pollination and fertilization are required for endosperm development.

Restitution nucleus: A cell nucleus with double chromosome number or unreduced chromosome number that develops due to failure of division.

Selfish offspring: Selfish offspring is a developing seed (offspring) which is the result of first fertilization in a multi-ovulate ovary and uses most available resources, leaving a paltry amount of the maternal resource supply to subsequently forming seeds.

Sexual twinning: The development of multiple embryos by splitting of sexual embryo/zygote.

Sporophytic apomixis: Type of apomixis in which embryo develops directly from a sporophytic cell of ovule, i.e., nucellus, integuments.

Key Questions

Q11.1 Differentiate between:
   a. Gametophytic apomixis and sporophytic apomixis
   b. Apospory and Diplospory
   c. Panicum-type and Hieracium-type of apospory
   d. Vegetative reproduction and Agamospermy

Q11.2 Write notes on:
   a. Methods to screen apomixis
   b. Potential applications of apomixis
   c. Genetics of polyembryony
   d. Classifications of polyembryony

Q11.3 Explain current understanding of genetic regulation of apomixis with examples.

Q11.4 Provide insights on the different types of polyembryony existing in angiosperms with suitable examples and illustrations?

Q11.5 Fill in the blanks:
   a. .... is an example of facultative apomict taxon.
   b. Polyembryony can be commonly observed in.................
   c. Neochrome theory was proposed by.................to explain phenomenon of ....
   d. In general the *Panicum* type of apomixis is characterized by absence of............... in organized female gametophyte.
   e. .... is an example of obligate apomict taxon.
   f. Most common examples of apomicts occur in the families ....
   g. ....was first scientist to discover phenomenon of polyembryony.
   h. ....coined the term apomixis.
   i. Restitution nuclei formation occurs in ....type of gametophytic apomixis.

Q11.6 State whether following statements are true or false. Justify your answer with suitable reason(s).
   a. Vegetative reproduction is not a type of asexual reproduction.
   b. Agamospermy involves gametes for formation of embryo in seed.
   c. Polyembryony has potential for crop improvement.
   d. The gametophytic apomixis does not involve parthenogenesis.
   e. Apomixis can be screened by use the of fluorescence microscopy.
   f. Apomeiosis plays a role in manifestation of sporophytic apomixis.
   g. Panicum-type of apomixis can be recognized by the presence of two or more and 4-celled megagametophyte/s in an ovule.

Practicals

Exercise 11.1 To screen apomixis in ovules using histo-clearing technique

Cyto-histological analysis of embryo sac development is a commonly used method to study female gametophyte development, apomixis and to trace the embryo development. This is achieved through histological sectioning, or ovule clearing (histo-clearing), and fluorescence staining. Nowadays, the method of histo-clearing is usually employed over histological sectioning due to several reasons; histo-clearing is quick and less time consuming, with no chance of loss of any part of the section, does not even require dissecting the ovules of plants with small ovaries (e.g., grasses). Thus, a large number of ovules can be studied and development can be traced quickly. Nonetheless, the availability of differential interference microscope (or Nomarski's microscope) is a chief limitation in this methodology. For studying female gametophyte development through histo-clearing; pistils can be fixed at different developmental stages (on basis of size), and after certain time intervals of bagging treatment. Most commonly used fixatives for fixation of ovules for clearing are: FAA (5ml formaldehyde: 5ml acetic acid: 90ml; 70% ethanol; v/v), FPA (5ml formaldehyde: 5ml Propionic acid: 90ml 70% ethanol; v/v) and Acetic alcohol (Acetic acid 3 parts: Ethanol 1 part, v/v). These acidic fixatives preserve cytoplasmic structures in a net-like pattern (formed by formaldehyde). Such fixatives dissolve certain cellular components like mitochondria. This choice of acidic fixatives is advantageous in ovule clearing technique as it helps in preservation of shape and position of the nuclei, vacuoles and cytoplasm (in part). Such fixatives also limits the appearance of artifacts during observation and image capture such as shrinkage caused by alcohol in tissue/organelles; which, is overcome by presence of acetic acid (Ruzin 1999). A wide choice of combinations of tissue clearing agents is available. For ovule clearing, the most common histoclearing media are Herr's solution, Hoyer's solution, and Methyl salicylate. Preparation of aforementioned solutions and methodology have been described below.

Material Required

Suggested flowers
Flowers of different development stages (bud to mature flowers) *Cenchrus* sp., *Cajanus cajan*, *Pennisetum* sp., and *Brassica* sp.

Chemicals
Lactic acid, chloral hydrate, clove oil, phenol, xylene, gum arabic, glycerol ethanol, methyl salicylate, double distilled water

Equipment
Stereo-zoom microscope/dissecting microscope, weighing balance, differential interference microscope, magnetic stirrer

Miscellaneous
Needle, forceps, glass vials, amber color glass bottles, teflon magnetic beads, glass droppers/pipettes, spatula

A) Ovule clearing with Herr's solution

Procedure

1. Dissect the flowers to separate pistils from other whorls. Fix pistils in FAA for about 24–48 hours and store them in 70% alcohol.
2. Dissect out ovules from fixed pistils under a stereo-zoom/dissecting microscope.
3. Keep the dissected ovules in lactic acid (saturated with chloral hydrate) for 24 hours.
4. Wash the ovules thrice in 70% ethanol (5 min each) and keep them in Herr's solution for 7 days (at room temperature).
5. Mount the cleared ovules on a glass-slide with the Herr's solution and observe under a microscope equipped with DIC optics.

Note: The duration of treatment of Herr's solution may vary from species to species depending upon the number and thickness of integuments, and nucellus. Thus, it needs to be standardized. Generally, the treatment for minimum 24–48 hours is required. For some species, treatment can be given at 50–60°C by keeping the glass vials with ovules immersed in Herr's solution in a hot air oven.

B) Ovule clearing with Hoyer's solution

The use of this solution for ovule clearing is advantageous over other methods of ovule clearing as it eliminates the need of fixation and treatment with other chemicals. This method is variously employed to trace female gametophyte development and embryo development in (fertilized ovules) *Brassica* sp., *Arabidopsis*.

Procedure

1. Dissect out ovules from fresh/fixed pistils under a stereozoom/dissecting microscope.
2. Keep the dissected ovules in Hoyer's solution for 24–48 hours.
3. Mount the cleared ovules on a glass-slide with the Hoyer's solution and observe under a microscope equipped with DIC optics.

C) Ovule clearing with methyl salicylate

This method of ovule clearing is more suitable for ovules with a thin nucellus. The cleared tissue/ovule might also be observed under phase contrast microscope which is the main advantage of this protocol.

Procedure

1. Dissect out ovules from fixed pistils under a stereozoom/dissecting microscope.
2. Pass the ovules through alcohol series: 80%, 90%, and 100% for dehydration. At each concentration keep the ovules for about 20–30 minutes.
3. After dehydration in 100% ethanol, treat the ovules with ethanol:methyl salicylate (3:1, v/v) and ethanol:methyl salicylate (1:3 v/v) mix for about 30 minutes in each (*duration of treatment requires standardization*).
4. Transfer the ovules in 100% methyl salicylate for further clearing (30 minutes–2 hours, ovules may be kept in methyl salicylate till observation).
5. Mount the cleared ovules on a glass-slide with methyl salicylate and observe under a microscope equipped with DIC optics or with phase contrast optics.

Result

Observations as in Fig. 11.8. You may also refer to Figure 5.23 in Chapter 5.

- *Preparation of lactic acid saturated with chloral hydrate:* Take 20 ml lactic acid in a glass bottle. To that add about 2 gm of chloral hydrate crystals. Put a teflon magnetic bead in the glass bottle and place it over magnetic stirrer for mixing for about 30 minutes. Add few more crystals (0.5 gm) if a clear solution is observed. Keep on adding and mixing until chloral hydrate crystals do not dissolve anymore.

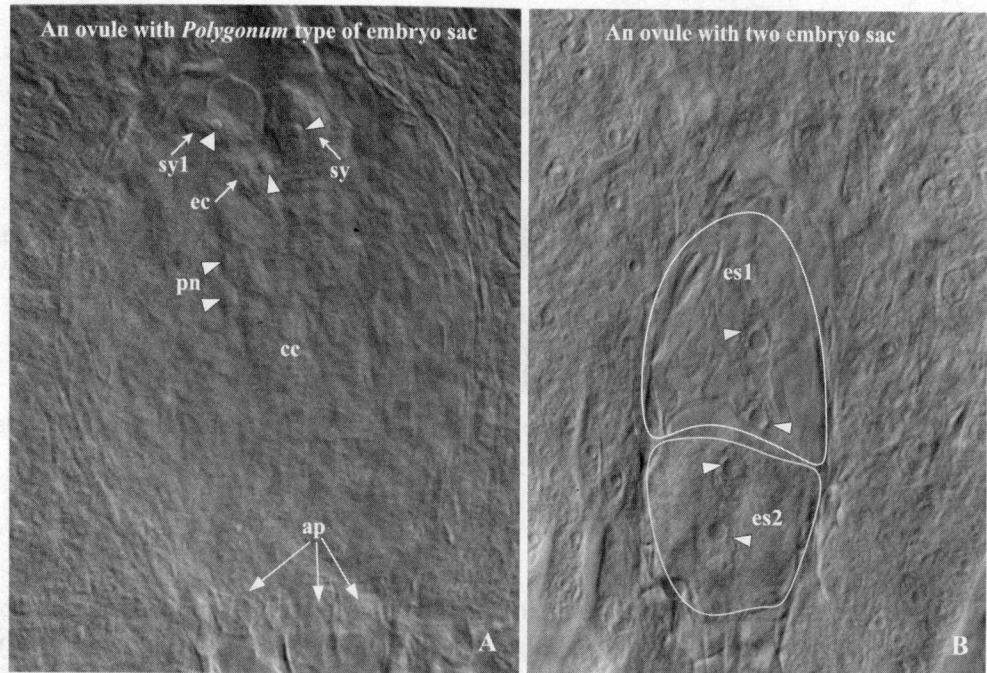

Figure 11.8 Ovules as observed after histoclearing under DIC microscope. Arrowhead indicates a nucleus. **A.** Ovule (enlarged view) showing Polygonum-type of embryo sac. **B.** An ovule with two embryo sacs (es1 and es2; outlined with white lines). **ec:** egg cell; **sy:** synergid; **pn:** polar nuclei, **cc:** central cell, **ap:** antipodals.

- *Preparation of Herr's solution* (Herr 1971): To prepare Herr's solution place a glass vial on weighing balance. Add lactic acid: chloral hydrate: clove oil: phenol: Xylene in the ratio of 2:2:2:2:1 respectively, (v/w/v/v/v/v). Fasten the cap of the vial and mix the contents well. One may use magnetic stirrer for mixing. Complete mixing of the contents may take about 30 minutes. The prepared solution can be stored in dark bottle at 4–10°C in refrigerator.
- *Preparation of Hoyer's solution* (Liu & Meinke 1998): Add 7.5 g gum arabic, 100 gm chloral hydrate, 5 ml glycerol in 30 ml water. Mix well and the prepared solution can then be stored in dark bottle at room temperature.

Precautions

1. Pistils must not be over fixed. Store them in 70% ethanol after 24–48 hours of fixation.
2. Phenol, chloral hydrate are corrosive chemicals. Wear gloves while handling such chemicals.
3. Always use glass pipette or glass dropper to pour in/out solutions.
4. Store Herr's and Hoyer's solution in amber color glass bottles.
5. Use fume hood or switch on exhaust fan of laboratory while using methyl salicylate.
6. Do not keep damaged ovules for clearing.
7. Do not let ovules dry while dissecting.

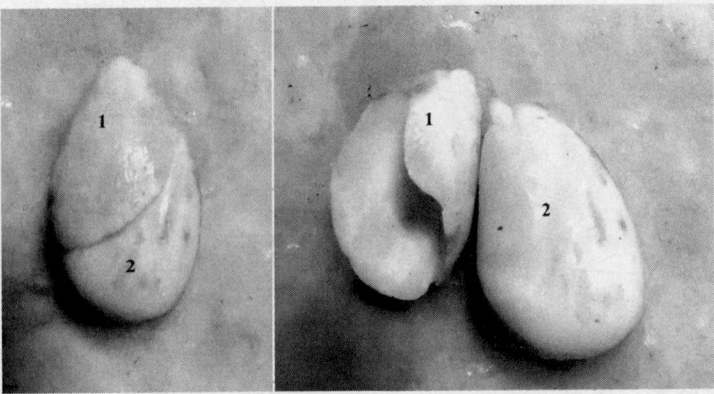

Figure 11.9 An almond seed with two embryos (labelled as 1 and 2). Also refer to Fig. 11.4 for similar observations in *Citrus* sp.

Exercise 11.2 To study the polyembryony in seeds

Polyembryony is of common occurrence in citrus, mango, almonds, jamun. For studying polyembryony, one may soak some seeds such as almond, in water. After removal of the seed coat multiple embryos can be observed, if present (Fig. 11.4 and 11.9) and seeds can be further dissected under the stereo-zoom microscope.

   If these seeds are sown in soil and emergence of multiple seedlings per seed may be observed, then it is an indication of polyembryony (refer to Fig. 11.5). Perform these two sets of experiments to calculate the percentage incidences of polyembryony in the above-mentioned plants species or any other species of your choice. You may also study female gametophyte development with clearing method to support your observation. Record your observations in Table 11.1 and discuss.

Table 11.1 Observation table to record incidences of polyembryony with two strategies: *seed sowing and seed dissection.*

Number of seeds studied		Number of seeds with two or more seedlings (n3)	Number of seeds showing more than one embryo (n4)	Percentage incidence of polyembryony= n3+n4/n1+n2 X 100
Number of seeds sown (n1)	Number of seeds dissected (n2)			
Citrus sp.				
Mangifera indica				
Prunus dulcis				
Syzygium cumini				
Seed of your choice				

Exercise 11.3 To determine the type of embryo by enumerating key features

Fig. 11.10 shows longitudinal sections of ovules. Can you identify which section shows the presence of a sexual embryo and which one is showing an adventitious embryo? (*Hint: arrowhead indicates the micropylar end*).

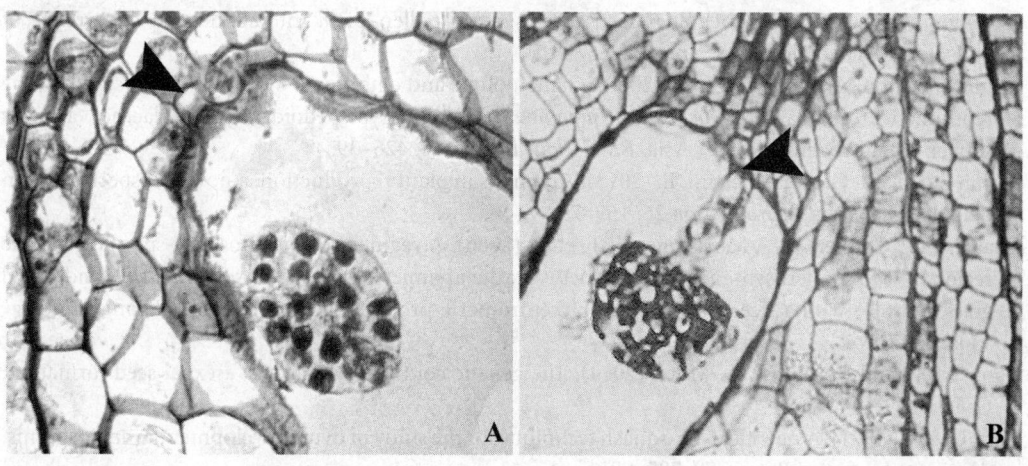

Figure 11.10 A.................................... B..

Bibliography

Albertini, E., Barcaccia, G., Carman, J. G. and Pupilli, F. (2019). Did apomixis evolve from sex or was it the other way around? *Journal of Experimental Botany* 70: 2951–64.

Aleza, P., Juarez, J., Ollitrault, P. and Navarro, L. (2011). Polyembryony in non-apomictic citrus genotypes. *Annals of Botany* 106: 533–45.

Alves, M. F., Duarte, M. O., Bittencourt, N. S. et al. (2016). Sporophytic apomixis in polyembryonic *Handroanthus serratifolius* (Vahl) S.O. Grose (Bignoniaceae) characterizes the species as an agamic polyploid complex. *Plant Systematics Evolution* 302: 651–9.

Baker, H. G. (1955). Self-compatibility and establishment after 'long-distance' dispersal. *Evolution* 9: 347–9.

Batygina, T. B. and Vinogradova, G. Y. (2007). Phenomenon of polyembryony: genetic heterogenity of seeds. *Russian Journal of Developmental Biology* 38: 126–51.

Bhat, V., Dwivedi, K. K., Khurana, J. P. and Sopory, S. K. (2005). Apomixis: an enigma with potential applications. *Current Science* 89: 1879–93.

Bicknell, R. A. and Koltunow, A. M. G. (2004). Understanding apomixis: recent advances and remaining conundrums. *The Plant Cell* 16: S228–S245.

Brukhin, V. and Baskar, R. (2019). A brief note on genes that trigger components of apomixis. *Journal of Biosciences* 44: 45.

Erdelská, O. and Vidovencová, Z. (1994). Cleavage polyembryony in vivo and in vitro. *Biologia Plantarum* 36: 329–34.

Firetti-Leggieri, F., Lohmann, L. G., Alcantara, S. et al. (2013). Polyploidy and polyembryony in *Anemopaegma* (Bignonieae, Bignoniaceae). *Plant Reproduction* 26: 43–53.

Ganeshaiah, K. N., Shaanker, R. U. and Joshi, N. V. (1991). Evolution of polyembryony: consequences to the fitness of mother and offspring. *Journal of Genetics* 70: 103–27.

Garcia-Aguilar, M., Michaud, C., Leblanc, O. and Grimanelli, D. (2010). Inactivation of a DNA methylation pathway in maize reproductive organs results in apomixis-like phenotypes. *Plant Cell* 2010: 3249–67.

Geetha, K. A., Kawane, A., Bishoyi, A. K. et al. (2013). Characterization of mode of reproduction in *Commiphorawightii* [(Arnot) Bhandari] reveals novel pollen–pistil interaction and occurrence of obligate sexual female plants. *Trees* 27: 567–81.

Ghimire, B., Son, S., Kim, J. H. et al. (2020). Gametophyte and embryonic ontogeny: understanding the reproductive calendar of *Cypripedium japonicum* Thunb. (Cypripedoideae, Orchidaceae), a lady's slipper orchid endemic to East Asia. *BMC Plant Biology* 20: 426–39.

Gianni Barcaccia, G. and Albertini, E. (2013). Apomixis in plant reproduction: a novel perspective on an old dilemma. *Plant Reproduction* 26: 159–79.

Gualtieri, G., Conner, J. A., Morishige, D. T. et al. (2006). A segment of the apospory-specific genomic region is highly microsyntenic not only between the apomicts *Pennisetum squamulatum* and buffel grass, but also with a rice chromosome 11 centromeric-proximal genomic region. *Plant Physiology* 140: 963–71.

Hand, M. L. and Koltunow, A. M. G. (2014). The genetic control of apomixis: asexual seed formation. *Genetics* 197: 441–50.

Herr, J. M. Jr. (1971). A new clearing-squash technique for the study of ovule development in angiosperms. *American Journal of Botany* 58: 785–90.

Hojsgaard, D. and Hörandl, E. (2019).The rise of apomixis in natural plant populations. *Frontiers in Plant Sciences* 10: 1–13.

Hörandl, E. (2006). The complex causality of geographical parthenogenesis. *New Phytologist* 171: 525–38.

Johnson, S. D., Eichmann, E. and Koltunow A. M. G. (2017). Genetic analyses of the inheritance and expressivity of autonomous endosperm formation in *Hieracium* with different modes of embryo sac and seed formation. *Annals of Botany* 119: 1001–11.

Kaushal, P., Dwivedi, K. K., Radhakrishna, A. et al. (2019). Partitioning apomixis components to understand and utilize gametophytic apomixis. *Frontiers in Plant Sciences* 10 (256): 1–17.

Kishore, K. (2016). Polyembryony. In K. G. Ramawat, J. M. Merillon and K. R. Shivanna, eds., *Reproductive Biology of Plants*. Boca Raton: CRC Press.

Koltunow, A. M. and Grossniklaus, U. (2003). Apomixis, a developmental perspective. *Annual Reviews of Plant Biology* 54: 547–74.

Koltunow, A. M. G., Johnson, S., D. Rodrigues, J. C. et al. (2011). Sexual reproduction is the default mode in apomictic *Hieracium* subgenus *Pilosella*, in which two dominant loci function to enable apomixis. *The Plant Journal* 66: 890–902.

Kumar, S. and Bhat, V. (2014). Application of omics technologies in forage crop improvement. In D. Barh, ed., *Omics Applications in Crop Science* Barh. Boca Raton: CRC Press, pp. 523–48.

Lakshmanan, K. K. and Ambegaokar, K. B. (1984). Polyembryony. In B. M. Johri, ed., *Embryology of Angiosperms*. Berlin: Springer-Verlag, pp. 445–74.

León-Martínez, G. and Vielle-Calzada, J. P. (2019). Apomixis in flowering plants: developmental and evolutionary considerations. *Current Topics in Developmental Biology* 131: 565–604.

Liu, C. M. and Meinke, D. W. (1998). The titan mutants of Arabidopsis are disrupted in mitosis and cell cycle control during seed development. *Plant Journal* 16: 21–31.

Mangla, Y., Chaudhary, M., Gupta, H. et al. (2015). Facultative apomixis and development of fruit in a deciduous shrub with medicinal and nutritional uses. *AoB PLANTS* 7: 1–12.

Matzk, F., Meister, A. and Schubert, I. (2000). An efficient screen for reproductive pathways using mature seeds of monocots and dicots. *The Plant Jouranl* 21: 97–108.

Nogler, G. A.(1984). Gametophytic apomixis. In B. M. Joshi, ed., *Embryology of Angiosperms*. Berlin: Springer-Verlag, pp. 475–518.

Olmedo-Monfil, V., Durán-Figueroa, N., Arteaga-Vandázquez, M. et al. (2010). Control of femalegamete formation by a small RNA pathway in *Arabidopsis*. *Nature* 2010: 628–32.

Ozias-Akins, P. and Conner, J. A. (2012). Regulation of apomixis. *In Plant Biotechnology and Agriculture: Prospects for the 21st Century*. Cambridge, MA: Academic Press.

Podio, M., Cáceres, M. E., Samoluk, S. S. et al. (2014). A methylation status analysis of the apomixis specific region in *Paspalum* spp. suggests an epigenetic control of parthenogenesis. *Journal of Experimental Botany* 65: 6411–24.

Ruzin, S. E. (1999). *Plant Microtechnique and Microscopy*. Oxford: Oxford University Press.

Wang, X., Xu, Y., Zhang, S. et. al. (2017). Genomic analyses of primitive, wild and cultivated citrus provide insights into asexual reproduction. *Nature Genetics* 49: 765–74.

Woo, J. K., Park Y. C., Lee, J. W. et. al. (2019). Evaluation of polyembryony for genetic resources and efficacy of simple sequence repeat markers for the identification of nucellar and zygotic embryo-derived individuals in *Citrus*. *Applied Biological Chemistry* 62: 1–11.

Zappacosta, D. Ochogavía, A. Rodrigo, J. M. et al. (2014). Increased apomixis expression concurrent with genetic and epigenetic variation in a newly synthesized *Eragrostis curvula* polyploid. *Scientific Reports* 4: 4423.

12

Seed

In ancient times seeds of Abrus precatorius (known as Ratti) were used to weigh gold.

12.1 Introduction

The protective seed habit is a significant feature in the evolutionary success of angiosperms. The seed, encloses an undeveloped miniature plant '*the embryo*' and acts as a functional unit which links the successive generations. Developmentally, seed is a fertilized ovule and, a typical angiospermous seed consists of an embryo, some storage tissue (mostly endosperm) and a seed-coat. While the embryo and the endosperm are the products of double fertilization, the seed-coat develops from the integument/s of the ovule. Embryos accomplish their early development before seed germination, protected by the surrounding seed coats and sustaining on the stored food in the endosperm. Protection provided to embryo by seed coat increases its chances of survival, and establishment of subsequent generations. Generally, seeds develop as discrete units attached to the inside of the fruit wall through a stalk called the funiculus. However, in many plants, seeds are associated with some other structures that help in their dispersal. In such cases, a single entity of the seed and the structure assisting in dispersal are together described as **dispersal units** or **diaspores**. For example, in the members of Asteraceae, the outer integument of the ovule is completely fused with the ovary wall and the diaspore is called a *cypsella*. Some other examples of diaspores are the seeds with the elaiosomes, achene (dry indehiscent fruits), and caryopsis (fruit type seen in grasses).

Here, one must acknowledge that all structures and processes associated with reproduction in angiosperms are directly responsible for the formation of seed. Seeds perform a wide variety of functions including dispersal, perennation (surviving seasons of stress such as winter), dormancy (a state of arrested development), and most importantly, perpetuation of a plant species.

A huge variation in the size, shape, color, seed coat, weight and dispersal mechanism can be observed among angiospermous seeds. The smallest known seeds are those of orchids which are about 85 micrometers in size and weigh about 0.8 micrograms, thus appearing similar to dust particles. Double coconut or *Lodoicea maldivica* has the largest (nearly 0.5 meter) and the heaviest (weighing up to 25 kg) known seeds in the world (Fig. 12.1). The size of the seed in a plant depends on the size of the embryo and also on whether the seed at maturity is endospermous or non-endospermous. Seeds in Orchids are small as the endosperm formation is completely suppressed, and also, the embryo is highly reduced. In plants like beans, lettuce, and peanuts, despite being non-endospermous, the seeds are relatively big because of the significant amount of reserve materials stored in the cotyledons of the embryo.

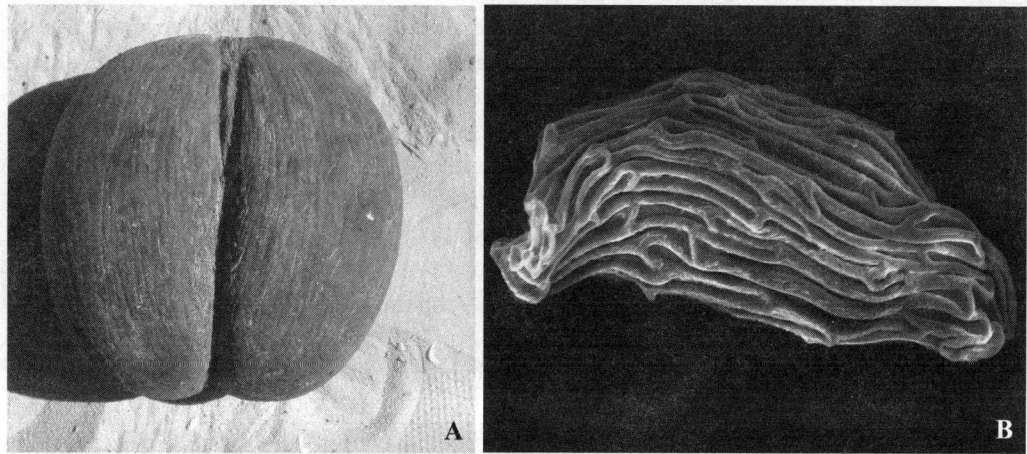

Figure 12.1 A. Seed of double coconut or *Lodoicea maldivica*. **B.** Scanning electron micrograph of an orchid seed (~ 250 μm in size). (*Source:* Creative commons search, *Mary Gillham Archive Project*, published under CC BY 2.0 License.)

For the perpetuation of a plant species, effective dispersal and recruitment of seeds is essential. Seed dispersal can be accomplished by either abiotic or biotic agencies accounting for the diversity of seed dispersal mechanisms that are seen in angiosperms. Seed dispersal mechanisms, are the key to maintain genetic diversity and population structure in all species.

Seeds being a major source of food in the form of grains and pulses are also vital for the survival of humans. Thus, the study of seed development, its structure, dispersal and natural recruitment encompasses an integral part of the reproductive biology of a plant. The study of different aspects of seed development, dispersal and establishment in a habitat is known as **seed biology**. In the present scenario of climate change, exploring seed biology has become essential as ineffective seed dispersal and limited natural recruitment is one of the reasons for many species becoming endangered (especially trees). In the current chapter, the variations in seed types, seed coat structure, seed development, and dispersal mechanisms with relevant examples have been discussed.

12.2 Structure and Morphology of Seeds

A typical angiospermous seed is a fertilized ovule enclosing embryo and endosperm protected by a seed coat. Thus, the ovule is the starting point in the development of a seed. The development of the embryo and the endosperm begins following fertilization of the egg cell and the central cell. Endosperm enlarges and occupies most of the volume of the embryo sac which also corresponds to the volume of seed. The embryo is the last to increase in volume as its rapid growth is at the expense of the endosperm. By the time the seed attains maturity, a major part of the endosperm has already been consumed by the embryo, which eventually fills the interior of the seed completely. The embryo in the seed may be very small as in buttercups, or, it may fill the seed almost completely as in members of Brassicaceae or legumes.

The seed coat is derived from the integument/s of the ovule. Bitegmic ovules give rise to seeds with two-layered seed coats, outer layer is called the **testa** (derived from outer integument) and the inner layer is called **tegmen** (derived from inner integument). Unitegmic ovules develop into seeds with a single layered seed-coat, the testa alone.

Different parts of testa are recognized on the basis of parts of outer integument from which they are derived. The outer epidermis and inner epidermis of outer integument differentiate as exotesta and endotesta respectively. The intervening layers of outer integument gives rise to mesotesta. Similarly, different regions can also be recognized in the tegmen viz. exotegmen, endotegmen or mesotegmen. The exotesta may consist of one or more rows of elongated and pallisade like cells as present in Fabaceae. The distinctive feature of seed-coat anatomy is the position of the mechanical layer/s. On the basis of the position of the mechanical layer, seeds can be designated as exotestal, mesotestal, endotestal, exotegmic, mesotegmic, endotegmic, or combinations of these. Seed coat may also be contributed by the diploid tissues of the mother plant like the chalaza and the raphe, or even by the nucellus. The outer layer of the raphe and the chalaza also undergo more or less a similar differentiation as observed in the integuments, e.g., Leguminaceae, Rhamnaceae. Endotestal like layers too develop on the inner side of the raphe, as seen in Onagraceae, Rutaceae.

Box 12.1

Differences between Testa and Tegmen

Testa	Tegmen
• It is the outermost protective coat of a seed.	• It is the inner protective coat of a seed.
• It is generally derived from the outer integument of the ovule.	• It is derived from the inner integument of the ovule.
• It is dark in color, generally thick and variously patterned.	• Thinner than the testa, it is generally white and membranous.
• It is impermeable and hence protects the seed to tide over unfavorable conditions and also during dormancy.	• It protects the embryo from dehydration.

The nutrients to the developing seeds are transported through a vascular system consisting of the placental, funicular, and raphal bundles. The massive or compound raphal bundles with well-differentiated xylem and phloem elements in the chalazal region may even extend as far as the micropyle (Meliaceae, Connaraceae, Dichapetalaceae). Seeds of primitive families are generally large and have extensively developed vascular systems. On the other hand, seeds in more advanced taxa (e.g., Sympetalae) are small, often with simple and poorly differentiated vascular bundles, or none at all.

The small scar resulting from the disconnection of the seed from the funicle is called the *hilum* and marks the place where the seed was formerly attached to the placenta. In many plants, as the seed matures, hilum enlarges and the seeds possessing the hilum are known as hilar seeds, e.g., Meliaceae, Fabaceae, Erythrina, *Aesculus*. The hilum of a seed is the channel for water uptake and efflux during germination. In Fabaceae members (*Cajanus* sp., *Phaseolus* sp.) a raised growth on hilum is present which is known as *strophiole* (Fig. 12.2). In such species movement of water inside of the seed through the testa is facilitated by the strophiole (Dell 1980). In many seeds, the micropyle of the ovule also persists as a small opening in the seed coat.

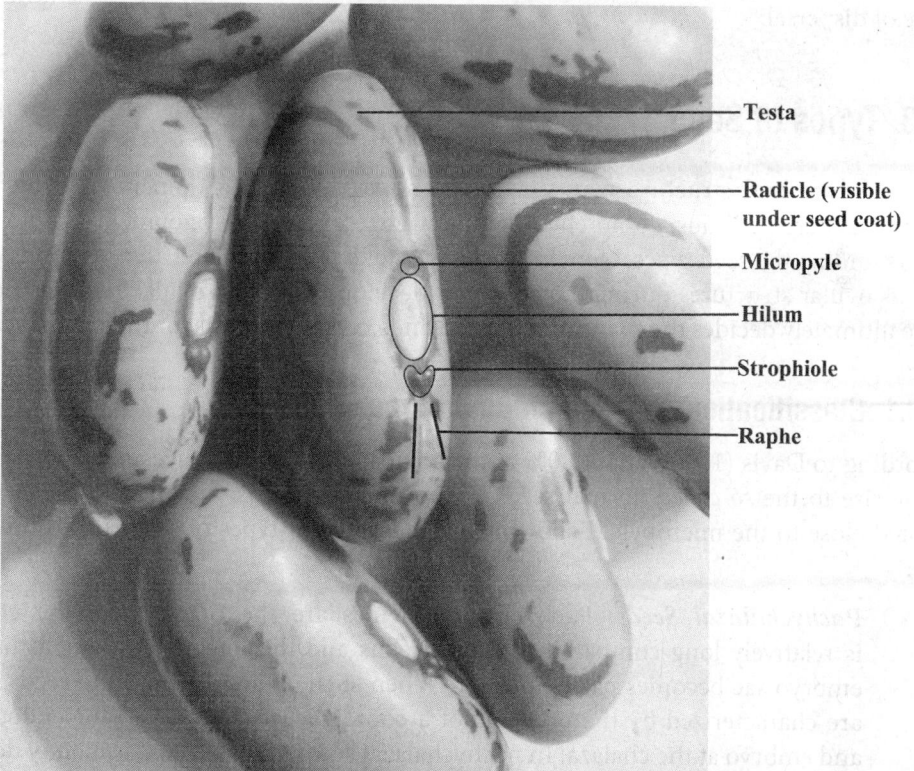

Figure 12.2 A mature seed showing different parts. See Color Plates (page 498).

Morphological seed traits include various features of a seed like its shape, color, texture, mass, seed extensions (wings, hooks), and associated structures (aril). Seeds may be classified based on their shapes into many types like discoid (laterally flattened), orbicular, reniform, ovate, obovate, cylindrical, conical, elliptic, and more. While discussing about seed shapes, one may recall the seeds we come across in daily-life such as different grains, pulses, seeds of different fruits. Seeds also show an astonishing variation in their color and may be white, black, brown, red and even green (due to the presence of chlorophyll, as in Loranthaceae, Santalaceae). Seed color is primarily because of the seed coat and in combination with odor, is important for the dispersal of diaspores. Color of the seed coat is also a commercially important trait for legumes, e.g., in common beans, lentils, vetch (*Vicia* sp.), or peas. Seed coat epidermis also displays variations; from being trichomatous in cotton to possessing stomata in *Hymenocallis occidentalis*, Magnoliaceae, Bombaceae and Euphorbiaceae. Seed size, considered an important aspect of plant reproduction and perpetuation, influences many stages in the life-history of plants. Species producing small seeds with shorter life spans are known to have low rates of seedling survival, and hence produce more seeds in order to survive. In many crops, seed size is an important agronomic trait (e.g., legumes, *Brassica* sp.). Seed mass also plays a vital role in the dispersal and population establishment of seeds. Usually, lighter seeds disperse to longer distances while heavier seeds have short range of dispersal.

12.3 Types of Seed

The seed type and structure may vary with the type and development of ovule from which it is derived. Thus, comprehensive knowledge of ovule development is crucial for understanding the seed types. Numerous changes take place in the relative positions of the various ovular structures during the course of development of an ovule. The shape of the ovule ultimately decides the shape of the seed which can be the basis for their classification.

12.3.1 Classification 1: Based on Ovule Type

According to Davis (1966), about 80% of the angiosperm families have anatropous ovules giving rise to the *so-called* normal seeds, where the embryo is straight and the hilum is situated close to the micropyle. Following are the different types of seeds based on ovule type:

- *Pachychalazal Seeds:* Pachychalazal ovules are the ones where the chalaza is relatively long compared to the nucellus and the integuments, such that the embryo sac becomes partly inferior. When such ovules develop into seeds, these are characterized by the presence of a container (pachychalaza) for endosperm and embryo at the chalaza. In pachychalazal seeds, the seed coat is mainly derived from the chalaza. In extreme cases the integumentary part of the ovule hardly undergoes any development, such that the seed coat is entirely chalazal in origin, and exotestal (e.g., mango).

- *Perichalazal Seeds:* In perichalazal ovules, the chalaza is longer in the median plane than the transverse plane. Seeds derived from such ovules are called perichalazal seeds and are rather rare, occuring mainly in family Annonaceae and Meliaceae.
- *Endopachychalazal Seeds:* In members of Euphorbiaceae, the region between the nucellus and the base of inner integument undergoes fusion. This results in shortening of inner integument or tegmen and formation of a massive chalaza also known as endopachychalaza. Seeds developing from such ovules are known as endopachychalazal seeds. Example: *Ricinus, Delchampia, Euphorbia.*
- *Campylotropous Seeds:* Such seeds are with a bent anti-raphal side (opposite of raphe) which exerts a one-sided pressure on the embryo sac. Therefore, the embryo sac (and later the embryo) becomes either U or V-shaped. Campylotropous seeds occur in Papaveraceae, Capparidaceae, Cactaceae, Leguminosae, Geraniaceae, and Malvaceae.
- *Obcampylotropous Seeds:* This type of seeds are the converse of the campylotropous seeds. Here, the raphal side extends, and the anti-raphe remains small. These can be seen in Leguminaceae, Vitaceae.
- *Orthotropous Seeds:* Such seeds develop from orthotropous ovules which do not show any curvature during development. Therefore, the hilum is situated opposite the micropyle. Orthotropous seeds are mostly found in one-seeded fruits, and occur in families Piperaceae, Polygonaceae, Urticaceae.

12.3.2 Classification 2: Based on Embryo Size, Shape and Position

The embryo in the seed consists of a root primordium (also called a radicle), a prospective shoot (also named a plumule or epicotyl), one or two cotyledons and a hypocotyl (a region connecting the radicle with the plumule). Martin (1946) classified the seeds on the basis of size and position of the embryo. He divided seeds into three major categories of basal, peripheral and axile which were further divided into 12 types. However, his classification included only very small seeds; referred to as dwarf (seed interior 0.3–2.0 mm long), and micro seeds (interior less than/equal 0.2 mm); and excluded seeds bigger than 2 mm. Nonetheless, Martin's seed classification was very exhaustive, the outline of which is given below (Fig. 12.3):

- **Basal:** Embryo is placed at the base
  Rudimentary: Embryo as wide as long (but undifferentiated)
  Broad: Embryo globular, broader than length
  Capitate: Embryo capitate (cap-like)
  Lateral: Embryo baso-lateral (on one side)
- **Peripheral:** Embryo dicot, elongated or large present on the periphery of the seed
- **Axile:** Embryo central (axile), straight, curved, coiled, bent or folded
  Linear: Embryo longer than wide
  Micro: Seeds minute, usually < 0.2 mm long, embryo globular in shape
  Dwarf: Embryo variable in relative size, small, usually oval to elliptic or oblong; cotyledons inclined to be poorly developed; seeds small, generally 0.3 mm to 2 mm

Spatulate: Embryo erect, stalk apparent, thin to thick and slightly expanded to broad

Investing: Embryo erect, stalk invested by half or more of its length by cotyledons

Bent: Embryo bent-like jack-knife

Folded: Embryo with thin cotyledons, usually extensively expanded and folded

In 2007, Baskin & Baskin modified Martin's classification. They named all seed types on the basis of the embryo and endosperm characteristics. The name 'dwarf' was removed and 'micro' was replaced by 'undifferentiated' to indicate that the embryos in fresh seeds lack organs.

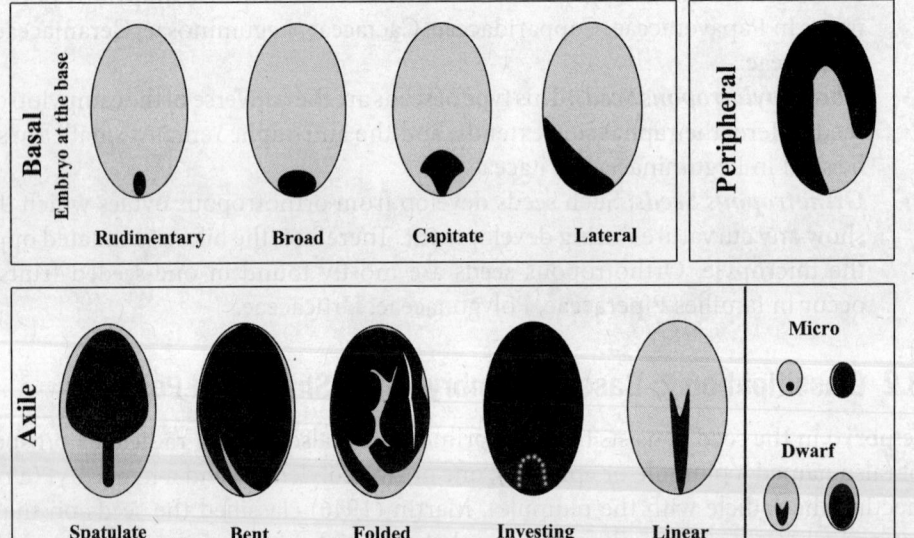

Figure 12.3 Diagrammatic representation of types of seeds based on embryo size, shape and position. (*Source*: Based on Martin 1946; Baskin & Baskin 2007.)

12.3.3 Other Types

- *On the basis of nutrient storage:* When the developing embryo consumes most of the endosperm to fulfill its nutritive requirement, no endosperm remains in the mature seed. Such seeds are called *non-endospermous* or *ex-albuminous seeds*, e.g., *Moringa olifera, Glycine max, Pisum sativum*. In some plants, the growth of the embryo starts much later than that of the endosperm. Thus, the embryo remains relatively small and the endosperm is not exhaustively consumed until maturity. The endosperm persists in mature seed as a storage tissue and such seeds are called *endospermous* or *albuminous seeds*, e.g., *Cocos nucifera, Phoenix roebelenii, Cucumis melo, Brassica* sp. In the albuminous seeds, the stored nutrients are used up only during or after germination. However, a strict distinction between these types is difficult as even in non-endospermous seeds, the remnants of some endosperm tissues can be seen.

- *On the basis of physiological behavior:* Based on physiological features like water content, ability to germinate, temperature tolerance; seeds can be of two types, viz. *orthodox seed*, and *recalcitrant seed*. The terms orthodox and recalcitrant seeds were first used by Roberts (1973) suggesting the physiological behavior of seeds. **Orthodox seeds** may be defined as seeds that could be dried to low moisture content (2–5%) and are able to tolerate freezing temperatures (0–4°C) (e.g., *Citrus* sp., *Capsicum* sp., *Lantana camera*, *Psidium guajava*, *Anacardium occidentale*, most grains, and legumes). Viability of such seeds can be extended in a predictable manner by moisture reduction and lowering of storage temperature. **Recalcitrant seeds** are those which cannot be dried below a relatively critical moisture content and are unable to tolerate freezing temperatures (e.g., mango, avocado, cacao, coconut, and jackfruit). Recalcitrant seeds lose viability once they are dried to a moisture-content below a critical value. The information on desiccation, temperature tolerance of seeds is important for horticulture practices, ex-situ seed storage and conservation.
- *On the basis of seed surface patterns:* Scanning electron microscopy has advanced the study of seed surface or seed coat sculpturing. Using SEM, shapes of various cells of seed coat, protuberances, projections can be studied and a study encompassing these aspects is called *seed micromorphology*. Such studies have proven to be an important tool for taxonomic purposes as they help in diagnosis of phylogenetic relationships between taxa (as seen in Brassicaceae, Acanthaceae, Asteraceae). A comprehensive list of terminology for the seed surfaces is given in Box 12.2 (also see Fig. 12.4).

Box 12.2
Description of Different Seed Surface Patterns

Micromorphology	Description
Foveate	Pitted or having depressions marked with little pits
Foveolate	Marked with little shallow pits
Ocellate	Having eye-like depressions, each with a raised circular border
Reticulate	Having a raised network of narrow and sharply angled lines; frequently presenting a geometric appearance, each area outlined by a reticulum
Reticulate-Foveate	Intermediate between reticulate and foveate type
Rugose	Wrinkled, irregular elevations making up the wrinkles and running mostly in one direction
Ruminate	Penetrated by irregular channels giving an eroded appearance and running in different directions
Sulcate	Grooved or furrowed with long V-formed depressions
Undulate	Having a wave-like pattern
Verrucate	Irregular projections or knobs

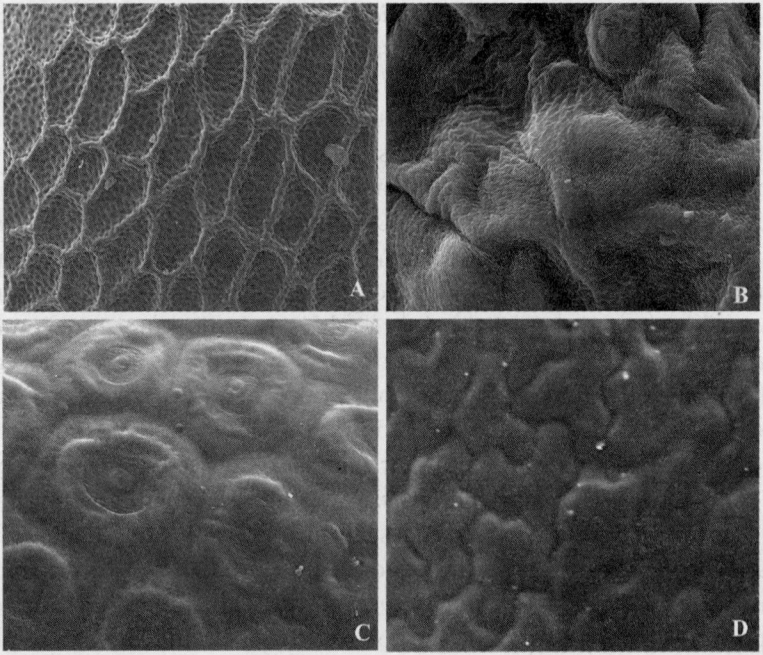

Figure 12.4 Scanning electron micrographs of different seeds exhibiting seed coat patterns.
A. Reticulate. **B.** Undulate. **C.** Ocellate. **D.** Rugose (Source: A–C: Gahr 2018, published under CC BY 4.0 License.)

12.4 Seed Coat

The development of seed coat from integument/s is triggered post-fertilization. The integument/s undergo divisions in periclinal and/or anticlinal plane. The periclinal divisions result in increase in the number of layers of the concerned integument. These divisions may be diffused or restricted to certain layers/elements of integument/s. The anticlinal divisions are mainly responsible for increase in the length. Complex seed-coats are mostly the result of both anti- and periclinal divisions whereas simple seed-coats develop only through anticlinal divisions. Within the seed-coat, remnants of nucellar tissue may persist, often separated from the seed-coat by a more or less conspicuous cuticle.

12.4.1 Structure and Development

The seed coat differentiation in some important families and taxa is described below:

- **Brassicaceae:** The family Brassicaceae includes a large number of important crops where seed and the seed coat characteristics are associated with important agronomic traits (e.g., *Brassica* for oil). Here, the ovules are bitegmic with both integuments differentiating into four distinct layers of seed coat. The outer three layers are derived from the outer integument of the ovule forming the testa. The fourth layer or tegmen is derived from the parenchyma cells of inner integument. The first layer is the epidermal layer of the outer integument which is generally one cell layer thick. Below it lies a sub-epidermal layer that may consist of one or more cell layers and is typically parenchymatous; sometimes it may be collenchymatous, or sclerotic. In some species, including *Arabidopsis*, this cell layer is entirely absent. The innermost layer of the outer integument is one-cell thick and forms a sclerotic layer, sometimes known as the palisade layer. The palisade layer develops secondary thickenings on their inner tangential and radial cell walls during development. Cells of the inner integument become compressed and accumulate flavonoids at maturity (a typical feature of Brassicaceae). It is interesting to note that the outer layer of the endosperm remains closely associated with the inner integument and together they form the aleurone layer (Bouman 1975) (Fig. 12.5 A). The epidermis of seed coats of Brassicaceae may contain mucilage (a pectic polysaccharide) that contributes to seed hydration, seed dispersal and seed germination (as seen in *Lepidium, Arabidopsis, Capsella bursa-pastoris*).

- **Fabaceae:** The seed coats of legumes are complex. Though the ovules are bitegmic; only the outer integument contributes to the formation of seed coat as the inner integument degenerates after fertilization. Instead, the endosperm and the maternal tissue are known to become a part of seed coat in legumes. In general, the outermost layer of testa is the epidermis which is a single layer of elongated palisade cells (macrosclereids) that are perpendicular to the surface of the seed. Next to the palisade layer is an hourglass like cell layer, which is composed of thick-walled osteosclereids. The innermost portion of the seed coat proper is made of parenchyma cells and is multi-layered. These parenchyma cells become partially flattened during development. Apart from these three layers, there is an aleurone layer (derived from endosperm), immediately next to the inner parenchyma. The aleurone and the crushed endosperm too are a part of the seed coat and the cells of this layer become tightly compressed in the seed coat due to expansion of the cotyledons (Fig. 12.5 B–C). An important feature of the seed coat of legumes is the vascular system which plays a key role in the lateral transfer of assimilates and nutrients to seed coat. The seed coat vascular systems in legumes vary structurally from extensive to simple. While the extensive systems anastomose to form reticulated networks throughout the entire seed coat (e.g., *Phaseolus vulgaris, Glycine max*), the simpler ones only have a single chalazal vascular bundle with two lateral branches extending into the seed coats (e.g., *Pisum sativum, Vicia faba, Medicago truncatula*). These characters of seed coats in Fabaceae help in providing physical dormancy and also can be used for taxonomy, physiology and biochemical investigations.

A: *Brassica* **sp.**

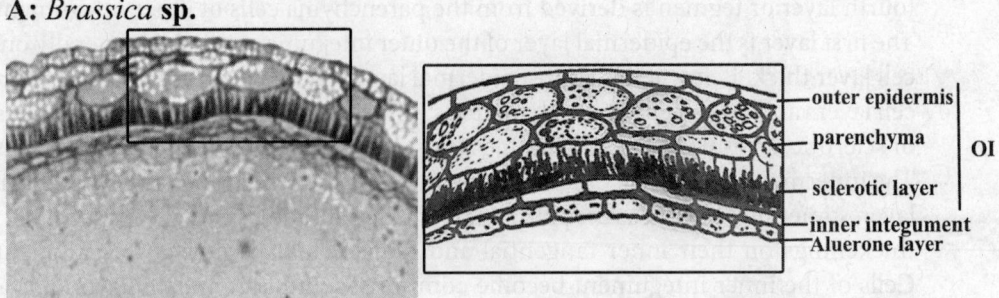

outer epidermis
parenchyma
sclerotic layer
inner integument
Aluerone layer

OI

B: *Glycine max*

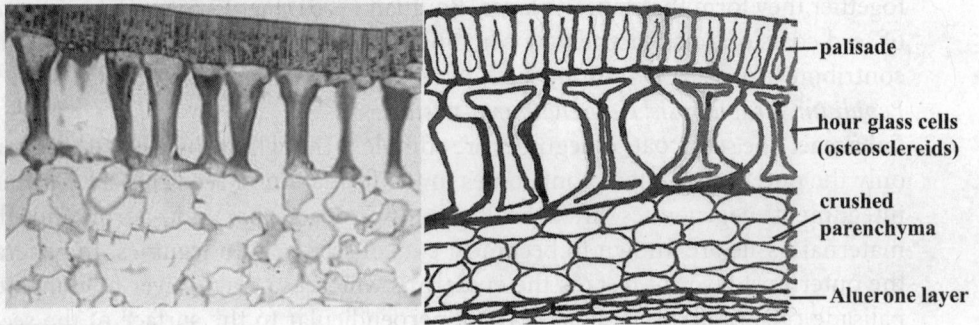

palisade

hour glass cells
(osteosclereids)

crushed
parenchyma

Aluerone layer

C: *Pisum sativum*

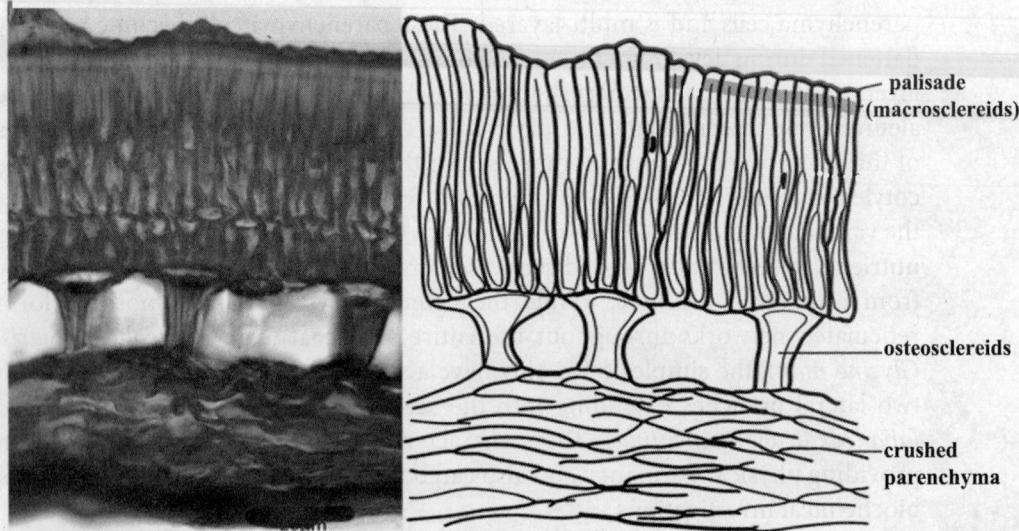

palisade
(macrosclereids)

osteosclereids

crushed
parenchyma

Figure 12.5 Seed coat in Brassicaceae (A) and Fabaceae (B–C). A. *Brassica* sp. B. Soybean C. *Pisum sativum*. **See Color Plates (page 498)**. (*Source*: A: Miller et al. 1999, *reproduced with permission*; B: Smykal et al. 2014, published under CC BY license.)

- **Cucurbitaceae:** A huge variation exists in the seed morphology, structure of seed coat and its development in Cucurbitaceae. The ovules are bitegmic in Cucurbitaceae, with 7–8 layered outer integument and the 2-layered inner integument. Only the outer integument develops into the seed coat as the inner integument degenerates soon after fertilization. In most of the cucurbits, the mature seed coat comprises five distinct zones (outside to inside):

1. **Epidermis:** The outer epidermal cells vary in shape as well as in degree of cuticularization and lignification. The cells may be elongated (e.g., *Benincasa hispida, Ruthalicia longipes),* isodiametric to polygonal (*Acanthosicyos horridus, Bryonia dioica, Lagenaria* sp.), irregular to polygonal, upright or even squarish (*Corallocarpus boehmii*).

2. **Hypodermis:** The hypodermis is generally 2–10 layered and is present only along the edges and at the chalazal end. In *Corallocarpus boehmii,* it is 9 to 11-layered, present throughout with usually small, polygonal, thick-walled and lignified cells with simple pit connections. In *Marah* sp. the hypodermis is 8–12 layers thick, made up of round and lignified cells with simple pit connections. Hypodermal cells accumulate brown pigments (Singh & Dathan 1974) (Fig. 12.6).

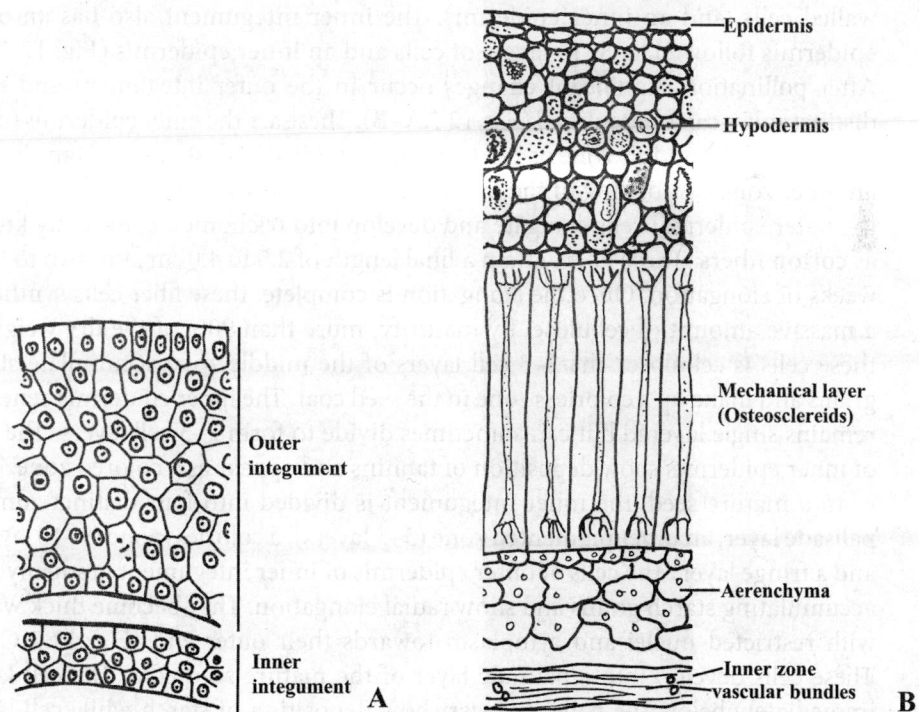

Figure 12.6 Diagrammatic sketch of seed coat in Cucurbitaceae (*Marah* sp.). **A.** Integuments of ovule. Note outer integument is thicker than the inner integument. **B.** Mature seed coat is derived from the outer integument and the mechanical layer is exotestal (*Source:* Singh & Dathan 1972).

3. ***Sclerenchyma or Mechanical layer:*** This is a single layer consisting of broad osteosclereids with dilated and branched ends. The size, breadth and branching of the sclereids show minor variations among species. For instance, the sclereids are narrow and palisade-like in *Ibervillea* but large astrosclereids are present in *C. boehmii.*

4. ***Aerenchyma:*** Next layers of cells are aerenchymatous. The aerenchymal region is 5–7 layered in *Corallocarpus* and *Kedrostis,* and is 3–5 layered in *Ibervillea.*

5. ***Inner zone or Chlorenchyma:*** The remaining cell layers of the outer integument enlarge and develop air-spaces and chloroplasts. The cells of the outer layers of the chlorenchyma are large with prominent airspaces while those of the inner layers are small and more compact as seen in *Luffa* sp.

• **Malvaceae:** Malvaceae is of great economic value because several of its members yield fiber. In some species, fiber comes from the seed coat, e.g., *Gossypium* sp. (cotton); the world's most important fiber plant. Two examples of seed coat development from the family are given below:

Gossypium **sp. (Cotton):** The seed and seed coat development in cotton has received much attention due to its commercial value. In cotton, the ovules are bitegmic and both the integuments participate in the formation of seed coat. At maturity the outer integument possesses an outer epidermis, 4–6 layers of thin-walled cells, and an inner epidermis. The inner integument also has an outer epidermis followed by 8–15 layers of cells and an inner epidermis (Fig. 12.7 A). After pollination, significant changes occur in the outer integument and three distinct zones can be identified (Fig. 12.7 A–B). These are the outer epidermis (outer most zone), the middle zone of 2–5 layers of cells filled with starch and tannins and an inner zone comprising of the inner epidermis. With subsequent development, the outer epidermal cells elongate and develop into trichomes, commonly known as **cotton fibers**. These fibers attain a final length of 2.5 to 4.0 cm after two to three weeks of elongation. Once the elongation is complete, these fiber cells synthesize a massive amount of cellulose. By maturity, more than 90% of the dry weight of these cells is cellulose. The 2–3 cell layers of the middle zone accumulate starch grains and make up a colorless zone in the seed coat. The inner epidermis generally remains single layered but can sometimes divide to form 2–3 cell layers. The cells of inner epidermis show deposition of tannins and appear as a colored zone.

In a mature seed, the inner integument is divided into four distinct zones: a palisade layer, an inner pigmented zone (3–5 layers), a colorless zone (9–10 layers), and a fringe layer. The cells of outer epidermis of inner integument gradually start accumulating starch grains and show radial elongation. They become thick walled with restricted nuclei and cytoplasm towards their outer end (Fig. 12.7 C–D). These cells develop into a palisade layer of the mature seed coat. The cell layers immediately below the palisade layer show deposition of starch while cell layers towards the inside accumulate tannins. This pigmented layer is followed by a colorless zone. The cells of inner epidermis of inner integument also show minor radial elongation and develop plate-like thickenings on their walls and form a fringe layer (Fig. 12.7 C–D).

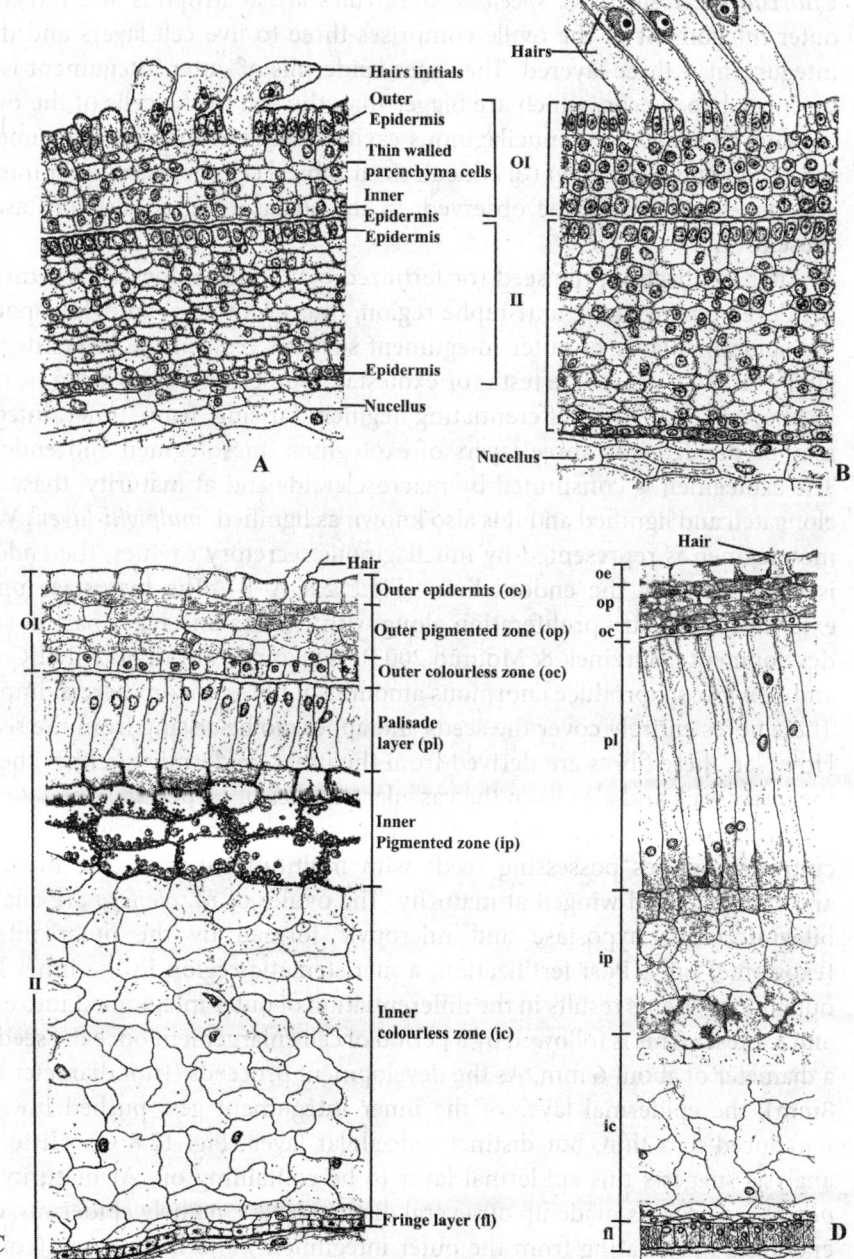

Figure 12.7 Seed coat development in *Gossypium* sp. A. Fertilized ovules after 3 days of pollination. **B.** After 6 days. **C.** After 10 days. **D.** After 20 days; showing seed coat differentiation. **A–B.** Initiation and elongation of hairs from the epidermis of the outer integument (**OI**). Three zones can be distinguished in the outer integument. **C.** Four distinct zones developed from the inner integument (**II**) are marked and apparent. During development the palisade layer differentiated from epidermis of inner integument shows considerable elongation (**D**). All the zones are marked with their abbreviations in C. (*Source:* Joshi et al. 1967, *reproduced with permission from INSA.*)

***Chorisia speciosa*:** In *C. speciosa*, the ovules are anatropous and bitegmic. The outer integument of the ovule comprises three to five cell layers and the inner integument is three layered. The outer epidermis of outer integument is formed by cube-shaped cells, which are bigger than the rest of the cells of the ovule and contain phenolics. The mucilaginous cavities between the two integuments and the endothelium like layer (of cuticularized cubical cells between inner integument and nucellus) can also be observed. In the chalazal region, a hypostase is also present.

After fertilization, the seed (or fertilized ovule) undergoes a curvature due to the development of the anti-raphe region, thus becoming campylotropous. Only the epidermal layer of outer integument survives (rest of the layers degenerate) giving rise to a uniseriate testa (or exotesta). Numerous stomata can be observed in the exotesta. The differentiating tegmen (arising from inner integument) may be divided into three layers of exotegmen, mesotegmen and endotegmen. The exotegmen is constituted by macrosclereids and at maturity, these cells are elongated and lignified and it is also known as lignified '*malpighi-layer*'. While the mesotegmen is represented by mucilaginous secretory cavities, the endotegmen is constituted by the endothelium (Fig. 12.8 A–B). The hypostase undergoes extensive secondary proliferation along with chalaza and takes part in seed coat development (Marzinek & Mourão 2003). It is important to note that *C. speciosa* and *Bombax* sp. produce enormous amounts of fibres of commercial importance. These fibres entirely cover the seeds and appear to be arising from the seed itself. However, these fibres are derived from the inner epidermis of either the fruit or the ovary wall; and also from the vascular bundles of carpels.

- **Moringaceae:** *Moringa oleifera* (Moringaceae; drumstick or wonder tree) is a cultivated species possessing seeds with multitude of uses. The mature seeds are triangular and winged at maturity. The ovules of *M. oleifera* are anatropous, bitegmic, with hypostase and micropyle formed by the outer integument (exostome) only. Post fertilization, a meristematic region in the inner layers of outer integuments results in the differentiation of outer integument into endotesta and exotesta. This is followed by a period of cell enlargement once the seed reaches a diameter of about 6 mm. As the development proceeds (fruit diameter reaching 8mm), the epidermal layer of the inner integument gets pushed inwards and is reduced to a thin, but distinct unicellular layer (Fig. 12.8 C). Histochemical analysis suggests this epidermal layer to be containing oil. At maturity, the *M. olifera* seed coat is made up of several distinct layers, namely epidermis, exotesta, endotesta (originating from the outer integument) and the tegmen (comprising the remnants of the inner integument). The exotesta is characterized by an epidermis with prominent papillae, followed by a layer or two of very small sub-epidermal cells ending in much larger, thin-walled reticulate cells. These cell layers have large pits, intercellular spaces and extend into the three wings of the seed. The distinction between exotesta and the endotesta is marked by reticulate cells as well as the vascular bundles underlying the seed wings as seen in a mature seed (Muhl et al. 2016).

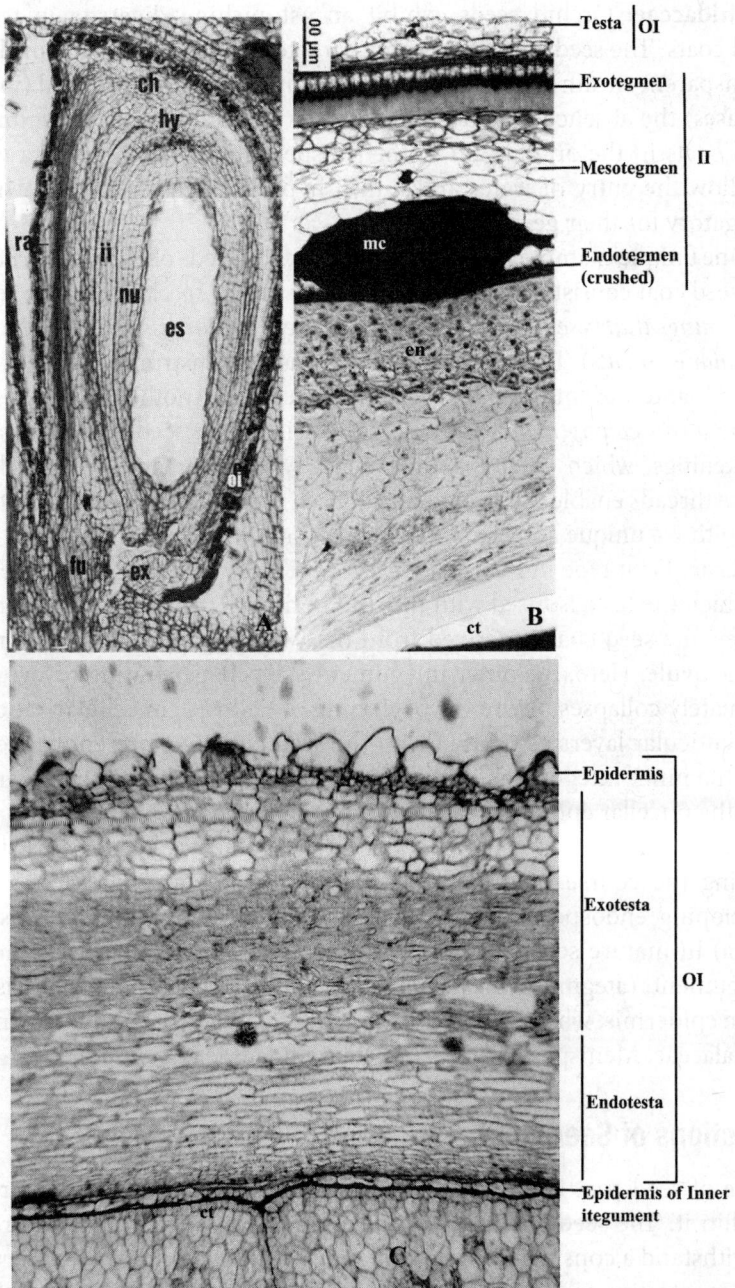

Figure 12.8 A–B. Seed coat in *Chorisia speciosa*. A. Longitudinal section of bitegmic ovule. **B.** In mature seed coat the outer integument (OI) contributes only single layer testa, while inner integument (II) forms well differentiated tegmen. Mucilagenous cavities (**mc**) are present in mesotegmen. Important to note that hairs covering the seed; ontogentically are the part of the fruit. **C. Seed coat in *Moringa oleifera*.** Ovules are bitegmic. In mature seed coat, the testa is well differentiated and the inner integument remains as single layered tegmen. (*Source*: A–B: Marzinek & Mourão 2003, *reproduced with permission*; C: Muhl et al. 2016, published under CC BY 2.0 License).

- **Orchidaceae:** Orchid seeds exhibit an astonishing diversity in terms of their seed coats. The seeds in orchids are characterized by a thin balloon-like seed coat (transparent sheath) with uniform cells of different cell types, and (in the majority of cases) the absence of endosperm (and even undifferentiated embryo). Sac-like seed coats in the orchids aid in their dispersal and also provide a wide opening to allow the entry of water and fungal hyphae. Such association with hyphae is obligatory for their germination.

  One unique example of seed coat is seen in the seeds of *Sobralia dichotoma*, where the seed coat consists of tracheoidal idioblasts (i.e., *special cells with secondary wall thickenings that resemble the tracheids of the vascular tissue system but are not true vascular elements*). These tracheoidal idioblasts are instrumental for uptake of water in seeds and prevent the embryo from desiccation. Another example is *Chiloschista lunifera (an epiphytic orchid),* where the cells of the seed-coat possess helical wall thickenings, which extend upon contact with water and produce long threads. These threads enable the seed to attach itself to moist barks of several types of trees, exhibiting a unique and elaborate mechanism of seed attachment and dispersal.

- **Poaceae:** Fruit type in Poaceae is caryopsis. **Caryopsis** is a single-seeded dry fruit in which the testa is fused with the thin pericarp (fruit wall) (Black et al. 2006). In barley, the seed coat is derived from the two cell layers of the inner integument of the ovule. Here, the outer integument starts degenerating before anthesis and ultimately collapses before the beginning of endosperm cellularisation. There are two cuticular layers associated with the seed coat; one present between the outer and the inner integument, and the second present between the inner integument and the nucellar epidermis originating from the latter.

- **Other Examples:** In the members of Acanthaceae, the seed coat is single layered. During the course of seed development, the integument is consumed by the developing endosperm, leaving just the epidermis which serves as a seed coat (testa) in mature seeds. In Loranthaceae, the seeds are naked as the ovule lacks integuments (ategmic ovules). Here, the endosperm is persistent and is surrounded by an epidermis, which covers the seed mass. A seed coat is also lacking in families Santalaceae, Menispermaceae, and Amaryllidaceae (*Crinum*).

12.4.2 Functions of Seed Coat

The chief role of seed coat is to provide protection to embryo and the nourishing tissue enclosed within it. The seed coat enables the embryo to pass through a dormant period and also to withstand a considerable amount of desiccation without losing its viability. The presence of a protective seed coat is the primary reason for successful establishment of seed-bearing plants in hostile terrestrial environments. Apart from protection, other roles performed by the seed coat are summarized below:

1. Various studies have highlighted the regulatory role of seed coat in embryo development. Although the seed coats of different species vary greatly in their structure and composition, they all undergo similar phases of development in relation to the embryo and the endosperm. In most species, seed coat and endosperm development precede the development and maturation of the embryo. Several mutant analyses in *Arabidopsis thalliana* have shown that deformity in seed coat development leads to deformation or abortion of embryo indicating that seed coats have a role in embryo development.

2. In legumes, enzyme invertases of seed coat play an essential role in the maternal control of seed development. Cell wall invertases facilitate nutrient collection by increasing the sucrose gradient in the unloading zones of the legume seed coat. The high hexose-to-sucrose ratios help in embryo growth by increasing cell division.

3. Seed coat plays an important role in seed dormancy. Abscisic acid present in seed coat is a determinant of dormancy, especially in species with seed coat-imposed dormancy.

4. Seed coat also helps in maintaining optimal environmental conditions for the viability of embryo and its germination. In radish and soybean, reactive oxygen species like superoxide radicals, hydrogen peroxide, hydroxyl radicals are synthesized by the seed coat and the embryo, as by-products of metabolic processes. These reactive oxygen species accompany germination and presumably provide resistance to pathogens during seedling emergence. Oxidases such as peroxidases, oxylate oxidases or amine oxidases can be potential sources of hydrogen peroxide. Peroxidases are commonly found in seed coats of barley.

5. A variety of storage compounds exist in the seed coat (e.g., phenolics in Leguminaceae, *Butia capitata*) performing multiple functions. Among legumes, the seed and the seed coat phenolics support nodulation by acting as chemo-attractants and promote rhizobial growth. Phenolics also induce transcription of nodulation genes in symbiotic bacteria subsequently supporting nodulation. Legumes with dark seed coats, as in soybean, broad beans, faba beans, lentils and peas, possess high antioxidant activity mainly due to the pro-anthocyanidins (*they are natural sources of antioxidant*). Seed coat stored tannins in sorghum are associated with a marked degree of resistance to pre-harvest seed germination. Tannins are also responsible for retarding seedling growth by decreasing the rate of starch and protein degradation in germinating high-tannin seeds, possibly by inactivating hydrolytic enzymes.

Box 12.3

Seed Dormancy

Seed dormancy is an essential morpho-physiological feature of seeds. It can be visualized as a temporary failure of a viable seed to germinate under a particular set of environmental conditions that normally allow germination. Depending on the species, seed dormancy may last from a few days to weeks to several years. It is an important survival strategy of plants that provides additional time for seed dispersal over greater distances and also maximizes seedling survival by avoiding germination under unfavorable conditions. However, negative consequences of high level of dormancy may include failure of germination and seedling establishment.

Induction, maintenance, and breaking of seed dormancy are complex physiological processes that are influenced by a wide range of endogenous and environmental factors. Plant hormones, mainly abscisic acid (ABA) and gibberellins (GA), are the major endogenous factors that act antagonistically to control seed dormancy and germination. ABA positively regulates the induction, and maintenance of dormancy, while GA enhances germination (Tuan et al. 2018). Temperature is a key environmental factor that regulates both dormancy and germination. The role of light as a regulator of dormancy is still debatable. Artificially, seed dormancy can be released by different treatments like after-ripening, cold stratification, acid/alkali treatment, scarification, light and temperature treatment. Nikolaeva (2004), and Baskin & Baskin (2004) have provided a comprehensive classification of seed dormancy, dividing it in five classes:

a. *Physiological dormancy:* It is the most abundant form of dormancy, induced by the endogenous physiological characters of a seed. It is prevalent in both gymnosperms and angiospermous seeds of temperate seed banks. This can be overcome by GA treatment, scarification, e.g., *Arabidopsis, Helianthus, Nicotiana.*

b. *Morphological dormancy:* It is present in seeds where embryos are underdeveloped (in terms of size), but differentiated. Such embryos need/take time to germinate, e.g., Arecaceae.

c. *Morpho-physiological dormancy:* This type includes seeds which have underdeveloped embryos along with physiological dormancy, e.g., *Trollius, Fraxinus excelsior.*

d. *Physical dormancy:* Such dormancy is evident in species where seeds have water impermeable layers of palisade cells in seed or fruit coat. Mechanical or chemical scarification alleviates such dormancy, e.g., *Melilotus, Trigonella.*

e. *Combinational dormancy:* This type of dormancy is manifested by the presence of thick seed coat along with physiological dormancy, e.g., *Trifolium.*

Morphological dormancy is thought to be the most primitive dormancy class. Morphological and morpho-physiological types of dormancy are present in basal angiosperms as well as in gymnosperms. Physical and combinational dormancy are restricted in distribution, and are thought to be an adaptation to specialized life strategies or habitats. Seeds with invested, bent and folded type of embryos frequently exhibit physical dormancy along with physiological dormancy (Savage & Metzger 2006). Seed dormancy also has an evolutionary relationship with the embryo and the seed size. Seed dormancy is thought to be an associated feature in the evolutionary trend towards increase in the relative size of embryo in seeds, both in angiosperms and gymnosperms (Baskin & Baskin 2004; Nikolaeva 2004).

Box 12.4

Xenia and Metaxenia

Different ovules in an ovary and also different flowers in a plant are fertilized by different pollen grains. Pollen from different sources have different effects on the characteristics of seeds and fruits. This phenomenon of direct or indirect effect of the pollen grains, or male gametes on tissues of seed and fruits other than the embryonic tissue was termed as **Xenia** by Wilhelm Focke (1881). It takes into account the differences in the size, shape, color, developmental timing, and chemical composition of seeds and fruits as a result of fertilization by different pollen/ male gametes. Xenia does not include effects associated with hybridization as expressed in the embryo. However, the term 'Xenia' was coined before the discovery of double fertilization and at that time, the endosperm was considered to be purely maternal in origin, and so the term xenia became associated mostly with effects seen in endosperm. Later, another term **metaxenia** was coined to account for specific cases of xenia occurring only in the maternal tissue, rather than on parts resulting from syngamy (Swingle 1928). In other words, xenia represents the morphological and biochemical features of seeds of the same fruit/plant due to different alleles brought by different pollen grains. While metaxenia is the phenomenon that defines effects/expression of male genotype, only in the morphology of fruit, like fruit length and diameter.

Both the phenomena of xenia and metaxenia have applications not only in genetic and physiological research but also in plant breeding and crop production. In plant genetics and breeding, the concept of xenia and metaxenia is used to identify the best pollenizer parents to shorten the fruit development period, and increase yield in mixed cultivar plantings. Both the phenomena have been used to improve yield in maize (Weingartner et al. 2002), and in several fruit crops, such as pecan nuts, pistachio nuts, avocado (Robbertse et al. 1996; Sedgley & Griffin 1989) and date palms (Nixon 1928; Shaheen et al. 1989).

12.5 Specialized Seed Associated Structures/Appendages

The shedding of seeds from the mother plants followed by successful germination is the ultimate goal of all plant species. Many species of angiosperms have devised several mechanisms for the effective dispersal and germination of seeds. These structures generally differentiate post-fertilization and may serve as the diagnostic feature of a family or a species. A brief description of some of the seed associated structures with their function is provided below.

- **Sarcotesta:** The term sarcotesta is applied to pulpy, colorful and sometimes edible parts of the seed coat which originate from the outer epidermis of testa. It stores sugars and lipids. The seeds with sarcotesta are adapted for endozoochory, a type of seed dispersal mechanism involving animals. The occurrence of pulpy layer followed by a hard-mechanical layer in such seeds; protects the seeds against the injuries which may happen in the gut, when consumed by animals. The most common example of sarcotestal seeds is *Punica granatum* (pomegranate, Lythraceae), where the pulpy layer originates from the epidermis of outer integument. The other examples are in families Magnoliaceae (*Magnolia*), Rutaceae (*Melicope glaberrima*), Sapindaceae (Fig. 12.9 A–C).

- **Operculum:** The operculum is a dehiscent cap like covering of a seed or fruit at the micropyle that detaches during germination. The operculum is also called a little seed lid and has a varied ontogeny. It may originate from the nucellar apex, exostome, endostome or even from the hilar region of a fertilized ovule. The operculum is of common occurrence in monocots (e.g., Arecaceae, Commelinaceae, Lemnaceae, Musaceae) but is also seen in some dicots (e.g., Begoniaceae). It plays an important role in maintaining seed dormancy in palms. In *Butia capitata* (Arecaceae), the seed coat is thin and seeds have an operculum at the micropyle. A basal-linear embryo is present in the cavity formed by the endosperm and remains connected to the operculum. The micropylar endosperm remains in contact with the lateral endosperm through cells with thin walls – establishing a line of weakness favoring dislocation of the operculum by growing radicle at germination. In Lemmnaceae, the operculum develops from the inner integument and its cells become dome-shaped and thick-walled. The other examples of opercula are *Victoria* sp. and *Elaeis guineensis* (oil palm) (Fig. 12.9 D).

- **Aril:** The term aril refers to a fleshy structure originating from the funiculus as an envelope which covers the seed partially or completely. There are many other fleshy structures found associated with seeds like sarcotesta described above. However, any structure developing from parts of the ovule other than funiculus is usually called *arillode or false aril or aril-like structure*. For instance, in *Catha edulis* (Celastraceae), an aril-like structure originates from the micropyle and is termed 'caruncula'. It later develops into the seed wings associated with seed dispersal (Zhang et al. 2014). The true aril which originates from funiculus is also designated as a supernumerary integument as it covers the seeds partially or completely. Aril is of wide occurrence among different families of angiosperms like Leguminaceae (*Eriosema glaziovii, Cytisus striatus, C. multiflorus*), Sapindaceae, Myristicaceae, Passifloraceae (Fig. 12.9 F–H).

  During the development of seed, aril accumulates various substances conferring olfactory cues and scents to seeds. Aril is of great economic importance as it is a good source of flavor-and aroma-rich compounds (as in *Myristica fragrans*; nutmeg), nutrients and even sugars (e.g., *Passiflora edulis, Litchi chinensis*).

- **Elaiosomes:** The term elaiosomes was introduced by the Sernander (1906) for the fleshy appendages of the seeds. Elaiosomes can be defined as fleshy, oily outgrowths of the seed coat. The elaiosomes can vary in color from generally being orange to yellow, or even white. Some examples of species with elaiosomes are *Acacia* sp. (Mimosaceae), *Sanguinaria canadensis* (Papavaraceae), *Polygala* (Fabaceae), *Turnera* (Turneraceae) (Fig. 12.9 I–J). Ants are attracted to the oil content of elaiosomes (a source of nutrition) and in turn disperse the seeds. During the collection process, some seeds being hard get detached from elaiosomes. Such seeds are not eaten by ants and act as a seed bank. Apart from aiding seed dispersal, elaiosomes also perform other functions like facilitating dehydration and hydration of the seed (e.g., in *Chelidonium majus*), inducing seed dormancy (e.g., *Calendula arvensis, Euphorbia cyparissias* and *Mercurialis annua*), and reserving water by absorbing it from the soil and subsequently transferring it to the rest of the seed during germination.

Figure 12.9 Seed associated structures/appendages. A–C. Sarcotesta in **A.** *Punica granatum.* **B.** *Magnolia* sp. and **C.** *Melicope glaberrima.* **D.** Longitudinal section of fruit of oil palm showing operculum below which embryo is placed. **E.** Caruncle in *Ricinus communis* seeds. **F–H.** Aril in: **F.** *Myristica fragrans.* **G.** *Litchi chinensis* **H.** *Passiflora edulis* fruit showing seeds covered with yellow-orange aril (ar). **I–J.** Elaiosomes: **I.** *Acacia* sp. opened pods with seeds showing yellow appendage 'elaiosomes' and a seed in enlarged view. **J.** Seed of *Philotheca difformis* with remnants of endocarp (arrow) in mature seed and function as elaiosome. **See Color Plates (page 499).** (*Source:* Creative commons search, B: Editor B; F and I: John Tann CC BY 2.0, D: Professor S. Natesan; C and J: Bayly et al. 2013; H: Silveira et al. 2016, published under CC BY Attribution License.)

- **Caruncle:** It is a plug like outgrowth that develops from the micropylar region of the integument due to proliferation or outgrowths of the exostome cells. It is a feature of family Euphorbiaceae. The caruncle in *Ricinus communis* accumulates up to 40% oil by weight. Like elaiosomes, it attracts ants and this eventually helps in the dispersal of seeds. It also absorbs water and helps in seed germination (Fig. 12.9 E).

Box 12.5

Genes Governing Seed Morphology and Seed Associated Structures

Seed coat color and seed associated structures are known to play a significant role in seed dispersal and represent a relevant character for systematic classification. These traits also impart agronomic value to certain species like *Passiflora, Magnolia* and *Brassica.* Recently, seed traits (like shape, size, seed coat color) have gained much attention and information on development/ genetics aspects of seed coat color and seed associated structures have appeared in literature. In earlier chapters (7, 9, 10), several genes have been listed which play a role in embryo-endosperm development and also affect seed size. Genes affecting seed coat color and seed associated structures include candidate gene *TRANSPARENT TESTA GLABRA 1 (TTG1)* in *Brassica* (Ren et al. 2017). This gene is related to flavonoid biosynthesis. Similarly, several *TTG* genes have been characterized in *Arabidopsis.* In *Magnolia grandiflora,* gene expression analyses showed that *AGAMOUS, AGL6* (C- and E-function MADS-box genes), and *TM8*-like genes; are involved in the development and ripening of fleshy structure 'sarcotesta' (Lovisetto et al. 2015). In *Arabidopsis thalliana,* homeobox gene *WUSCHEL* (WUS) expresses exclusively in the nucellus but in *Passiflora* sp. it has been observed that, WUS ectopic expression at funiculus is induced by *AINTEGUMENTA* (*ANT*) promoter, and such induction causes formation of extra/supernumerary integument, i.e., aril (Silveira et al. 2016).

12.6 Seed Dispersal

Seed dispersal is the movement of seeds from the immediate environment of their parents to establish themselves in unknown distant destinations. It is a key mechanism for gene flow and establishment of new populations. The seed dispersal mechanism and its ecology are the topics of much interest to biologists as they are one of the key steps in the process of plant regeneration. There are a variety of mechanisms through which seeds disperse involving both biotic and abiotic agencies. Seed dispersal mechanism has been divided into several categories based on the agency involved in dispersal. Gravity is considered to be the simplest dispersal method. Not only do coconuts fall from trees under the influence of gravity, but they also roll, and can even be carried great distances by ocean currents. Many plants show ballistic seed dispersal where seeds sit inside a pod till the pod dries out. This creates a tension within and causes fruit to burst and fling considerable distance. Quite a number of seeds are carried aloft by the wind or animals. These methods reflect the survival strategies, adaptations and optimization for establishment of progeny.

Studying seed dispersal is an important part of reproductive biology of a species. It is also important for strategizing conservation plans in the scenario of climate change. The fruiting phenology of multiple tree species is changing. This shift/change is destroying the mutualistic association between the plant and its seed dispersal agent, leading to lesser and ineffective seed dispersal. This association is essential for frugivore birds and endozoochorous species, as they generally rely on high energy food (primarily fruits) during their reproductive phase. Hence, changing phenology not only affects the plant communities due to lack of effective seed dispersal but also has an impact on life strategies of animals.

12.6.1 Advantages of Dispersal

The primary benefit of seed dispersal is their movement or departure away from the parent plant thereby negating competition amongst the siblings. This decreases chances of seed or seedling mortality due to lack of resources and also saves them from attacks by predators or pathogens. Another advantage of dispersal is the colonization of new sites. As the seeds spread, they may colonize newer habitats. However, the density of their distribution usually decreases with distance from the parent plant; in turn increasing their chances of germination and survival (Travest 2018). Seed dispersal also drives gene flow, plant population dynamics and functional connectivity along landscapes. Long distance dispersal strategies are not just effective in colonization of species in new places but also maintain species and genetic diversity in a geographical area.

12.6.2 Mechanisms of Dispersal

Seed dispersal can take place by both abiotic and biotic mechanisms. Interestingly, in some angiosperms, dispersal of seeds can take place without the involvement of any external agency either by explosive opening of the fruits or by the springing of a trip lever. Such species are called autochorous species and seed dispersal in such species is called as **autochory**. This mechanism reflects the adaptations for seed dispersal in scarcity of biotic means or weakened abiotic factors. Abiotic modes of seed dispersal include wind, water and even gravity. Dispersal aided by wind is called as **anemochory**. Seeds dispersed by water remain buoyant and keep themselves afloat on the surface of water, and this mechanism of dispersal is called as **hydrochory**. A huge number of seeds are transported to different areas by different animal vectors as well; a process called **zoochory**. There are certain types of seeds which possess hooks or barbs on their outer surface by means of which they can easily adhere to the exteriors of animal vectors. Such dispersal is called **exozoochory or epizoochory**. Many seeds possess fleshy appendages or coverings which are consumed by animals, to be ejected later at a different place. The dispersal of seeds occurs after passing through the gut of these animals and is known as **endozoochory**. A lesser known agents of seed dispersal are ants and the mechanism is known as **myrmecochory**. Ants feed on fleshy appendages like elaiosomes, caruncle and in turn disperse them.

There are some plants which lack a defined dispersal mechanism or even disperser rewards, such species are called barochorous species and the phenomenon is called **barochory**. Seeds in some species produce mucilage upon being wetted and such species are called **ombrohydrochorous species**, e.g., *Lepidium* sp. The mucilage helps in maintaining prolonged seed dormancy and prevents herbivory. Likewise seeds of Podostemaceae members also produce mucilage on coming in contact with water which enables the seeds to stick firmly to rocks in fast flowing river waters. The mucilage also supports the germination of these seeds.

Recent studies have highlighted that some seeds may have more than one opportunity of being effectively dispersed and the dispersal may occur in two phases. The first phase consists of the initial movement of seeds away from the mother plant (i.e., **primary dispersal**), followed by a second phase (i.e., **secondary dispersal**) in which seeds are further dispersed, usually by another mechanism or agent. This is common in plants which are first dispersed by endozoochory or ballistically, and are subsequently moved by ants, dung beetles, rodents, birds or even predators (frugivores) that carry seeds in their digestive tracts. For example, *Viola* seeds are first dispersed ballistically and then dispersed by ants. Similarly, in genus *Ruellia*, primary seed dispersal is by explosion in capsular fruit and secondary dispersal is achieved by hydrochory, zoochory and even anemochory. In this scenario, the primary explosive mechanism helps in short distance dispersal while secondary mechanism accomplishes the long-distance dispersal. It is also not uncommon to observe plants combining two or three modes/agents for seed dispersal, e.g., *Myrtus* seeds are dispersed by ants, birds, and mammals all together.

Each dispersal mechanism comes with its own adaptations in seeds. The morphological devices that help in seed dispersal are usually quite apparent (e.g., aril, caruncle). In addition, seed dispersal mechanisms are usually influenced by the features/seed traits like seed color, shape and others. For instance, species dispersed ballistically have significantly larger seeds as compared to those dispersed by anemochory; species with mammal-dispersal have significantly larger seeds than those with bird-dispersal. Such a correlation between seed traits and dispersal vectors is called **dispersal syndrome**.

12.6.2.1 *Autochory*

Incredibly, a number of plant species utilize explosive force to fling their seeds away. Explosive seed dispersal is heavily reliant on mechanical phenomena. Generally, elastic energy stored in fruit tissue aids autochory induced by dehydration (e.g., Geraniaceae fruits), or in some cases by excessive hydration which builds up turgor pressure to cause an explosion (e.g., *Impatiens capensis*). Many mistletoes have explosive fruits with sticky seeds to propel their parasitic offspring high into neighboring trees. *Hura crepitans* (sandbox tree) has exploding capsules that can launch seeds up to 100 meters away with speed of up to 70 meters per second. The sound of the explosion can be heard echoing through the forest.

The seed dispersal from the fruit of *Oxalis* sp. is an example of catapult where each seed is ejected like an individual shot from a gun. The fruit in *Oxalis* is comprised of five valves distributed around a central axis, and each valve contains several seeds, each covered with an aril. In *Oxalis,* the aril is specialized and performs the function of a catapult. Here, the aril is two layered with its outer layer being lipidic in nature and hydrophobic, while the inner layer is hydrophilic. The absorption of water by the inner layer causes its expansion and exerts a pressure on the outer layer. The expanded inner layer curls and exerts mechanical pressure; which cracks open the aril and expels the seed (Stage S1–S3). The ejection velocity of the seed is around 4.7 m/s. The seeds in *Oxalis* also bear a crack and a protrusion towards the next seed (Fig. 12.10 A). Therefore, once a seed is triggered, its curly aril will contact the next seed's protrusion and induce its firing. Thus, a chain reaction starts and the whole valve gets dehisced (Fig. 12.10 A).

Autochory in *Arceuthobium* (commonly known as dwarf mistletoe; Santalaceae), is an example of fluid pressure catapult. The ripe fruit of dwarf mistletoes consists of broadly fusiform-spheric seeds attached on short stems (pedicels). An abscission zone, representing the weakest region of the fruit, develops between the stems and the base of the fruit. Inside the fruit, a layer of viscin tissue surrounds each seed. During swelling of the fruit, the viscin tissue expands and starts to exert a hydrostatic force on the seeds and tensile stress in the cell walls. Once critical pressure is reached, the cell walls of the pedicel break at the abscission zone, thereby discharging the seeds (Fig. 12.10 B). The seed discharge happens in approximately 4.4×10^{-4} seconds and liquid cell content exerts a launch velocity of 13.7 m/s in *A. vaginatum*. The highest measured launch distance recorded in genus is 14.6 m in *A. cryptopodum* (Sakes et al. 2016).

In *Impatiens* sp. the explosive seed dispersal is aided by cell layers of the fruit wall. The fruits in *Impatiens* have five fused carpels at columella (in centre) throughout their length (Fig. 12.10 C). The valves contain a bilayered structure: an inner cell-layer that shortens by water absorption and an outer cell-layer that expands by water absorption (also known as expansion tissue). These two contrasting cell layers cause deformation of the adjacent valves due to inflow of water as the inner cell-layer compresses and the outer cell-layer expands. This in turn results in storage of elastic energy. When the fruit is fully ripe and the conditions are dry, then the tension within the fruit is released even with a mild breeze or a jerk, leading to separation of valves from the columella. Due to sudden release of high tension in the fruit, an inward curvature in the valves occurs. Thus, the five carpel walls immediately curl up and throw out the seeds (Fig. 12.10 C).

Another interesting example of an autochorous species is *Ecballium elaterium* (squirting cucumber). Squirting cucumber has exploding fruits which are borne on long and a curved stalk. The tip of the stalk fits into the fruit like a stopper. At maturity, seeds develop a mucilaginous layer and exert a turgor pressure on the fruit wall. Due to this mechanical pressure, the fruits pop off from the end of their stalk, and liberate the seeds forcibly. It is notable that during explosive seed dispersal, the magnitude of force, and the manner/angle at which seeds are launched are major factors.

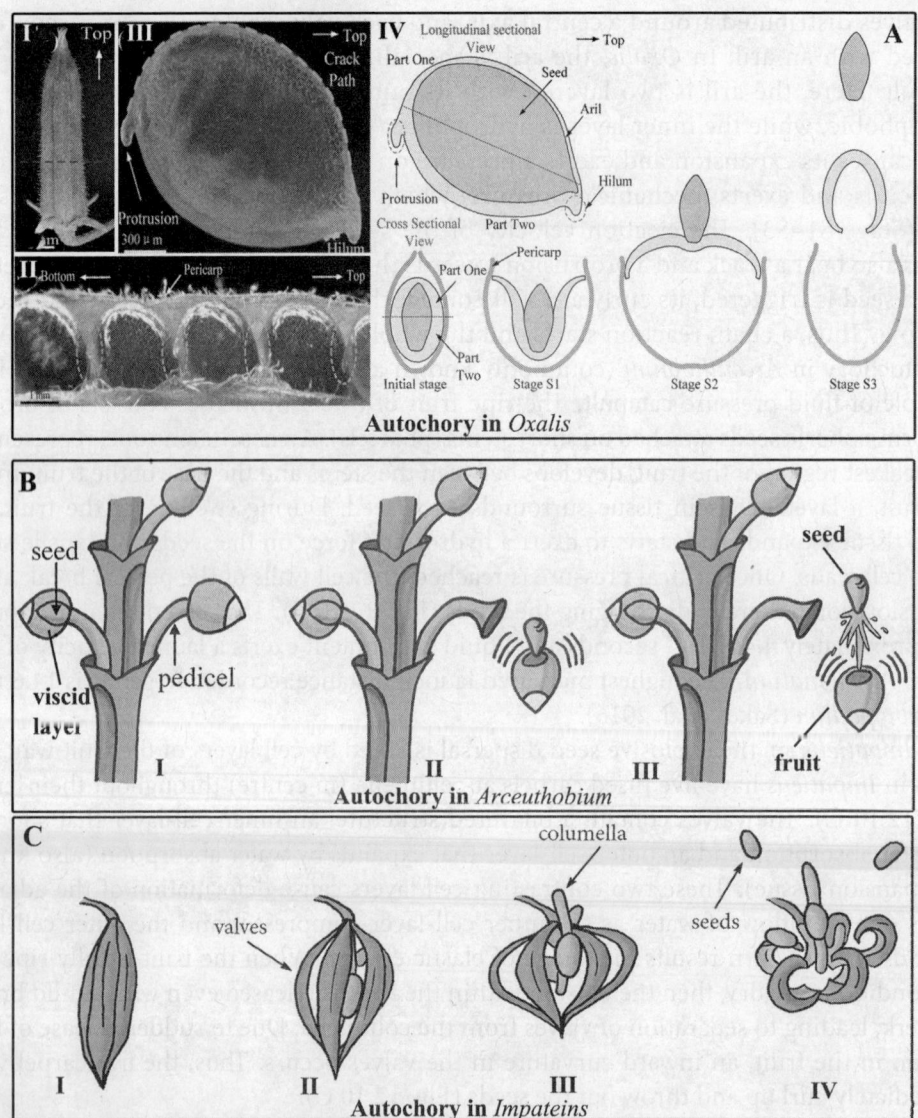

Figure 12.10 Diagrammatic representation of mechanisms of autochory. **A.** *Oxalis* sp. I. Fruit, an upright pod. II. A mature seed. III. Fruit split open, showing seed arrangement in one valve. Note the position of the protrusion (encircled) is located near the front seed, which plays a key role in the continuous ejection. IV. Representation of stages of seed ejection (in cross sectional view of fruit). **B.** *Arceuthobium* sp. I. The ripe fruits comprise of fusiform-spheric seeds and are attached to short pedicels. The seeds remain covered with a viscid layer. II–III. Viscid layer expands due to absorption of water and exerts a pressure. When a critical pressure is reached, the fruit breaks free from the pedicel and discharges the seed together with the liquid cell content. **C.** *Impateins* sp.: I. A five valved mature seedpod. II–IV. Elastic energy is stored in the seedpod by the absorption of water in the valves. When a critical pressure is reached, the valves separate from the columella and subsequently the coiling disperse/release the seeds from valves. **See Color Plates (page 500).** (*Source*: A: Li et al. 2020; B–C: Sakes et al. 2016, published under CC BY Attribution License.)

12.6.2.2 *Anemochory*

Seed dispersal by wind or anemochory is the most common type of seed dispersal, given the ubiquitous presence of wind in virtually all environments. So, anemochory even may occur in seeds non-specialized for wind dispersal (without dispersal syndrome). In general, the seeds specialized for anemochory are; produced in large numbers, small in size, light in weight and possess membranous wings or appendages or fluffy hairs on their surface (Fig. 12.11). The common groups and families showing anemochory as primary dispersal mechanism are birch, elms, some orchids, Asteraceae (diaspore is cypsela), Apocyanaceae, Dipterocarpaceae.

The seeds with wings can be found in *Oroxylum indicum, Holoptelea integrifolia, Ulmus* sp., and *Acer* sp. Such winged seeds glide easily in the air. Certain diaspores also show adoption of an optimum wing curvature to maximize descent time (*period between detachment from mother till touching the ground*). Recent investigations have revealed that each type of seed appendage possessed by an anemochorous seed is associated with a specific flight mechanism. For example, winged maple seeds auto-rotate and generate lift as they fall by leading-edge vortex mechanism. Such a feature is possible by virtue of highly optimized shape, size, and angle of wings. In contrast to the double/single winged maple diaspores, Dipterocarpaceae members possess multiple-winged diaspores which exhibit auto-rotating flight during dispersal. Spinning of the fruit occurs by a helicopter-like mechanism in which the horizontal force on the wings promotes greater forward lift than reverse drag. Many members of Aceraceae and Oleaceae have helicopter seeds with a rigid or membranous wing at one end. The wing typically has a slight pitch (like a propeller or fan blade), causing the seed to spin as it falls. Depending on the wind velocity and distance above the ground, helicopter seeds can be carried to considerable distances away from the parent plant. On the other hand, tumbleweeds roll across the plains, also using wind to disperse their seeds. One of the most-amazing adaptation for anemochory can be seen in *Alsomitra macrocarpa* (Javan cucumber), whose seeds are borne with translucent aerodynamic gliders of ~12 cm (Fig. 12.11 B). Another unique dispersal mechanism occurs in the Vanilloideae–Vanilleae, where seeds are dispersed by wind, aided by large sclerified wings. This type of anemochory is very different from the wind-dispersal of the tiny 'dust-seeds' prevalent in the rest of the family Orchidaceae.

In smaller or lighter diaspores, there is the presence of hairs, bristles or the pappus as seen in diaspores (cypsela) of Asteraceae (Figure 12.11 D) which help in their dispersal by wind. While small seeds in dandelion are covered by a tuft of hair, seeds in *Calotropis* have parachute-like sails (Fig. 12.11 C). Seeds in members of Malvaceae like *Ceiba pentandra* and *Chorisia speciosa* are covered by cottony hair which help in dispersal by the wind. *Papaver* and *Aristolochia* plants too exhibit wind assisted seed dispersal where the mature fruits are capsules possessing several valves at maturity (Fig. 12.11 F). A gentle blow of air shakes the fruits, liberating seeds with each jerk. This mechanism helps in gradual dispersal of seeds in batches.

Figure 12.11 Seeds/diaspores/fruits exhibiting features of anemochory. **A–B.** Winged seeds of *Holoptelea integrifolia* and *Alstromeria* sp. **C.** A tiny seed of *Calotropis* with tuft of plumose hairs. **D.** Head with cypselae in *Taraxacum* sp. Note the pappus are arranged like bristle over a stalk which help in flight of diaspore. **E.** Winged samara in *Acer* sp. **F.** Mature capsule of poppy with valves. **See Color Plates (page 501).** (*Source:* B and F: Creative Commons Search.)

12.6.2.3 *Hydrochory*

It is generally thought that seeds of aquatic plants are dispersed by water. However, this notion is not always true as there are many other ways in which water plants disperse their seeds. Also, several non-aquatic plants like the ones growing besides streams, rivers, oceans, rely on water for dispersal of their seeds. Seeds dispersed by water are light enough to float on the surface of water and possess structures that help them to remain buoyant. Most trees found on tropical beaches are dependent on water for their seed dispersal. Their seeds are woody and with waterproof coverings enabling them to float in the salty water for long periods.

Hydrochory in coconut is supported by characteristics of its fruit wall. Coconut fruit is a one seeded drupe, having a waxy epicarp impermeable to water, a fibrous mesocarp and a stony endocarp. While the fibrous mesocarp provides buoyancy to coconut, the thick endocarp protects the embryo. The massive cellularized endosperm provides nutrients to the growing embryo and keeps it alive for months (~110 days) in the sea. This makes coconut a plant to have successfully colonized tropical islands around the world. The other plants and families where hydrochory can be observed are *Sagittaria, Nyphaea, Nelumbo* sp. (Fig. 12.12).

Though mangroves exhibit vivipary, they may also be considered as a group where water assists in dispersal of propagules. The embryo starts germinating in the seed while the seed/ fruits are still on the mother plant. The embryo does not undergo any dormancy. These

Figure 12.12 **A.** Coconut fruit dispersed by water currents. **B.** Coconut germinating at sea shore. **C.** Mangrove propagules dispersed in sea water. **D.** Fruit of *Nelumbo nucifera* with seeds. **See Color Plates (page 501).** (*Source*: Creative common search and *Encyclopedia Britannica*.)

propagules are oblong-elliptic to long narrow structures encompassing seedlings, further enclosed in a buoyant pericarp (e.g., *Rhizophora* sp., *Avicennia* sp.). After germination, the propagules fall/break from the parent tree and float on surface of sea water (Fig. 12.12 C). These long narrow propagules can survive for up to a year in salt water floating upright. They may develop roots rapidly when they encounter a suitable substrate. Some orchids also exhibit hydrochory for seed dispersal e.g., *Epipactis gigantea*, *Disa* sp. The seed coat in these species is highly lignified and reflects an important feature associated with hydrochory.

12.6.2.4 *Zoochory*

Zoochory or the dispersal of seeds by animals is of two types: exozoochory and endozoochory. Different groups of animals aid in the dispersal of seeds to variable distances, either by carrying them (exozoochory), or by ingesting and dispersing through fecal matter (endozoochory). Seeds dispersed by exozoochory (or ectozoochory or epizoochory), generally cling to the fur or feathers of animals, or produce a fruit or nut that the animals eat or store (e.g., nuts by squirrels). Many plants produce fruits or individual seeds covered by hooks or spines which get attached to the fur or feathers of the animals or, sometimes even to our clothes or bags. The seeds are eventually carried away from the parent plant to new places providing them additional space to grow. Some plants like dandelion, use exozoochory as an alternative or additional means of seed dispersal. In dandelions, in addition to the hairy pappus, there also are microscopic barbs that facilitate secondary dispersal by getting stuck on to the animals. Other examples of exozoochory are *Acyranthus aspera* and grasses, viz. *Setaria,* which produce hooked diaspores. The fruits of *Xanthium strumarium* bear tenacious teeth and hooks which stick to the fur or feathers of animals, clothes of human beings and disperse (Fig. 12.13). Seeds of plants like *Prunella vulgaris* and *Plantago* sp. produce mucus aiding them to stick to surfaces, considered an adaptation for exozoochory.

Figure 12.13 Diaspores dispersed by exochory. A. Fruits of *Xanthium strumarium* **B.** Diaspores of *Acyranthus aspera* clinging to cloth. **See Color Plates (page 502).**

Endozoochory is seed dispersal by animals where fruits or seeds are ingested and later ejected through regurgitation or defecation. In endozoochory, the fruits and seeds fulfill the metabolic needs of the animals and in turn get a specific treatment while passing through the gut, for their enhanced germination. For example, seeds of blackberries and cherries actually require to be softened by the stomach acids and digestive enzymes to overcome dormancy. This treatment makes the seeds permeable to water and enhances chances of germination. Moreover, fecal matter serves as an organic medium for the germination of seeds.

To attract seed dispersing animals, many plants have edible fruits covering the seeds. Fruits which have a colorful pulp or some kind of fragrance (like guava and figs) serve as food for animals (frugivory) which in turn eject seeds in suitable conditions to germinate. Among the different groups of animals, birds are the most effective agents of endozoochory (ornithochory). They effect dispersal to longer distances and are attracted by the various characters of fruits and seed appendages (aril, sarcotesta). Fruits of *Passiflora* sp. have sweet flesh in the form of an egg-like rind, which attracts the birds and is considered important for seed dispersal (Castillo et al. 2020). However, these generalizations may not be true for all the species (refer to Box 12.7 for details).

Although neglected in the past, recent studies have indicated that large animals (like monkeys, sambhar, and elephants) play a very significant role in seed dispersal. Monkeys are instrumental in seed dispersal by both exozoochory and endozoochory. They are known to disperse the seeds of *Aegle marmelos* (Bael) in forests. The two largest frugivores, tapirs (*Tapirus terrestris*) and muriquis (*Brachyteles arachnoides*) in the neotropics (Brazil), are known to be the dispersal agents of about 35 plant species. Likewise in tropical evergreen forests of Thailand, seeds of large-fruited *Platymitra macrocarpa* (Annonaceae) are effectively dispersed by mega-frugivores like gibbons, elephants and sambar deer (McConkey et al. 2018; Fig 12.14 E). In Mauritius and other oceanic islands, Giant Aldabra tortoise feeds on *Syzygium mamillatum* fruits which have a fleshy endocarp. The seeds are dispersed with the faecal matter of the tortoise (Hansen et al. 2008). Lizards too are effective seed dispersers in terms of fruit removal, seed dispersal in suitable habitats, and seed germination, e.g., seeds of *Capparis spinosa* are effectively dispersed by the lizard *Teratoscincus roborowskii*, through endozoochory (Yang et al. 2021). Frugivorous animals, mostly birds and mammals, can sometimes also act as seed predators by pecking or biting fruits for their pulp or seeds. For example, the Woodpigeon can disperse large seeds with a hard coat but its gut typically destroys smaller and weaker seeds (Snow & Snow 1988). A category of endozoochory which has not received much attention is **diploendozoochory**. It is secondary seed dispersal system whereby a secondary seed disperser, (carnivorous vertebrate) consumes a primary seed disperser (usually a rodent or bird) with undamaged seeds in their digestive tract which are subsequently eliminated with its feces.

Squirrels, ants, and rodents are mostly seed predators but under some conditions can also act as agents of dispersal. Squirrels and ants have the habit of collecting seeds for their food. In the process of seed gathering, squirrels collect variety of nuts and carry them to a cache. Often, these animals forget to eat these seeds, which have been kept securely in their burrow, eventually resulting in their dispersal. Ants are very important seed hoarders and get attracted to oily seeds and other such structures. Gathering these oily structures, ultimately promotes seed dispersal (Fig. 12.15). Myrmecochory or dispersal of seeds by ants is

Figure 12.14 Zoochory. A–B. Ornithochory (dispersal by birds). **A.** A bird feeding on berries. **B.** A Hornbill eating a nut. Hornbills are known to eat hard nuts and help in their dispersal. **C–D.** Dispersal by squirrel: **C.** Squirrel eating a berry (C) and a nut (D). **E.** *Platymitra macrocarpa* fruit eaters, major agency for seed dispersal by both endozoochory and exozoochory. **See Color Plates (page 502).** (*Source:* A–D: *Encyclopedia Britannica* and Creative Commons; E: after McConkey et al. 2018, published under CC BY License.)

commonly observed in *Strombocactus* sp. (and many cactus genera), *Lamium amplexicaule*, *Croton sonderianus, Rhamnus, Euphorbia, Borderea chouardii, Acacia neurophyll.* The seeds of most cacti are hard and bear elaiosomes and aril, serving as a source of food to ants. The ants eat away these sweet, oily structures but pose no harm to hard seeds. Recent studies in *Lamium amplexicaule* (Tanaka et al. 2015) have highlighted that ants are not only effective seed dispersal agents, but also safeguard seeds of many species against predation by beetles. (Fig. 12.15).

Box 12.6

Differences between Exozoochory and Endozoochory

Exozoochory	Endozoochory
• Animals mediated seed dispersal where seeds are either carried away or get attached to the body and fur of animals.	• Animals mediated seed dispersal where seeds are ingested by animals, passed through their gut and ejected either through regurgitation or defecation.
• The adaptation for such dispersal includes appendages like a hook and the diaspore is generally dry.	• Mostly, such seeds are borne with fleshy appendages (e.g., aril) which are a source of food and attractant for animals.
• Exozoochorous seeds/diaspores are generally small in size and light in weight.	• Endozoochorous seeds/diaspores are generally large.
• Exozoochorous seeds/diaspores generally do not require specific treatment/s for germination.	• Species dispersed by endozoochory need scarification treatment for germination which is accomplished in gut of animals.
• Exozoochory is a generalized phenomenon.	• Endozoochory usually exhibits mutualistic association/interaction.
• Examples: *Xanthium* sp., *Setaria*.	• Examples: *Pittosporum* sp., *Aegle marmelos*, *Passiflora* sp., Figs.

Box 12.7

Lack of Flesh and Adaptation to Endozoochory: An Investigation

Endozoochory in angiosperms is often associated with diaspores possessing fleshy, attractive tissues like aril and sarcotesta. Possession of fleshy appendages is seen as an adaptation to prevent digestion in the animal gut. However, new studies have put forth significant evidence that endozoochory in nature is not limited to frugivory; diaspores without 'external flesh and attractive cues' are also commonly dispersed, over long distances, via birds and mammals by granivory. A recent study investigated 11 angiosperm taxa (*Allium angulosum*, *Astragalus contortuplicatus*, *Bolboschoenus planiculmis*, *Cirsium brachycephalum*, *Cuscutalu puliformis*, *Cyperus flavescens*, *Echinochloa crus-galli*, *Elatine hungarica*, *Elatine hydropiper*, *Lychnis coronaria*, *Glycyrrhiza echinata*) for their dispersal. Investigations have revealed that these taxa are typically endozoochorous, but deviate from this syndrome sometimes. Their diaspores are achene, caryopsis and cypsela. The dry diaspores possess the mechanical protection layer(s) in seed coats or pericarp which is often one-celled and lignified (e.g., the achenes of *Bolboschoenus* and *Cyperus*) or may be multi-layered with sclereids or fibers (e.g., *Cirsium* cypsela, *Echinochloa*). Although the diaspores were ingested by birds, still a high percentage of intact diaspores were found in their droppings. The seeds passed from bird gut showed significant scarification of seed coat but no damage to the embryo is observed leading to successful germination. This study highlights that the seed coat or pericarp of the vast majority of angiosperm plants possess sufficient mechanical endurance to withstand – with different degrees of success – passing through different types of digestive tracts (e.g., of birds, ungulates or fish). Furthermore, and critically, there is no major difference in seed and seed coat structure between plants with a fleshy fruit and plants with dry types of diaspores. Authors have also pointed out that the evolution of external flesh surely represents a specialization, but not a specialization for diaspore architectures and its dispersal mutualistic interactions. Granivory can also provide dispersal benefits to the plants and thus is comparable to rugivory (Costea et al 2019, and references therein).

Figure 12.15 Myrmecochory. A. Seed of *Borderea chouardii* with elaiosomes and an ant carrying a seed. **B.** A worker ant (*Tetramorium tsushimae*) carrying a seed of *Lamium amplexicaule* into its nest. **C.** An ant (*Rhytidoponera metallica*) transporting a seed of *Acacia neurophylla* by holding it by elaiosome. **See Color Plates (page 503).** (*Source*: A: Garci et al. 2012; B: Tanaka et al. 2015; C: Lengyel et al. 2009; A–C: published under CC BY 4.0 License.)

Glossary

Anemochroy: Seed/diaspore dispersal facilitated by wind.

Auotchory: Seed dispersal by mechanical energy derived from an inherent mechanism present in the fruit, e.g., by explosion, or catapult as seen in *Oxalis, Arceuthobium* sp.

Caruncle: Plug like outgrowth containing oil that develops from the micropylar region of the integument in family Euphorbiaceae, associated with dispersal of seeds by ants.

Caruncula: An aril like structure in seed, originating from micropyle which can develop wings at a later stage for dispersal.

Caryopsis: A one-seeded fruit in which the fruit wall is united with the seed coat, e.g., cereals.

Diaspore: It is a dispersal unit consisting of a seed with an appendage.

Dispersal syndrome: Morphological characters of fruit and seed that are associated with a dispersal mechanism.

Elaiosomes: Fleshy, starch or lipid rich bodies attached at the anterior region of seeds acting as nutrient rich reward for seed dispersing ants.

Endostome: Region of the micropyle formed by the inner integument of the ovule.

Endotesta: The inner layer(s) of the testa formed by the inner epidermis of outer integument of ovule.

Endozoochroy: Animal mediated seed dispersal where the animals ingest the fruits and seeds. The seeds get dispersed after their release from animal gut.

Exostome: Part of the micropyle formed by the outer integument of the ovule.

Exotegmic seeds: Seeds in which mechanical layer is present in the exotegmen, the layer of tegmen that develops from the outer epidermis of inner integument.

Exotesta: The outer layer of the testa that develops from the outer epidermis of outer integument.

Exozoochroy: Animal mediated seed dispersal where seeds get dispersed by sticking or attaching to the body, fur or other parts of an animal.

Flight time or descent time: In anemochory, the duration between detachment of seed from mother plant and the time it falls on the ground is known as flight time. Longer flight time suggests greater distance of seed dispersal.

Hourglass cell layer: Pillar-like osteosclerids that constitute the single layer of hypodermis in seed coat of some species.

Sarcotesta: The pulpy, colorful and sometimes edible part of seed coat that develops from the outer epidermal cell layers of testa.

Secondary Seed Dispersal: Two-phase seed dispersal involving a second dispersal phase through which seeds are carried to a new location, e.g., the primary dispersal by winds and secondary by the ants.

Seed dispersal: Mechanism of transfer of seeds to newer sites for germination and establishment.

Seed longevity: The period for which seeds retain viability after harvesting. It is generally evaluated by the germination ratio, which decreases with the loss of seed viability during storage.

Strophile: A raised outgrowth in seeds where the raphe ends, seen in members of Fabaceae like *Cajanus* sp. and *Phaseolus* sp. It facilitates the movement of water inside the seed through the testa.

Tegmen: Inner layer of the seed coat derived from the inner integument of ovule

Testa: Outer layer of the seed coat derived from the outer integument of ovule.

Key Questions

Q12.1   **Fill in the blanks:**
   a. ....is an example of an albuminous seed.
   b. The members of family .... have green seeds due to presence of chlorophyll in seed coat.
   c. Myrmecochory is the seed dispersal by .... and .... is an example for the same.

    d. Elaiosomes are...............rich, colorful appendages and help in seed dispersal by
       ....

    e. The mature seeds of *Brassica* sp. are comprised of endosperm which is rich
       in................

    f. Aerenchymatous cells are present in seed coats in members of family ....

    g. The generalized function of sclereids in the seed coat is....................

Q12.2 Differentiate between:
    a. Exozoochory and Endozoochory
    b. Testa and Tegmen
    c. Anemochory and Hydrochory

Q12.3 Seed coat in Fabaceae is diverse. Justify the statement with suitable examples and diagrams.

Q12.4 Write a detailed note on seed coat development in cotton.

Q12.5 Discuss the various roles performed by seed coat with suitable examples.

Q12.6 Enumerate the various seed associated structures present in flowering plants with their function, composition and origin.

Q12.7 List the various adaptations that are displayed by a seed to be able to be dispersed by water and wind.

Q12.8 Define seed dispersal. What are the advantages of seed dispersal?

Q12.9 Autochory represents an effective form of seed dispersal that enables the plants to disperse seeds efficiently. This statement is true or false. Justify your answer.

Q12.10 Give an example (family, genus, or species) where the following are present:

a. Albuminous seeds	j. Aril
b. Myrmechory	k. Seeds with extensively developed vascular system
c. Anemochory	
d. Sclereids in seed coat	l. Hydrochory
e. Endozoochory	m. Exotegmic seeds
f. Exozoochory	n. Seed with strophile
g. Winged seeds	o. Operculum
h. Tracheoidal idioblasts in seed coat	p. Sarcotesta
i. Dwarf/micro seeds	

Practicals

Exercise 12.1 To calculate the mass and determine the shape (seed traits) of different seeds

Seed mass is an important aspect of plant reproduction as it influences different stages in the life-history of plants. Differences in seed mass among species are because of varying levels of starch and endosperm nutrients in the seeds. Seed mass may influence germination percentage. For instance, small-seeded species which produce a large number of seeds have a short life span. In all such species, the seedlings have low rates of survival. On the other hand, it is widely accepted that large seeds generally have a higher germination percentage than small seeds (Xu et al. 2014; Wang et al. 2016). Seed mass estimation is of commercial importance for agronomic species as it indicates the yields and vigor of the crop.

Seed shape is another important trait that affects germination. Theoretical studies have predicted that species with elongate and flat seeds germinate more frequently than those with round seeds. Elongate and flat seeds are more likely to experience post-dispersal predation than round seeds, and should therefore show prompt germination to avoid risks of mortality. However, strong empirical data supporting this prediction are lacking (Wang et al. 2016).

Seed Mass: Seed mass, is also a measure of its size. It may be calculated by the average weight of a single seed. For each plant species, seed mass may preferably be calculated by weighing 100 air-dried seeds. If the seed size is big, then one may weigh a single seed or weigh 10 seeds for calculations. Replication of weighing must be done for statistical analysis.

Seed Shape: Seed dimension variance is a standard index to describe seed shape (Thompson et al. 1993; Moles et al. 2000; Wang et al. 2016). It takes into account three main perpendicular dimensions, i.e., length, width, and height. The variance can be calculated as follows (after Wang et al. 2016), using the average of 10 seeds of any species:

$$\text{variance} = \sqrt{\frac{\left(1-\dfrac{\text{width}}{\text{length}}\right)^2 + \left(1-\dfrac{\text{height}}{\text{length}}\right)^2 + \left(1-\dfrac{\text{height}}{\text{width}}\right)^2}{3}}$$

The given equation provides a value that expresses the extent to which seed shape varies from sphericity. For example, spherical seeds have a variance of 0, and elongated or flattened seeds have a variance up to 0.33 (Thompson et al. 1993). In other words, larger variance values are associated with flatter seeds and smaller values with rounder seeds. Thus, seeds may be classified as: spheroid or nearly spheroid, cylindrical, tubular, conical, ellipsoid (broad-ellipsoid & narrow-ellipsoid), ovoid, obovoid, trigonous or prismatic, oblate- discoid, lenticular, fusiform or linear (Wang et al. 2016).

Procedure

Seed mass:

1. Select a species of your choice.
2. Collect the seeds and clean them for soil and extragenous material.
3. Air-dry the seeds for about two days.
4. Weigh 100 seeds (or 10, if seed size is large) and note their weight.
5. Calculate the average weight of a single seed.

Seed shape:

1. Select a species of your choice.
2. Collect the seeds and clean them for soil and extragenous material.
3. Air-dry the seeds for about two days.

4. Record the length, width, and height of seeds using digital scale/vernier caliper.
5. Calculate the variance with the given equation.
6. Predict the seed shape with the values obtained.

Exercise 12.2 To classify different types of seeds

One important aspect of seed biology is the classification of seeds which can be done on the basis of the number of cotyledons, type of ovule from which seed has developed, presence or absence of endosperm, seed-embryo size ratio, embryo positioning. Another aspect in seed biology is the study of seed traits, viz. seed coat color and its development and such. These characters vary in different species. The number of cotyledons may be noted after removing seed coats of soaked seeds or one may cut free hand sections of the seed (transverse or longitudinal) using a blade. Size of seed and embryo can be measured using a common scale or a digital scale or a vernier caliper. You may study the seeds present in your kitchen and in your surroundings for observing variation in seed traits between species. Record your observation in Table 12.1.

Exercise 12.3 To study seed dispersal mechanisms in various species

An elaborate account has been given under section 12.5 for each type of seed dispersal mechanism with several examples. You may attempt to fill Table 12.2 based on your understanding and observations of the plants in your vicinity or college campus.

Exercise 12.4 To calculate percentage seed viability, germinability and seed vigor

Seed viability is a measure of the duration during which a seed is still capable of germinating and giving rise to a seedling. Seeds lose their viability after storage depending upon temperature and moisture conditions (Black et al. 2006). Testing seed viability is of particular importance to forestry and agriculture, as both rely on seed germination. Seed vigor measures the properties of the seed which determine the level of activity or physiological capability of a seed during germination and seedling emergence. Loss of seed vigor reduces physiological capabilities of a seed, e.g., enzyme activity, protein synthesis. Seeds lose vigor before they lose the ability to germinate. Therefore, seeds that have similar germinability may still differ in their vigor and the seedlings may display differential performance. Loss of seed vigor starts before harvest and continues during harvest, processing and storage. Seed vigor testing is an important practice in seed storage programs especially for crop species like pulses, cereals, as they are subjected to prolonged storage. A seedling has four morphological parts: root system, hypocotyl, cotyledons and epicotyl. Seedings are said to be vigorous if all parts are well developed. If any deficiency like a lesion or necrosis is observed in the seedling, then it is classified as weak. The **Paper Piercing Test** is a simple test to check seed vigor. The principle of the paper piercing test is that vigorous seeds are expected to produce strong seedlings (all four parts intact) which can pierce a particular type of paper, while seedlings of poor vigor may not be able to do so. Therefore, the seedlings which emerge by piercing the paper are more vigorous than those which are not able to emerge through the paper (Bhanuprakash & Umesha 2015).

Table 12.1 Observation table to study seed traits

Species	Observations						Results	
	Seed coat color	Size of seed	Number of cotyledons	Position of embryo	Size of embryo	Embryo size:seed size ratio	Dicot/Monocot	Seed class (as per Martin's classification)
Phaseolus vulgaris (kidney bean)								
Glycine max (soyabean)								
Triticum aestivum (wheat)								
Vinca rosea								
Acacia sp. (babul)								
Tagetes sp. (marigold)								
Azadirachta indica (neem)								
Zea mays (maize)								

Table12. 2 Observation table for seed dispersal mechanism

Species	*Oroxylum indicum, Bombax celba, Chorisia sp., Tridax procumbens*	*Cocos nucifera, Nelumbo sp., Trapa sp., Eichhornia sp*	*Setaria sp., Achyranthus aspera*	*Morus alba,* Guava, Mango	*Aegle marmelos, Magnolia sp., Withania somnifera, Oxalis sp.*
Type of diaspore/seed (Large/big)					
Type of appendage/s					
Type of fruit/seed or diaspore					
Specific mechanism observed/studied					
Any other observation					
Possible seed dispersal mechanism					

Seed Viability: TTC test (Black et al. 2006)
For testing seed viability 0.2%–1% solution of TTC may be prepared in 10–15% sucrose solution.

Material Required
Seeds, TTC solution, glycerol, lactic acid, Stereo-zoom microscope, beaker, petriplates, glass slides, needles, forceps, blade, distilled water.

Procedure

1. Soak the seeds in water for hydration for about 6–10 hours. The duration may be increased if the seed coat is very hard. Seeds in water may be kept at 40°C for enhanced imbibing.
2. Take the soaked seeds and gently blot them dry.
3. Immerse the soaked seeds in TTC solution for 30 minutes–2 hours. The duration may be increased/decreased depending on the seed coat thickness in the species. Some workers suggest piercing in cotyledons or the removal of seed coat can reduce this time significantly. But precaution must be taken not to damage the embryo.
4. Wash the seeds in distilled water once after staining.
5. Dissect the embryo under dissecting microscope/stereo zoom microscope and mount in glycerol or lactic acid.
6. Viable embryo turns magenta-deep pink.

Seed Germinability
Seed germinability tests are easier for the species where dormancy is not much prevalent or can be easily overcome by acid/alkali treatment or by simple scarification. Such tests are difficult to employ for the species which show endozoochory and have pulpy fruits. One must completely remove the pulp/flesh present over the seed to eliminate the chances of fungal/bacterial growth during germination.

Material Required
Seeds, germination paper, water, petriplates sets (9 inches), soilrite mix, pots (6 inches)

Procedure

1. Take 100 seeds in batches of 10 each. Clean them for extraneous dust and debris using tissue paper. Seeds may be washed once in tap water for cleaning.
2. Spread the germination paper on the larger petriplate and moisten it using water. One may also sow the seeds in a peat-soilrite mix for testing germinability. If germination paper is not available Whatman sheets or blotting paper may be used instead.
3. Keep 10 seeds of each batch at same distance from each other and cover the petriplates. Such a setup can be kept in the dark for germination. Make replicates.
4. Check after intervals of minimum 24 hours for germination and sprinkle water to keep paper moist.

5. Record the duration and number of seeds which exhibit the emergence of radicle. Also record the number of seeds from which radicle and plumule does not appear.
6. Calculate the percentage for germinability.

Seed Vigor: Paper Piercing Test
This test can be conducted in similar way as the germinability test except that the pots where seeds have been sown are covered with thin paper.

Observations
Compare and discuss the results obtained after seed viability test, germinability test, and vigor test.

Exercise 12.5 To calculate the ovule: seed ratio

Ovules are the progenitor of seeds. A successful event of fertilization of ovule is necessary for the formation of embryo and endosperm and hence the seed itself. As evident from previous chapters (Ovule, Pollination, Pollen–pistil Interaction, and Fertilization), there are several reasons to suggest that all ovules produced by a species do not convert into the seed. These include:

1. All ovules produced in an ovary may not be receptive at the time of pollination.
2. Ovules may be aborted on occasions.
3. All ovules in the ovary may not be fertilized especially in the species with large number of ovules (e.g., in Bignoniaceae) due to insufficient pollen load or incompatible pollen grains.
4. Embryo may be aborted sooner or later (hence seed) in case of fertilization by self-pollens.
5. Seeds may be aborted due to sibling rivalry or limited resource allocation.

This can be understood in another perspective, i.e., reproductive strategy of the plants. In general, insect and bat pollinated species produce large number of ovules so that even low pollination events can lead to seed set. In case of sufficient pollination all ovules may get fertilized but ovules fertilized earlier in the sequence attempt to draw resources towards themselves and so ovules fertilized later receive less resources and get aborted (*sibling rivalry*) (*observe a pea pod with seeds for same*). In wind-pollinated plants, a few ovules (mostly one) are produced in an ovary, as the chances of pollination are low. Thus, by producing large number of flowers, plants attempt to maximize their reproductive success.

Thus, for assessing reproductive success of the plant simply counting the number of ovules (per flower in plant) will not be sufficient. One must consider the seed set as well. Together, ovules and seeds are good indicators of reproductive success of a species. Hence, one may consider the ovule to seed ratio as a measure of reproductive success. One may also compare the ovule-seed ratio in self vs. cross pollinated species.

Suggested species: *Oroxylum indicum*, *Delonix regia*, *Cassia* sp., *Vinca rosea*, *Solanum nigrum*, *Phlox sp.*

Procedure

1. Estimate the ovule production per flower in a selected species (refer to protocol 5.6, Chapter 5).
2. Collect the fruits carefully in fruiting season of species.
3. Dissect the fruits (n=10) and count the number of seeds per fruit. If the seeds are small dissection may be done under stereo-zoom microscope.
4. Calculate the ovule to seed ratio which indicates the number of ovules that have been successfully fertilized and converted into seeds.

Exercise 12.6 Observation of seed dispersal mechanism

Someone observed the *Calotropis* sp. plant with mature fruits. After two days, the person was surprised to see hair-like appendages on seeds in the opened fruit (Fig. 12.16). What is the explanation? What could be the seed dispersal mechanism?

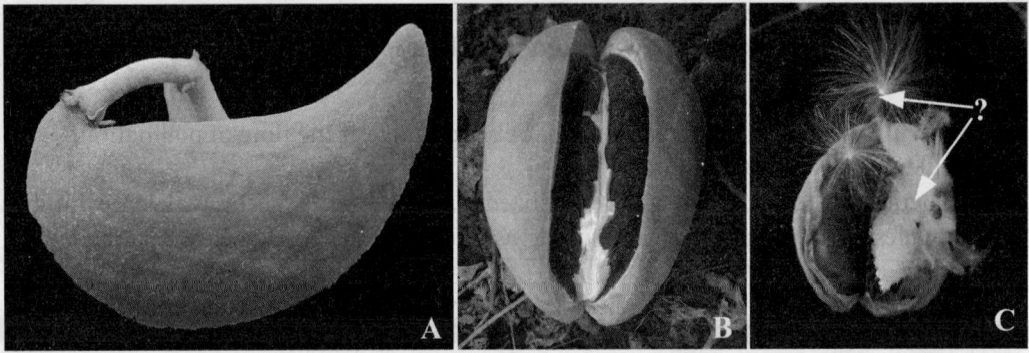

Figure 12.16 A. Mature fruit **B–C.** Opened fruit showing seeds with hair like appendages. **See Color Plates (page 503).**

Bibliography

Abraham, Y. and Elbaum, R. (2013). Hygroscopic movements in Geraniaceae: the structural variations that are responsible for coiling or bending. *New Phytologist* 199: 584–94.

Ames, O. (1963). The significance of the angiosperm seed. *Economic Botany* 17: 3–10.

Baskin, C. C. and Baskin, J. M. (2007). A revision of Martin's seed classification system, with particular reference to his dwarf-seed type. *Seed Science Research* 17: 11–20.

Baskin, J. M. and Baskin, C. C. (2004). A classification system for seed dormancy. *Seed Science Research* 14: 1–16.

Bayly, M. J., Holmes, G. D., Forster, P. I. et al. (2013). Major Clades of Australasian Rutoideae (Rutaceae) Based on rbcL and atpB Sequences. *PLoS ONE* 8(8): e72493. doi:10.1371/journal.pone.0072493.

Bhanuprakash, K. and Umesha, B. (2015). Seed biology and technology. In Bahadur et al., eds., *Plant Biology and Biotechnology: Volume I: Plant Diversity, Organization, Function and Improvement*. New Delhi: Springer-India, pp. 469–97.

Black, M., Bewley, J. D. and Halmer, P. (2006). *The Encyclopedia of Seeds: Science, Technology and Uses*. Trowbridge, UK: Cromwell Press.

Boesewinkel, F. D. and Bouman, F. (1984). The seed: structure. In B. M. Johri, ed., *Embryology of Angiosperms*. Berlin: Heidelberg: Springer.

Bouman, F. (1975). Integument initiation and testa development in some Cruciferae. *Botanical Journal of Linnean Society* 70: 213–29.

Castillo, N. R., Melgarejo, L. M. and Blair, M. W. (2020). Seed structural variability and germination capacity in *Passiflora edulis* Sims f. edulis. *Frontiers of Plant Sciences* 11: 498–508.

Corner, E. J. H. (1976). *The Seeds of Dicotyledons*. Cambridge: Cambridge University Press.

Costea, M., El Miari, H., Laczkó, L. et al. (2019). The effect of gut passage by waterbirds on the seed coat and pericarp of diaspores lacking 'external flesh': Evidence for widespread adaptation to endozoochory in angiosperms. *PLoS ONE* 14(12): e0226551. https://doi.org/10.1371/journal.pone.0226551.

Davis, G. L. (1966). *Systematic Embryology of the Angiosperms*. New York: Wiley.

Dell, B. (1980). Structure and function of the strophiolar plug in seeds of *Albizia lophantha*. *American Journal of Botany* 67: 556–63.

Fauli, R. A., Rabault, J. and Carlson, A. (2019). Effect of wing fold angles on the terminal descent velocity of double-winged autorotating seeds, fruits, and other diaspores. *Physical Review* 100: e013108.

Gabr, D. G. (2018). Significance of fruit and seed coat morphology in taxonomy and identification for some species of Brassicaceae. *American Journal of Plant Sciences* 9: 380–402.

García, M. B., Espadaler, X. and Olesen, J. M. (2012). Extreme reproduction and survival of a true cliffhanger: the endangered plant *Borderea chouardii* (Dioscoreaceae). *PLoS ONE* 7(9): e44657. doi:10.1371/journal.pone.0044657.

Hansen, D. M., Kaiser, C. N. and Muller, C. B. (2008). Seed dispersal and establishment of endangered plants on oceanic islands: the janzen-connell model, and the use of ecological analogues. *PLoS ONE* 3(5): e2111. doi:10.1371/journal.pone.0002111.

Hayashi, M., Feilich, K. L. and Ellerby, D. J. (2009). The mechanics of explosive seed dispersal in orange jewelweed (*Impatiens capensis*). *Journal of Experimental Botany* 60: 2045–53.

Heneidak, S. and Khalik, K. A. (2015). Seed coat diversity in some tribes of Cucurbitaceae: implications for taxonomy and species identification. *Acta Botanica Brasilica* 29: 129–42.

Henrique, I., Freitas, A. F. and de Moraes, R. (2019). Seed morphology of Ruellieae species (Acanthaceae) in Brazil and its taxonomic implications. *Systematic Botany* 44: 631–51.

Joshi, P. C., Wadhwani, A. M. and Johri, B. M. (1967). Morphological and embryological studies in *Gossypium* L. *Proceedings of Indian National Science Academy B*. 33: 37–93.

Kloth, R. H. and Turley, R. B. (2010). Physiology of seed and fiber development. In J. M. Stewart, D. M. Oosterhuis, J. J. Heitholt and J. R. Mauney, eds., *Physiology* of Cotton. Dordrecht: Springer.

Lengyel, S., Gove, A. D., Latimer, A. M. et al. (2009) Ants sow the seeds of global diversification in flowering plants. *PLoS ONE* 4: e5480. doi:10.1371/journal.pone.0005480.

Lentink, D., Dickson, W. B., van Leeuwen, J. L. and Dickinson, M. H. (2009). Leading-edge vortices elevate lift of autorotating plant seeds. *Science* 324: 1438–40.

Li, S., Zhang, Y. and Liu, J. (2020). Seed ejection mechanism in an *Oxalis* species. *Scientific Reports* 10: 8855–63.

Lisci, M., Bianchini, M. and Pacini, E. (1996). Structure and function of the elaiosome in some angiosperm species. *Flora* 191: 131–41.

Lovisetto, A., Masiero, S., Rahim, M. A. et al. (2015). Fleshy seeds form in the basal Angiosperm *Magnolia grandiflora* and several MADS-box genes are expressed as fleshy seed tissues develop. *Evolution and Development* 17: 82–91.

Martin, A. C. (1946). The comparative internal morphology of seeds. *The American Midland Naturalist* 36: 513–660.

Marzinek, J. and Mourão, K. S. M. (2003). Morphology and anatomy of the fruit and seed in development of *Chorisia speciosa* A. St.-Hil. – Bombacaceae. *Brazilian Journal of Botany* 26: 23–34.

McConkey, K. R., Nathalang, A., Brockelman, W. Y. et al. (2018). Different megafauna vary in their seed dispersal effectiveness of the megafaunal fruit *Platymitra macrocarpa* (Annonaceae). *PLoS ONE* 13: e0198960. https://doi.org/10.1371/journal.pone.0198960.

Miller, S. S., Bowman, L. A., Gijzen, M. and Miki, B. L. A. (1999). Early development of the seed coat of soybean (*Glycine max*). *Annals of Botany* 84: 297–304.

Moise, J. A., Hani, S., Gudynaite-Savitch, L. et al. (2005). Seed coats: structure, development, composition, and biotechnology. *In Vitro Cellular and Developmental Biology-Plant* 41: 620–44.

Moles, A. T. and Westoby, M. (2004). Seedling survival and seed size: a synthesis of the literature. *Journal of Ecology* 92: 372–83.

Muhl, Q. E., du Toit, E. S., Steyn, J. M. and Robbertse, P. J. (2016). The embryo, endosperm and seed coat structure of developing *Moringa oleifera* seed. *South African Journal of Botany* 106: 60–6.

Nikolaeva, M. G. (2004). On criteria to use in studies of seed evolution. *Seed Science Research* 14: 315–20.

Nixon, R. (1928). Immediate influence of pollen in determining the size and time of ripening of the fruit of the date palm. *Journal of Heredity* 19: 241–55.

Oliveira, N. C. C., Lopes, P. S. N., Ribeiro, L. M. et al. (2013). Seed structure, germination, and reserve mobilization in *Butia capitata* (Arecaceae). *Trees* 27: 1633–45.

Petruzzello, M. (2020). Falling far from the tree: 7 brilliant ways seeds and fruits are dispersed. *Encyclopaedia Britannica*.

Rabault, J., Fauli, R. A. and Carlson, A. (2019). Curving to fly: synthetic adaptation unveils optimal flight performance of whirling fruits. *Physical Review Letters* 122: e024501.

Rabinowitz, D. (1978). Dispersal Properties of Mangrove Propagules. *Biotropica*10: 47–57.

Ramchandani, S., Joshi, P. C. and Pundir, N. S. (1966). Seed development in *Gossypium* Linn. *Indian Cotton Journal* 20: 97–106.

Ren, Y., He, Q., Ma, X. and Zhang, L. (2017). Characteristics of color development in seeds of brown and yellow-seeded heading chinese cabbage and molecular analysis of BRSc, the candidate gene controlling seed coat color. *Frontiers in Plant Sciences* 8:1410. doi: 10.3389/fpls.2017.01410.

Robbertse, P. J., Coetzer, L. A., Johannsmeier, M. F. and Swart, D. J. (1996). Has yield and fruit size as influenced by pollination and pollen donor: a joint progress report. *South African Avocado Growers' Association Yearbook* 19: 63–7.

Roberts, E. H. (1973). Predicting the storage life of seeds. *Seed Science and Technology* 1: 499–514.

Sakes, A., van der Wiel, M., Henselmans, P. W. J. et al. (2016). Shooting mechanisms in nature: a systematic review. *PLoS ONE* 11(7): e0158277. doi:10.1371/journal.pone.0158277.

Savage, W. E. F. and Metzger, G. L. (2006). Seed dormancy and the control of germination. *New Phytologist* 171: 501–23.

Schupp, E. W., Jordano, P. and Maria-Gomez, J. (2010). Seed dispersal effectiveness revisited: a conceptual review. *New Phytologist* 188: 333–53.

Seale, M. and Nakayama, N. (2019). From passive to informed: mechanical mechanisms of seed dispersal. *New Phytologist*. doi:10.1111/nph.16110.

Sedgley, M. and Griffin, A. R. (1989). *Sexual Reproduction in Tree Crops*. London: Academic Press.

Sernander, R. (1906) Entwurfeiner Monographie der europäischen Myrmekochoren. *K Sven Vetensk Akal Handl* 41: 1–407 (original not seen).

Shaheen, M. A., Bacha, M. A. and Nasr, T. A. (1989). Effect of male type on fruit chemical properties in some date palm cultivars. *Annals of Agricultural Sciences* 34: 265–81.

Silveira, S. R., Dornelas, M. C. and Martinelli, A. P. (2016). Perspectives for a framework to understand aril initiation and development. *Frontiers in Plant Sciences* 7: https://doi.org/10.3389/fpls.2016.01919.

Singh, D. and Dathan, A. S. R. (1972). Structure and development of the seed coat in Cucurbitaceae. VIII. seeds of *Marah* Kell. *Bulletin of the Torrey Botanical Club* 99: 239–42.

Singh, D. and Dathan, A. S. R. (1974). Structure and development of the seed coat in Cucurbitaceae. IX. Seeds of *Corallocarpus, Kedrostis* and *Ibervillea. Bulletin of the Torrey Botanical Club* 101: 78–82.

Singh, D. (1966). Structure and development of seedcoat in cucurbitaceae. I. Seeds of *Biswarea* Cogn., *Edgaria* Clarke, and *Herpetospermum* Hook. *Proceedings of Indian Academy of Sciences, Biological Sciences* 115: 267–75.

Smýkal, P., Vernoud, V., Blair, M. W. et al. (2014). The role of the testa during development and in establishment of dormancy of the legume seed. *Frontiers in Plant Sciences* 5: 351. doi:10.3389/fpls.2014.00351.

Swingle, W. T. (1928). Metaxenia in date palm, possibly a hormone action by the embryo or the endosperm. *Journal of Heredity* 19: 257–68.

Tackenberg, O., Poschlod, P. and Bonn, S. (2003). Assessment of wind dispersal potential in plant species. *Ecological Monographs* 73: 191–205.

Tanaka, K., Ogata, K., Mukai, H. et al. (2015). Aadaptive advantage of myrmecochory in the ant-dispersed herb *Lamium amplexicaule* (Lamiaceae): predation avoidance through the deterrence of post-dispersal seed predators. *PLoS ONE* 10: e0133677. doi:10.1371/ journal.pone.0133677.

Thompson, K., Band, S. R. and Hodgson, J. G. (1993). Seed size and shape predict persistence in soil. *Functional Ecology* 7: 236–41.

Traveset, A. and Rodríguez-Pérez, J. (2018). Seed dispersal. *Reference Module in Earth Systems and Environmental Sciences* 3: 592–99. *Encylopedia of Ecology* (Second edition), Elsevier.

Tuan, P. A., Kumar, R., Rehal, P. K. et al. (2018). Molecular mechanisms underlying abscisic acid/gibberellin balance in the control of seed dormancy and germination in cereals. *Frontiers in Plant Sciences* 9: 668. doi: 10.3389/fpls.2018.00668.

Wang, Z., Wang, L., Liu, Z. et al. (2016). Phylogeny, seed trait, and ecological correlates of seed germination at the community level in a degraded sandy grassland. *Frontiers in Plant Sciences* 7: 1532–42.

Wehncke, E. V. (2010). Seed dispersal and conservation. *Encyclopedia of Animal Behavior* 119–24. doi:10.1016/b978-0-08-045337-8.00327-2.

Weingartner, U., Kaeser, O., Long, M. and Stamp, P. (2002). Combining cytoplasmic male sterility and xenia increases grain yield of maize hybrids. *Crop Science* 42: 1848–56.

Xu, J., Li, W. L., Zhang, C. H. et al. (2014). Variation in seed germination of 134 common species on the Eastern Tibetan Plateau: phylogenetic, life history and environmental correlates. *PLoS ONE* 9:e98601. doi:10.1371/journal.pone.0098601.

Yang, Y. Lin, Y., and Shi, L. (2021). The effect of lizards on the dispersal and germination of *Capparis spinosa* (Capparaceae). *PLoS ONE* 16(2): e0247585. https://doi.org/10.1371/journal.pone.0247585.

Zhang, K., Zhang, Y., Ji, Y. et al. (2020). Seed Biology of *Lepidium apetalum* (Brassicaceae), with particular reference to dormancy and mucilage development. *Plants* 9: 333–46.

Zhang, X., Zhang, Z. and Stützel, T. (2014). Ontogeny of the ovule and seed wing in *Catha edulis* (Vahl) Endl. (Celastraceae). *Flora* 209: 179–84.

13

Plant Germline Transformation

Transformation of germline offers several advantages over transformation of somatic cell/tissue in the development of transgenics.

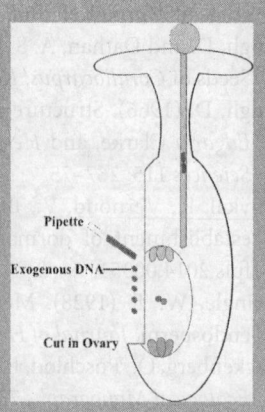

13.1 Introduction

Plant breeding is an age-old practice of genetic improvement of plants so that they become more useful to humans. It involves combining selected parental plants to obtain the next generation with an improved genetic potential of disease-resistance, stress-tolerance or better yields. The process includes manual crosses or controlled pollination followed by an artificial selection of progeny. Plant breeding, for its enormous benefits to mankind is widely practiced. However, the process is labor intensive and it takes many years for the integration of the required gene and development of a desired progeny. Another major limitation of conventional breeding is that the gene transfer can be achieved only in genetically related species/genera. Even in conspecific plants, the incompatibility of crosses becomes a major barrier. In the last few decades, these limitations of plant breeding have largely been overcome by modern plant genetic engineering/transformation techniques which allow insertion of foreign genes from one organism into another organism. Introduction of foreign genes is relatively less time consuming and does not require recipient and donor organisms to be genetically related to each other. For example, Bt-cotton which was created by genetically altering the cotton genome using genes from the soil bacterium *Bacillus thuringiensis* (Pursell & Perlak 2004).

The technique of developing transgenic plants has become an integral part of crop improvement programs across the world (Eapen 2011). Although, transgenic approach for crop improvement has several advantages over conventional breeding, it suffers from some drawbacks as well. The transformation techniques used are expensive, genotype-dependent and involve time taking procedures such as *in vitro* culture and regeneration of explants. Therefore, alternate methods of quick and easy transformation are being developed. One such approach which offers several advantages is transformation of germ cells of plants instead of somatic cells/tissues. Germ cells of plants include sperm cells in the pollen grain (male germ cell) and egg cell in the female gametophyte (female germ cell). The process of introduction of desired genes into germ cells is known as **Germline Transformation**. This method of transformation has the potential to produce genetically modified plants within less time as it bypasses some tedious steps of *in vitro* regeneration. The genetically transformed germ cells can also be used directly for fertilization to recover transgenics without tissue culture and the procedure is known as *in planta* transformation. The transformed haploid germ cells can also be used for production of double haploids for recovery of homozygous lines of transgene, eliminating the need of laborious and time-consuming back crosses. So far, techniques of germline transformation have employed a large number of permutation and combinations of gene delivery systems and some methods such as floral dip (Clough & Bent 1998) and MAGELITR (Touraev et al. 1997) have yielded positive results. According to Mohanty et al. (2016) male germline transformation studies have been conducted in more than 15 dicot and 11 monocot taxa and about 23 dicot and four monocot taxa have been investigated for female germline transformation.

Plant transformation systems have changed immensely over the years in search of simpler and efficient methods, and germline transformation is a promising technique in this regard. Some of the important methodologies used for transformation of female and male germline have been discussed in this chapter.

13.2 Plant Germline Transformation

The first successful genetic transformation in plants was reported in 1983 (Herrera-Estrella et al. 1983; Fraley et al. 1983). Since then, numerous genetic transformation methods have been used to introduce new traits into various plant species. The process of plant genetic transformation can be broadly classified into three steps: (1) introduction of foreign DNA into the selected tissue (explant) of target plant; (2) *in vitro* regeneration of the transformed explant into a complete plant via either embryogenesis or direct organogenesis; (3) selection of transgenic plants. To recover a fertile transgenic plant, standardization of each step of the process is required. Main optimizations include selection of gene delivery method and a responsive explant.

For transfer of a new gene, either *Agrobacterium*-mediated transformation, or direct DNA transfer through Biolistic method are mostly preferred. The former method involves the use of a bacterium, *Agrobacterium tumefaciens* which has the ability to infect plant cells with a piece of its DNA, into which a foreign gene can be inserted and transferred. While direct delivery of DNA does not require any vector, and a plasmid DNA carrying gene of interest is directly inserted into cells. Most commonly used direct DNA delivery method is microprojectile bombardment or Biolistic method. There are several other methods of direct DNA delivery which include introduction of DNA molecules into protoplasts mediated by polyethylene glycol (PEG), sonication, electroporation or liposome treatment (Hansen & Wright 1999).

To develop efficient transformation protocols different explants are used for different species. Mostly somatic tissues like cotyledons, leaves, stems, and roots are used as explant for transformation followed by regeneration. Irrespective of the gene delivery method used, a tissue culture period is an essential requirement for developing transgenic plants when somatic tissues are used as explants (Eapen 2011). This makes the whole process tedious, genotype-dependent with prolonged cycles of regeneration which may result in somaclonal variation and production of chimeras (Li et al. 2002). Occasionally the transgenic plants also need several generations of back crosses to bring the inserted gene in a homozygous state so that chances of segregation in subsequent progenies (seeds) are minimal.

Considering difficulties encountered in conventional transformation methods, alternate ways which are quick and without tissue culture are sought after by scientific community. One such technique of plant transformation which can potentially circumvent the constraints associated with conventional transformation techniques is plant germline transformation. It is the method of introduction of gene of interest into the male or female germ cells of a plant. For transforming male gamete, either microspore or pollen grains are used and for transforming female gamete, either unfertilized ovules or unfertilized ovaries and occasionally complete flower buds are used. Based on the germ cell being transformed, plant germline transformation can be classified as male or female germline transformation. From a transformed germ cell, transgenic plants can be recovered either through *in vivo* pollination or *in vitro* regeneration pathway. In the former, introduction of foreign gene into either male or female gamete prior to fertilization can give rise to transgenic zygote and eventually plants, completely bypassing the requisite for *in vitro* culture. This method is also known as *in planta* transformation. Alternatively, the transformed gametic tissues can be cultured under *in vitro* conditions to induce androgenesis or gynogenesis for development of much valuable double haploid plants. Regardless of the technique used, recovering transgenics through germline transformation is easier and quicker, especially for *in vitro* recalcitrant crops. Some examples of techniques which have been used for male and female germline transformation are MAGELITR (Touraev et al. 1997), floral dip method (Clough & Bent 1998), pollen transformation (Wang et al. 2001), pistil transformation (Chumakov et al. 2006; Mamontova et al. 2010) and ovary-drip transformation (Yang et al. 2009). However, all these techniques are restricted to few species and reproducibility is still obscure.

Box 13.1

Advantages and Disadvantages of Plant Germline Transformation

Advantages	Disadvantages
• Fast and regeneration independent method • A method of *in planta* transformation • Not prone to chimerism • Avoids somaclonal variation • Cheap method • Easy production of double haploids	• Not demonstrated in many species • Fails if genes are not introduced into germ cell at the correct stage • Frequency of transformation is low

13.3 Male Germline Transformation

The male germline transformation is the process of introduction of foreign gene into either the immature microspores or mature pollen grains. Based on the stage used for transformation, male germline transformation can be distinguished into two pathways (Resch & Touraev 2011; Fig. 13.1).

1. *Gametophytic pathway*: It involves either transformation of microspores or direct transformation of mature pollen grains which are subsequently used for pollination to produce transgenic seeds. If microspores are used for transformation, they need to undergo *in vitro* maturation before use in pollination, and it is recognized as microspore maturation-based transformation. Whereas, if transformed pollen grains are directly employed for pollination, it is known as mature pollen based transformation.

2. *Sporophytic pathway*: It involves transformation of immature pollen grains (microspores, uninucleate stage) followed by their induction towards the embryogenesis (androgenesis). This strategy is also known as microspore or immature pollen-embryogenesis-based transformation and in part employs pollen and anther culture (refer to Box 13.2 for details).

In the gametophytic pathway, the resultant seeds and transgenic plants can be selected using selectable markers. This pathway eliminates tissue culture and makes the technique genotype independent. Both microspore and pollen grain are an ideal target for transformation because of *en-masse* availability, synchrony in their development and easy isolation from anthers. A 2-celled mature pollen grain carries a vegetative and a generative cell, the latter gives rise to two sperm cells. Pollen-based transformation aims to introduce a desired foreign gene into the DNA of generative or the sperm cell. The introduction of DNA into the sperm cell which is destined to fuse with the egg cell is essential or else no transgenic will be produced. Often, it is difficult to introduce genes into mature pollen grains because of the thick exine and therefore, immature pollen or uninucleate microspore is considered an ideal stage for germline transformation. Besides, when transgene is introduced at uninucleate stage it is passed on to both the sperm cells.

The sporophytic pathway, reprograms the microspores or immature pollen grains towards embryogenesis to give rise to haploid plants via embryos (androgenesis). Alternatively, microspores may first be induced to undergo embryogenic development and then transformed. Irrespective of the procedure, recovered haploid plants can undergo spontaneous or microtubular chemicals induced diploidization to develop homozygous transgenic plants (double haploids) in one generation (Touraev et al. 2001). Production of double haploids ensures there is no segregation of transgene and it is stably inherited to future generations. This fixation of a trait in homozygous condition otherwise requires years of backcrosses in breeding programs.

Methods Used for Transformation: The term 'pollen transformation' was used for the first time by Dieter Hess in 1974. A technique called 'gametic transformation' which involved transferring a single gene or short genetic segments of pollen to the genome of female plant

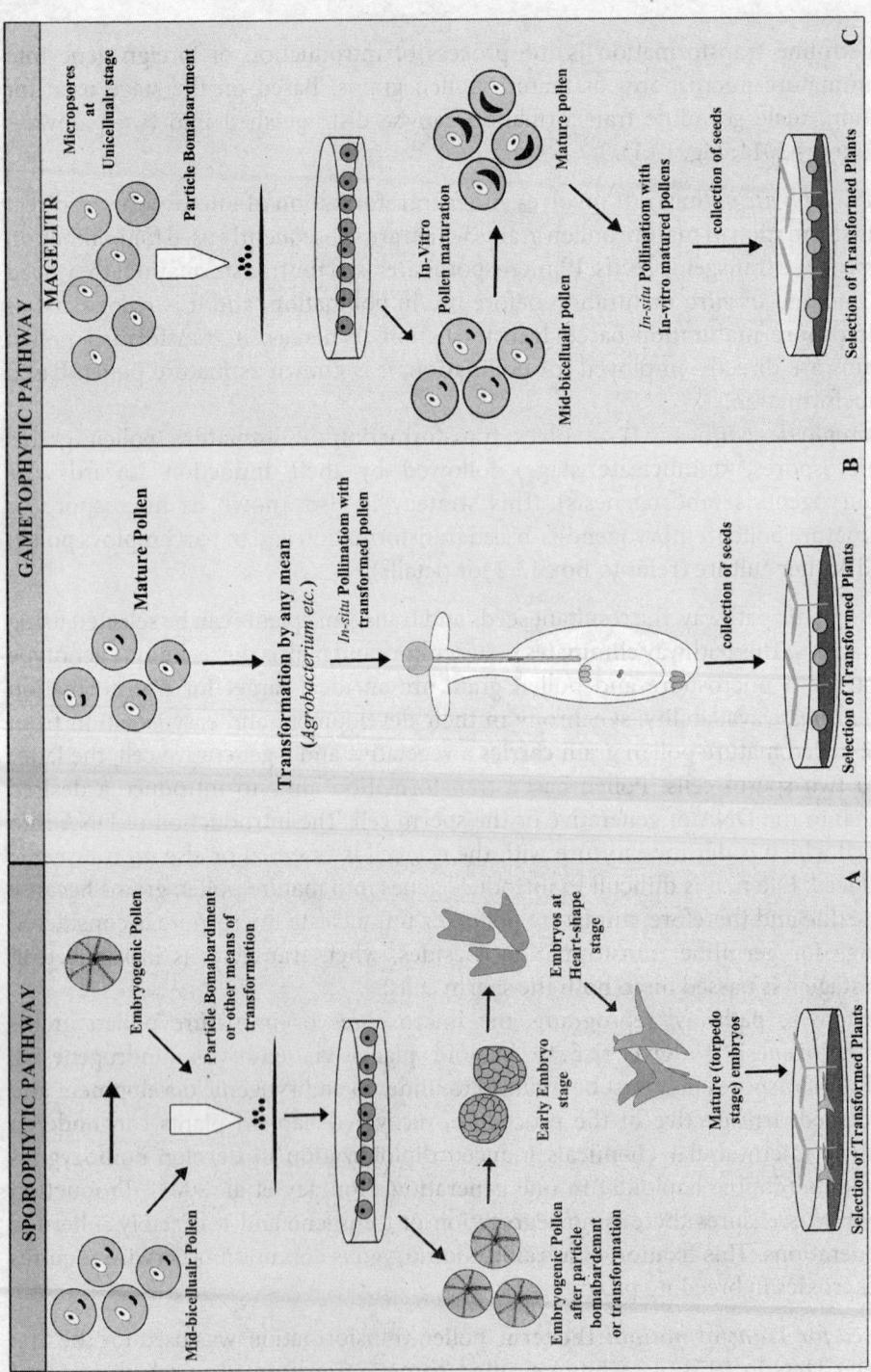

Figure 13.1 Diagrammatic representation of different methods used for male germline transformation. A. Sporophytic pathway: This method uses immature pollen for transformation followed by androgenesis. B-C. Gametophytic pathway. B. Transformation of mature pollen grains followed by in situ pollination with transformed pollen grains. C. MAGELITR approach targets the unicellular pollen grains for transformation which undergo in vitro maturation and are then used for in vivo pollination. See Color Plates (page 504). (Source: Adapted and modified after Resch & Touraev 2011.)

Box 13.2

Androgenesis: Anther and Pollen Culture

Anther and pollen culture aim at raising haploid plants from microspores and pollen grains. The first successful anther culture and production of haploid plants from the pollen grains was demonstrated in 1964 by Shipra Guha and Satish Chandra Maheshwari. Since then, anther and pollen culture have been used extensively in crop improvement programmes. For raising haploids, anther or pollen grains are cultured at the appropriate stage of pollen development by placing in an appropriate nutrient medium, which is known as androgenesis. It is an integral step of sporophytic pathway of male germ-line transformation.

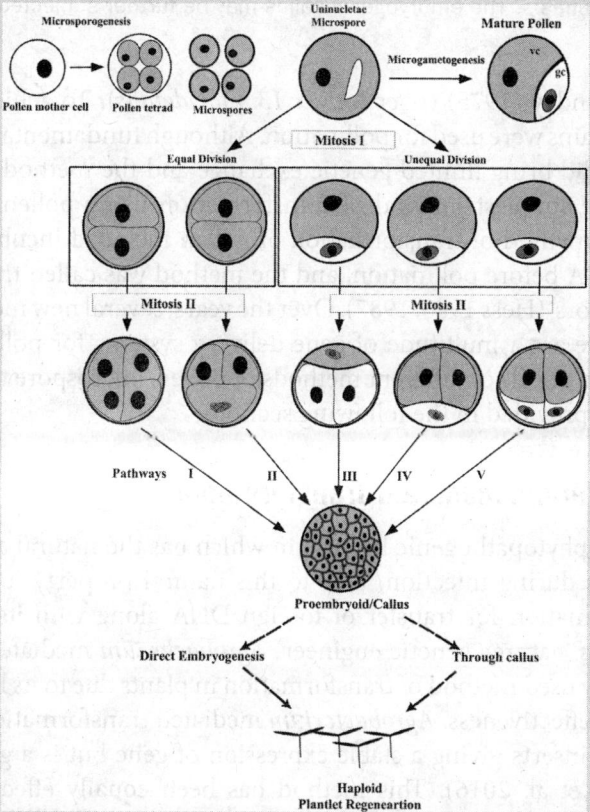

Figure 13.2 Diagrammatic representation of development of embryogenic callus from microspore cultured at uninucleate stage. Under *in vitro* conditions, uninucleate microspore are reprogrammed towards the sporophytic divisions (mitosis). I–V depict the different pathways in formation of proembryoid/callus. I. Equal division in the microspore occurs. Distinct vegetative and generative cells do not form. Both daughter cells contribute to callus formation. II–IV Distinct vegetative and generative cells form. II. Only vegetative cell divides and forms the callus. III. Only generative cell divides and forms the callus. IV. Both vegetative and generative cells divide and their derivatives form the callus (e.g., *Datura innoxia*). V. The first division is symmetrical, and the pollen embryos develop exclusively from the vegetative cell (e.g., *B. napus*). (*Source:* Based on Bhojwani et al. 2015; Bhaskara 2017.)

For the induction of androgenesis pollen development stage is a critical factor. The most amenable stage for induction of androgenesis is the uninuclear stage of pollen and early binuclear stage (just after first pollen mitosis). In the later stages, the genetic machinery of pollen regulates it to develop into gametes only, and once committed, reversal is not possible. Various types of stress (dark period, temperature, starvation treatment) can induce sporophytic development in pollen at early stages. After the start of sporophytic divisions in pollen grains a multicellular structure develops, enclosed within the pollen wall. The pressure exerted by increasing cell mass leads to its liberation from pollen wall. The contribution of young vegetative cell and generative cell in callus varies and may be species specific (Fig. 13.2). Depending on the culture conditions and aim, two modes of pollen embryo development can be distinguished: embryogenesis via callus and direct embryogenesis. The embryogenic callus may be further subjected to transformation.

was reported by Pandey (1975) (*refer to Box 13.3 for details*). To achieve transformation, irradiated pollen grains were used for pollination. Although fundamentally it was not genetic transformation, it did bring limited genetic exchange and the method can be regarded as the earliest and the simplest method of transformation using pollen (Resch & Touraev 2011). First actual method of transformation of pollen included incubation of dry pollen in a solution of DNA before pollination, and the method was called transformation using pollen as 'supervectors' (Hess 1980; 1987). Over the years, several new methods have evolved and at present there are a multitude of gene delivery systems for pollen and microspore transformation. The details of different methods of pollen/microspore transformation with some examples are provided in the following sections.

13.3.1 *Agrobacterium* Mediated Transformation

Agrobacterium is a phytopathogenic bacterium which has the natural ability to transfer its DNA to plant cells during infection. Due to this natural property it is used as a vector in genetic transformation for transfer of foreign DNA along with its own DNA, and is popularly known as 'nature's genetic engineer'. *Agrobacterium* mediated transformation is the most commonly used method of transformation in plants due to its high transformation efficiency and cost effectiveness. *Agrobacterium* mediated transformation usually results in single or low copy inserts giving a stable expression of gene but is a genotype dependent method (Mohanty et al. 2016). This method has been equally effective in transferring exogenous DNA or gene of interest into male germline. For transformation, the pollen grains or pollen derived embryos are incubated with the *Agrobacterium* culture known as co-cultivation of pollen grains with *Agrobacterium* (Hess & Dressler 1989). Application of *Agrobacterium* prior to or after pollination has been reported to give rise to transgenic zygotes in maize and rice (Luo & Wu 1989; Ohta 1986). However, the method did not yield a constant gene integration and results were also non-reproducible in other species. Another method for transformation of pollen grains is vacuum infiltration with *Agrobacterium*. In this method, vacuum pulls out the air from intercellular spaces in the plant tissue and allows better penetration of *Agrobacterium* present in infiltration medium into those inter cellular spaces (Bechtold et al. 1993). Higher transformation efficiency has been reported in

Box 13.3
Gametic Transformation

While attempting to overcome interspecific incompatibility in *Nicotiana* spp., gene transfer by irradiated pollen grains known as mentor pollen grains (mentor pollination) was reported by Pandey (1975). The irradiation caused shearing in chromatin of generative nucleus of mentor pollen and such pollen failed to form sperm cells. However, such pollen grains retained the ability to give rise to pollen tube. Pandey (1975) suggested that some chromatin (may be a gene or few genes) of mentor pollen get discharged within egg cell by pollen tube. This event was called pseudo-fertilization of egg cell. In the study, the transfer of a flower color allele in the interspecific cross was provided as phenotypic evidence of transformation. However, molecular or biochemical evidences of transformation and mechanism of allele transfer were not investigated. This method was later named as **gametic transformation** (Pandey 1978). After this report, several groups world-wide attempted this methodology. However, usage of the term gametic transformation was restricted and instead **egg transformation** was used (Chin and Gordon 1989). Moreover, subsequent studies showed contrasting inheritance and expression of transferred paternal alleles. For instance, in *Brassica* sp., pollination with irradiated pollen resulted in progeny with maternal-like plants (Banga et al. 1984; Chin & Gordon 1989). The random and untargeted transfer of chromatin or alleles rendered it as a method for genome modification than genetic transformation since latter involves the transfer of targeted gene/s of one species to another species (Banga et al. 1984).

Arabidopsis thaliana with application of vacuum infiltration of *Agrobacterium* into pollen grains (Ye et al. 1999). Tjokrokusumo et al. (2000) also reported use of vacuum infiltration to transform pollen of *Petunia hybrida* which were later used for pollination to bring about *in planta* transformation. Recently, Kumari et al. (2017) used another technique in *Albuca nelsonii* and *Tulbaghia violacea*, where germinated pollen grains were incubated with the *Agrobacterium* culture and were later transferred to the stigma. This method is known as pollen tube pathway of male germline transformation.

Microspore derived embryos are also used as target for *Agrobacterium* mediated transformation. Sangwan et al. (1993) demonstrated production of transgenic plants in *Datura inoxia* and *Nicotiana tabacum* after introduction of DNA into immature pollen embryos via *Agrobacterium*. However, using *Agrobacterium*-mediated transformation for microspore derived embryos is not successful if androgenic embryos remain enclosed in the microspore wall which makes infiltration of *Agrobacterium* difficult. Therefore, the most effective stage of transformation is when embryogenic structures are coming out of the microspore wall (Kumlehn et al. 2006). Using this strategy Kumlehn et al. (2006) confirmed that 6–11 days old embryogenic cultures can be transformed using *Agrobacterium* in barley.

To overcome the barriers in penetration of *Agrobacterium* into plant tissues and consequently increase the transformation efficiency, *Agrobacterium* mediated male germline transformation has been combined with several other approaches (Mohanty et al. 2016). These include pre-treatments such as physical wounding of tissue by ultrasonication, use of silicon carbide fibres, sand, ammonia and aluminium oxide (Trick & Finer 1997; Kim et al. 2006) and combining *Agrobacterium* with particle bombardment ('Agrolistics'; Abdollahi et al. 2009).

13.3.2 Particle Bombardment Method

In this method, DNA is initially coated on metal microparticle (gold or tungsten) and these DNA coated particles are then bombarded on cell/tissue by helium gas under high pressure. The force and acceleration of coated particles facilitate them to cross the physical barriers (e.g., cell wall). This method of ballistic delivery is often called *biolistics* and the device for shooting particles is called *Gene Gun*. It was developed by Sanford et al. (1987) and is a good alternative for gene delivery in plants that are unresponsive to *Agrobacterium* mediated transformation like monocots. It is a genotype independent method and both mature and immature pollen grains can be used for this kind of transformation. However, it can result in multi-copy inserts of transgene. The biolistic-mediated entry of foreign DNA into pollen were first reported in *Lycopersicon esculentum* and *Nicotiana* sp. (Twell et al.1989). However, transgene showed transient expression in pollen only, and was not transferred to the progeny. Particle bombardment method inducing transformation of immature pollen grains (microspores, at early bicellular stage) was reported first in *N. tabacum* by Stöger et al. (1995). Transformed microspores, however, exhibited poor *in vitro* regeneration and low transformation efficiencies. Stable microspore transformants using particle bombardment was achieved in *Brassica napus* by modifying the microspore culture conditions (Fukuoka et al. 1998). Successful transformation in monocot species have been reported using variants of the particle bombardment method. For example, use of a pneumatic particle gun in lily (Tanaka et al. 1995), using arabinogalactans and bombardment of cells at S-phase of cell cycle in barley (Shim et al. 2009) and bombarding pollen with magnetic particles and followed by selection on the basis of magnetic force in maize and asparagus (Kakuta et al. 2001).

13.3.3 MAGELITR

The method of MALE GERM LINE TRANSFORMATION (MAGELITR), was developed by Touraev et al. (1997). It was the first successful demonstration of introduction of transgene into the unicellular microspores of tobacco. The transformed microspores were then matured *in vitro* and used for pollination, fertilization and finally development of transgenic plants. Touraev et al. performed microprojectile bombardment at the G1 phase of the cell cycle of unicellular microspores. However, they obtained very low transformation efficiency in their study (5 resistant seeds out of 30000 total seeds produced). Nonetheless, molecular and genetic analysis showed stable transmission of genes in next three generations. Aziz & Machray (2003), using the same approach, optimized tobacco microspore transformation to achieve high transformation frequency of 15%. They were able to obtain stable transformants with low copy number using this procedure. The method exploits the gametophytic pathway of male germline transformation and is a fast and genotype-independent technique. The progeny also does not show chimerism and somaclonal variation. However, the technique of *in vitro* maturation of unicellular microspores after introduction of transgene has to be standardized for each species. Another disadvantage of the method is low transformation efficiency as reported in several species like *Antirrhinum majus*, wheat, *Arabidopsis* (Resch & Touraev 2011).

13.3.4 Microinjection

It is a technique of delivery of DNA into cells using microcapillaries and microdevices. It is a skill-based procedure which entails minimal damage to cell so that it is able to survive and proliferate after treatment. The successful male germ line transformation of plants via microinjection of foreign DNA has been reported in few cases. De La Pena et al. (1987) injected *Agrobacterium* harboring the *aminoglycoside II gene* (which confers kanamycin resistance) into the young floral tillers of rye plants. They hypothesized that the DNA will be transported through the vascular system to germ cells, which can take up the DNA only when they are in a competent stage (i.e., when the germ cells are not surrounded by a callose wall after meiosis). Although the method eliminated use of tissue culture procedures, transformation efficiency was very low in their study; only 3 seeds out of 1000 seeds from 37 plants expressed the target DNA. The microinjection of DNA directly into pollen grains was first demonstrated in maize and barley using fluorescent dye (lucifer yellow) (Bolik & Koop 1991). In *Brassica*, the DNA solution was injected into the pollen derived embryos at 4–12 cell stage. However, the DNA integration in each cell was not obtained and chimeric embryos were produced under *in vitro* conditions (Neuhaus et al. 1987). Microinjection is associated with several technical disadvantages which limits its wider application. Bypassing the pollen exine for delivery of DNA and maintenance of the osmoticum of pollen cytoplasm to prevent bursting of pollen grains are some of the challenges. Moreover, application of this strategy needs sophisticated instrumentation and precision.

13.3.5 Sonication

Sonication is the application of sound energy to agitate particles in a sample or solution. This method has been used to aid uptake of DNA by plant cells. DNA can be directly introduced into pollen by subjecting them to ultrasound during incubation (Wang et al. 2007; 2008). During sonication, the hydrostatic pressure imposed on the pollen membrane helps in exchange of solutes from the medium and the cytoplasm (Muorten and Janne 1990). Transformation of pollen grains using sonication and subsequent use in pollination and development of transgenic plants was first reported in maize (Wang et al. 2001). Sonication-assisted introduction of *aroAM1* gene in the pollen grains and successful production of transgenic plants has been reported in *Brassica juncea* (Wang et al. 2008). Sonication has also proved to be beneficial in enhancing *Agrobacterium* mediated transformation of pollen grains.

13.3.6 Electroporation

Electroporation is the technique of using electric pulse to create temporary pore/s in a membrane such that movement of molecules across membrane can be facilitated. It is a method where foreign DNA is directly delivered into plant tissue and has proved effective for pollen transformation. Electroporation has been used to introduce DNA into pollen grains of tobacco (Matthews et al. 1990; Abdul-Baki et al. 1990). Production of transgenic plants using electroporation mediated pollen transformation has been demonstrated in *Nicotiana tabacum* (Smith et al. 1994). Electroporation is a genotype independent procedure but the range and frequency of voltage needs to be standardized for each system.

Box 13.4

Advantages and Disadvantages of Various Gene Delivery Methods Used for Male Germline Transformation

Method	Advantages	Disadvantages
Agrobacterium mediated transformation	Easy, high transformation efficiency, cost effective, single or low copy insertion for stable expression	Genotype dependent method, many species are recalcitrant to *Agrobacterium* mediated transformation
Microprojectile Bombardment	Alternative method for delivery of foreign DNA into plants which are recalcitrant to *Agrobacterium* mediated transformation, genotype independent	Multi-copy insertion of transgene
MAGELITR	Fast, genotype independent, stable transmission of genes, progeny does not show chimerism and somaclonal variation	Very low transformation efficiency, *in vitro* maturation of unicellular microspores needs standardization for each species
Microinjection	Genotype independent	Difficult to bypass the pollen exine, bursting of pollen grains, needs sophisticated instrumentation and skilled manpower
Sonication	Genotype independent	Not demonstrated in many species
Electroporation	Genotype independent	Range and frequency of voltage need to be standardized for each species

13.4 Female Germline Transformation

In flowering plants, ovules carry the female gametophyte or the embryo sac which in turn houses the female gamete; the egg cell. Introduction of desired genes into the female germ cell prior to fertilization can give rise to transgenic zygote and subsequently transgenic plant. Like male germline transformation, genetic alteration of female germline cells limits the chances of somaclonal variation associated with *in vitro* culture-mediated genetic transformation and regeneration. However, unlike pollen grains, isolation of ovules and subjecting them to transformation methods is extremely strenuous and also inefficient. Therefore, either unfertilized ovaries or complete flower buds, flowers, inflorescences and sometimes even complete plants are used for transformation. Such *in planta* transformations are non-destructive, cost effective, genotype independent and many transgenics can be recovered from a single transformation experiment (Mohanty et al. 2016). However, when compared to somatic tissues, the transformation efficiency achieved through female germline is very low and also the biological mechanisms governing the transformation are not completely understood (Mohanty et al. 2016). The first report of *in planta* transformation was in maize by Graves and Goldman (1986). Later, Feldmann and Marks (1987) achieved a stable transformation in *Arabidopsis* and selfing of resultant plants led to the development of transformed seeds. Like other plant tissues, *Agrobacterium*-mediated transformation is the most commonly employed method for gene delivery in female germline transformation also. There are several methods which utilize *Agrobacterium* mediated gene delivery in combination with *in planta* transformation strategy. Some of the methods of female germline transformation are discussed below.

13.4.1 *In Planta* Agroinfiltration

As mentioned earlier, agroinfiltration is the infiltration of exogenous DNA-carrying Agrobacteria into the intercellular space of the plant tissue by applying vacuum. The transformation of whole plants using agroinfiltration is an easy and simple method of *in planta* transformation. It was first time used for transformation in *Arabidopsis* (Bechtold et al. 1993). The method can be divided into following steps: growing plant up to flowering stage, uprooting of plants, inoculation of whole plant in infiltrating medium comprising *Agrobacterium* under vacuum, replanting, collection of seeds, and selection of transgenics using selectable markers. The *Agrobacterium* suspension medium for infiltration usually comprises of sucrose and suitable hormones. Success with *in planta* agroinfiltration was previously limited to plants in Brassicaceae (e.g., *Arabidopsis*, *Brassica rapa*, *B. napus*, *B. juncea*, *Camelina sativa*) which has now been extended to other taxa like *Eustoma grandiflorum*, *Setaria viridis* and *Solanum lycopersicum* (Saha & Blumwald 2016, Sharda et al. 2017; Fang et al. 2018). This method relies on standardization of factors like composition of infiltration medium, developmental stage of plant, duration and pressure of vacuum applied. A major disadvantage of this method is uprooting and replanting of target plant which makes it untenable for large plants.

13.4.2 Floral Dip

The floral dip method overcomes the disadvantages of *in planta* agroinfiltration by avoiding the damage caused to plant during vacuum infiltration and replantation. In this method, only the developing inflorescences attached to the plant are used, eliminating the need for uprooting and replanting. The inflorescences are immersed into the buffered infiltration medium containing *Agrobacterium* and a surfactant. The success of this method is also stage dependent. For instance, in *Arabidopsis thaliana*, floral dip was accomplished by inoculating flowers with *Agrobacterium* containing

Box 13.5

Differences between Male Germline and Female Germline Transformation

Male Germline Transformation	Female Germline Transformation
• Introduction of foreign gene into male germ cell i.e., generative cell or sperm cells.	• Introduction of foreign gene into female germ cell, i.e., egg cell.
• Targeted delivery of foreign gene into microspore or pollen grains.	• Gene delivery is not targeted as mostly flowers and inflorescences are used for transformation intended to deliver foreign gene to female germ cells.
• Various methods of gene delivery are used.	• Mostly *Agrobacterium* mediated gene delivery is used.
• May require tissue culture for *in vitro* maturation of microspores or pollen embryogenesis	• Tissue culture is completely eliminated.
• Pollen grains or microspore are isolated from anther, transformed and then can be used for pollination.	• Method of *in planta* transformation.

the GUS marker gene at different developmental stages of female gametophyte. It was reported that transformation occurred in flowers which were at third mitotic division of megagametogenesis (Bechtold et al. 2000). Floral dip method has been successful in generating transgenic plants in *Raphanus sativus*, *Cicer arietinum* and wheat. In these studies, authors suggested that the plant development stage and use of surfactant are important parameters to get optimum transformation (*reviewed in* Kaur & Devi 2019). The major disadvantages of floral dip method are requirement of many flowers because there is no specific target site of transformation, handling of large volume of bacterial suspension and the difficulty in inoculating large flowers or inflorescence. A modified version of floral dip involves dropping *Agrobacterium* inoculum using the 1 mL syringe on individual spikelets in rice (Ratanasut et al. 2017).

13.4.3 Floral Spray

The floral spray method is employed for *in planta* transformation of plant species which are not suitable for vacuum infiltration or floral dip either due to their size or big flowers. This method involves spraying an inoculum of *Agrobacterium* containing the transgene on the immature floral buds once or multiple times until the suspension begins to drip off. This is followed by covering flower/inflorescence to maintain humidity. The transformation using floral spray was first demonstrated in *Arabidopsis* by Chung et al. (2000). In their study, floral buds at very young stage of development were sprayed with *Agrobacterium* and about 2.4% transformation efficiency was reported.

13.4.4 Pollen Tube Mediated Transformation

Another method which essentially is an alternative to genetic transformation is the pollen-tube pathway mediated transformation (PTT). For PTT, stigma is severed shortly after pollination and an exogenous DNA solution is applied onto the cut end of the style of the recipient plant (Luo and Wu 1989). The exogenous DNA is then transported through pollen tube to the ovary where it may integrate either with unfertilized egg cell or undivided zygote (Ali et al. 2015) (Fig. 13.3A). It was first proposed by Zhou et al. (1983) for transformation in cotton (*Gossypium hirsutum*) and later confirmed in rice by Luo and Wu (1989). Since then, this method has been used for transformation in numerous species like cotton, rice, maize, soybean, wheat, *Petunia*, tomato and watermelon. This technique has several advantages as it eliminates difficult and time-consuming tissue culture process and is independent of genotype. In spite of recent reports of successful transformation using PTT in soyabean (Liu et al. 2009) and cotton (Bibi et al. 2013), there are reservations about universal applicability and advantages of PTT over other gene delivery techniques. Here it is noteworthy that, both 'Gametic Transformation' and 'Pollen-tube Pathway Transformation' use pollen tube mediated gene delivery but are different in methodology (refer to Box 13.6).

Box 13.6

Differences between Gametic and Pollen-tube Pathway Mediated Transformation

Gametic Transformation	Pollen-tube Pathway Mediated Transformation
• Method uses the irradiated mentor pollen grains as source of DNA.	• Method directly uses exogenous DNA for transformation.
• Transfer of random chromatin (or alleles) to egg cell.	• Transfer of desired gene of interest to egg cell.
• Pseudo-fertilization and parthenogenesis of egg is involved.	• There is fusion between egg cell and sperm cell nucleus.
• No ablation done in pistil.	• A part of style is excised after pollination and exogenous DNA solution is dropped on the cut end of style.

13.4.5 Ovary/Pistil Drip Transformation

This method involves opening of ovary by making a cut, followed by the direct dripping of the exogenous DNA solution into the ovary near ovules (Fig. 13.3B). This method was devised by Liu et al. (2009) in soybean where an increased transformation efficiency was observed (~3%) as compared to pollen tube transformation method (~1%). This method is a variation of PTT and also known as modified pollen-tube based transformation method (Ali et al. 2014). Both PTT and ovary drip have been successful for transformation in maize, with better transformation efficiency seen in ovary drip method (Yang et al. 2009). The

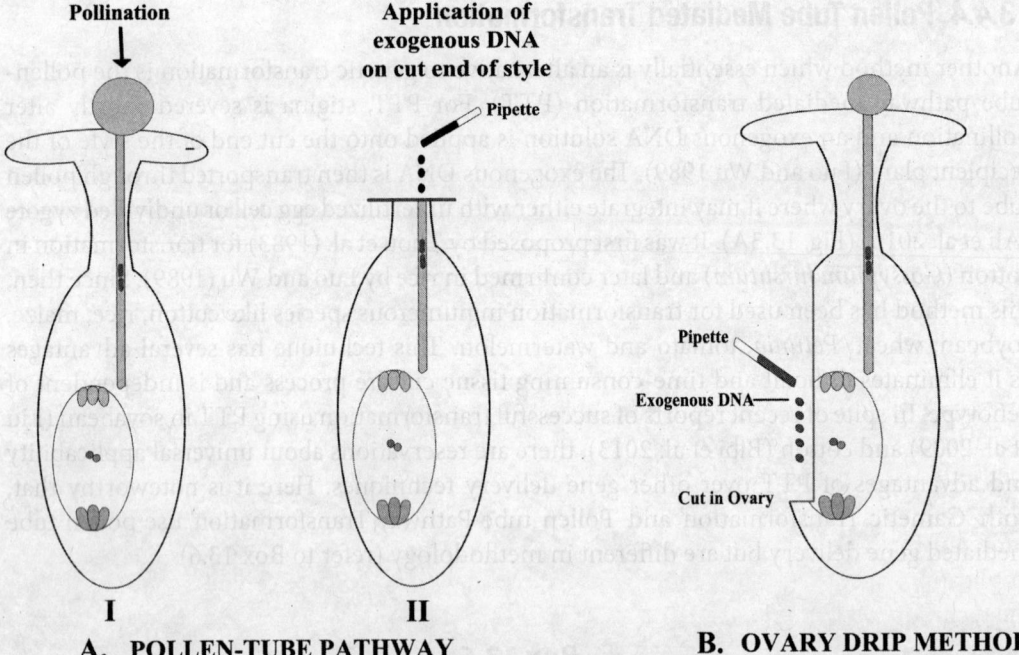

Figure 13.3 **A.** Diagrammatic representation of pollen-tube pathway mediated transformation. **B.** ovary/pistil drip transformation. Exogenous DNA is transferred directly near the ovule by incising a part of ovary. (*Source*: Modified after Ali et al. 2014.)

ovary drip method has advantage over the pollen tube transformation method probably because the distance for migration of the exogenous DNA to reach the ovule is short and therefore, chances to reach embryo sac are higher. However, this method needs skill for the excision and opening of the ovary such that mechanical injury to ovules can be controlled (Shi et al. 2012).

13.5 Factors Influencing Germline Transformation

There are several factors which affect the success of germline transformation. Some of them are mentioned below:

- The developmental stage and physiology of the plant has a great impact on success of transformation and its efficiency. Transformation both in male and female germline is amenable in early stages of gametophyte development, e.g., pollen grains are difficult to transform after bicellular stage.
- The selection of right strain of *Agrobacterium* is critical because it exhibits host specificity. The compatibility between genotype of the plant and bacterial strain is required for obtaining high transformation efficiency.
- In DNA delivery mediated by *Agrobacterium*, the concentration and type of surfactant used in inoculum has a direct role in success and efficiency of transformation.

- Similarly, the duration of treatments like co-cultivation, vacuum infiltration, incubation also has a bearing on the success of procedure. Under-incubation may prevent transformation and over-incubation may prove lethal for the explant.
- When using microprojectile bombardment for transformation, pressure of the helium gas, microprojectile travel distance, duration of exposition to an osmotic agent and the concentration of osmotic agent plays an important role.
- Concentration and amount of foreign DNA used for transformation also has an impact on overall success of transformation.
- The germline transformation studies can only be performed during the flowering period of a species. Thus, physical factors affecting flowering play an important role. Overall, use of these techniques need a systematic knowledge of reproductive structures of a species.

Germline transformation is a promising technique which can be an alternate and easy method for developing transgenic plants. Minimization or elimination of tissue culture and regeneration steps and developing transgenic plants directly by using transformed pollen and ovules can bring down the cost of developing transgenic plants. This will have a major implication for crop improvement and can prove beneficial to plant breeders. However, the technique needs to be standardized to obtain reproducible results and be extended to a wider range of plants.

Glossary

Androgenesis: Regeneration of plants through isolated microspore/anther cultured under *in vitro* conditions.

Co-cultivation: Culturing of *Agrobacterium* and plant cell populations together with some degree of contact between them.

Electroporation: Technique that uses electrical pulse to create pores in cells through which exogenous DNA can enter the cell.

Explant: Small pieces of plant tissues which are cultured in a nutrient medium under sterile conditions to regenerate a plant.

Floral dip: Method of transformation in which developing inflorescence are submerged into the infiltration medium containing exogenous DNA-carrying Agrobacteria.

Gametic Transformation: Transfer of a single gene or short genetic segments of pollen to the egg cell via pollen tubes coming out of irradiated pollen grains.

Germline: The specialized group of cells containing and transmitting genetic information from generation to generation.

Gynogenesis: Regeneration of plants through unfertilized ovaries or ovules cultured under *in vitro* conditions.

In planta gene transformation: Direct transformation of the intact plant without involving any tissue culture.

Microinjection: The technique of delivering foreign DNA into a living cell through a glass micropipette.

Particle Bombardment: Direct penetration of genetic material into the cells using high density micron-size particles accelerated to high speeds.

Transformation: Genetic alteration of a cell either by introduction or direct taking up of exogenous DNA.

Vacuum Infiltration: The infiltration of exogenous DNA-carrying Agrobacteria into the intercellular space of the plant tissue by applying vacuum.

Key Questions

Q13.1 Differentiate Between:
a. Male germline transformation and Female germline transformation
b. Gametic transformation and Pollen-tube pathway mediated transformation

Q13.2 Fill in the blanks:
a. Regeneration of plants through microspore culture under *in vitro* conditions is known as................
b. .... is the regeneration of plants through unfertilized ovules under *in vitro* conditions.
c. Application of sound energy to agitate particles in a sample is known as ....
d. .... uses vacuum for infiltration of DNA into plant cells.
e. A combination of particle bombardment and *Agrobacterium* mediated gene delivery is known as ....

Q13.3 Expand the following:
a. MAGELITR
b. PTT

Q13.4 What is plant germline transformation? Enumerate the advantages and disadvantages of the procedure.

Q13.5 Discuss methods used for gene delivery in male germline transformation, citing the advantages and disadvantages of each.

Q13.6 Discuss the methods used for female germline transformation, citing the advantages and disadvantages of each.

Practicals

Exercise 13.1 To identify the uninucleate stage of pollen grains and demonstration of technique of anther culture

The regeneration of haploid or di-haploid plants from anthers or isolated microspores under *in vitro* conditions is known as anther culture or microspore culture. The original technique of anther culture was developed by Guha and Maheshwari (1964; 1966) and has since been modified by various workers. While culturing anther, the target is generation of plants from the microspores and not anther tissue which is diploid. Under optimal conditions, anther or microspores are stimulated to undergo embryogenesis. The embryos are produced via either of the two pathways i.e., 1. Through callus 2. Direct embryogenesis (Fig. 13.2). It is established that late uninucleate to early binucleate microspores are most suitable for induction of embryogenesis. The somatic embryos developed from microspores generate haploid plants, from which double haploid plants can be obtained by chromosome-doubling techniques, e.g., application of colchicine. Thus, microspore culture enables the production of homozygous (similar allele at every locus) plants in a comparatively shorter duration as compared to conventional breeding practices. These homozygous plants are useful tools in plant breeding and genetic studies. In addition, haploid embryos are used in mutant isolation, gene transfer, studies of storage product biochemistry, and physiological aspects of embryo maturation (Palmer & Keller 1997; Bhojwani & Razdan 1996).

Till now several nutrient media have been formulated for anther culture. The popular ones are Nitsch and Nitsch medium (Nitsch & Nitsch 1969), N6 medium (Chu 1975), and modification of B5 and MS media as suggested by Keller & Armstrong in 1977. In general, a medium is a nutrient mix of inorganic salts that consists of macroelements, microelements, vitamins, amino acid and carbohydrate essential for plant growth. These salts are vital for plant metabolism and improve the quality of anther callus. The studies aiming at anther culture have shown that no generalization can be made for medium and concentration of ingredients, e.g., Nitsch and Nitsch medium was developed for tobacco anther culture. N6 Medium has been developed for the anther culture of rice (*Oryza sativa*). The concentration of inorganic salts in some media are given in provided in Box 13.7. Here it is suggested that the reader may select any medium of choice. To exemplify the process of medium preparation, protocol for preparation of Nitsch and Nitsch medium is provided in Box 13.7. One may also purchase a ready-mix of all these media which are easily available in the market.

Suggested material: Datura sp., *Nicotiana* sp., *Brassica* sp.

A. Identification of the floral buds with anthers having uninucleate stage of pollen grains

The identification of uninucleate stage of pollen grains is prerequisite for start of anther/microspore culture. For this floral bud of different developmental stages are collected and the size of floral buds are recorded using digital scale or scale. Also record the size of anthers for each stage.

Material Required: 1% Acetocarmine solution, needle, forceps, hotplate, glass slide, coverslips, blotting sheet, compound microscope

Procedure

1. Dissect out the anthers (at least one) from each stage on a glass slide.
2. Add a drop of acetocarmine and crush the anther using forceps/needle to liberate microspores.
3. Remove the extra tissue of anther from the slide.
4. Place the coverslip and observe the preparation under compound microscope for the nuclear stage of microspore.

(Note: If needed add some more acetocarmine solution from the sides of coverslips and gently warm over hot plate. This step aids in better staining of microspore nucleus.)

5. By examining pollen preparation from each stage, one may mark the floral development stage of anthers with uninucleate stage of pollen.

B. Anther Culture

Material Required

Chemicals: Inorganic salts for medium (as listed in Table) or ready-mix for medium, sucrose, agar-agar, 1N NaoH/ HCl, double distilled water, ethanol

Glassware: 4-inch sterilized petriplates, glass slides, measuring cylinder, conical flask, droppers, micropipette,

Equipment: Weighing balance, hot plate, laminar flow, autoclave, filters for sterilization (with syringes), spirit lamp

Miscellaneous: Forceps (12-inch size for inoculation), dissecting forceps, needle, tissue paper

Procedure

Step 1 Preparation of medium

1. If using ready mix for medium preparation then constitute medium as suggested by the manufacturer.

OR

If you are making the medium using salts separately then prepare stock solutions, as it is not feasible to weigh the required amount of some of the components in final solution. The stock solutions can be stored in refrigerator for about 1 month (except iron salt solution).

After stock solution preparation, constitute the medium by adding required volume of each solution (only heat stable supplements) and sucrose. Dissolve the mix in little volume of double distilled water and then make up the final volume (e.g., for 1 liter) by adding remaining amount of double distilled water.

2. Adjust the pH of the medium to 5.75 ± 0.5 using 1N NaOH/ HCl.

3. Add the agar-agar (0.8% *w/v*) and heat the medium to boiling till complete dissolution of gelling agent.

4. Sterilize the medium by autoclaving at 15 lbs and 121°C for 15 min.

5. Cool the autoclaved medium to about 45°C before adding heat labile supplements (vitamins#).

   (*#Note: Heat labile supplements must be filter sterilized under aseptic conditions before adding them into the autoclaved medium.*)

6. After adding all supplements pour the medium (~20ml) into test tubes (of 50 ml) or pour the medium in the sterilized petriplates in a laminar flow.

7. Allow the medium to solidify. Cover the petriplates and plug the test tube using cotton plug. (The petriplates and test tubes with solidified medium can be stored for later use.)

Step 2 Pretreatment

1. Select the floral buds of appropriate stage for culture.
2. Floral buds/inflorescence (e.g., in case of cereals), may be given a cold pretreatment at 4°C or dark pretreatment. The pretreatment and its duration are species specific, e.g., pretreatment of panicles in grasses at 4°C for a week increases the percentage of anthers that produce callus. This step is optional

Step 3 Surface sterilization and dissection of anthers from floral buds

1. Give a quick rinse (~30 sec) to floral buds in 70% ethanol. One may also give a wash to floral buds in few drops of surfactant (e.g., Triton-X, Tween-80) followed by thorough washing with running water (for ~20 minutes).
2. Immerse the excised floral buds in 20% commercial Sodium hypochlorite solution for 10–15 minutes. Treated floral buds must be thoroughly rinsed (3–4 times) with sterile distilled water.
3. Dissect out anthers from the sterilized floral buds without damage to anthers; using sterilized forceps and needles. (*Dissection must be done under laminar flow using pre-sterilized glass slides and forceps.*)

Step 4 Inoculation: Using sterilized forceps inoculate the excised anthers onto the agarified solid medium. Cover the petriplates immediately after inoculation.

Step 5 Maintenance of Cultures: The anther cultures are usually maintained in alternating periods of light (12–18 h) and darkness (12–6 h). The temperature may be kept between 22–28°C (Vasil 1973). However, different species require standardization of optimum conditions for maintenance of cultures. For example, the anther cultures of *Brassica* species are best and must be maintained throughout in the dark.

Observation: The responsive anthers gradually show color change (10–15 days after inoculation, also species dependent) and anther wall tissues turn brown. Depending on the species, after 4–8 weeks, anthers may burst open due to the pressure exerted by the growing pollen callus or pollen embryos. This can be observed under compound microscope.

Precautions

1. Stir the medium while dissolving the gelling agent, otherwise it will char at bottom of conical flask.
2. Appropriate pH of the medium must be set before addition of agar-agar; as acidic pH will lead to decreased gelation resulting in semi-solid gel while alkaline pH will lead to formation of a hardened gel.
3. Always use double distilled water/tissue culture grade water for media preparation.
4. Maintain aseptic conditions during inoculation and for maintenance of culture.
5. Use Sodium hypochlorite carefully as it is toxic.
6. Decontaminate work surfaces with 70% ethanol or 10% bleach (freshly prepared) before and after work.

Box 13.7

Nutrient Media for Anther Culture

Ingredients	milligrams/litre Nitsch and Nitsch Medium (1969)	milligrams/litre N6 (Chu 1975; 1978)
Potassium nitrate	950	2830
Ammonium nitrate	720	-
Magnesium sulphate. $7H_2O$	185	185
Potassium phosphate monobasic	68	200
Ammonium sulphate	-	463
Calcium chloride. $2H_2O$	166	166
Manganese sulphate.$4H_2O$	25	4.4
Boric acid	10	1.6
Potassium iodide	-	0.8
Molybdic acid (sodium salt). $2H_2O$	0.25	-
Zinc sulphate.$7H_2O$	10	1.5
Copper sulphate.$5H_2O$	0.025	-
Ferrous sulphate.$7H_2O$	27.8	27.8
EDTA disodium salt.$2H_2O$	37.25	37.3
myo - Inositol	100	-
Thiamine hydrochloride	0.50	1
Pyridoxine hydrochloride	0.50	0.5
Nicotinic acid (Free acid)	5.00	0.5
Folic acid	0.50	-
Biotin	0.05	-
Glycine (Free base)	2	2
2-4 D	-	2
Kinetin	-	0.5
Sucrose %	2	5–12
Agar-agar%	0.8	0.8

Exercise 13.2 To demonstrate the floral dip method of germline transformation

Floral dip is an *in planta* transformation technique which bypasses the need for *in vitro* culture conditions. In this method, developing inflorescences attached to the plant are immersed into the buffered infiltration medium containing *Agrobacterium* and a surfactant. It is quick and easy to perform procedure with simple laboratory facilities. A simple protocol of floral dip was first given by Bechtold et al. (1993) for *Arabidopsis* transformation. Among the factors affecting the success of procedure, appropriate amount of surfactant and health of plant are most important. Species with small flowers can only be used in this method, thus *Arabidopsis* or *Brassica* or rice may be used for the procedure. Before start, one must be acquainted with the methods and safety protocols involved in tissue culture and disposal of waste bacterial culture.

Material required: Soilrite mix, 4 inches pots, *Agrobacterium tumefaciens* strain carrying gene of interest, LB/YEB culture medium, antibiotics (Kanamycin, Rifampicin, etc.), Silwet L-77, Falcon tubes (50 ml)
Instruments: Incubator shaker, Spectrophotometer, Centrifuge
Glasswares: Test tubes, cuvettes

Procedure (Based on Bechtold et al. 1993; Clough & Bent 1998)

Step 1 Plant Material

1. Grow plants until they are flowering. Plants may be grown in soilrite mix in 4 inches pots.
2. Select plants which have many immature flowers (inflorescences) and a few fertilized young siliques.

Step 2 Agrobacterium tumefaciens culture and dip

1. Prepare *Agrobacterium tumefaciens* strain carrying gene of interest on a binary vector.
2. Set up a liquid culture of *A. tumefaciens* at 28 °C in Luria Broth medium (or any other medium like YEB) with antibiotics (Kanamycin, Rifampicin, etc.) to select for the binary plasmid.
3. Centrifuge (at 5000 rpm, 5–10 minutes) the cultures at mid-log cells or a recently stationary phase.
4. Decant the supernatant and resuspend *Agrobacterium* pellet in 5% Sucrose solution (~100 ml, freshly prepared or autoclaved if prepared earlier). Measure the OD (600 nm) of suspension and read it near 0.8. About 50–100 ml bacterial suspension is required for two or three small pots, or 100–200 ml for two or three 3.5" (9 cm) pots. (*The bacterial suspension can also be prepared in sterilized falcon tubes/test tubes depending on the scale of experiment.*)
5. Add Silwet L-77 to a concentration of 0.05% (500 ul/L) in bacterial suspension and mix well. Due to L-77 toxicity, one may use 0.02% or as low as 0.005% of surfactant.
6. Dip above-ground parts of plant including inflorescences (without uprooting) in bacterial suspension for 2 to 3 seconds, with gentle agitation. One may observe a film of liquid that coat the plants.
7. After dipping, keep plants covered for 16 to 24 hours to maintain high humidity. Use saran wrap/cling film to cover the plants.
8. Plants can be watered and grown normally. Seeds may be harvested at maturity and transformants may be selected using antibiotic or herbicide selectable marker.

Precautions

1. Select healthy plants for experiment. Plants may be grown under controlled conditions in plant growth rooms.
2. Do not expose the plants to excessive sunlight after treatment/dip.
3. Wear gloves while performing the experiment.
4. Decontaminate work surfaces with 70% ethanol or 10% bleach (freshly prepared) before and after work.

Bibliography

Abdollahi, M. R., Corral-Martínez, P., Mousavi, A. et al. (2009). An efficient method for transformation of pre-androgenic, isolated Brassica napus microspores involving microprojectile bombardment and *Agrobacterium*-mediated transformation. *Acta Physiologiae Plantarum* 31: 1313.

Abdul-Baki, A. A., Saunders, J. A., Matthews, B. F. and Pittarelli G. W. (1990). DNA uptake during electroporation of germinating pollen grains. *Plant Science* 70: 181–90.

Ali, A., Bang, S. W., Chung, S. M. and Staub, J. E. (2014). Plant Transformation via Pollen Tube-Mediated Gene Transfer. *Plant Molecular Biology Reporter* 33: 742–7.

Aziz, N. and Machray, G. C. (2003). Efficient male germline transformation for transgenic tobacco production without selection. *Plant Molecular Biology* 51: 203–11.

Banga, S. S., Shashi, K., Banga, S. K. and Labana K. S. (1984). Gametic gene transfer in indian mustard (*Brassica juncea* (l.) coss.). *Heredity* 53: 293–7.

Bechtold, N., Ellis, J. and Pelletier, G. (1993). In planta *Agrobacterium*-mediated gene transfer by infiltration of adult *Arabidopsis thaliana* plants. *Life Science* 316: 1194–9

Bechtold, N., Jaudeau, B., Jolivet, S. et al. (2000). The maternal chromosome set is the target of the T-DNA in the in-planta transformation of *Arabidopsis thaliana*. *Genetics* 155: 1875–87.

Bechtold, N., Jolivet, S., Voisin, R. and Pelletier G. (2003). The endosperm and the embryo of *Arabidopsis thaliana* are independently transformed through infiltration by *Agrobacterium tumefaciens*. *Transgenic Research* 12: 509–17.

Bhaskara, G. B. (2017). Basic principles and recent advances in anther/pollen culture for crop improvement. In B. D. Prasad, S. Sahni, P. Kumar and M. W. Siddiqui, eds., *Plant Biotechnology, Volume 1: Principles, Techniques, and Applications*. New Jersey: Apple Academic Press, pp. 88–123.

Bhojwani, S. S. and Razdan, M. K. (1996). *Plant Tissue Culture: Theory and Practice*. Amsterdam: Elsevier Science Pub.

Bibi, N., Fan, K., Yuan, S. et al. (2013). An efficient and highly reproducible approach for the selection of upland transgenic cotton produced by pollen tube pathway method. *Australian Journal of Crop Science* 7: 1714–22.

Bolik, M. and Koop H. U. (1991). Identification of embryogenic microspores of barley (*Hordeum vulgare* L.) by individual selection and culture and their potential for transformation by microinjection. *Protoplasma* 162: 61–8.

Brew-Appiah, R. A. T., Ankrah, N., Liu, W., Konzak, et at. (2013). Generation of doubled haploid transgenic wheat lines by microspore transformation. *PLOS ONE* 8: e80155.

Chin, S. F. and Gordon, G. H. (1989). Pollination with irradiated pollen in rice *Oryza sativa* L. First (Ml) generation. *Heredity* 63: 163–70.

Chu, C. C. (1978). The N6 medium and its applications to anther culture of cereal crops. In *Proc. Symp. Plant Tissue Cult.*, pp. 45–50. Peking: Science Press.

Chumakov, M. I., Rozhok, N. A., Velikov, V. A. et al. (2006). *Agrobacterium*-mediated in planta transformation of maize via pistil filaments. *Russian Journal of Genetics* 42: 893–7.

Chung, M. H., Chen, M. K. and Pan, S. M. (2000). Floral spray transformation can efficiently generate *Arabidopsis.* transgenic plants. *Transgenic Research* 9: 471–86.

Clough, S. J. and Bent, A. F. (1998). Floral dip: a simplified method for *Agrobacterium*-mediated transformation of *Arabidopsis thaliana. Plant Journal* 16: 735–43.

De La Pena, A., Lörz, H. A. and Schell, J. (1987). Transgenic rye plants obtained by injecting DNA into young floral tillers. *Nature* 325: 274–6.

Duan, X. and Chen, S. (1985). Variation of the characters of rice (*Oryza sativa*) induced by foreign DNA uptake. *China Agricultural Science* 3: 6–9.

Eapen, S. (2011). Pollen grains as a target for introduction of foreign genes into plants: an assessment. *Physiology and Molecular Biology of Plants* 17: 1–8.

Fang, F., Oliva, M., Ehi-Eromosele, S., Zaccai, M., Arazi, T. and Oren-Shamir, M. (2018). Successful floral-dipping transformation of post-anthesis lisianthus (*Eustoma grandiflorum*) flowers. *The Plant Journal* 96 (4): 869–79.

Feldmann, K. A. and Marks, M. D. (1987). *Agrobacterium* mediated transformation of germinating seeds of *Arabidopsis thaliana*: a non-tissue culture approach. *Molecular Genomics and Genetics* 208: 1–9.

Fraley, R. T., Rogers, S. G., Horsch, R. B. et al. (1983). Expression of bacterial genes in plant cells. *Proceedings of the National Academy of Sciences* 80: 4803–7.

Graves, A. C. F. and Goldman, S. L. (1986). The transformation of *Zea mays* seedlings with *Agrobacterium tumefaciens. Plant Molecular Biology* 7: 43–50.

Guha, S. and Mahcshwari, S. C. (1964). In vitro production of embryos from anthers of *Datura. Nature* 204: 497.

Guha S. and Maheshwari S. C. (1966). Cell division and differentiation of embryos in the pollen grains of *Datura* in vitro. *Nature* 212: 97–8.

Hansen, G. and Wright, M. S. (1999). Recent advances in the transformation of plants. *Trends in Plant Sciences* 4: 226–30.

Herrera-Estrella, L., Depicker, A., Van Montagu, M. et al. Expression of chimaeric genes transferred into plant cells using a Ti-plasmid-derived vector. *Nature* 303: 209–13.

Hess, D. and Dressler, K. (1989). Tumor transformation of *Petunia hybrida* via pollen cultured with *Agrobacterium tumefaciens. Acta Botanica Brasilica* 102: 202–7.

Hess, D. (1980). Investigations on the intra- and inter-specific transfer of anthocyanin genes using pollen as vectors. *Zeitschrift für Pflanzenphysiologie* 98: 321–37.

Hess, D. (1987). Pollen-based techniques in genetic manipulation. *International Review of Cytology* 107: 367–95.

Huang, G., Dong, Y. and Sun, J. (1999). Introduction of exogenous DNA into cotton via pollen-tube pathway with GFP as a reporter. *China Science Bulletin* 44: 698–701.

Kakuta, H. (2003). Magnetic particles used for genetic transformation of pollen. *Agricell Report* 40: 26.

Kaur, R. P. and Devi, S. (2019). In planta transformation in plants: a review. *Agricultural Reviews* 40: 159–74.

Keller, W. A. and Armstrong, K. C. (1977). Embryogenesis and plant regeneration in *Brassica napus* anther cultures. *Canadian Journal of Botany* 55: 1383–8.

Kim, T. H. and Lee, Y. Y. (2006). Fractionation of corn stover by hot-water and aqueous ammonia treatment. *Bioresource Technology* 97: 224–32.

Kumari, A., Baskaran, P. and van Staden, J. (2017). Gene transfer utilizing pollen-tubes of *Albuca nelsonii* and *Tulbaghia violacea. Crop Breeding and Applied Biotechnology* 17: 228–34.

Kumlehn, J., Serazetdinova, L., Hensel, G., Becker. D. and Loerz, H. (2006). Genetic transformation of barley (*Hordeum vulgare* L.) via infection of androgenetic pollen cultures with *Agrobacterium tumefaciens. Plant Biotechnology Journal* 4: 251–61.

Li D, Shi, W. and Deng, X. (2002). Agrobacterium-mediated transformation of embryogenic calluses of Ponkan mandarin and the regeneration of plants containing the chimeric ribonuclease gene. *Plant Cell Reporter* 21: 153–6.

Liu, J., Su. Q., An, L. and Yang, A. (2009). Transfer of a minimal linear marker free and vector-free smGFP cassette into soybean via ovary-drip transformation. *Biotechnology Letters* 31: 295–303.

Luo, Z. X. and Wu, R. (1989). A simple method for the transformation of rice via pollen tube pathway. *Plant Molecular Biology Reporters* 6: 165–74.

Mamontova, E. M., Velikov, V. A., Volokhina, I. V., et al. (2010). Agrobacterium-mediated in planta transformation of maize germ cells. *Russian Journal of Genetics* 46: 501–4.

Matsumoto, T. K. and Gonsalves, D. (2012). Biolistic and other non-Agrobacterium technologies of plant transformation. *Plant Biotechnology and Agriculture: Prospects for the 21st century*. Cambridge, MA: Academic Press, pp. 117–29.

Matthews, B. F., Abdul-Baki, A. A. and Saunders, J. A. (1990). Expression of a foreign gene in electroporated pollen grains of tobacco. *Sexual Plant Reproduction* 3: 147–51.

Ming, L., Yang, J., Cheng, Y. Q. and Li-Jia. (2009). An optimization of soybean (*Glycine max* (L.) Merrill) *in planta* ovary transformation using a linear minimal gus gene cassette. *Journal of Zhejiang University* 10: 870–6.

Mishra, K. P., Joshua, D. C. and Bhatia, C. R. (1987). In vitro electroporation of tobacco pollen. *Plant Science* 52: 135–9.

Mohanty, D., Chandra, A. and Tandon, R. (2016). Germline Transformation for Crop Improvement. In V. R. Rajpal, R. S. Rao, S. N. Raina, eds., *Molecular Breeding for Sustainable Crop Improvement, Sustainable Development and Biodiversity*. Cham: Springer, pp. 343–95.

Neuhaus, G., Spangenberg, G., Scheid, O. M. and Schweiger, H. G. (1987). Transgenic rapeseed plants obtained by the microinjection of DNA into microspore-derived embryoids. *Theoretical and Applied Genetics* 75: 30–36.

Niazian, M., Noori, S. A. S., Galuszkam, P. and Mortazavian, S. M. M. (2017). Tissue culture-based *Agrobacterium*-mediated and *in planta* transformation methods. *Czech Journal of Genetics and Plant Breeding* 53: 133–43.

Nitsch, J. P. and Nitsch, C. (1969). Haploid plants from pollen grains. *Science* 163: 85–7.

Ohta, Y. (1986). High-efficiency genetic transformation of maize by a mixture of pollen and exogenous DNA. *Proceedings of National Academy of Science USA* 83: 715–19.

Palmer, C. D. and Keller, W. A. (1997). Pollen Embryos. In K. R. Shivanna and V. K. Sawhney, eds., *Pollen Biotechnology for Crop Production and Improvement*. Cambridge: Cambridge University Press.

Pandey, K. K. (1975). Sexual transfer of specific genes without gametic fusion. *Nature* 256: 310–13.

Pandey, K. K. (1978). Novel techniques of gene transfer and plant improvement: an appraisal of transformation in eukaryotes. *New Phytologist* 81: 685–704.

Pechan, P. (1989). Successful cocultivation of *Brassica napus* microspores and proembryos with *Agrobacterium*. *Plant Cell Reports* 8: 387–90.

Purcell, J. P. and Perlak, P. J. (2004). Global impact of insect-resistant (Bt) cotton. *AgBioForum* 7: 27–30.

Ratanasut, K., Rod-In, W. and Sujipuli, K. (2017). In planta *Agrobacterium*-mediated transformation of rice. *Rice Science* 24: 181–6.

Resch, T. and Touraev, A. (2011). Pollen transformation technologies. In C. N. Stewart Jr., A. Touraev, V. Citovsky, T. Tzfira, eds., *Plant transformation technologies*. Oxford: Wiley-Blackwell, pp. 83–92.

Saha, P. and Blumwald, E. (2016). Spike-dip transformation of *Setaria viridis*. *The Plant Journal* 86: 89–101.

Sanford, J. C., Skubik, K. A. and Reisch, B. I. (1985). Attempted pollen-mediated plant transformation employing genomic donor DNA. *Theoretical and Applied Genetics* 69: 571–4.

Sangwan, R. S., Ducrocq, C. and Sangwan-Norreel, B. (1993). *Agrobacterium*-mediated transformation of pollen embryos in *Datura innoxia* and *Nicotiana tabacum*: production of transgenic haploid and fertile homozygous dihaploid plants. *Plant Science* 95: 99–115.

Sharada, M. S., Kumari, A., Pandey, A., et al. (2017). Generation of genetically stable transformants by *Agrobacterium* using tomato floral buds. *Plant Cells Tissue and Organ Culture* 129: 299–312.

Saunders, J. A. and Matthews, B. F. (1995). Pollen transformation in tobacco: pollen electro-transformation methods. In J. A. Nickoloff, ed., *Methods in Molecular Biology (vol 55): Plant Cell Electroporation and Electrofusion Protocols*. Totowa: Humana Press Inc., pp. 81–8.

Shi, X. X., Du, G., Wang, X. and Pei, D. (2012). Studies on gene transformation via pollen-tube pathway in walnut. *Acta Horticulture Sinica* 39: 1243–52.

Shim, Y. S., Pauls, K. P. and Kasha, K. J. (2009). Transformation of isolated barley (*Hordeum vulgare* L.) microspores: II. Timing of pretreatment and temperatures relative to results of bombardment. *Genome* 52: 175–90.

Smith, C. R., Saunders, J. A., Wert, S. V. et al. (1994). Expression of GUS and CAT activities using electrotransformed pollen. *Plant Science* 104: 49–58.

Stöger, E., Fink, C., Pfosser, M. P. and Heberle-Bors, E. (1995). Plant transformation by particle bombardment of embryogenic pollen. *Plant Cell Reports* 14: 273–8.

Tanaka, T., Nishihara, M., Seki, M. et al. (1995). Successful expression in pollen of various plant species of in vitro synthesized mRNA introduced by particle bombardment. *Plant Molecular Biology* 28: 337–41.

Tjokrokusumo, D., Heinrich, T., Wylie, S., Potter, R. and McComb, J. (2000). Vacuum infiltration of *Petunia hybrida* pollen with *Agrobacterium tumefaciens* to achieve plant transformation. *Plant Cell Reports* 19: 792–7.

Touraev, A., Pfosser, M. and Heberle-Bors, E. (2001). The microspore: a haploid multipurpose cell. *Advances in Botanical Research*: 53–109.

Touraev, A., Stgert, E., Voronin, V. and Heberle-Bors, E. (1997). Plant male germ line transformation. *The Plant Journal* 12: 949–56.

Trick, H. N. and Finer, J. J. (1997). SAAT: sonication-assisted *Agrobacterium*-mediated transformation. *Transgenic Research* 6: 329–37.

Twell, D., Klein, T. M., Fromm, M. E. and McCormick, S. (1989). Transient expression of chimeric genes delivered into pollen by microprojectile bombardment. *Plant Physiology* 91: 1270-4.

Wang, J., Li, Y. and Liang, C. (2008). Recovery of transgenic plants by pollen mediated transformation in *Brassica juncea*. *Transgenic Research* 17: 417–24.

Wang, J. X., Sun, Y., Cui, G. and Hu, J. (2001). Transgenic maize plants obtained by pollen-mediated transformation. *Acta Botanica Sinicia* 43: 275–9.

Wang, R., Li, R., Xu, T. and Li, T. (2017). Optimization of the pollen-tube pathway method of plant transformation using the Yellow Cameleon calcium sensor in *Solanum lycopersicum*. *Biologia* 72: 1147–55.

Wang, W., Wang, J., Yang, C., et al. (2007). Pollen-mediated transformation of *Sorghum bicolor* plants. *Biotechnology and Applied Biochemistry* 48: 79–83.

Yang, A., Su, Q. and An, L. (2009). Ovary dip transformation: A simple method for directly generating vector and marker free transgenic maize (*Zea mays* L.) with a linear GFP cassette transformation. *Planta* 229: 793–80.

Ye, G. N., Stone, D., Pang, S. Z., et al. (1999). *Arabidopsis* ovule is the target for *Agrobacterium* in planta vacuum infiltration transformation. *Plant Journal* 19: 249–57.

Zhou, G. Y., Weng, J., Zhen, Y. S., et al. (1983). Introduction of exogenous DNA into cotton embryos. In R. Wu, L. Grossman, K. Moldave, eds., *Methods in Eenzymology. Vol. 101. Recombination DNA, Part C.* New York: Academic, pp 433–81.

Index

abiotic pollination, 2, 182–188
  anemophily, 182–185
  hydrophily, 185–187
abscisic acid (ABA), 312, 315, 407–408
adventitious/adventive embryony, 361–363,
  367, 380
agamospermy, 357–358
Agrobacterium-mediated transformation, 437,
  443, 447
aleurone layer, 306–310, 315, 399
Allium, 65, 133, 139–140, 370, 423
allogamy, 175, 284
alveoli formation, 303–307
ambophily, 36, 202
Amici, Giovanni Battista, 28
amoeboid tapetum/periplasmodial tapetum/
  invasive tapetum, 52–53, 56
amphitropous ovule, 125, 157
amylogenesis, 91
amylolysis, 91
anatropous ovule, 122, 124, 156, 394
androdioecious/androdioecy, 17, 179
androecium, 9, 11–12, 82
androgenesis, 438–442
andromonoecious, 17
anemochory, 413, 417–418
anemophily, 182–185
Angraecum sesquipedale, 28
aniline blue, 249, 266, 279
anther, 2, 6–7, 12–13
  bithecous, 42, 96
  development, 42–45
  events, 57
  locular fluid, 59
  monothecous, 13, 41
  structure, 41–42
  thecae, 12, 41

anther culture, 439, 441, 452–456
anther dehiscence, 2, 43, 45, 48–49, 52, 57–60,
  61, 69, 176, 180, 182–184, 192
  dehydration, 57–60, 66, 86, 225, 380, 384,
    392, 410, 414
  time, 206
  types, 57
anther primordium, 43, 46
anther wall layers, 2, 45–56, 96, 98
  endothecium, 41–44, 48–52, 57, 61, 91, 131
  epidermis, 41, 45–48, 57, 96
  middle layers, 41, 44, 51–52
  tapetum, 41, 44, 52–56, 60, 63, 66, 68, 75–77,
    80, 97, 99, 264–265
anther wall ontogeny, 44
antipodal cells/antipodals, 32, 137–144, 148,
  150–151
ant pollination/myrmecophily, 192
apical cell (ap/ac), 327–331
apomixis, 357–358, 365
  applications, 375
  genetics, 373
  study methods/screening, 379
  role, 373
  types, 366
apical embryo domain, 332, 334
apomeiosis, 357, 366
Arachis hypogea, 295
archesporial cell, 43, 45, 49–50, 96, 126, 132,
  361, 366
aril, 410, 414–415
Aristolochia, 63, 86, 109, 133, 180, 191–193, 417
autochory, 413–415
autogamy, 175–176, 180, 271
auxins, 312, 337–338, 364

Color Plates

Figure 2.1 [Page 10]

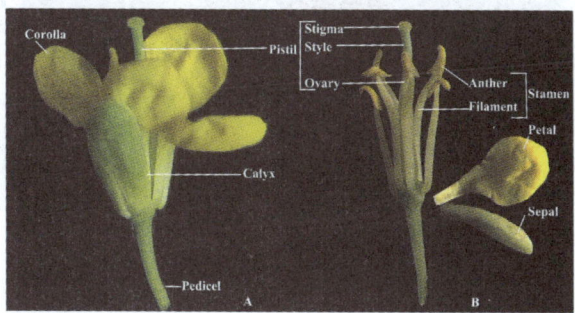

Figure 2.2 [Page 11]

Figure 2.3 [Page 12]

Figure 2.4 [Page 13]

Figure 2.5 [Page 14]

Figure 2.6 [Page 15]

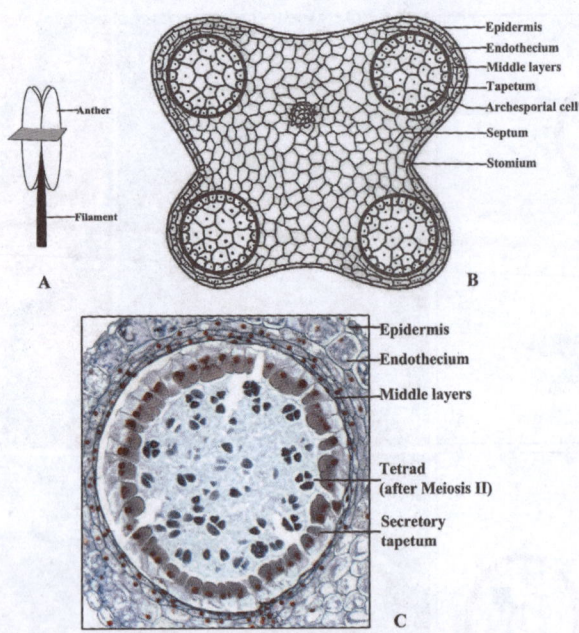

Figure 4.1 [Page 42]

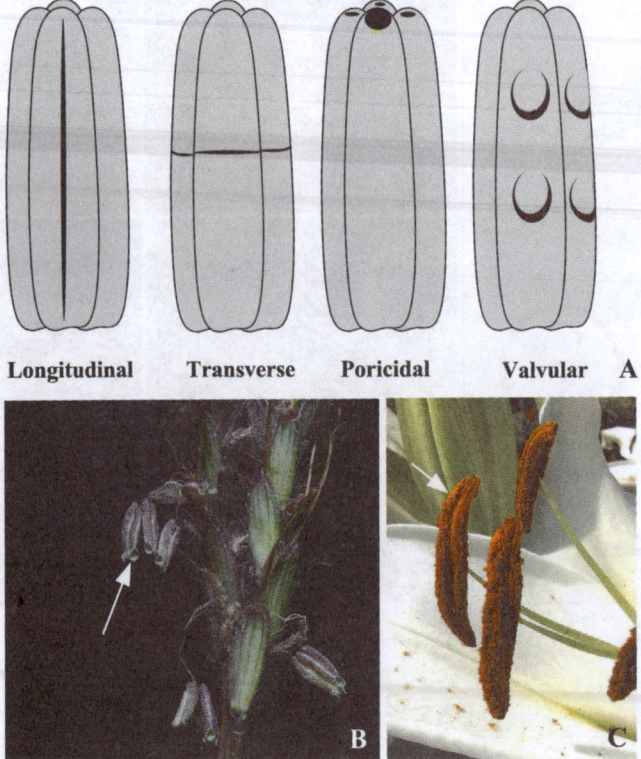

Figure 4.5 [Page 58]

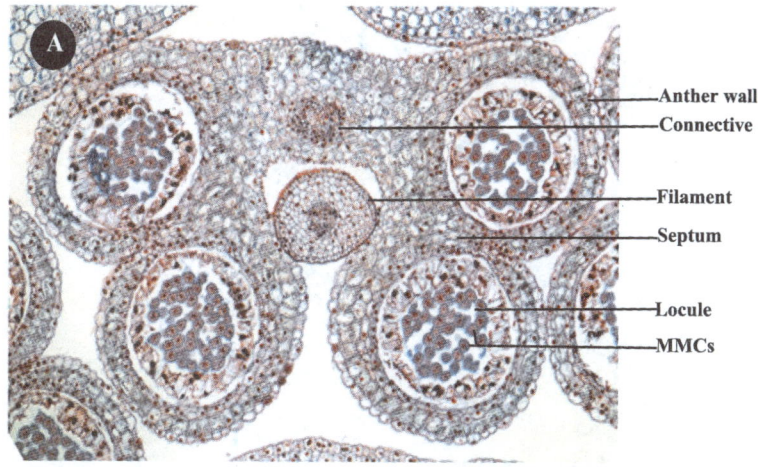

A bithecous teralocular (or tetrasporangiate) anther with microspore mother cells (MMCs)

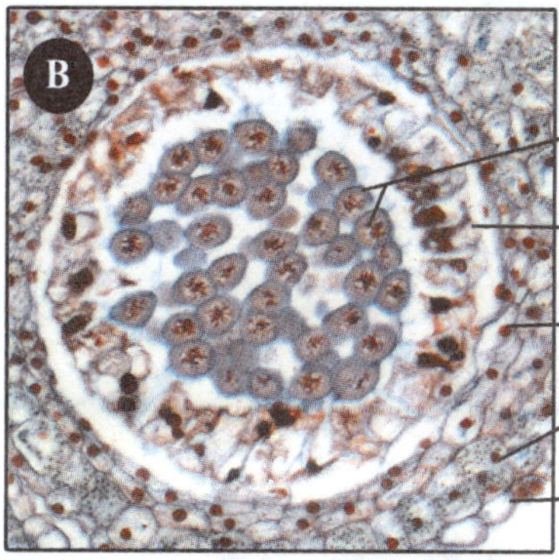

An anther locule showing microspore mother cells at meiosis I

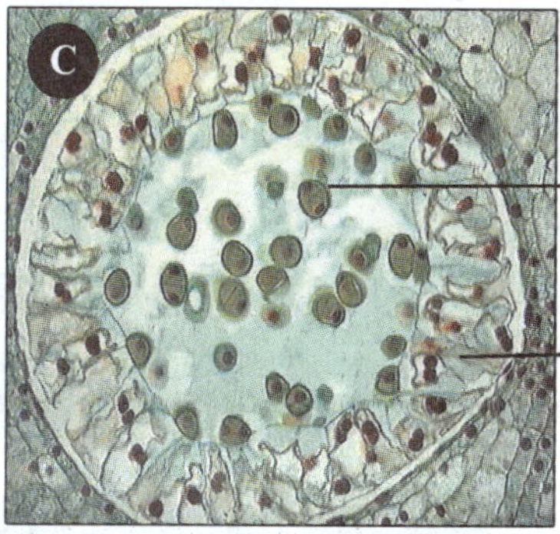

An anther locule showing microspore mother cells after meiosis I, dyad stage

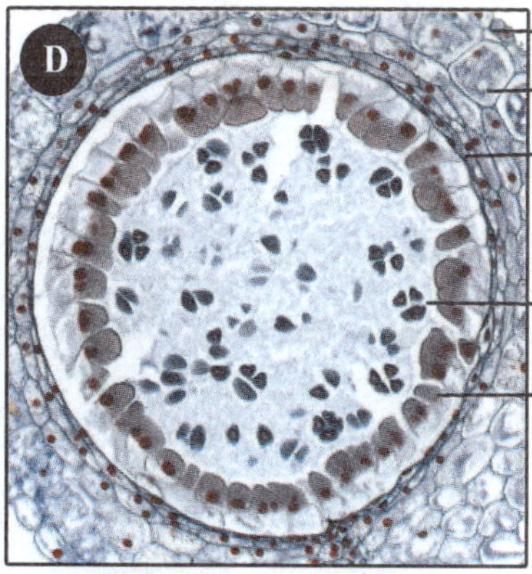

An anther locule showing microspore mother cells after meiosis II, (microspore tetrad)

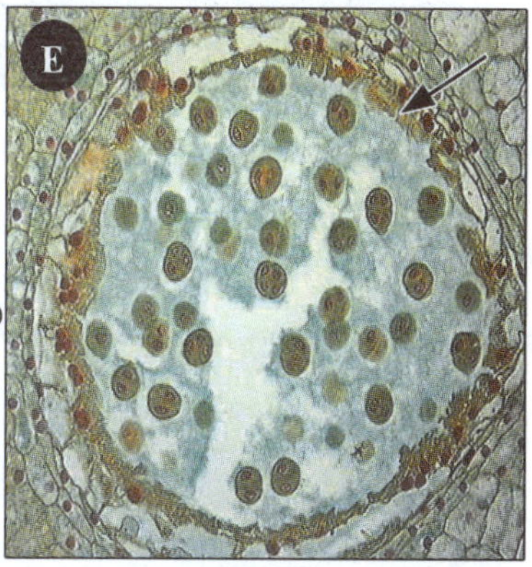

An anther locule showing microspores in tetrad and degeneartion of secretory tapetum (arrow)

Figure 4.18 [Page 98]

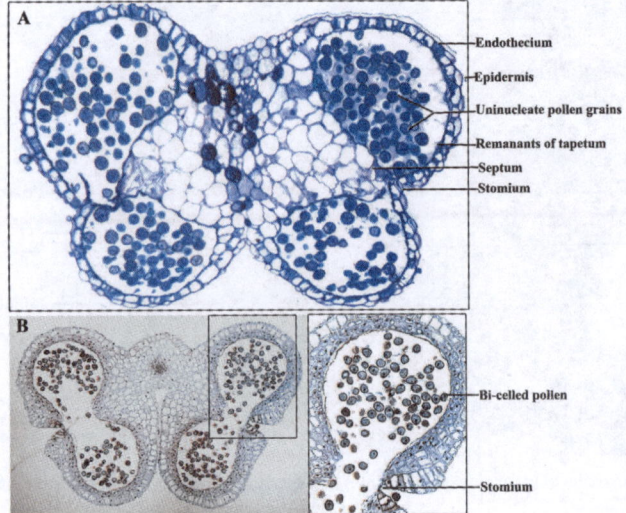

Figure 4.19 [Page 99]

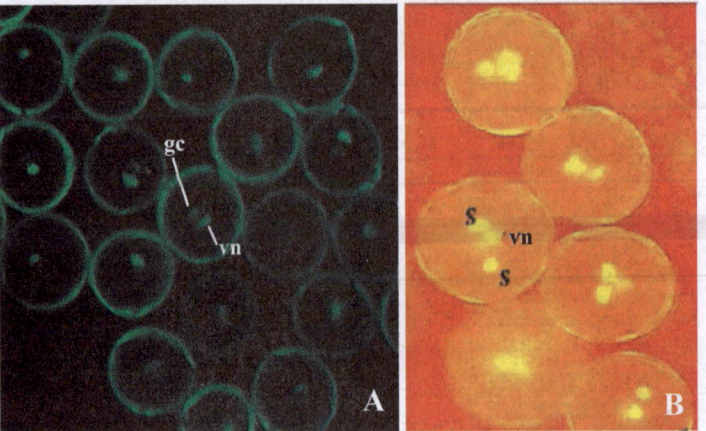

Figure 4.21 [Page 102]

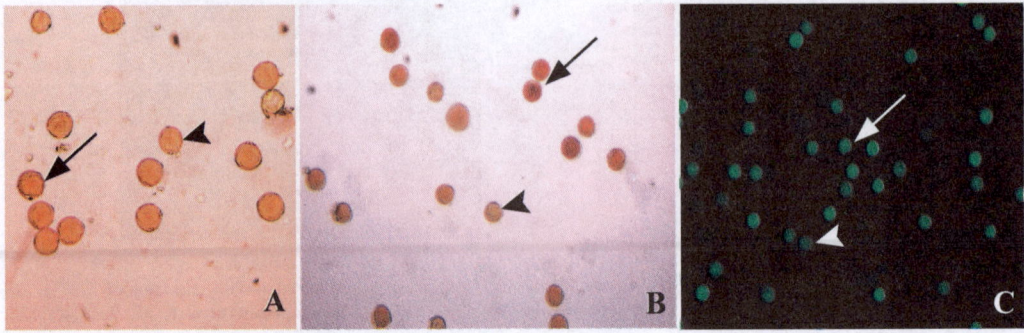

Figure 4.22 [Page 106]

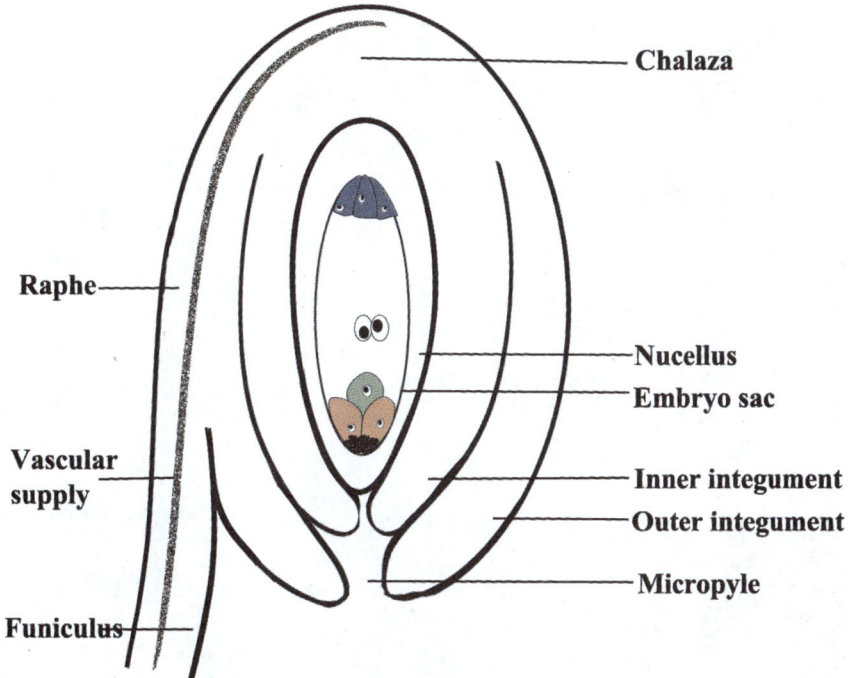

Figure 5.1 [Page 120]

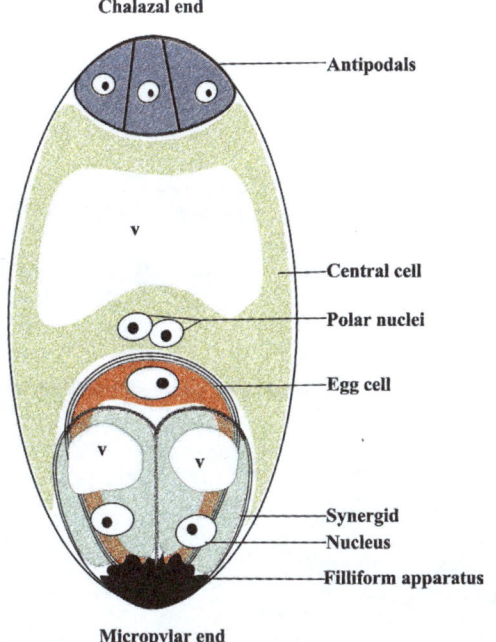

Figure 5.10 [Page 137]

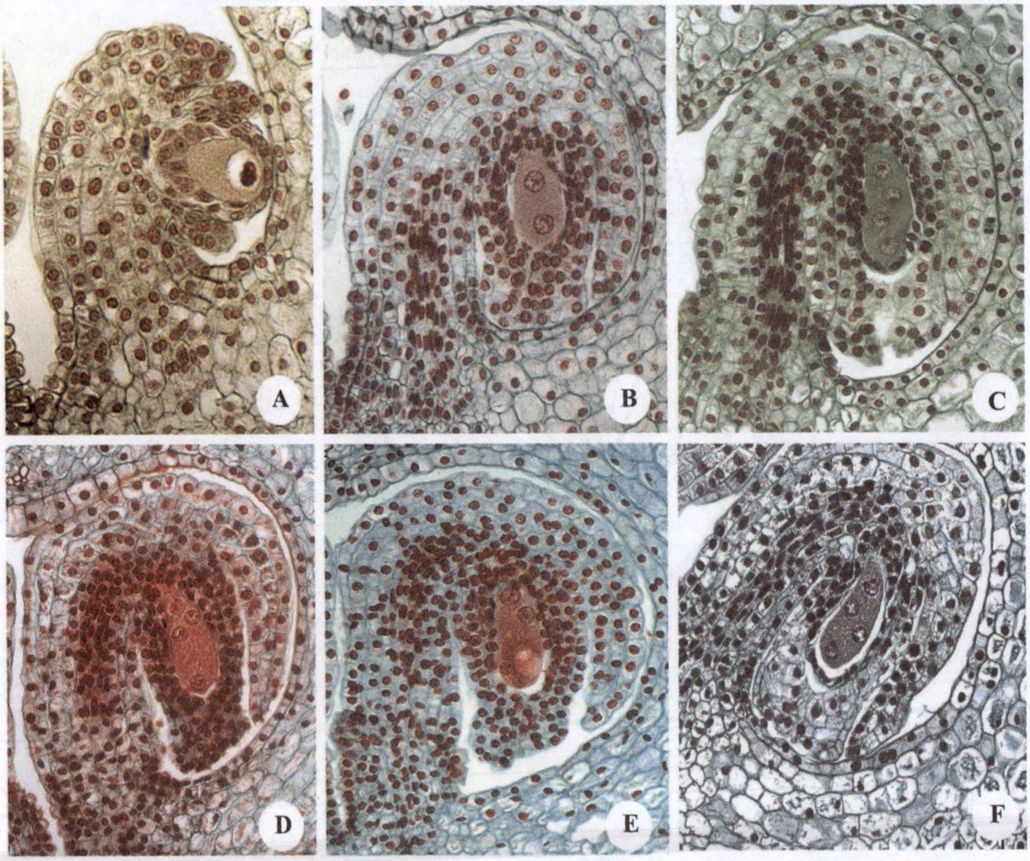

Figure 5.22 [Page 163]

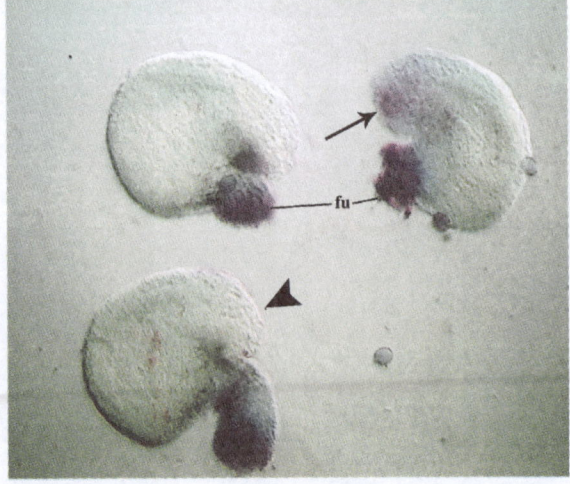

Figure 5.24 [Page 169]

Figure 6.3 [Page 180]

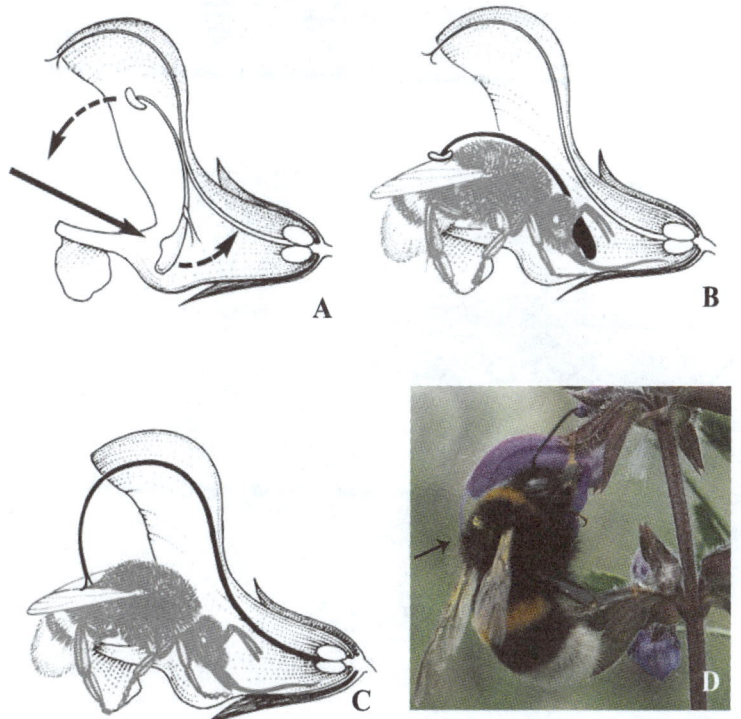

Figure 6.6 [Page 190]

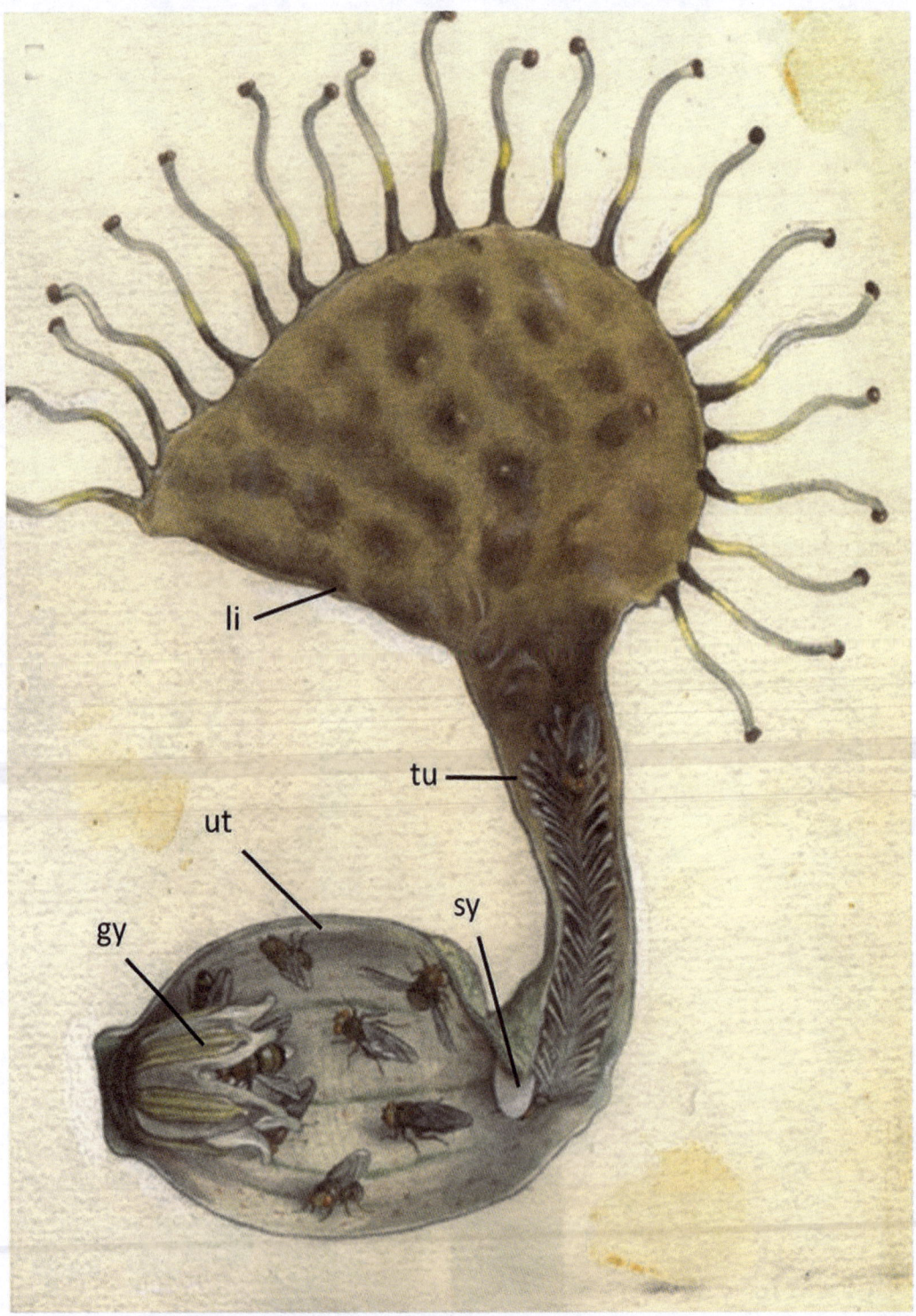

li

tu

ut

sy

gy

Figure 6.7 [Page 191]

Figure 6.8 [Page 194]

Figure 6.9 [Page 197]

Figure 6.10 [Page 199]

Figure 6.11 [Page 199]

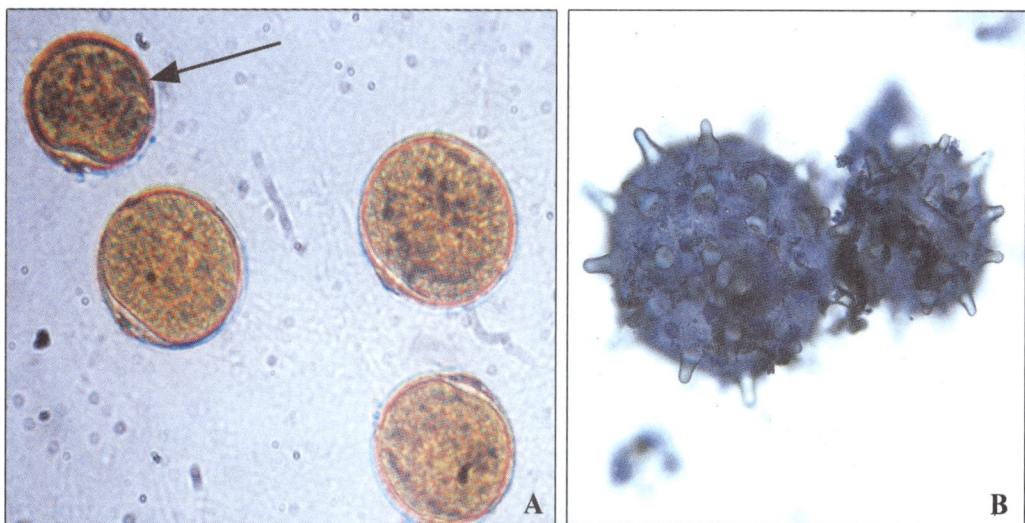

Figure 6.12 [Page 208]

Figure 7.1 [Page 219]

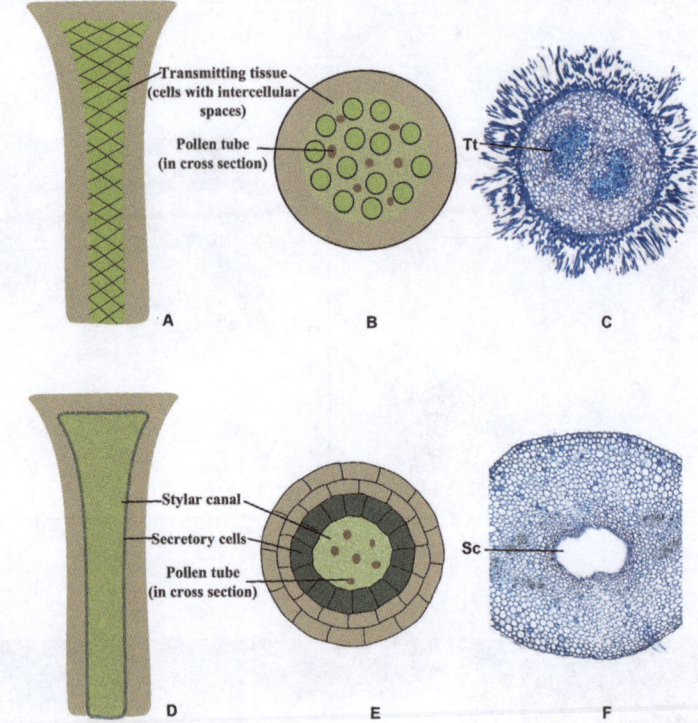

Figure 7.3 [Page 220]

Figure 7.5 [Page 224]

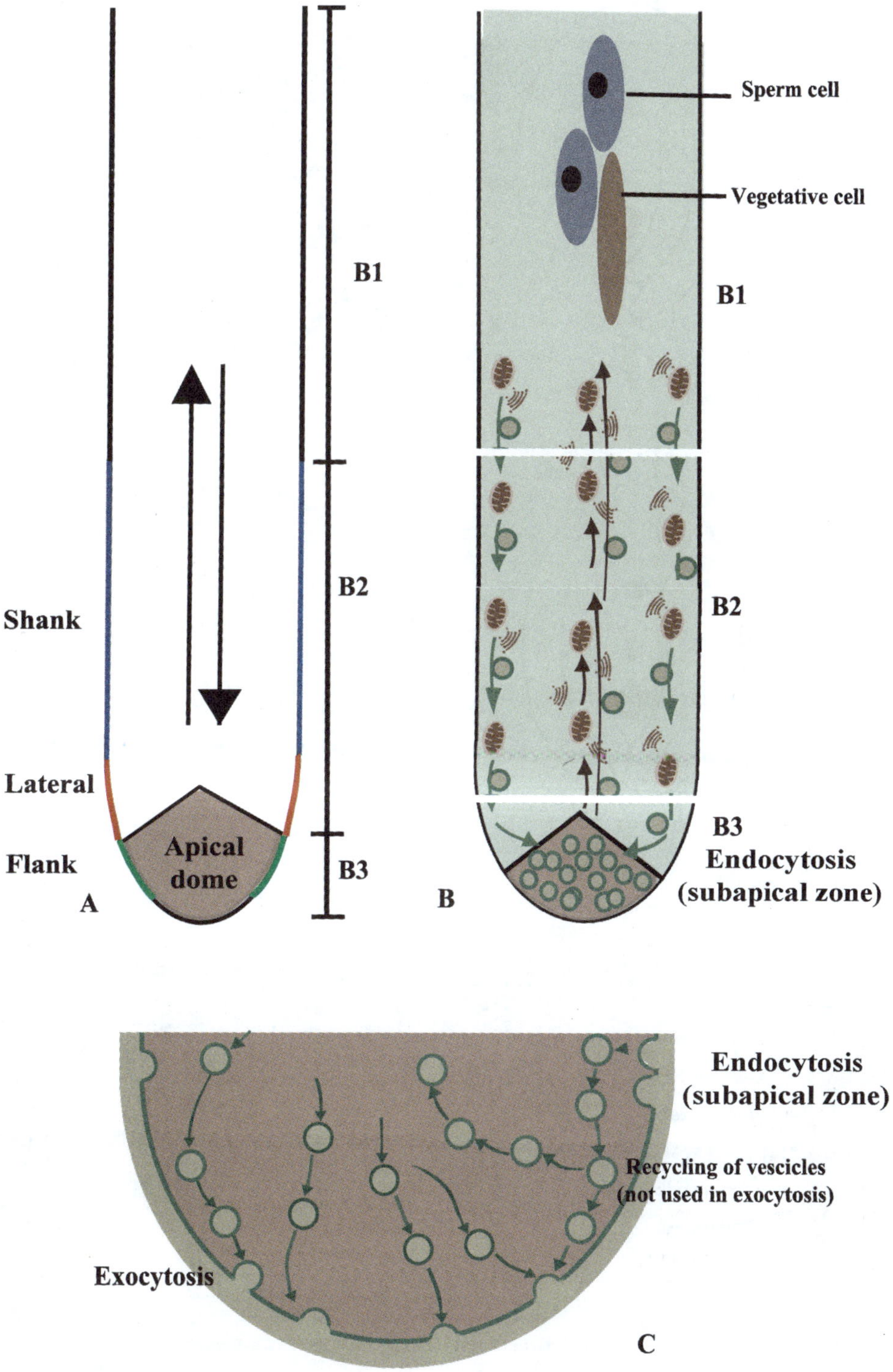

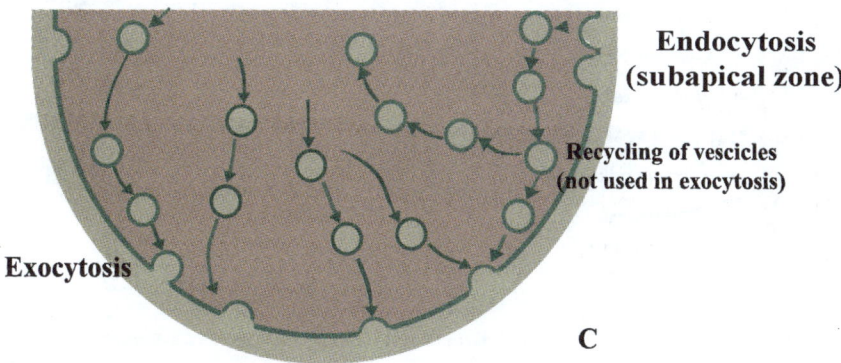

Figure 7.7 [Page 230]

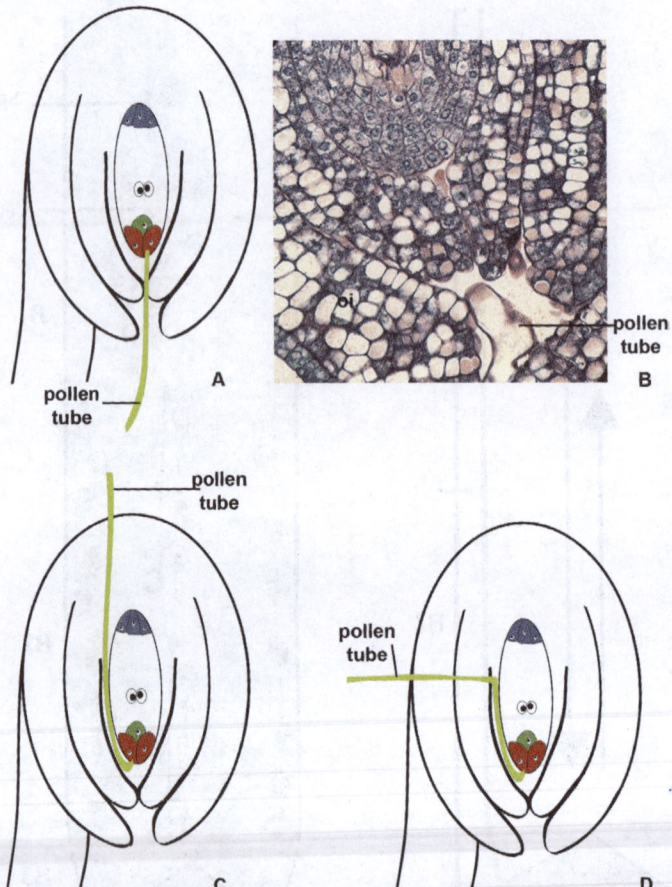

pollen tube

A

B

pollen tube

pollen tube

C

pollen tube

D

oi

pollen tube

Figure 7.8 [Page 234]

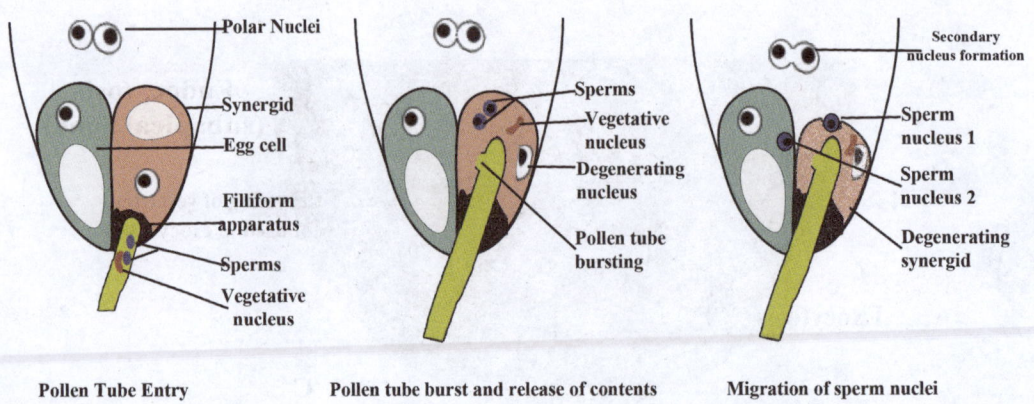

Polar Nuclei

Synergid

Egg cell

Filliform apparatus

Sperms

Vegetative nucleus

Sperms

Vegetative nucleus

Degenerating nucleus

Pollen tube bursting

Secondary nucleus formation

Sperm nucleus 1

Sperm nucleus 2

Degenerating synergid

Pollen Tube Entry
A

Pollen tube burst and release of contents
B

Migration of sperm nuclei
C

Figure 7.9 [Page 235]

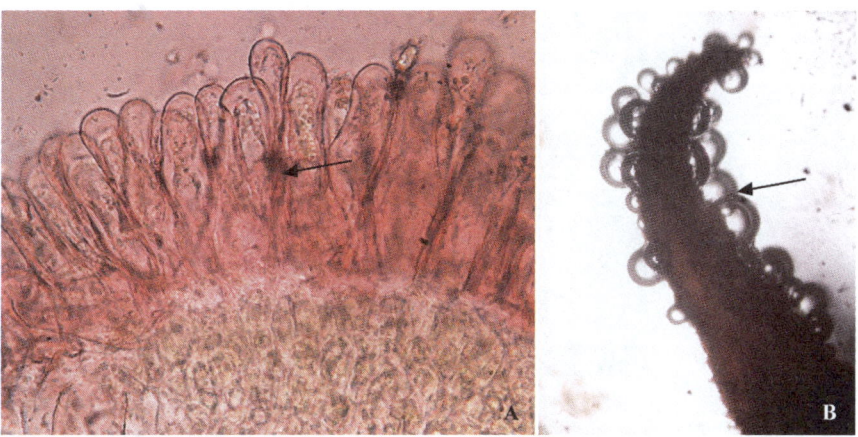

Figure 7.10 [Page 245]

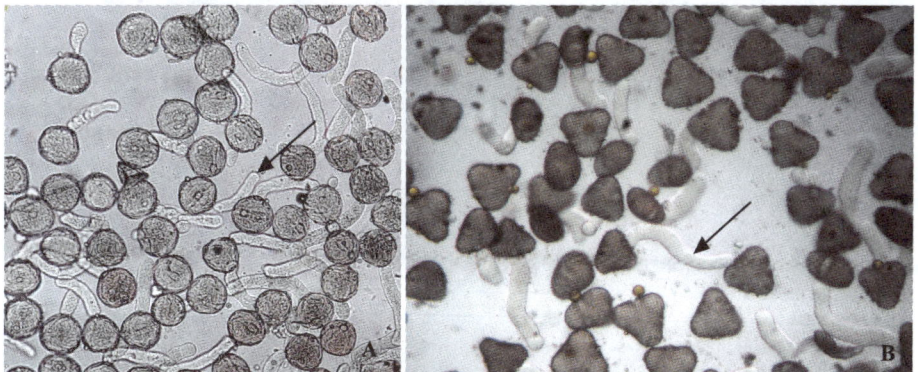

Figure 7.11 [Page 248]

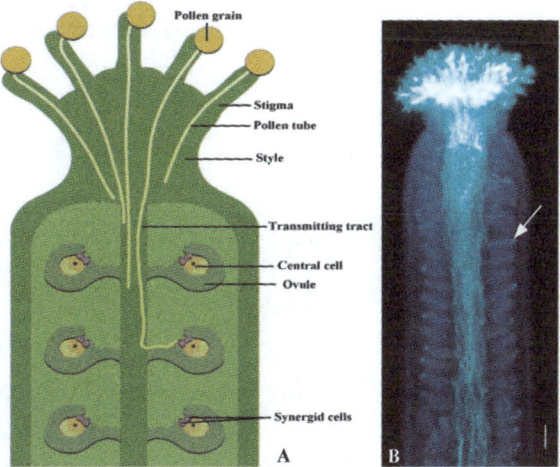

Figure 7.12 [Page 250]

Figure 8.1 [Page 259]

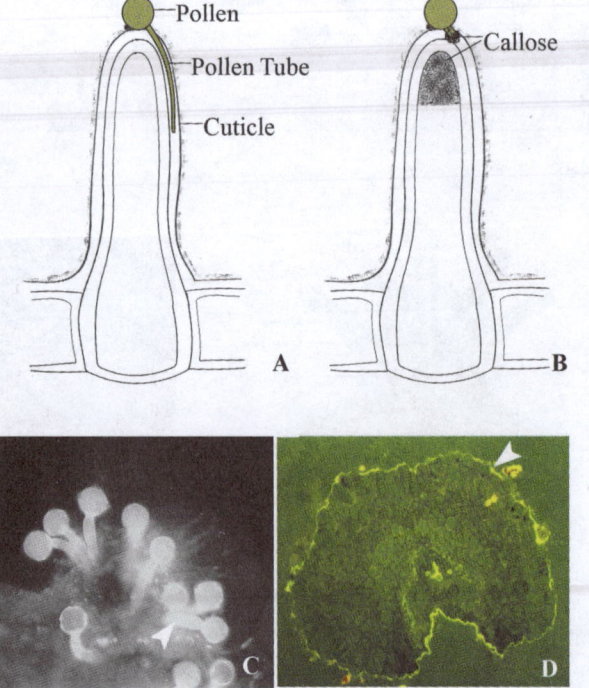

Figure 8.8 [Page 266]

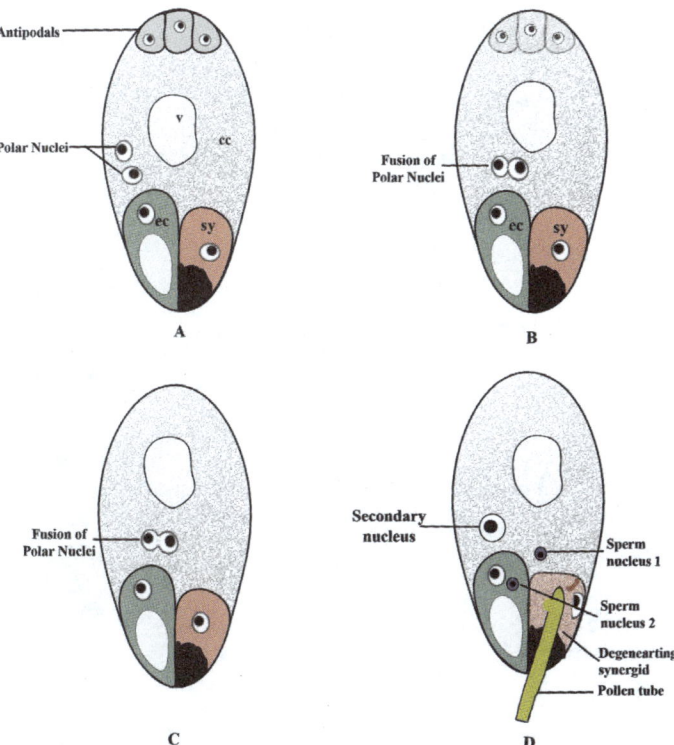

Figure 9.1 [Page 291]

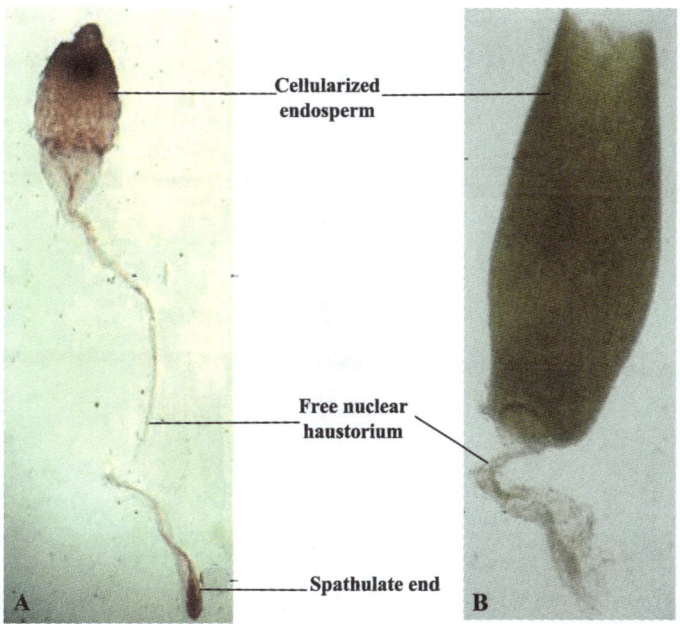

Figure 9.19 [Page 320]

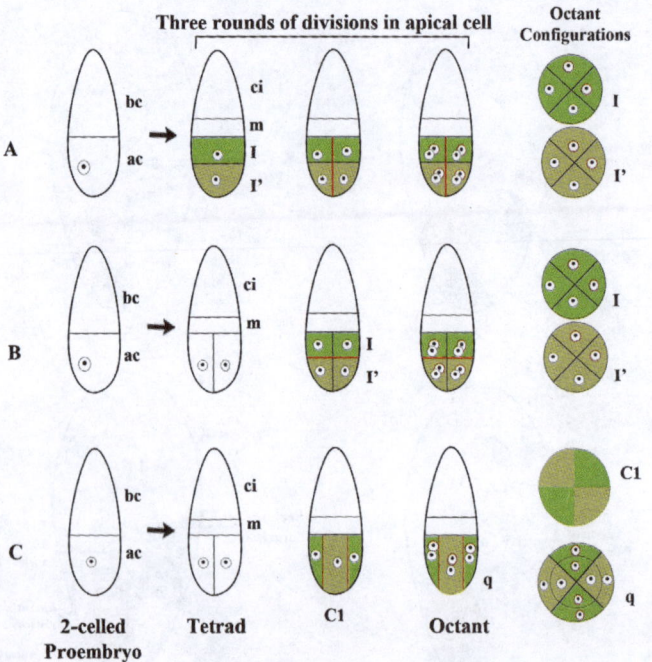

Figure 10.4 [Page 331]

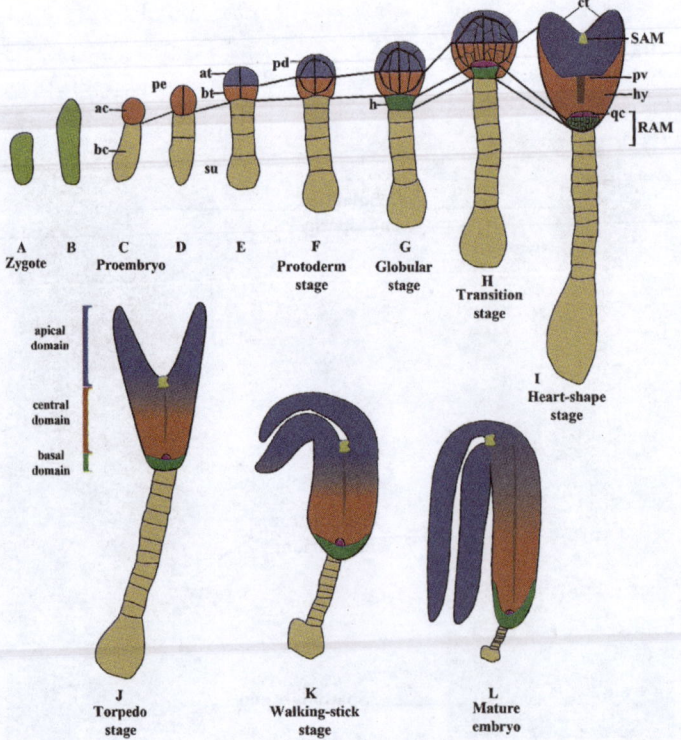

Figure 10.5 [Page 333]

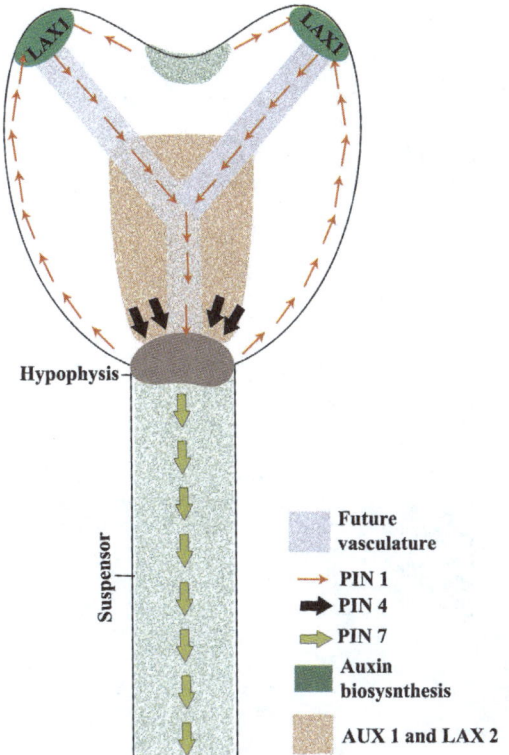

Figure 10.8 [Page 337]

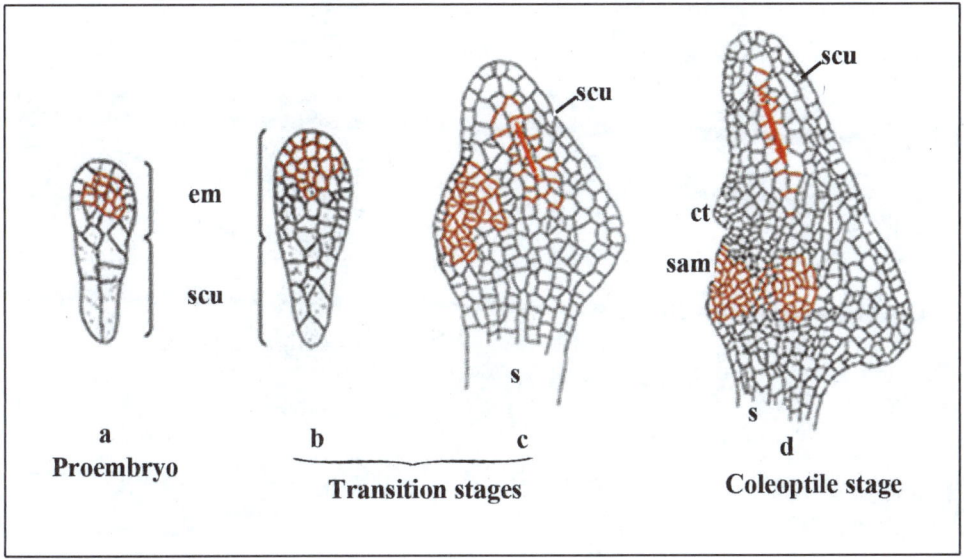

Figure 10.9 [Page 338]

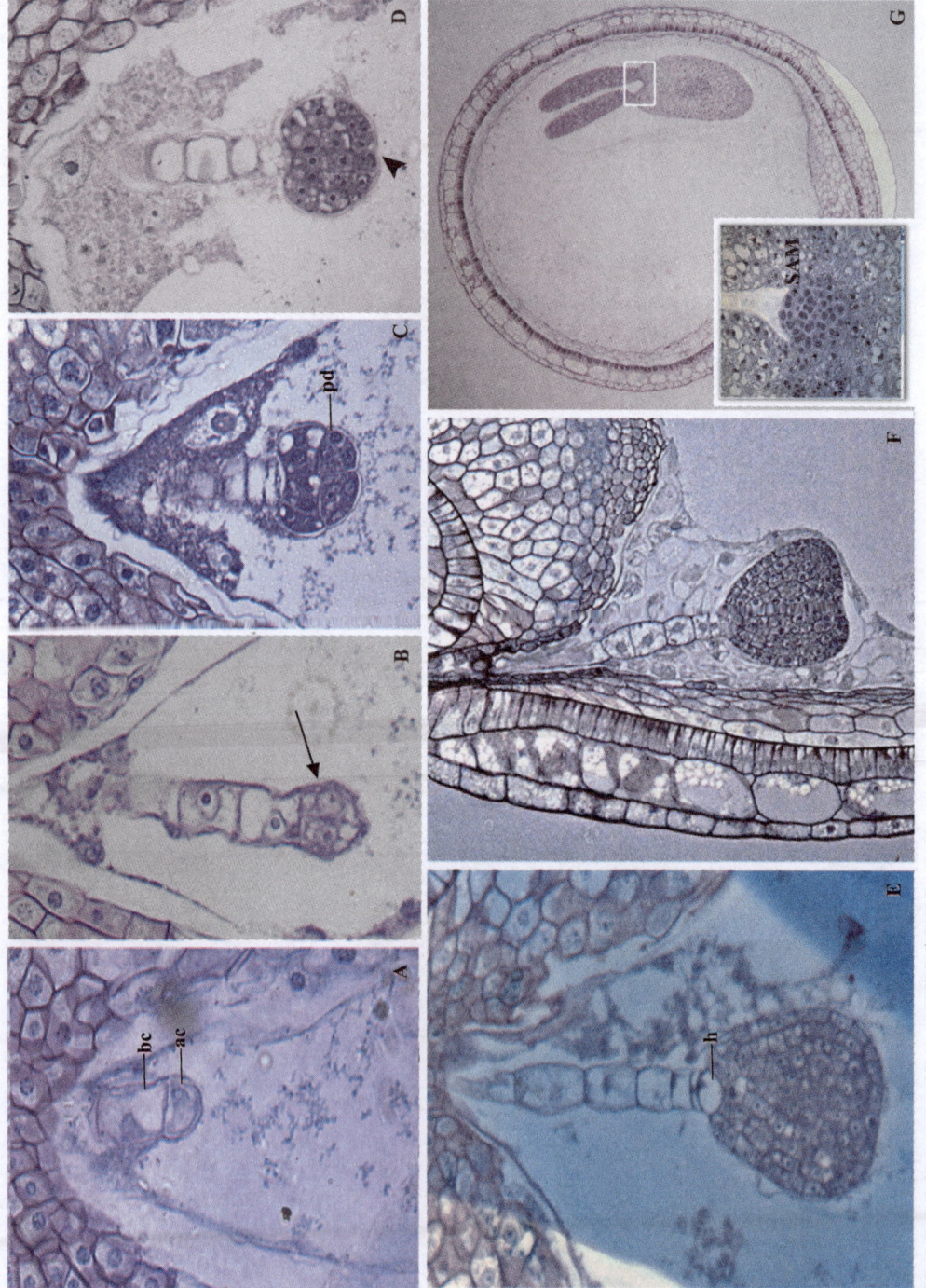

Figure 10.15 [Page 352]

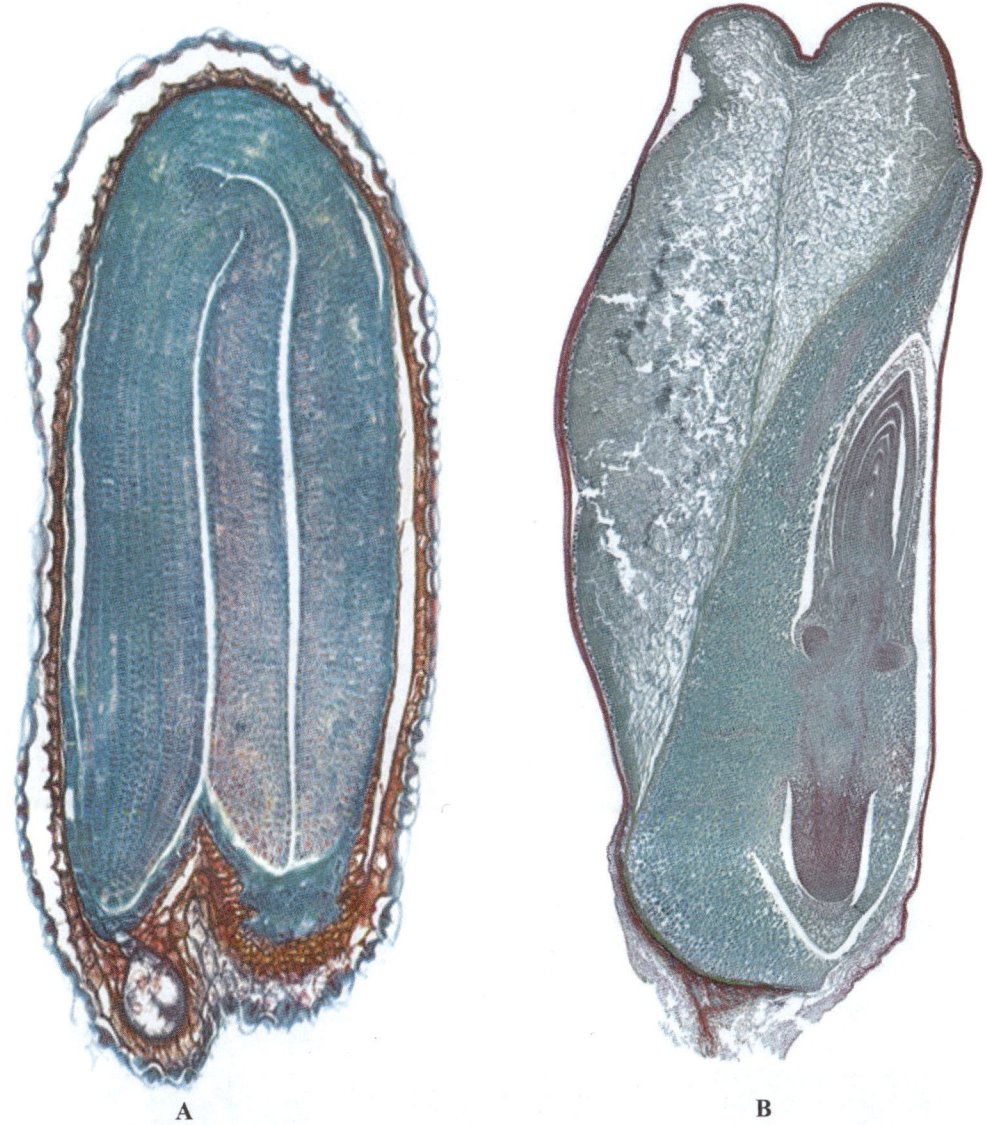

A   B

Figure 10.17 [Page 354]

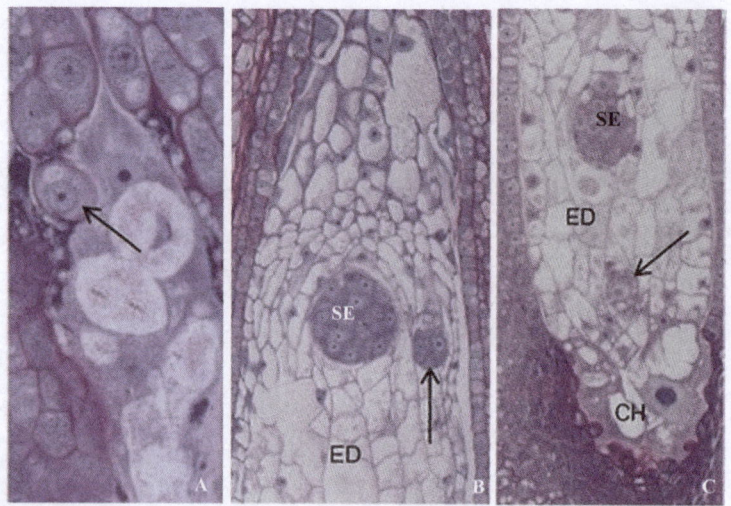

Figure 11.2 [Page 362]

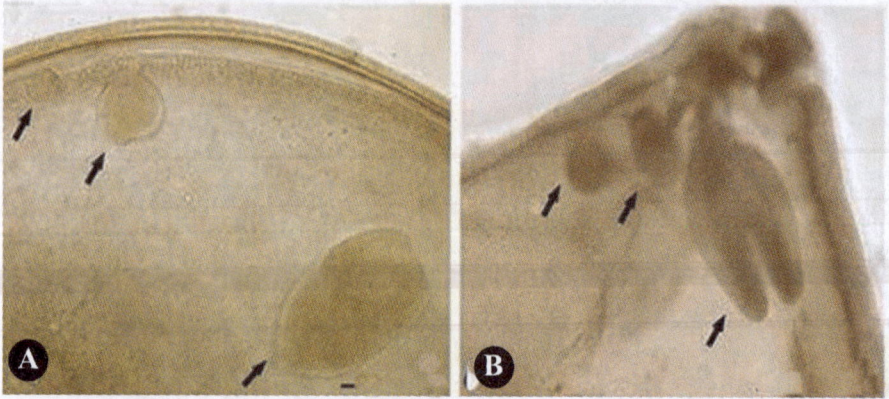

Figure 11.3 [Page 362]

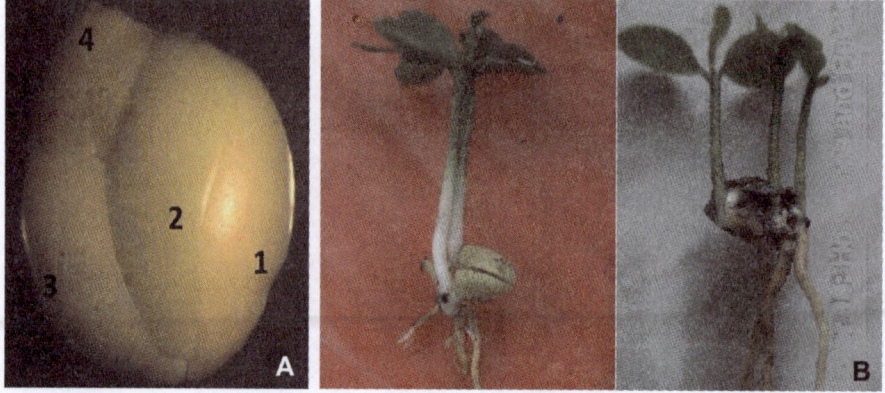

Figure 11.4 [Page 363]

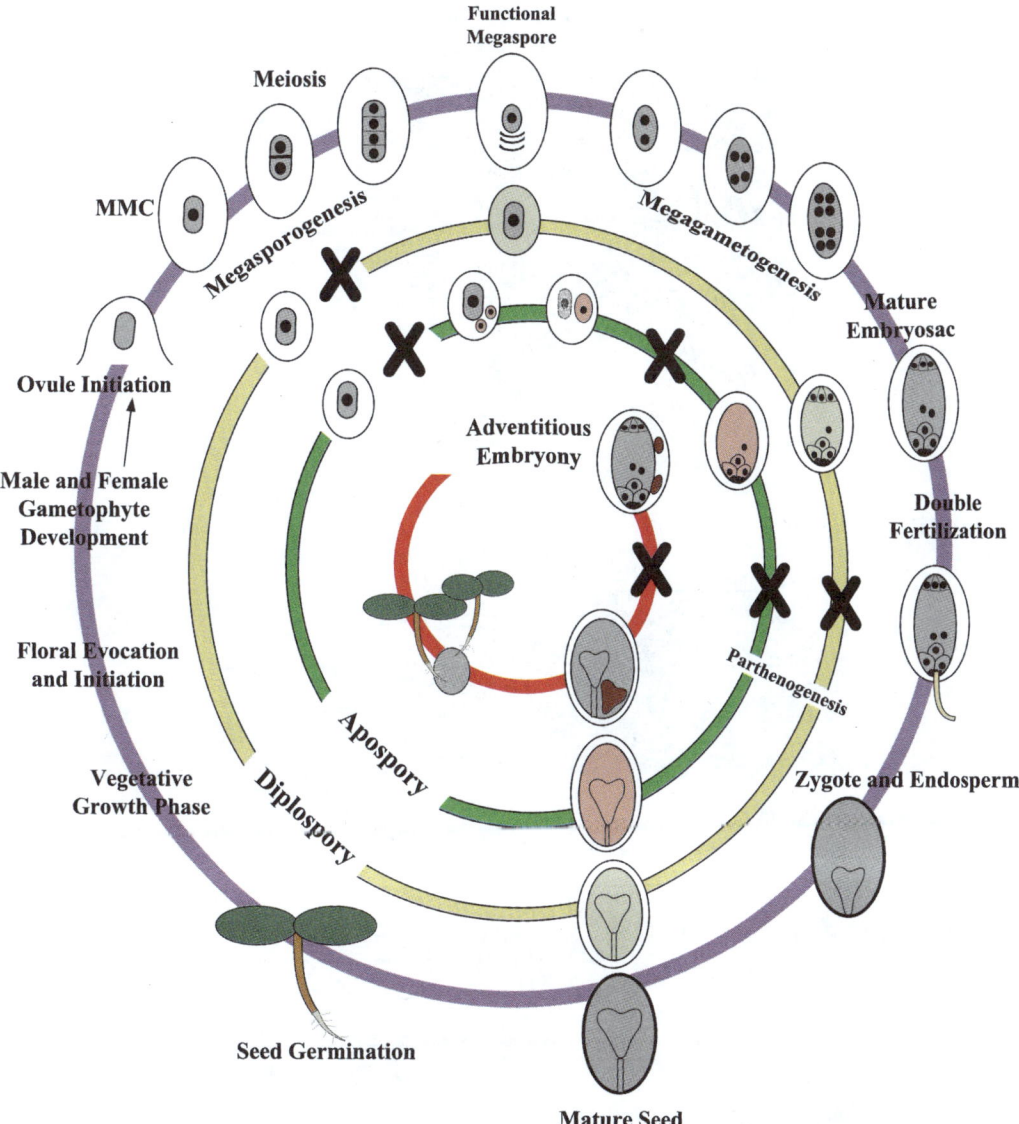

Figure 11.5 [Page 368]

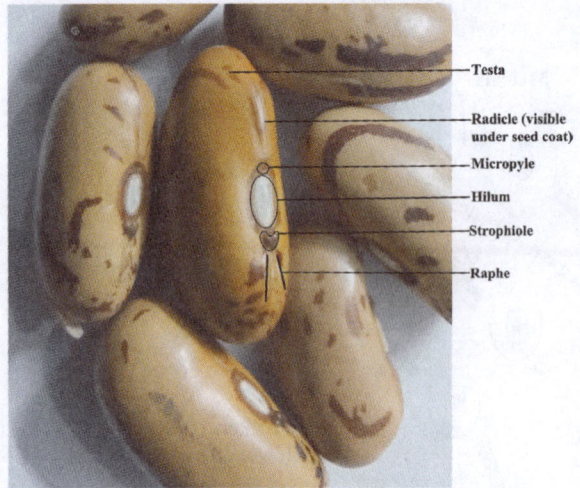

Figure 12.2 [Page 393]

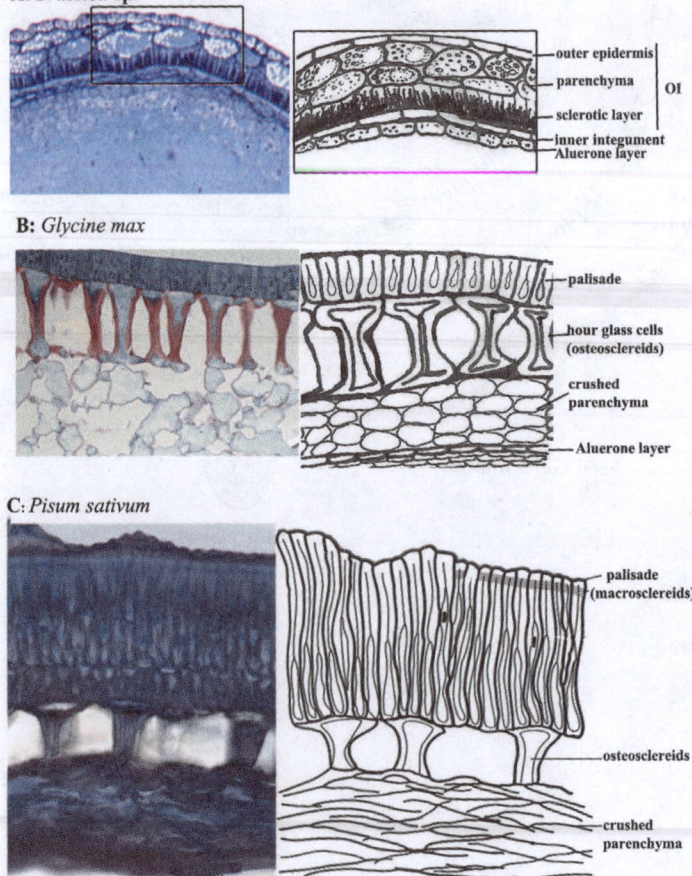

Figure 12.5 [Page 400]

Figure 12.9 [Page 411]

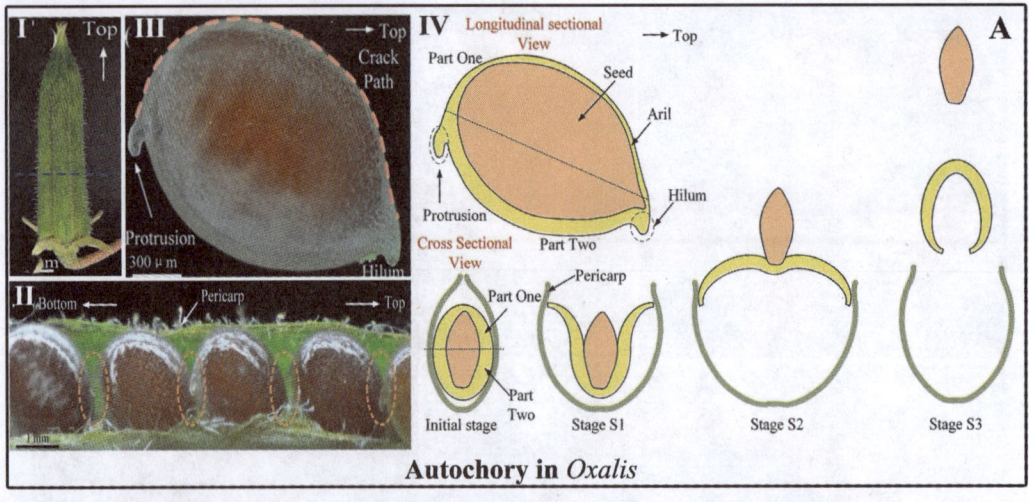

Autochory in *Oxalis*

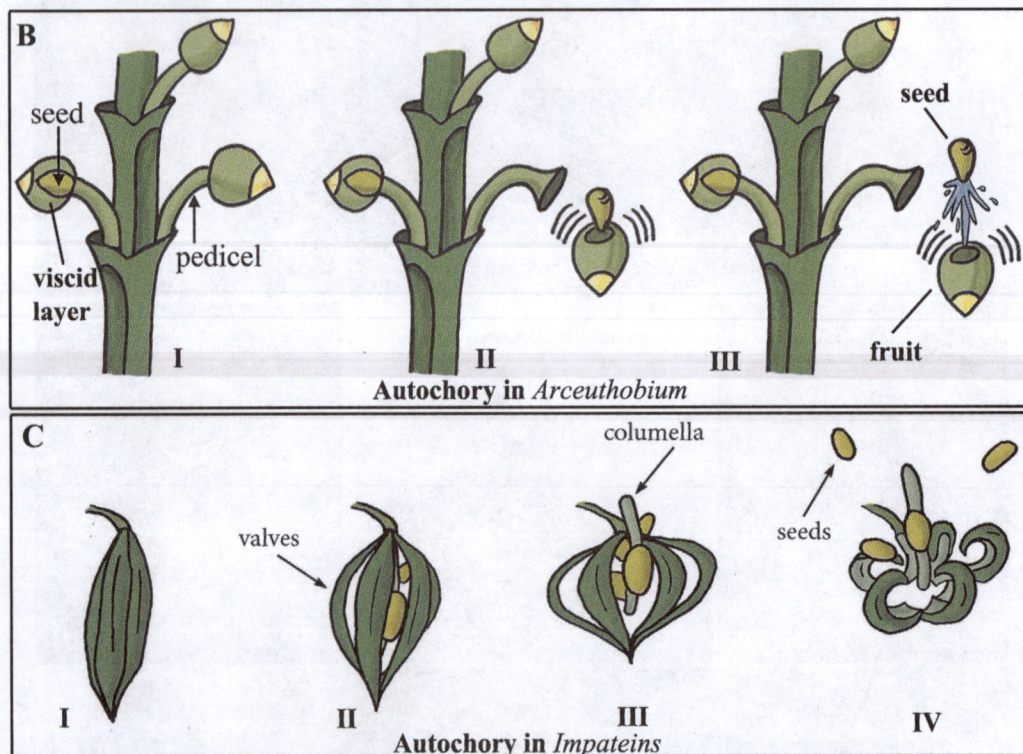

Autochory in *Arceuthobium*

Autochory in *Impateins*

Figure 12.10 [Page 416]

Figure 12.11 [Page 418]

Figure 12.12 [Page 419]

Figure 12.13 [Page 420]

Figure 12.14 [Page 422]

Figure 12.15 [Page 424]

Figure 12.16 [Page 432]

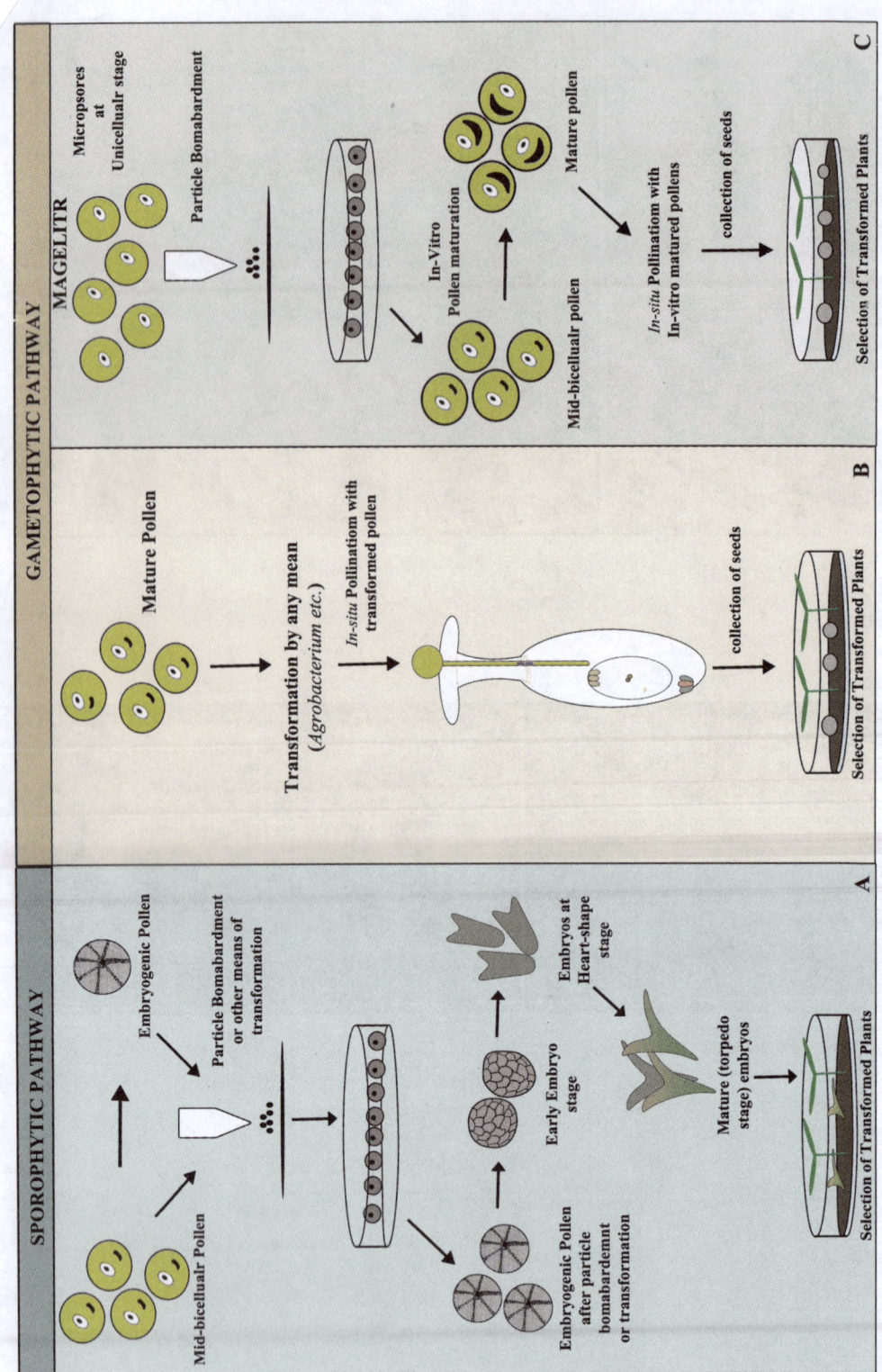

Figure 13.1 [Page 440]